Genomics Handbook

Volume I

Contents

Preface

The simplest way to define Genomics is by understanding it to be the study of the genomes of organisms. Genomics is the branch of medical science that studies and researches the genome, which is the genetic material or map of a living organism, be it human or other species like animals, plants or microbes that is contained in its DNA. Such study is done to better understand the workings and mechanisms of the organism, and what happens when certain genes or genetic materials interact with each other and the environment in which it exists. This discipline also focuses on the study of genes, and chromosomes as well as their functions in search of aspects that could be helpful in spotting, preventing or controlling disease through drug treatments or vaccinations. This field of course also automatically is the one that is making efforts to determine the entire DNA sequence of organisms and fine-scale genetic mapping. Genomics can also be said to be an interdisciplinary field that includes studies of intra-genomic phenomena such as epistasis, heterosis, and pleiotropy along with other interactions between various parts within the genome. By studying genomics and the related interaction with the environment, medical professionals will be able to better understand the reason that why some people get sick from certain environmental factors, infections, and behaviors, while others do not. Therefore this field has various important applications in the medical field. This also heightens the already rapidly increasing demands for skilled researchers and professionals in this field.

This book is an attempt to compile and collate all available research on genomics under one aegis. I am grateful to those who put their hard work, effort and expertise into these research projects as well as those who were supportive in this endeavour.

Editor

1

The Novelty of Human Cancer/Testis Antigen Encoding Genes in Evolution

Pavel Dobrynin,[1,2] Ekaterina Matyunina,[1] S. V. Malov,[2] and A. P. Kozlov[1,2]

[1] *The Biomedical Center, Saint Petersburg 194044, Russia*
[2] *Dobzhansky Center for Genome Bioinformatics, Saint Petersburg State University, Saint Petersburg 190004, Russia*

Correspondence should be addressed to Pavel Dobrynin; pdobrynin@gmail.com

Academic Editor: Ancha Baranova

In order to be inherited in progeny generations, novel genes should originate in germ cells. Here, we suggest that the testes may play a special "catalyst" role in the birth and evolution of new genes. Cancer/testis antigen encoding genes (CT genes) are predominantly expressed both in testes and in a variety of tumors. By the criteria of evolutionary novelty, the CT genes are, indeed, novel genes. We performed homology searches for sequences similar to human CT in various animals and established that most of the CT genes are either found in humans only or are relatively recent in their origin. A majority of all human CT genes originated during or after the origin of Eutheria. These results suggest relatively recent origin of human CT genes and align with the hypothesis of the special role of the testes in the evolution of the gene families.

1. Introduction

In order to be inherited in progeny generations, novel genes should originate in germ cells. Available data suggest that the generation of novel genes in germ cells is ongoing process, for example, the promiscuity of gene expression in spermatogenic cells [1, 2]. Novel genes may originate through different mechanisms (retrogenes, segmental duplicates, chimeric, and *de novo* emerged genes), but all of them are uniformly expressed in the testis ([3–8]; reviewed in [9]). These observations led us to suggest that testes may play a "tissue catalyst" role in the birth and evolution of new genes [9]. Previously, we proposed the expression of evolutionarily novel genes in tumors [10].

Cancer/testis or cancer/germline antigen genes are a class of genes with predominant expression in testis and in a variety of tumors, with a significant exclusion of some CT antigens also expressed in the brain. Here we set forth to test the hypothesis that cancer/testis antigen genes should be composed of evolutionarily new or young gene family. We performed homology searches for sequences similar to human CT in various animals. Additionally, as an extensive traffic of novel genes has been described for mammalian X chromosome [3, 6, 11], we also performed this analysis separately for genes located on this chromosome only.

2. Methods

The list of CT antigens gene was retrieved from CT Database (http://www.cta.lncc.br) and included 265 genes. Among them, there are 105 CT antigens that are encoded by the X chromosome (CT-X genes) and 105 that are located on various autosomes (autosome CT genes, or non-X CT genes). Eight CT antigen encoding genes are located on the Y chromosome.

To assess the evolutionary novelty of the studied group of CT genes by searching orthologues for each of CT genes, the HomoloGene.release 66 (http://www.ncbi.nlm.nih.gov/homologene/) tool from NCBI website was used. HomoloGene is a database of both curated and computed gene orthologs and orthologues and now covers 21 organisms. Curated orthologs include gene pairs from the Mouse Genome Database (MGD) at the Jackson Laboratory, the Zebrafish Information (ZFIN) database at the University of Oregon, and from published reports. Computed orthologs

Table 1: Distribution of all human CT genes according to the origin of their orthologues in different taxa of human lineage.

Taxa	Chromosome names																							
	1	2	3	4	5	6	7	8	9	10	11	12	13	14	15	16	17	18	19	20	21	22	X	Y
Eukaryote	1	1					1			2			1							1	1		1	
Opisthokonta	1									1														
Bilateria		1				1		1									2						2	
Coelomata						2		2	1												1			
Euteleostomi	4	4	2	1	1	2		2		3	1	2		1	1	1			1				3	
Amniota	3														2		2							
Eutheria	2		3	3	1	2		4	3		3	2			1	1	3	1	6	2		1	21	8
Euarchontoglires		1												1		1		1	1	1		1	4	
Catarrhini																		1		1	1		28	
Homininae								1										1					13	
Homo sapiens															1								33	
Total	11	7	5	4	2	7	1	10	4	6	4	4	1	2	5	3	7	4	8	5	3	2	105	8

and orthologues, which are considered putative, are identified from BLAST nucleotide sequence comparisons between all UniGene clusters for each pair of organisms [12]. As an input, the program uses gene name and/or taxon name, and the output is clusters of orthologues. For this study, the search was performed in several completely sequenced eukaryotic genomes, including *H. sapiens*, *P. troglodytes*, *M. mulatta*, *C. lupus*, *B. taurus*, *M. musculus*, *R. norvegicus*, *G. gallus*, *D. rerio*, *D. melanogaster*, *A. gambiae*, *C. elegans*, *S. cerevisiae*, *K. lactis*, *A. gossypii*, *S. pombe*, *M. oryzae*, *N. crassa*, *A. thaliana*, *O. sativa*, and *P. falciparum*.

According to the origin of their orthologues in different taxa of human phylogeny, the CT genes and all human genes were distributed into 11 groups. The differences in distribution of CT genes and all human genes were assessed using the chi square test [13]. Sheffe's S method of multiple estimation ([14, 15]; for counts see also [16]) was applied to define the difference and to show stochastically that the origin of human CT genes is substantially more recent than that for all human genes.

3. Results

The results obtained using HomoloGene tool applied to human CT genes are presented in Table 1. The full list of studied CT genes is present in Supplementary material (see Supplementary Material available online at http://dx.doi.org/10.1155/2013/105108. HomoloGene assigned each gene to a certain homology group which includes orthologues from different taxa within human lineage. Of 265 genes represented in CT Database, 47 did not match any homology group, probably because of the differences in the gene names making matches with HomoloGene database difficult. Human CT genes orthologues are widely distributed throughout the human lineage. For example, for one CT-X gene (*FAM133A*), the orthologues were found in all Eukaryota, and for two CT-X genes (*MAGEC1* and *SPANXN4*), the orthologues were first found in Bilateria, and for three CT-X genes (*ARX*, *IL13RA*, and *FAM46D*), the time of origin was placed in Euteleostomi. There were substantially larger numbers of CT-X genes with orthologues emerging in Eutheria, Catarrhini, and Homininae and of CT-X genes that were found exclusively in humans. Interestingly, there was a Eutheria-specific subfamily TSPY1 composed of 8 CT genes and located on chromosome Y.

Similarly searches for the orthologues were performed for all CT-X genes, all autosomal CT genes, all human CT genes, and all annotated protein coding genes in human genome (assembly GRCh37) (Table 2 and Figure 1).

The results show that the proportion of autosomal CT genes that has orthologues originated in Euteleostomi and in Eutheria (24.8% and 36.2%, accordingly) is greater than that on chromosome X. Only a few of autosomal CT genes are exclusive for humans. We found that CT gene *POTEB* (prostate, ovary, testis-expressed protein on chromosome 15, Ensembl: ENSG00000233917) has a poorly characterized homologue (LOC100287399, Ensembl: ENSG00000230031) that is according to HomoloGene criteria is exclusive to *H. sapiens*. This newly described homolog (LOC100287399, Ensembl: ENSG00000230031) has not been previously annotated as a gene of CT family.

Among all annotated human protein coding genes, the proportion of genes specific to humans only is very small (0.85%). The list of these human-specific genes includes 163 entries, 33 of which are CT-X genes.

For CT-X genes, the distribution was different: 31.4% of CT-X genes (five *CT45A* genes, twelve *CT47A* genes, fifteen *GAGE* genes, and four *XAGE* genes) are present in humans only, while 39.1% of CT-X genes have orthologues that emerged in *Catarrhini* or *Homininae*. This means that the majority (70.5%) of CT-X genes present in human genome are either novel or relatively recent. At the same time, distribution of all genes located on X chromosome is similar to that for all human genes (see Supplementary Table IV).

The distribution of all human CT genes shows that 30.73% of CT genes have orthologues that originated in *Eutheria*. This proportion is larger than the proportion of all human genes with pan-*Eutherian* orthologues (16.41%). Importantly, 36.7% of all human CT genes originated in *Catarrhini*, *Homininae*,

TABLE 2: CT-X genes, autosomal CT genes, all human CT genes, and all annotated human genes with orthologues originated in different taxa of *H. sapiens* lineage.

Taxa	CT-X genes		Autosome CT genes		All CT genes		All human genes	
Eukaryota	1,0%	1	7,6%	8	4,13%	9	15,19%	2900
Opisthokonta			1,9%	2	0,92%	2	3,21%	613
Bilateria	1,9%	2	4,8%	5	3,21%	7	8,12%	1549
Coelomata			5,7%	6	2,75%	6	8,27%	1579
Euteleostomi	2,9%	3	24,8%	26	13,30%	29	32,77%	6256
Amniota			6,7%	7	3,21%	7	8,39%	1601
Eutheria	20,0%	21	36,2%	38	30,73%	67	16,41%	3132
Euarchontoglires	3,8%	4	6,7%	7	5,05%	11	1,75%	334
Catarrhini	26,7%	28	2,9%	3	14,22%	31	2,66%	507
Homininae	12,4%	13	1,9%	2	6,88%	15	2,38%	454
Homo sapiens	31,4%	33	1,0%	1	15,60%	34	0,85%	163
Total	100,0%	105	100,0%	105	100,00%	218	100,00%	19088

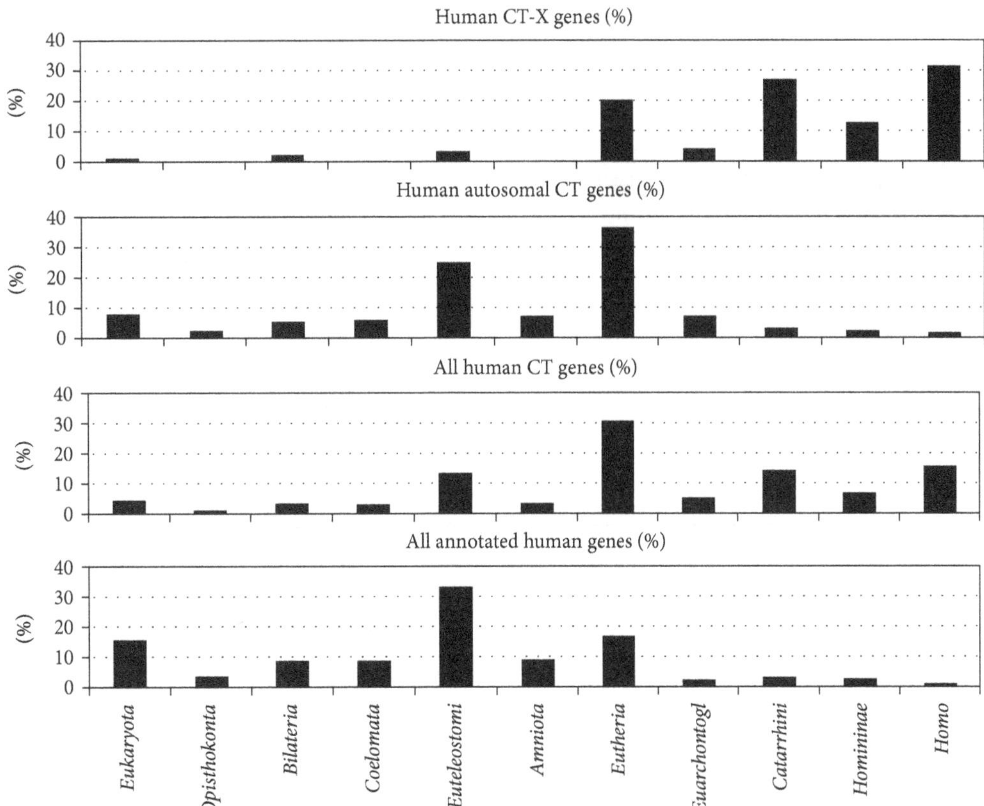

FIGURE 1: The proportions of CT-X genes, autosomal CT genes, all human CT genes, and all annotated human genes with orthologues originated in different taxa of *H. sapiens* lineage.

or humans. Thus, the majority of human CT genes (72.48%) originated during or after the emergence of Eutheria. On the other side, the majority of annotated human genes (75.95%) were older than *Eutheria*.

A significance of the difference between distribution of all human genes and all human CT genes according to the origin of their orthologues in different taxa was confirmed bychi square test (P value less than 10^{-6}). Moreover, 95% confidence region for the cumulative distribution function of CT human genes displays that CT genes are stochastically younger as compared to all human genes. In other words, the probability that a gene randomly chosen from all human genes is younger than some fixed time T is less than the probability that a randomly chosen CT gene is younger than T. Therefore, there is a significant bias in time of origin for human CT genes as compared to all human genes. If human CT genes would be

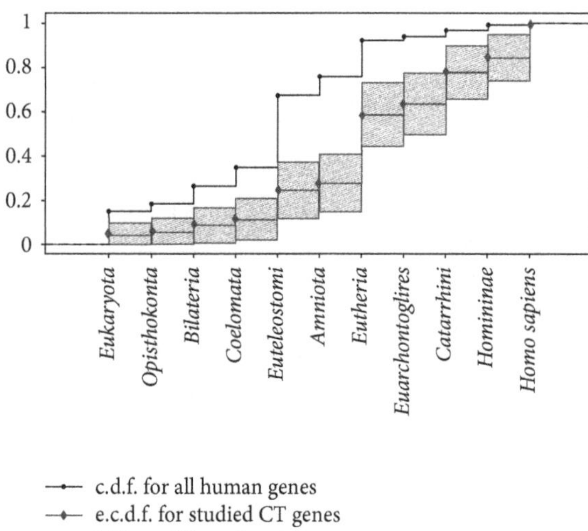

FIGURE 2: Cumulative distribution function for all human genes and empirical distribution function for all CT human genes, in accordance with the origin of their orthologues in different taxa, with 95% confidence bands. c.d.f.—cumulative distribution function. e.c.d.f.—empirical cumulative distribution function.

obtained as a sample from some probabilistic distribution, the probability that CT human genes originated not earlier than *Catarrhini* or *Eutheria* would be significantly higher than the respective probability for census of all human genes (Figure 2). This statistical trial confirms that the origin of human CT genes is relatively recent.

4. Conclusion

Cancer/testis antigen genes (CTA or CT genes) encode a subgroup of tumor antigens expressed predominantly in testis and various tumors. CT antigens may be also expressed in placenta and in female germ cells [17–20]. In addition, some CT antigens are expressed in the brain [21].

Experimentally, human CT genes were discovered by a variety of immunological screening methods [22], serological identification of antigens by recombinant expression cloning (SEREX) [23], expression database analysis [24, 25], massively parallel signature sequencing [26], and other approaches. The fact that many CT antigens have been identified using SEREX suggests that they are highly antigenic [23, 27].

The first CT gene discovered was *MAGEA1* that encodes for an antigen of human melanoma [22]. This gene belongs to a family of 12 closely related genes clustered at Xq28. A second cluster of *MAGE* genes, *MAGEB*, was discovered at Xp21.3, and the third, encoding *MAGEC* genes, is located at Xq26-27. The expression of *MAGEA-MAGEC* genes (*MAGE-I* subfamily) is restricted to testis and cancer, whereas more distantly related clusters *MAGED-MAGEL* (subfamily *MAGE-II*) are expressed in many normal tissues. *MAGE-I* genes are of relatively recent origin, and *MAGE-II* genes are relatively more ancient. For example, *MAGE-D* genes are conserved between man and mouse. One of these genes corresponds to the founder member of the family, and the other *MAGE* genes are retrogenes derived from the common ancestral gene [19, 28, 29].

To date, CTD atabase (http://www.cta.lncc.br/) includes 265 CT genes. More than half of them are located on X-chromosome (CT-X genes) [21]. The analysis of the DNA sequence of the human X chromosome predicts that approximately 10% of the genes on the X chromosome are of the CT antigen type [30]. Non-X CT genes are distributed throughout the genome and are represented mainly by single-copy genes [19, 27, 31].

In normal testis, CT-X genes are expressed in proliferating germ cells (spermatogonia). Non-X CT genes are expressed during later stages of germ-cell differentiation, that is, spermatocytes [19]. Among human tumors, CT antigens are expressed in melanoma, bladder cancer, lung cancer, breast cancer, prostate cancer, sarcoma, ovarian cancer, hepatocellular carcinoma, hematologic malignancies, and so forth [21, 27, 31, 32]. Genome-wide analysis of 153 cancer/testis genes expression has led to their classification into testis-restricted ($N = 39$), testis/brain-restricted ($N = 14$) and testis-selective ($N = 85$) groups of genes, the latter group showing some expression in nongermline tissues. The majority of testis-restricted genes belong to CT-X group (35 of total 39 testis-restricted groups), while non-X CT genes are expressed in a less restrictive way [21].

Multiple CT antigens are often coexpressed in tumors suggesting that this expression program is coordinated for entire family [19, 23, 33]. CT gene expression is controlled by epigenetic mechanisms which include DNA methylation and histone posttranslational modifications [31]. Other mechanisms of CT gene regulation include sequence-specific transcription factors and signal transduction pathways such as activated tyrosine kinases [34].

The functions of CT-X genes are largely unknown. On the contrary, more is known about functions of non-X CT genes which are associated with meiosis, gametogenesis, and fertilization. Non-X CTs are also more conserved during evolution [21, 27, 31, 32].

CT-X genes tend to form recently expanded gene families, many with nearly identical gene copies [17–20, 26, 32, 35].

The prevalence of large, highly homologous inverted repeats (IRs) containing testes genes on the X- and Y-chromosomes was described in humans and great apes [36, 37]. CT-X gene families are also located in direct or inverted repeats [20].

The study of clusters of homologous genes originated by gene duplication roughly after the divergence of the human and rodent lineages discovered several families of CT genes among recent duplicates [38].

In the other paper, the authors also studied recent duplications in the human genome and found that CT genes were represented in this gene set, including the family of *PRAME* (preferentially expressed antigen of melanoma) genes located on chromosome 1 and expressed in the testis and in a large number of tumors [39]. Duplicated *PRAME* genes are hominid specific, having arisen in human genome since the divergence from chimps. *PRAME* gene family also expanded in other *Eutheria*. Chimp and mouse have orthologous *PRAME* gene clusters on their chromosomes 1 and 4, respectively [39, 40].

Rapid evolution of cancer/testis genes has been demonstrated on the X chromosome. In particular, the comparison of human: chimp orthologues of these genes has shown that they diverge faster and undergo stronger positive selection than those on the autosomes or than control genes on either X chromosome or autosomes [41].

SPANX-A/D gene subfamily of cancer/testis-specific antigens evolved in the common ancestor of the hominoid lineage after its separation from orangutan. Southern blot and database analyses have detected *SPANX* sequences only in primates [17]. The coding sequences of the *SPANX* genes evolved rapidly, faster than their introns and the 5′ untranslated regions, with accelerated rates of substitutions in both synonymous and nonsynonymous codon positions. The mechanism of *SPANX* genes expansion was segmental DNA duplications, with evidence of positive selection. *SPANX-N* is the ancestral form, from which the *SPANX-A/D* subfamily evolved in the common ancestor of hominoids approximately 7 MYA [35, 42]. *SPANX* genes are expressed in cancer cells and highly metastatic cell lines from melanomas, bladder carcinomas, and myelomas [35].

The *GAGE* cancer/testis antigen gene family contains at least 16 genes which are encoded by an equal number of tandem repeats. All *GAGE* genes are located at Xp11.23. *GAGE* genes are highly identical and evolved under positive selection that supports their recent origin [43, 44].

The *XAGE* family of cancer/testis antigen genes belongs to superfamily of *GAGE*-like CT genes. It is located on chromosome Xp11.21-Xp11.22. Three *XAGE* genes are described, as well as several splice variants of *XAGE-1* [45, 46].

CT45 gene family was discovered by massively parallel signature sequencing. It includes six highly similar (>98%) genes that are cluctered in tandem on chromosome Xq26.3. CT45 antigen is expressed in Hodgkin's lymphoma and in other human tumors [26, 47–49].

CT47 cancer/testis gene family is located on chromosome Xq24. Among normal tissues, it is expressed in the testis and (weakly) in placenta and brain. In tumors, its expression was found in lung cancer and esophageal cancer. The *CT47* family member is characterized by high (>98%) sequence homology. Chimp is the only other species in which a gene homologous to *CT47* was found by other authors [20].

Our work is the first systematic study of the evolutionary novelty of the whole class of CT genes. To assess the evolutionary novelty of CT genes, we applied the HomoloGene tool of NCBI. To construct the clusters of orthologues, the HomoloGene program uses information from blastp, phylogenetic analyses, and syntheny information when it is possible. Cutoffs on bits per position and Ks values are set to prevent unlikely "orthologs" from being grouped together. These cutoffs are calculated based on the respective score distribution for the given groups of organisms [12].

We searched for orthologues of each of CT genes among annotated genes in several completely sequenced eukaryotic genomes and built distributions of all CT-X genes, all autosomal CT genes, all human CT genes, and all annotated protein coding genes from human genome according to the origin of their orthologues in 11 taxa of human lineage.

We have shown that 31.4% of CT-X genes are exclusive for humans and 39.1% of CT-X genes have orthologues originated in *Catarrhini* or *Homininae*. Thereby, the majority of human CT-X genes (70.5%) are novel or recent in its origin. Our data are in good correspondence with evidence obtained by other groups on rapid expansion of certain CT-X gene families and high homology of their members which suggest their recent origin.

Altogether 36.7% of all human CT genes originated in *Catarrhini*, *Homininae*, and humans. We have also found that 30.73% of all human CT genes originated in *Eutheria*. These CT genes acquired functions in *Eutheria*. This indicates the importance of processes in which tumors and CT antigens were involved during the evolution of *Eutheria*. CT genes originated in *Eutheria* are located mostly on autosomes. CT genes originated in *Catarrhini*, *Homininae*, and humans are located predominantly on X chromosome. This difference is probably related to evolution of mammalian X chromosome since the origin of *Eutheria* [50], especially to the acquisition of its special role in the origin of novel genes [9].

Thus, the majority of CT-X genes are either novel or young for humans, and the majority of all human CT genes (72.48%) originated during or after the origin of *Eutheria*. These results suggest that the whole class of human CT genes is relatively evolutionarily new.

In its turn, this conclusion confirms our prediction about expression of evolutionary recent and novel genes in tumors [10]. The expression of cancer/testis genes in tumors is then a natural phenomenon, not aberrant process as suggested by many authors (e.g., [19, 27, 32, 34, 40]).

References

[1] E. E. Schmidt, "Transcriptional promiscuity in testes," *Current Biology*, vol. 6, no. 7, pp. 768–769, 1996.

[2] K. C. Kleene, "Sexual selection, genetic conflict, selfish genes, and the atypical patterns of gene expression in spermatogenic cells," *Developmental Biology*, vol. 277, no. 1, pp. 16–26, 2005.

[3] E. Betrán, K. Thornton, and M. Long, "Retroposed new genes out of the X in Drosophila," *Genome Research*, vol. 12, no. 12, pp. 1854–1859, 2002.

[4] C. A. Paulding, M. Ruvolo, and D. A. Haber, "The *Tre2* (USP6) oncogene is a hominoid-specific gene," *Proceedings of the National Academy of Sciences of the United States of America*, vol. 100, no. 5, pp. 2507–2511, 2003.

[5] X. She, Z. Cheng, S. Zöllner, D. M. Church, and E. E. Eichler, "Mouse segmental duplication and copy number variation," *Nature Genetics*, vol. 40, no. 7, pp. 909–914, 2008.

[6] M. T. Levine, C. D. Jones, A. D. Kern, H. A. Lindfors, and D. J. Begun, "Novel genes derived from noncoding DNA in Drosophila melanogaster are frequently X-linked and exhibit testis-biased expression," *Proceedings of the National Academy of Sciences of the United States of America*, vol. 103, no. 26, pp. 9935–9939, 2006.

[7] T. J. A. J. Heinen, F. Staubach, D. Häming, and D. Tautz, "Emergence of a new gene from an intergenic region," *Current Biology*, vol. 19, no. 18, pp. 1527–1531, 2009.

[8] H. Kaessmann, N. Vinckenbosch, and M. Long, "RNA-based gene duplication: mechanistic and evolutionary insights," *Nature Reviews Genetics*, vol. 10, no. 1, pp. 19–31, 2009.

[9] H. Kaessmann, "Origins, evolution, and phenotypic impact of new genes," *Genome Research*, vol. 20, no. 10, pp. 1313–1326, 2010.

[10] A. P. Kozlov, "The possible evolutionary role of tumors in the origin of new cell types," *Medical Hypotheses*, vol. 74, no. 1, pp. 177–185, 2010.

[11] J. J. Emerson, H. Kaessmann, E. Betrán, and M. Long, "Extensive gene traffic on the mammalian X chromosome," *Science*, vol. 303, no. 5657, pp. 537–540, 2004.

[12] E. W. Sayers, T. Barrett, D. A. Benson et al., "Database resources of the National Center for Biotechnology Information," *Nucleic Acids Research*, vol. 40, pp. D13–D25, 2012.

[13] R Development Core Team, *R: A Language and Environment for Statistical Computing*, R Foundation for Statistical Computing, Vienna, Austria, 2011, http://www.R-project.org/.

[14] H. Sheffe, *The Analysis of Variance*, Wiley, New York, NY, USA, 1959.

[15] H. Sheffe, "Multiple testing versus multiple estimation. Improper confidence sets. Estimation of directions and ratios," *The Annals of Mathematical Statistics*, vol. 41, no. 1, pp. 1–29, 1970.

[16] L. A. Goodman, "Simultaneous confidence intervals for contrasts among multinomial populations," *The Annals of Mathematical Statistics*, vol. 35, pp. 716–725, 1964.

[17] A. J. W. Zendman, J. Zschocke, A. A. Van Kraats et al., "The human SPANX multigene family: genomic organization, alignment and expression in male germ cells and tumor cell lines," *Gene*, vol. 309, no. 2, pp. 125–133, 2003.

[18] A. J. W. Zendman, D. J. Ruiter, and G. N. P. Van Muijen, "Cancer/testis-associated genes: identification, expression profile, and putative function," *Journal of Cellular Physiology*, vol. 194, no. 3, pp. 272–288, 2003.

[19] A. J. G. Simpson, O. L. Caballero, A. Jungbluth, Y. T. Chen, and L. J. Old, "Cancer/testis antigens, gametogenesis and cancer," *Nature Reviews Cancer*, vol. 5, no. 8, pp. 615–625, 2005.

[20] Y. T. Chen, C. Iseli, C. A. Yenditti, L. J. Old, A. J. G. Simpson, and C. V. Jongeneel, "Identification of a new cancer/testis gene family, CT47, among expressed multicopy genes on the human X chromosome," *Genes Chromosomes and Cancer*, vol. 45, no. 4, pp. 392–400, 2006.

[21] O. Hofmann, O. L. Caballero, B. J. Stevenson et al., "Genome-wide analysis of cancer/testis gene expression," *Proceedings of the National Academy of Sciences of the United States of America*, vol. 105, no. 51, pp. 20422–20427, 2008.

[22] P. Van Der Bruggen, C. Traversari, P. Chomez et al., "A gene encoding an antigen recognized by cytolytic T lymphocytes on a human melanoma," *Science*, vol. 254, no. 5038, pp. 1643–1647, 1991.

[23] U. Sahin, O. Tereci, H. Schmitt et al., "Human neoplasms elicit multiple specific immune responses in the autologous host," *Proceedings of the National Academy of Sciences of the United States of America*, vol. 92, no. 25, pp. 11810–11813, 1995.

[24] M. J. Scanlan, C. M. Gordon, B. Williamson et al., "Identification of cancer/testis genes by database mining and mRNA expression analysis," *International Journal of Cancer*, vol. 98, no. 4, pp. 485–492, 2002.

[25] M. J. Scanlan, A. J. Simpson, and L. J. Old, "The cancer/testis genes: review, standardization, and commentary," *Cancer Immunity*, vol. 4, article 1, 2004.

[26] Y. T. Chen, M. J. Scanlan, C. A. Venditti et al., "Identification of cancer/testis-antigen genes by massively parallel signature sequencing," *Proceedings of the National Academy of Sciences of the United States of America*, vol. 102, no. 22, pp. 7940–7945, 2005.

[27] Y. H. Cheng, E. W. P. Wong, and C. Y. Cheng, "Cancer/testis (CT) antigens, carcinogenesis and spermatogenesis," *Spermatogenesis*, vol. 1, pp. 209–220, 2011.

[28] P. Chomez, O. De Backer, M. Bertrand, E. De Plaen, T. Boon, and S. Lucas, "An overview of the MAGE gene family with the identification of all human members of the family," *Cancer Research*, vol. 61, no. 14, pp. 5544–5551, 2001.

[29] Y. Katsura and Y. Satta, "Evolutionary history of the cancer Immunity antigen MAGE gene family," *PLoS ONE*, vol. 6, no. 6, Article ID e20365, 2011.

[30] M. T. Ross, D. V. Grafham, A. J. Coffey et al., "The DNA sequence of the human X chromosome," *Nature*, vol. 434, pp. 325–337, 2005.

[31] E. Fratta, S. Coral, A. Covre et al., "The biology of cancer testis antigens: putative function, regulation and therapeutic potential," *Molecular Oncology*, vol. 5, no. 2, pp. 164–182, 2011.

[32] O. L. Caballero and Y. T. Chen, "Cancer/testis (CT) antigens: potential targets for immunotherapy," *Cancer Science*, vol. 100, no. 11, pp. 2014–2021, 2009.

[33] U. Sahin, O. Tureci, Y. T. Chen et al., "Expression of multiple cancer/testis antigens in breast cancer and melanoma: basis for polyvalent CT vaccine strategies," *International Journal of Cancer*, vol. 78, pp. 387–389, 1998.

[34] S. N. Akers, K. Odunsi, and A. R. Karpf, "Regulation of cancer germline antigen gene expression: implications for cancer immunotherapy," *Future Oncology*, vol. 6, no. 5, pp. 717–732, 2010.

[35] N. Kouprina, M. Mullokandov, I. B. Rogozin et al., "The SPANX gene family of cancer/testis-specific antigens: rapid evolution and amplification in African great apes and hominids," *Proceedings of the National Academy of Sciences of the United States of America*, vol. 101, no. 9, pp. 3077–3082, 2004.

[36] H. Skaletsky, T. Kuroda-Kawaguchl, P. J. Minx et al., "The male-specific region of the human Y chromosome is a mosaic of discrete sequence classes," *Nature*, vol. 423, no. 6942, pp. 825–837, 2003.

[37] P. E. Warburton, J. Giordano, F. Cheung, Y. Gelfand, and G. Benson, "Inverted repeat structure of the human genome: the X-chromosome contains a preponderance of large, highly homologous inverted repeated that contain testes genes," *Genome Research A*, vol. 14, no. 10, pp. 1861–1869, 2004.

[38] IHGSC, "International human genome sequencing consortium," *Nature*, vol. 431, pp. 931–945, 2004.

[39] Z. Birtle, L. Goodstadt, and C. Ponting, "Duplication and positive selection among hominin-specific PRAME genes," *BMC Genomics*, vol. 6, article 120, 2005.

[40] T. C. Chang, Y. Yang, H. Yasue, A. K. Bharti, E. F. Retzel, and W. S. Liu, "The expansion of the PRAME gene family in Eutheria," *PLoS ONE*, vol. 6, no. 2, Article ID e16867, 2011.

[41] B. J. Stevenson, C. Iseli, S. Panji et al., "Rapid evolution of cancer/testis genes on the X chromosome," *BMC Genomics*, vol. 8, article 129, 2007.

[42] N. Kouprina, V. N. Noskov, A. Pavlicek et al., "Evolutionary diversification of SPANX-N sperm protein gene structure and expression," *PLoS ONE*, vol. 2, no. 4, article e359, 2007.

[43] Y. Liu, Q. Zhu, and N. Zhu, "Recent duplication and positive selection of the GAGE gene family," *Genetica*, vol. 133, no. 1, pp. 31–35, 2008.

[44] M. F. Gjerstorff and H. J. Ditzel, "An overview of the GAGE cancer/testis antigen family with the inclusion of newly identified members," *Tissue Antigens*, vol. 71, no. 3, pp. 187–192, 2008.

[45] A. J. W. Zendman, A. A. Van Kraats, U. H. Weidle, D. J. Ruiter, and G. N. P. Van Muijen, "The XAGE family of cancer/testis-associated genes: alignment and expression profile in normal tissues, melanoma lesions and Ewing's sarcoma," *International Journal of Cancer*, vol. 99, no. 3, pp. 361–369, 2002.

[46] S. Sato, Y. Noguchi, N. Ohara et al., "Identification of XAGE-1 isoforms: predominant expression of XAGE-1b in testis and tumors," *Cancer Immunity*, vol. 7, article 5, 2007.

[47] Y. T. Chen, M. Hsu, P. Lee et al., "Cancer/testis antigen CT45: analysis of mRNA and protein expression in human cancer," *International Journal of Cancer*, vol. 124, no. 12, pp. 2893–2898, 2009.

[48] Y. T. Chen, A. Chadburn, P. Lee et al., "Expression of cancer testis antigen CT45 in classical Hodgkin lymphoma and other B-cell lymphomas," *Proceedings of the National Academy of Sciences of the United States of America*, vol. 107, no. 7, pp. 3093–3098, 2010.

[49] H. J. Heidebrecht, A. Claviez, M. L. Kruse et al., "Characterization and expression of CT45 in Hodgkin's lymphoma," *Clinical Cancer Research*, vol. 12, pp. 4804–4811, 2006.

[50] B. T. Lahn and D. C. Page, "Four evolutionary strata on the human X chromosome," *Science*, vol. 286, no. 5441, pp. 964–967, 1999.

2

High Levels of Sequence Diversity in the 5′ UTRs of Human-Specific L1 Elements

Jungnam Lee,[1] Seyoung Mun,[2] Thomas J. Meyer,[3] and Kyudong Han[1]

[1] *Department of Nanobiomedical Science & WCU Research Center, Dankook University, Cheonan 330-714, Republic of Korea*
[2] *Department of Microbiology, College of Advance Science, Dankook University, Cheonan 330-714, Republic of Korea*
[3] *Department of Biological Sciences, Louisiana State University, Baton Rouge, LA 70803, USA*

Correspondence should be addressed to Kyudong Han, kyudong.han@gmail.com

Academic Editor: Brian Wigdahl

Approximately 80 long interspersed element (LINE-1 or L1) copies are able to retrotranspose actively in the human genome, and these are termed retrotransposition-competent L1s. The 5′ untranslated region (UTR) of the human-specific L1 contains an internal promoter and several transcription factor binding sites. To better understand the effect of the L1 5′ UTR on the evolution of human-specific L1s, we examined this population of elements, focusing on the sequence diversity and accumulated substitutions within their 5′ UTRs. Using network analysis, we estimated the age of each L1 component (the 5′ UTR, ORF1, ORF2, and 3′ UTR). Through the comparison of the L1 components based on their estimated ages, we found that the 5′ UTR of human-specific L1s accumulates mutations at a faster rate than the other components. To further investigate the L1 5′ UTR, we examined the substitution frequency per nucleotide position among them. The results showed that the L1 5′ UTRs shared relatively conserved transcription factor binding sites, despite their high sequence diversity. Thus, we suggest that the high level of sequence diversity in the 5′ UTRs could be one of the factors controlling the number of retrotransposition-competent L1s in the human genome during the evolutionary battle between L1s and their host genomes.

1. Introduction

Transposable elements are a considerable component of the human genome, responsible for approximately 45% of the human genome sequence [1]. These elements are associated with genomic instability via *de novo* insertions, insertion-mediated deletions, and recombination events [2–8] and are responsible for a number of genetic disorders [9]. Almost all of the transposable elements belong to one of four types: long interspersed elements (LINEs), short interspersed elements (SINEs), long terminal repeat (LTR) retrotransposons, and DNA transposons [1, 10–12]. Among them, LINE-1s or L1s are one of the most successful retrotransposon families in the human genome, with 516,000 copies comprising 17% of the human genomic sequence [1]. A full-length functional L1 element is about 6 kb in length and contains a 5′ untranslated region (UTR) bearing an internal RNA polymerase II promoter, two open reading frames (ORF1 and ORF2), and a 3′ UTR terminating in a poly(A) tail [13]; ORF1 encodes an RNA-binding protein that has demonstrated nucleic acid chaperone activity *in vitro*, and ORF2 encodes a protein with both endonuclease (EN) and reverse transcriptase (RT) activities, which are required for L1 retrotransposition [14–16]. The generally accepted model for L1 retrotransposition is called target-primed reverse transcription (TPRT). In this mechanism, the L1 RNA forms a ribonucleoprotein complex with its own encoded proteins and then moves back to the nucleus where the L1 EN makes a single-stranded nick producing a free 3′-hydroxyl at the end of a poly(T) overhang to which the 3′ poly(A) tail of the L1 RNA anneals. The L1 RT then primes at this site and initiates reverse transcription. In addition to the newly inserted L1 sequence, TPRT typically generates 7-to-20 bp target site duplications flanking each side of the L1 insertion [17, 18].

Most previous studies of human-specific L1s focused on the 3′ UTRs rather than the 5′ UTRs because over 90% of L1 elements are 5′ truncated [19, 20], a result of the termination of reverse transcription, for unknown reasons, before the synthesis of a new L1 copy is complete. The 5′ UTR-deficient L1 is unable to propagate because it does not contain a promoter as well as transcription factor (TF) binding sites such as Yin Yang 1 (YY1), two putative SRY-related TF binding sites, and two putative RUNX3 TF binding sites [21]. These truncated insertions are therefore unlikely to influence the evolutionary history of L1 elements. It has been reported that several distinct lineages of L1s propagated in primate genomes simultaneously during the first 30 million years (myrs) of primate evolution, between 70 and 40 myrs ago. However, only a single lineage of L1 elements appears to have been retrotranspositionally active over the last 40 myrs. Although competition between distinct L1 lineages for host factors has been suggested as a reason why a single lineage came to dominance in the human genome for the last 40 myrs, a comprehensive explanation remains unclear [22].

The human genome contains ~2,000 copies of human-specific L1s [23, 24]. However, among them, only approximately 80 L1 copies are thought to be able to retrotranspose actively in the human genome. These are called retrotransposition-competent L1s and may be further divided into two groups, the "active" and "dead" L1s, according to whether they can be shown to retrotranspose *in vitro* [25]. To explore any connection between the rate of substitution and the regulatory elements residing within the L1 5′ UTR, we identified substitutions along the 5′ UTR sequences of the human-specific L1s. In addition, we estimated the mutation rate of each component of the human-specific L1s, based on the divergence among them. The comparison of the mutation rate of the L1 5′ UTRs with those of other L1 structural components (ORF1, ORF2, a partial of ORF2 and 3′ UTR, and the 3′ UTR) indicates that the 5′ UTR has accumulated mutations at a faster rate than the other components. Furthermore, an analysis of the CpG sites in all components found no correlation between the CpG contents and mutation rates of the L1 components.

2. Materials and Methods

2.1. Dataset. A total of 1,835 human-specific L1 elements were previously identified [23]. We manually inspected them to extract full-length L1 elements because this study focuses on the L1 5′ UTR. This manual inspection process yielded 443 full-length L1 elements including 897 bp of the 5′ UTR sequence. We had previously identified the genomic positions of the 443 human-specific L1 elements in hg17 (UCSC May 2004 freeze) of the human genome reference sequence [23]. We converted these hg17 positions into positions within the current hg19 (UCSC February 2009 freeze) assembly of the human genome reference sequence using the BLAST-Like Alignment Tool (BLAT) utility (http://genome.ucsc.edu/cgi-bin/hgBlat) [26]. The genomic loci of the set of full-length human-specific L1s are described in Table 1 in supplementary material available online at doi: 10.1155/2012/129416.

2.2. Estimation of Substitution Frequencies on L1 5′ UTRs. In a previous study of human-specific L1s, six L1 subfamilies were established based on the diagnostic mutations that are shared between all members of each subfamily [23]. In this study, we grouped the 443 full-length L1 elements into the six subfamilies, L1Hs-Ta1, L1Hs-Ta0, L1Hs-preTa, L1Hs-1AB, L1PA2, and L1PA3, based on the previously established taxonomy [23]. However, the previous study had examined only a partial segment of each L1 element, an 864 bp segment corresponding to the 3′ end of ORF2 and the entire 3′ UTR. It was therefore necessary for us to construct a full-length consensus sequence for each of the L1 subfamilies. Using the module MegAlign, available in the package DNAStar, we generated the consensus sequence for each L1 subfamily.

For the subfamily of L1Hs-preTa, we aligned the 5′ UTRs of all members and compared them with its consensus sequence using the biological sequence alignment editor BioEdit v.7.0 [27]. For the alignment, we discarded the first nine base pairs of the 5′ UTR because the most 5′ end of human L1s is known to be highly variable [20]. In Excel, we generated a matrix of substitutions relative to the consensus sequence at each nucleotide position within the 5′ UTR throughout all elements in the alignment. Using the same method, we examined the substitutions existing within the 5′ UTRs of the other L1 subfamilies represented in our dataset. Finally, we summed the substitution numbers counted from each subfamily to calculate the substitution frequencies per nucleotide position across all human-specific L1 5′ UTRs.

2.3. Age Estimates of Human-Specific L1s Based on L1 Components. All of the human-specific L1 elements were divided into their 5′ UTR, ORF1, ORF2, and 3′ UTR components using a combination of the L1Xplorer (http://line1.bioapps.biozentrum.uni-wuerzburg.de/l1xplorer.php) annotation tool [28] and manual inspection. The sizes of the 5′ UTR, ORF1, ORF2, and 3′ UTR were 897, 1017, 3828, and 202 bp, respectively. We grouped the L1 components based on the categories of subfamily and component and aligned the members of each group using the BioEdit v.7.0 software [27]. The relationship of each component within an L1 subfamily was reconstructed using a median-joining network [29, 30], as implemented in the NETWORK 4.6 (http://www.fluxus-engineering.com/sharenet.htm) software [29]. Then, the ages of the human-specific L1s were estimated based on the divergence among all the copies of each group. For this calculation, we used 0.15% per site per myr as the nucleotide mutation rate [31]. A previous study of human-specific L1s suggested that this method is useful in estimating the age of human-specific L1s [23]. To compare the observed mutation rates between the L1 components, we assumed that older apparent age was indicative of a higher mutation rate for the component.

2.4. Analysis of the CpG Contents of the Four L1 Components. Using an in-house Perl script, we counted the numbers of CpG and GpC sites on the consensus sequence of each L1 subfamily. To compare CpG contents among the four L1 components, we calculated the ratio of CpG to GpC for each component. Pearson's correlation coefficient (r) between

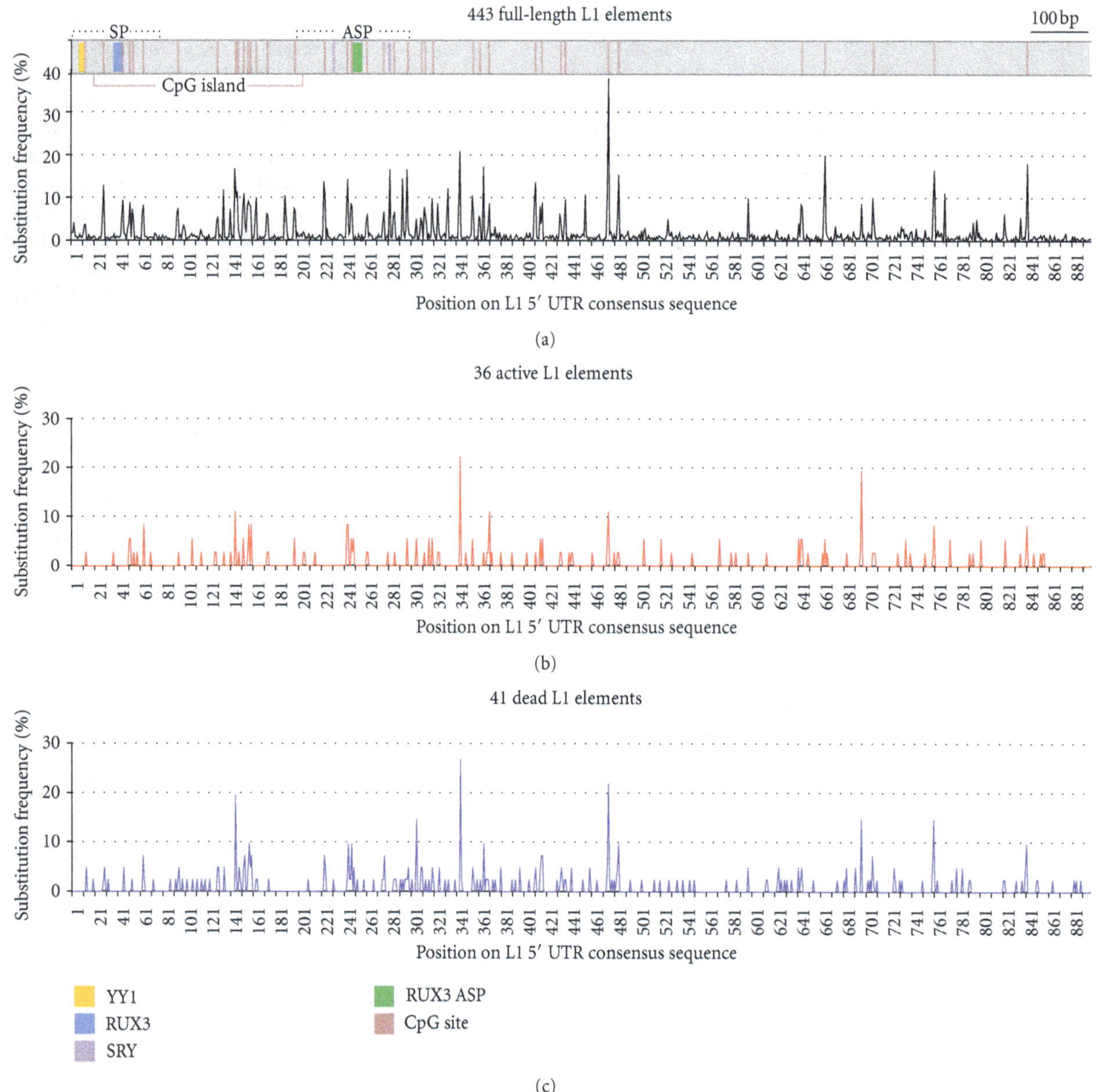

FIGURE 1: Substitution frequencies along the 5′ UTR sequences of human-specific L1 elements. The percentage of substitutions per nucleotide position along the 5′ UTR sequence was calculated. The structure of the L1 5′ UTR is shown on the top. The TF binding sites, sense promoter (SP), antisense promoter (ASP), CpG sites, and CpG islands are indicated by colored boxes and lines. (a) 443 full-length human-specific L1 elements. (b) 36 active L1 elements *in vitro*. (c) 41 dead L1 elements *in vitro*.

CpG contents and mutation rates of the L1 components was calculated. The two-tailed P value (i.e., statistical significance) was considered using the online freely available software, GraphPad QuickCalcs (http://www.graphpad.com/quickcalcs/index.cfm).

3. Results and Discussion

3.1. Substitutions within the 5′ UTRs of Full-Length Human-Specific L1s. L1 elements are transcribed by RNA polymerase II, but an internal promoter within the 5′ UTR is independent of the TATA-box. L1 transcription is initiated at variable positions within the L1 5′ UTR. Thus, it is clear that the 5′ UTR plays an important role in regulating the initiation of L1 transcription. Sequence differences within the 5′ UTR can result in different transcriptional activities because single or a combination of nucleotide differences within the L1 5′ UTRs influences the promoter activity [32]. From a previous study of human-specific L1s [23], we identified 443 full-length L1 elements and grouped them into six different L1 subfamilies using their diagnostic substitutions. By comparing the sequences of the L1 5′ UTRs with the respective consensus sequence for each subfamily, we identified mutations within the 5′ UTRs of the 443 human-specific L1s. Figure 1(a) shows

Table 1: Average frequency of substitutions across the promoter and TF binding sites of the L1 5′ UTR.

Region	Active L1 elements (%)	Dead L1 elements (%)	All L1 elements in the human genome (%)
Sense promoter	0.66	0.99	2.14
YY1	0.15	0.27	1.38
Runx3	0.00	0.26	0.53
SRY-1	0.00	0.49	0.81
Runx3 ASP	0.29	0.26	0.86
SRY-2	0.00	0.00	0.41
5′ UTR overall	0.53	0.78	1.62

the substitution frequency per nucleotide position among the L1 5′ UTRs. The colored boxes indicate TF binding sites: a YY1 binding site, two putative SRY-related TF binding sites, and two putative RUNX3 TF binding sites. As can be seen in Table 1, the average frequency of substitutions across the entire length of the L1 5′ UTR is 1.62%. It has been suggested that the coexistence of distinctive L1 5′ UTRs is unstable in the host genome because they compete with one another for host factor(s) which is/are required for L1 transcription. Thus, we expected that the TF binding sites would show different substitution patterns from the other regions within the L1 5′ UTR. The results of this analysis show that substitution frequencies in the TF binding sites were indeed far lower than the average frequency, except for the YY1 binding site, which showed a similar substitution frequency to the average frequency. This result makes some sense in light of the previous finding that the YY1 binding site is not required for L1 retrotransposition [21]. However, it also presents a bit of a paradox because the presence of the YY1 binding site is the only feature common to all of the seven distinctive types of L1 5′ UTR that emerged during last 70 myrs of L1 evolution [22].

Unlike other regulatory sequences, the promoter within the L1 5′ UTR showed a higher substitution frequency than the overall average, as can be seen in Table 1. We believe that the high mutation rate of L1 5′ UTR can be largely attributed to the high mutation rate of the L1 promoter. To initiate L1 transcription, RNA polymerase II needs to recognize and bind to the L1 promoter. Thus, mutations that accumulate within the L1 promoter have the potential to reduce L1 retrotransposition activity and the number of retrotransposition-competent L1s in the human genome. We suggest that any host defense system(s) against L1 elements may be among the factors contributing to the high substitution frequency observed within the L1 promoter region of the 5′ UTR.

In one recent study of the retrotransposition activity of human-specific L1s, 82 human-specific L1s with intact ORF1 and ORF2 were identified; the ORFs encode proteins that are essential for L1 retrotransposition. In the study, the L1s were cloned to test L1 retrotransposition in cultured cells. In the cell culture assay, 40 L1s were experimentally shown to be active, while the others appear retrotranspositionally inactive in human cells [25]. Because both of the two groups retain intact ORF1 and ORF2, we reasoned that the sequence difference between their 5′ UTRs is responsible for the difference between their retrotransposition abilities. As seen in Figures 1(b) and 1(c), we examined the substitution frequency per nucleotide position on the 5′ UTRs of these two L1 groups. No significant difference between the substitution patterns of the two groups was detected: the substitution frequencies of "active" and "dead" L1 5′ UTRs averaged 0.53% and 0.78%, respectively. While at first glance this result seems surprising, we believe that it is reasonable. As mentioned previously, the initial step in L1 retrotransposition is transcription, and the 5′ UTR is known to regulate this event. *In vivo*, only the L1 elements that have a functional 5′ UTR are able to retrotranspose. In contrast, in the cell culture assay of L1 retrotransposition, it did not matter whether the L1s had a functional 5′ UTR for the initiation of L1 transcription as the transcription of both "active" and "dead" Lls was initiated from the promoters of cloning vectors. In other words, the two L1 groups do not directly represent retrotransposition-competent L1s and retrotransposition-incompetent L1s *in vivo*. Thus, it is possible that some or most of the "active" L1s previously reported have no functional 5′ UTR *in vivo*. Regarding that all of the "dead" L1s are presumably inactive *in vivo*, it was not a surprising result that the two different L1 groups have similar substitution patterns within their 5′ UTRs.

3.2. The Mutation Rates of L1 5′ UTRs Compared with Those of Other L1 Components. After an L1 element inserts into the human genome, mutations accumulate at different rates at each nucleotide position within the L1 element. These mutation rates are usually measured in substitutions (fixed mutations) per base pair per generation. A full-length human L1 is divided into four components: the 5′ UTR, ORF1, ORF2, and the 3′ UTR. Using the NETWORK 4.6 (http://www.fluxus-engineering.com/sharenet.htm) software [29], we estimated the ages of the six L1 subfamilies, based on the divergence among the 5′ UTRs of all the copies of each subfamily [23, 29]. In addition, we calculated their ages based on the divergence among their 3′ UTRs using the same method. As can be seen in Table 2, the estimated ages of the 443 human-specific L1s averaged 9.90 myrs old, which was calculated based on the divergence among the 5′ UTRs. This estimate is older than 7.75 myrs old, which was calculated based on the divergence among the 3′ UTRs. Previous studies of L1 elements constructed the phylogenetic relationships among different L1 subfamilies using the combined sequences of a partial ORF2 (pORF2) and the 3′ UTR, as most L1 insertions are truncated. It was assumed that L1 elements accumulate mutations at the neutral rate after their insertion [33, 34]. The age estimates of L1 subfamilies based on the divergence among the combined L1 sequences did indeed show that mutations have occurred within human-specific L1s at the neutral rate after their insertion into the host genome [23]. However, as can be seen in Table 2, the mutation rate of the L1 5′ UTR is faster than the neutral rate and faster even than that of the 3′ UTR. This high mutation rate is likely detrimental for many L1 copies in terms of retrotransposition

Table 2: Age estimation of human-specific L1 elements based on each L1 component.

Subfamily[a]	Subfamily[b]	No. of full-length L1 elements	5′ UTR Age ± SD (myrs)	ORF1 Age ± SD (myrs)	ORF2 Age ± SD (myrs)	3′ UTR only Age ± SD (mys)	pORF2 and 3′ UTR Age ± SD (myrs)[b]
L1PA3	L1PA3-1A L1PA3-1Aa L1PA3-1B L1PA3-1Ba L1PA3-1Bb L1PA2-1A	106	20.30 ± 0.65	12.83 ± 0.71	11.31 ± 0.73	15.32 ± 2.11	12.71 ± 0.63
L1PA2	L1PA2-1B L1PA2-1C L1PA2-1D L1PA2-1Da L1PA2-1Db L1PA2-1E	147	13.85 ± 0.38	9.15 ± 0.59	9.63 ± 0.69	11.32 ± 1.14	7.62 ± 0.47
L1HS-1AB	L1HS-1A L1HS-1B	32	8.74 ± 0.53	5.50 ± 0.57	4.88 ± 0.63	10.48 ± 1.87	5.09 ± 0.56
L1HS-preTa	L1HS-preTa	62	7.22 ± 0.41	3.95 ± 0.41	3.79 ± 0.89	4.15 ± 0.84	3.13 ± 0.25
L1HS-Ta0	L1HS-Ta0	38	4.90 ± 0.33	3.25 ± 0.42	3.29 ± 0.47	2.96 ± 0.51	2.73 ± 0.22
L1HS-Ta1	L1HS-Ta1	58	4.38 ± 0.65	2.35 ± 0.43	2.28 ± 0.44	2.29 ± 0.41	1.94 ± 0.20
Average		443[c]	9.90 ± 0.49	6.17 ± 0.52	5.86 ± 0.64	7.75 ± 1.15	5.54 ± 0.39

[a]In this study.
[b]Source Lee et al. [23].
[c]Total number of L1 elements.

activity, but it could also offer a benefit to L1s if it increases the speed at which these elements can adapt to changing genomic circumstances.

We extended this analysis to ORF1 and ORF2 to compare their mutation rates with that of the 5′ UTR. The estimated ages of the 443 human-specific L1s averaged 6.17 and 5.86 myrs old, which were calculated based on divergences among ORF1s and ORF2s of all the copies of each L1 subfamily, respectively. These two age estimates are similar to the estimate of 5.54 myrs old, which was calculated based on the divergence among the combined sequences of the L1s. As shown in Figure 2, we compared the age estimates calculated for each of the four components of the L1s (the 5′ UTR, ORF1, ORF2, and the 3′ UTR) with the age estimate calculated for the combined sequence, using Welch's t-test [35]. The standard deviations of the age estimates are nonoverlapping. Especially, the pairwise comparison of the two age estimates based on the divergence among 5′ UTRs and the divergence among 3′ UTRs showed that they are significantly different to one another: the two-tailed P value was less than 0.0001. Given these findings, we can assert that the 5′ UTRs of L1s accumulate mutations at a rate faster than the neutral rate. Interestingly, despite this faster mutation rate within the 5′ UTRs, human-specific L1s have shared the same type of L1 5′ UTR since the divergence of humans and chimpanzees [23]. It has been suggested that L1s require host factor(s) for their replication, and that competition between different L1 subfamilies for these limiting resources may have led to the single L1 lineage observed in the human genome today [22]. Thus, it is possible that novel L1 5′ UTRs periodically emerge

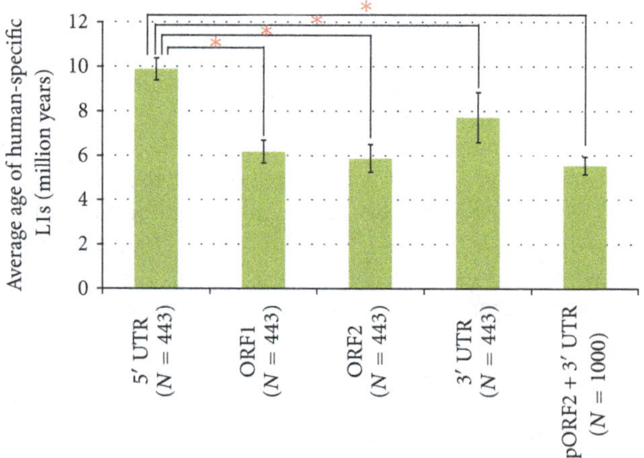

Figure 2: Average age estimates of human-specific L1 elements. N represents the number of samples. The pORF2 + 3′ UTR data are from Lee et al. [23]. * indicates significant differences of $P < 0.0001$ in Welch's t-test.

in the human genome via mutation, but they are not detected in this study as an L1 having a novel 5′ UTR is rapidly lost due to competition with preexisting L1s in the human genome.

3.3. The Comparison of Human-Specific L1 5′ UTRs with Chimpanzee-Specific L1 5′ UTRs. The structure of human-specific L1s is the same as that of chimpanzee-specific L1s, although the human-specific L1s have evolved in a single lineage while chimpanzee-specific L1s have evolved in two distinct lineages since the divergence of humans and chimpanzees [23]. Only 19 full-length chimpanzee-specific L1s have been detected in the chimpanzee genome [23]. We recovered these and examined the mutation rate of each component of the chimpanzee-specific insertions. Like their human-specific counterparts, the 5′ UTRs of the chimpanzee-specific elements showed the highest mutation rate (see Supplemetary Table 2). The overall mutation rate of the chimpanzee-specific L1s was by far higher than that of the human-specific L1s, which may have resulted from the relatively small sample size of chimpanzee-specific L1s available for analysis; unlike with the human-specific L1s, the chimpanzee-specific sample size was too small to divide into L1 subfamilies. The necessity of treating the chimpanzee-specific elements, which were likely comprised of members from several different subfamilies, as a single subfamily undoubtedly led to a higher divergence estimate among them and, subsequently, resulted in the high mutation rate of the chimpanzee-specific L1s. On the other hand, the finding that both groups of L1s show a higher mutation rate within their 5′ UTRs relative to the other L1 components or the L1 sequences as a whole is consistent with our other findings.

3.4. The Analysis of CpG Dinucleotides within Human-Specific L1 Elements. CpG dinucleotides occur at less than 10% of their expected frequency in the human genome because substitution at a CpG site, caused primarily by the methylation of cytosine, occurs 10- to 50-fold more frequently than other substitutions in primate genomes [36–38]. In spite of this high mutation rate from 5-methyl-CpG to TpG, many cellular functions such as gene expression rely on these CpG dinucleotides [39]. Any genomic region which contains a relatively high density of CpG sites is called a CpG island, and these islands are defined as regions of DNA of at least 200 bp with a GC percentage of greater than 50% and with an observed/expected CpG ratio of greater than 60% [40]. Under these criteria, we searched our L1 component sequences for CpG islands using the CpG island searcher utility (http://cpgislands.usc.edu) [41]. Among the four L1 components, only the 5′ UTR contains a CpG island. To determine whether CpG content is a major factor determining the mutation rates of the L1 components, we compared CpG contents among them. As shown in Figure 3, the 5′ UTR contains the highest density of CpG dinucleotides while ORF2 contains the lowest density. This observation of CpG content mirrored our observation of estimated mutation rates across L1 components, and we therefore wondered whether CpG content may be a major factor influencing the mutation rates of the L1 components. To investigate this hypothesis, we analyzed the correlation between the CpG contents and mutation rates of the L1 components and found that the mutation rate is not correlated with the CpG content ($r = 0.6665$, $P = 0.5356$). Thus, we can state that the relatively high frequency of CpG sites within the L1 5′ UTR is not a major source of the higher mutation rates observed in this component.

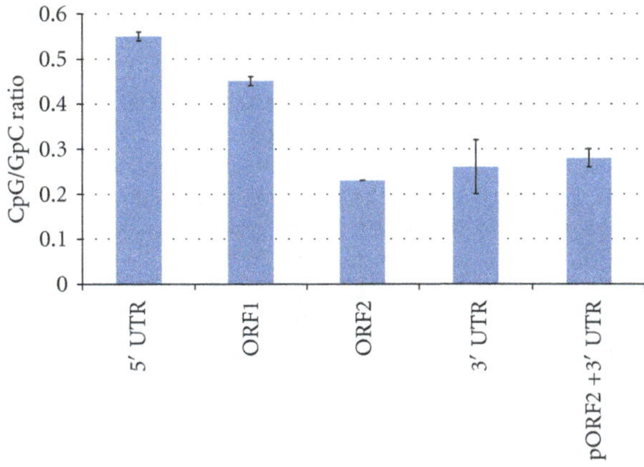

FIGURE 3: CpG-to-GpC ratio in L1 components. The vertical axis represents the ratio of CpG to GpC dinucleotide loci in each L1 component. The highest ratio is observed in the 5′ UTR.

4. Conclusions

In conclusion, the mutation rate of the L1 5′ UTR is higher than that of the other L1 components, ORF1, ORF2, and the 3′ UTR. However, the TF binding sites are relatively conserved among the human-specific L1 elements. We suspect that any host defense system(s) against L1 elements may be the cause of the higher mutation rate of the 5′ UTR and, in contrast, that L1 elements have kept the TF binding sites conserved against the host defense system(s) in order to survive; it could be the result of the evolutionary battle between L1s and their host. We believe that the increased frequency of substitutions within the 5′ UTRs could play a key role in regulating L1 retrotransposition activity and the number of retrotransposition-competent L1s in the human genome. However, not much is known about the factors causing the high level of sequence diversity in the L1 5′ UTRs. Although we suggest that the relationship between the L1 5′ UTR and other host factors, including the defense systems, causes the high mutation rate of the L1 5′ UTR, we could not rule out other possible factors such as the low fidelity of L1 reverse transcriptase. Thus, more research is needed to explore this intriguing possibility. Finding the factors that cause the increased mutation rates observed within L1 5′ UTRs will shed light on our understanding about L1 evolution and how the human genome may place controls on the retrotransposition rate of its resident L1 element population.

Acknowledgments

The authors thank two anonymous reviewers who helped to improve the manuscript. The present work was conducted by the Research Fund of Dankook University in 2010.

References

[1] E. S. Lander, L. M. Linton, B. Birren et al., "Initial sequencing and analysis of the human genome," *Nature*, vol. 409, no. 6822, pp. 860–921, 2001.

[2] P. L. Deininger and M. A. Batzer, "Alu repeats and human disease," *Molecular Genetics and Metabolism*, vol. 67, no. 3, pp. 183–193, 1999.

[3] P. A. Callinan, J. Wang, S. W. Herke, R. K. Garber, P. Liang, and M. A. Batzer, "Alu retrotransposition-mediated deletion," *Journal of Molecular Biology*, vol. 348, no. 4, pp. 791–800, 2005.

[4] D. E. Symer, C. Connelly, S. T. Szak et al., "Human L1 retrotransposition is associated with genetic instability in vivo," *Cell*, vol. 110, no. 3, pp. 327–338, 2002.

[5] N. Gilbert, S. Lutz-Prigge, and J. V. Moran, "Genomic deletions created upon LINE-1 retrotransposition," *Cell*, vol. 110, no. 3, pp. 315–325, 2002.

[6] K. Han, S. K. Sen, J. Wang et al., "Genomic rearrangements by LINE-1 insertion-mediated deletion in the human and chimpanzee lineages," *Nucleic Acids Research*, vol. 33, no. 13, pp. 4040–4052, 2005.

[7] S. K. Sen, K. Han, J. Wang et al., "Human genomic deletions mediated by recombination between Alu elements," *The American Journal of Human Genetics*, vol. 79, no. 1, pp. 41–53, 2006.

[8] J. Xing, H. Wang, V. P. Belancio, R. Cordaux, P. L. Deininger, and M. A. Batzer, "Emergence of primate genes by retrotransposon-mediated sequence transduction," *Proceedings of the National Academy of Sciences of the United States of America*, vol. 103, no. 47, pp. 17608–17613, 2006.

[9] J. M. Chen, P. D. Stenson, D. N. Cooper, and C. Férec, "A systematic analysis of LINE-1 endonuclease-dependent retrotranspositional events causing human genetic disease," *Human Genetics*, vol. 117, no. 5, pp. 411–427, 2005.

[10] P. L. Deininger and M. A. Batzer, "Mammalian retroelements," *Genome Research*, vol. 12, no. 10, pp. 1455–1465, 2002.

[11] H. H. Kazazian Jr., "Mobile elements: drivers of genome evolution," *Science*, vol. 303, no. 5664, pp. 1626–1632, 2004.

[12] A. F. Smit, "The origin of interspersed repeats in the human genome," *Current Opinion in Genetics and Development*, vol. 6, no. 6, pp. 743–748, 1996.

[13] H. H. Kazazian Jr. and J. V. Moran, "The impact of L1 retrotransposons on the human genome," *Nature Genetics*, vol. 19, no. 1, pp. 19–24, 1998.

[14] Q. Feng, J. V. Moran, H. H. Kazazian Jr., and J. D. Boeke, "Human L1 retrotransposon encodes a conserved endonuclease required for retrotransposition," *Cell*, vol. 87, no. 5, pp. 905–916, 1996.

[15] V. O. Kolosha and S. L. Martin, "In vitro properties of the first ORF protein from mouse LINE-1 support its role in ribonucleoprotein particle formation during retrotransposition," *Proceedings of the National Academy of Sciences of the United States of America*, vol. 94, no. 19, pp. 10155–10160, 1997.

[16] S. L. Mathias, A. F. Scott, H. H. Kazazian Jr., J. D. Boeke, and A. Gabriel, "Reverse transcriptase encoded by a human transposable element," *Science*, vol. 254, no. 5039, pp. 1808–1810, 1991.

[17] T. G. Fanning and M. F. Singer, "LINE-1: a mammalian transposable element," *Biochimica et Biophysica Acta*, vol. 910, no. 3, pp. 203–212, 1987.

[18] D. D. Luan, M. H. Korman, J. L. Jakubczak, and T. H. Eickbush, "Reverse transcription of R2Bm RNA is primed by a nick at the chromosomal target site: a mechanism for non-LTR retrotransposition," *Cell*, vol. 72, no. 4, pp. 595–605, 1993.

[19] A. F. Smit, G. Toth, A. D. Riggs, and J. Jurka, "Ancestral, mammalian-wide subfamilies of LINE-1 repetitive sequences," *Journal of Molecular Biology*, vol. 246, no. 3, pp. 401–417, 1995.

[20] S. T. Szak, O. K. Pickeral, W. Makalowski, M. S. Boguski, D. Landsman, and J. D. Boeke, "Molecular archeology of L1 insertions in the human genome," *Genome Biology*, vol. 3, no. 10, research 0052, 2002.

[21] J. N. Athanikar, R. M. Badge, and J. V. Moran, "A YY1-binding site is required for accurate human LINE-1 transcription initiation," *Nucleic Acids Research*, vol. 32, no. 13, pp. 3846–3855, 2004.

[22] H. Khan, A. Smit, and S. Boissinot, "Molecular evolution and tempo of amplification of human LINE-1 retrotransposons since the origin of primates," *Genome Research*, vol. 16, no. 1, pp. 78–87, 2006.

[23] J. Lee, R. Cordaux, K. Han et al., "Different evolutionary fates of recently integrated human and chimpanzee LINE-1 retrotransposons," *Gene*, vol. 390, no. 1-2, pp. 18–27, 2007.

[24] R. E. Mills, E. A. Bennett, R. C. Iskow et al., "Recently mobilized transposons in the human and chimpanzee genomes," *The American Journal of Human Genetics*, vol. 78, no. 4, pp. 671–679, 2006.

[25] B. Brouha, J. Schustak, R. M. Badge et al., "Hot L1s account for the bulk of retrotransposition in the human population," *Proceedings of the National Academy of Sciences of the United States of America*, vol. 100, no. 9, pp. 5280–5285, 2003.

[26] W. J. Kent, "BLAT—the BLAST-like alignment tool," *Genome Research*, vol. 12, no. 4, pp. 656–664, 2002.

[27] T. A. Hall, "BioEdit: a user-friendly biological sequence alignment editor and analysis program for Windows 95/98/NT," *Nucleic Acids Symposium Series*, vol. 41, pp. 95–98, 1999.

[28] T. Penzkofer, T. Dandekar, and T. Zemojtel, "L1Base: from functional annotation to prediction of active LINE-1 elements," *Nucleic Acids Research*, vol. 33, pp. D498–D500, 2005.

[29] H. J. Bandelt, P. Forster, and A. Röhl, "Median-joining networks for inferring intraspecific phylogenies," *Molecular Biology and Evolution*, vol. 16, no. 1, pp. 37–48, 1999.

[30] R. Cordaux, D. J. Hedges, and M. A. Batzer, "Retrotransposition of Alu elements: how many sources?" *Trends in Genetics*, vol. 20, no. 10, pp. 464–467, 2004.

[31] M. M. Miyamoto, J. L. Slightom, and M. Goodman, "Phylogenetic relations of humans and African apes from DNA sequences in the psi eta-globin region," *Science*, vol. 238, no. 4825, pp. 369–373, 1987.

[32] L. Lavie, E. Maldener, B. Brouha, E. U. Meese, and J. Mayer, "The human L1 promoter: variable transcription initiation sites and a major impact of upstream flanking sequence on promoter activity," *Genome Research*, vol. 14, no. 11, pp. 2253–2260, 2004.

[33] E. Pascale, C. Liu, E. Valle, K. Usdin, and A. V. Furano, "The evolution of long interspersed repeated DNA (L1, LINE 1) as revealed by the analysis of an ancient rodent L1 DNA family," *Journal of Molecular Evolution*, vol. 36, no. 1, pp. 9–20, 1993.

[34] C. F. Voliva, S. L. Martin, C. A. Hutchison III, and M. H. Edgell, "Dispersal process associated with the L1 family of interspersed repetitive DNA sequences," *Journal of Molecular Biology*, vol. 178, no. 4, pp. 795–813, 1984.

[35] B. L. Welch, "The generalisation of student's problems when several different population variances are involved," *Biometrika*, vol. 34, pp. 28–35, 1947.

[36] J. Sved and A. Bird, "The expected equilibrium of the CpG dinucleotide in vertebrate genomes under a mutation model," *Proceedings of the National Academy of Sciences of the United States of America*, vol. 87, no. 12, pp. 4692–4696, 1990.

[37] R. Anbazhagan, J. G. Herman, K. Enika, and E. Gabrielson, "Spreadsheet-based program for the analysis of DNA methylation," *BioTechniques*, vol. 30, no. 1, pp. 110–114, 2001.

[38] K. J. Fryxell and W. J. Moon, "CpG mutation rates in the human genome are highly dependent on local GC content,"

Molecular Biology and Evolution, vol. 22, no. 3, pp. 650–658, 2005.

[39] M. Li and S. S. Chen, "The tendency to recreate ancestral CG dinucleotides in the human genome," *BMC Evolutionary Biology*, vol. 11, article 3, 2011.

[40] M. Gardiner-Garden and M. Frommer, "CpG islands in vertebrate genomes," *Journal of Molecular Biology*, vol. 196, no. 2, pp. 261–282, 1987.

[41] D. Takai and P. A. Jones, "Comprehensive analysis of CpG islands in human chromosomes 21 and 22," *Proceedings of the National Academy of Sciences of the United States of America*, vol. 99, no. 6, pp. 3740–3745, 2002.

3

Generation and Analysis of Expressed Sequence Tags from *Chimonanthus praecox* (Wintersweet) Flowers for Discovering Stress-Responsive and Floral Development-Related Genes

Shunzhao Sui,[1] Jianghui Luo,[1] Jing Ma,[1] Qinlong Zhu,[2] Xinghua Lei,[3] and Mingyang Li[1]

[1] *College of Horticulture and Landscape, Chongqing Engineering Research Center for Floriculture, Key Laboratory of Horticulture Science for Southern Mountainous Regions of Ministry of Education, Southwest University, Chongqing 400715, China*
[2] *College of Life Science, South China Agricultural University, Guangzhou 510642, China*
[3] *Department of Botony, Chongqing Agricultural School, Chongqing 401329, China*

Correspondence should be addressed to Mingyang Li, limy@swu.edu.cn

Academic Editor: Jinfa Zhang

A complementary DNA library was constructed from the flowers of *Chimonanthus praecox*, an ornamental perennial shrub blossoming in winter in China. Eight hundred sixty-seven high-quality expressed sequence tag sequences with an average read length of 673.8 bp were acquired. A nonredundant set of 479 unigenes, including 94 contigs and 385 singletons, was identified after the expressed sequence tags were clustered and assembled. BLAST analysis against the nonredundant protein database and nonredundant nucleotide database revealed that 405 unigenes shared significant homology with known genes. The homologous unigenes were categorized according to Gene Ontology hierarchies (biological, cellular, and molecular). By BLAST analysis and Gene Ontology annotation, 95 unigenes involved in stress and defense and 19 unigenes related to floral development were identified based on existing knowledge. Twelve genes, of which 9 were annotated as "cold response," were examined by real-time RT-PCR to understand the changes in expression patterns under cold stress and to validate the findings. Fourteen genes, including 11 genes related to floral development, were also detected by real-time RT-PCR to validate the expression patterns in the blooming process and in different tissues. This study provides a useful basis for the genomic analysis of *C. praecox*.

1. Introduction

Chimonanthus praecox (L.) Link, wintersweet, belongs to the Calycanthaceae family. It is a perennial deciduous shrub and blossoms in winter, from late November to March. Its unique flowering time and long blooming period make it one of most popular ornamental plants in China [1]. *C. praecox* is mainly a garden plant that also provides cut flowers. The flower is strongly fragrant and may be used as a source of essential oil, which has received much attention in New Zealand [2]. *C. praecox* thrives in cold environments and blooms in low-temperature seasons with little rainfall. The plant is assumed to be rich in genes related to floral development and adversities, especially those responding to environmental stress factors. However, the molecular mechanism that regulates floral development and copes with stresses in *C. praecox* flowers remains unclear.

Expressed sequence tags (ESTs) have been proven to be an efficient and rapid means to identify novel genes (and proteins) induced by environmental changes or stresses [3–7]. Genes related to flower form, longevity, and scent from roses, *Phalaenopsis equestris*, and *Pandanus fascicularis* were identified by ESTs [8–10]. The present study used transcriptomic analysis of *C. praecox* flowers to identify novel genes induced by environmental changes or related to floral development and ultimately to better understand the physiological and genetic basis of cold acclimation in flowers of woody plants.

2. Materials and Methods

2.1. Complementary DNA (cDNA) Library Construction and Sequencing. The process of *C. praecox* blossoming includes

Figure 1: Stages of *C. praecox* blooming. Stage 1, sprout period: bud scales loosen; Stage 2, flower-bud period: flower buds turn green; Stage 3, display-petal period: flower buds enlarge and turn yellow; Stage 4, initiating bloom period: fragrance emerges; Stage 5, bloom period: flowers fully open with strong fragrance; Stage 6, wither period: petals begin withering.

the following: Stage 1, sprout period; Stage 2, flower-bud period; Stage 3, display-petal period; Stage 4, initiating bloom period; Stage 5, bloom period; Stage 6, wither period [11]. *C. praecox* flower buds or flowers at the six stages of development (Figure 1) were collected from the nursery at Southwest University, Chongqing, China, for cDNA library construction. The samples were immediately frozen with liquid nitrogen and then refrigerated at −80°C until RNA isolation. Total RNA samples were extracted from these flower buds or flowers using RNA isolation kits (W6771; Watson Biotechnologies Inc.), and RNA quality was detected by BioSpec-mini (Shimadzu). The final RNA sample for the cDNA library construction was bulked by pooling equal amounts of total RNA from each stage.

cDNA synthesis from the mixed total RNA and library construction through directional cloning of cDNAs into the λTriplEx2 vector was performed using a SMART cDNA Library Construction Kit (K1051-1) according to the manufacturer's instructions. The size of the insert fragment and the recombinant rate were measured by PCR, as described by Gao and Hu [12], using a random selection of 50 clones. All clones were sequenced from the 5′ end using ABI 3700 (Applied Biosystems).

2.2. Sequence Clustering, Annotation, and Functional Categorization. Poor-quality sequences, sequences with less than 100 bases, and vector sequences were trimmed from the raw sequences using SeqMan II (DNASTAR, Inc., Madison, WI) and manually. The trimmed cDNA sequences were assembled into clusters using the assembly program within SeqMan II set to default parameters. The assembly parameters were set to require a minimum match of 80% over 12 bp to initiate the assembly process. A consensus sequence for unigenes was exported from SeqMan II.

Unigenes (contigs and singletons) were annotated using BLASTX against the NCBI nonredundant protein database with a cut-off E value of the best hit of $\leq 10^{-5}$ [13]. Sequences without a reliable match ($>10^{-5}$) were subsequently compared with the NCBI nonredundant nucleotide database by performing BLASTN (score >100) for complementary annotation [14]. All well-annotated unigenes were then further classified and mapped to the three Gene Ontology (GO) categories (biological, cellular, and molecular) via AmiGO (http://amigo.geneontology.org/) [15].

2.3. Expression Analysis of Selected Genes Using Quantitative Real-Time RT-PCR. For real-time expression studies, *C. praecox* seeds were kept under a 16 h light/8 h dark cycle in a growth chamber at 25°C. Small seedlings were subsequently transferred to plastic pots containing a mixture of soilrite and vermiculite (1 : 1). The plantlets were raised to the sixth leaf stage and then subjected to cold treatment (4°C) for 15 min, 1 h, and 6 h. Control plantlets were maintained at 25°C. The tissues were harvested and snap-frozen in liquid nitrogen and kept at −80°C until further use.

The total RNA of flowers at the different developmental stages as well as floral organs dissected from Stage 4 flowers, roots, stems, and leaves were extracted with the above-mentioned method. Equal amounts of DNA-free RNA (5 μg) from different tissues were reverse-transcribed using a PrimeScript RT Reagent Kit with gDNA Eraser (TaKaRa). The primers used for real-time PCR (Table 1) were designed by Primer Premier 5.0 (PREMIER Biosoft, USA). Real-time PCRs were performed in triplicate on 10 μL reactions using

Table 1: Forward (F) and reverse (R) primers used in real-time PCR.

Gene code	Annotation (sequence homology)	Primer sequence (5′ → 3′)	Tm (°C) a	Cycles	Product length
Cp88	Membrane channel protein	F: ATGTGGACTTTTTGCTGGAGTTTTT R: GCAATACAAGCATTTACCCAGTCCT	59	40	90
Cp173	Low temperature and salt responsive protein	F: CTGACGCTCTTTGGATGGCTACC R: ACGAATCAACGGTCACAAACACG	59	40	172
Cp215	Putative abscisic acid-induced protein	F: TGCTCTTTACTTGCTGGCTCGTCT R: TCTCGCCCATTCTCTTCTATGTATTTC	59	40	193
Cp274	Glutathione S-transferase	F: ATGTGCTAAATCTCATTCCAGTCTTCC R: GCCGTGATATCCTGCGAAACTAG	59	40	85
Cp359	1,4-Alpha-glucan-maltohydrolase	F: TATTATCTGGCAAGGTTCCACTCGTT R: TGTGTTGTAGAACCCAGCAGTCAGC	59	40	86
Cp364	Cold acclimation Protein WCOR413-like protein beta form	F: TCCAATAAGCCGAATCAAAATACAGT R: GTCTGTCTTCATCGCCAAATACTCC	59	40	84
Cp375	3-Oxoacyl-[acyl-carrier-protein] synthase	F: GTAACTGCGCCGAAACGAGAAAAAG R: CCCAAAGACAGAAACCAGCCCC	59	40	78
Cp440	Putative multiple stress-responsive zinc-finger protein	F: GTTTCCCGATTTTGTGTGGCG R: AGTTGTTGATGCAGAGAATTGGTCCT	59	40	183
Cp465	Catalase	F: CATCGTTCGCTTCTCTACCGTTATT R: TGTTTCCGACCAGATCAAAGTTACC	59	40	118
Cp197	AGL9.2	F: TGAGAATAAGATAAATCGGCAGGTGAC R: TTTTCCTCTGCTGGAGAAGACAACG	59	40	128
Cp203	LLP-B3 protein	F: GCGATGTGAAACTTGTCAGTAGCCC R: AGTTTGGCACAATCAGGACGAGGTT	59	40	171
Cp268	Caffeoyl-CoA *O*-methyl-transferase-like protein	F: AACACAAATGGAGAAGAGAAAACCC R: CATCGGCAGAGGTCGTCATAAGA	59	40	193
Cp297	SRG1-like protein	F: TAAGAACACCGTCCCATCTCGCTAC R: GAGTGCAGTCTCTCCAATTCATCAG	59	40	143

Table 1: Continued.

Gene code	Annotation (sequence homology)	Primer sequence (5′ → 3′)	Tm (°C) a	Cycles	Product length
Cp328	STYLOSA protein	F: GGGGAACCAACAAGAACAACAGAT R: GCCCAGTGGCACCATTTAGAAGAT	59	40	161
Cp330	RAC-like G-protein Rac1	F: ACCGCTTCTACCGCCTTTTCTCC R: AGGTCTTTCCAACAGCACCATCTCC	59	40	166
Cp360	Peroxisomal fatty acid betao-xidation multifunctional protein AIM1	F: TATGCTGGTGTTTTGAAAGAGGCG R: CTATGCCAGACCCCATCAAACCTC	59	40	193
Cp383	Secondary cell wall-related glycosyltransferase family 47	F: CAGGCGACACCCCCTCCTCTAAC R: TCAAGGCATCAGAGTTACGCACAAAG	59	40	150
Cp423	Putative polyketide synthase	F: GAGTTCAGACACTGTATGCCCTTGG R: TTGCTTGTAAAATTCGTCCTTCGTT	59	40	165
Cp436	Allergen-like protein BRSn20	F: CTCTTATCGCTCTCTGCTTCCTTCC R: GTTGCTGTTAACGTCTCTGCATTCC	59	40	176
Cp458	MADS-box protein 9	F: TGGTTGAGTATGATGTAGAGAAGCGAC R: ACCCATCATCTGAAGGTGTGCTATT	59	40	98
Cp82	Dormancy related protein	F: TGGAATAGAAAGAGCAAATACAGCG R: TACAAAAGGCTCAATGGCGTCC	59	40	172
Cp24	No hits	F: GCAGTTTACATACTACGGGAGAGGCT R: CGGCTTACGGAATCGTCATCAC	59	40	106
Cp64	Putative protein	F: CCAATCACTCTCCCTGAGGATGTAT R: TTCACCGACTCCTTGTTCTTTTAGC	59	40	159
Internal control	Actin	F: GTTATGGTTGGGATGGGACAGAAAG R: GGGCTTCAGTAAGGAAACAGGA	59	40	199
Internal control	Tublin	F: TAGTGACAAGACAGTAGGTGGAGGT R: GTAGGTTCCAGTCCTCACTTCATC	59	40	139

[a] Annealing temperature.

$5\,\mu$L of SsoFast EvaGreen Supermix (Bio-Rad), $0.5\,\mu$L of each primer (final concentration of 500 nM), $3.5\,\mu$L of water, and $0.5\,\mu$L of cDNA template. Table 1 shows the primer pair-specific temperatures. PCR was carried out using a Bio-Rad CFX96 real-time system machine based on the manufacturer's instructions under the following conditions: denaturation at 95°C for 30 s; 40 cycles of 95°C for 5 s; 59°C for 5 s. Dissociation kinetics was performed using the real-time PCR system at the end of the experiment (60 to 95°C; continuous fluorescence measurement) to check the annealing specificity of the oligonucleotides. A comparative Ct (threshold cycles) method of relative quantification was used to analyze the real-time quantitative RT-PCR data using Bio-Rad CFX Manager Software Version 1.6. Actin and tubulin were used as housekeeping genes for the calculation of relative transcript abundance. The sizes of the amplified products were confirmed via gel electrophoresis. Negative controls with no templates were carried out concurrently.

3. Results

3.1. General Characteristics of the cDNA Library and the Ests. A cDNA library constructed from *C. praecox* flowers at different stages of development was used as a source of ESTs. The titers of the primary library and amplified library were 1.4×10^6 and 1.0×10^{10} pfu/mL, respectively, with a recombinant rate of 96% for the original library. The sizes of the inserts ranged from 0.5 to 2.5 kb, and the average insert size was estimated to be 1.1 kb by PCR amplification of inserts from 50 randomly selected clones. These results indicate that our cDNA library was qualified.

In total, 896 random cDNA clones were successfully sequenced to generate ESTs. Trimming of the short sequences (<100 bp), vector sequences, and poor-quality sequences resulted in 867 high-quality ESTs, constituting a total of 584,201 bases in the *C. praecox* sequence. The average read length of these ESTs was 673.8 bp. All 867 sequences were deposited in GenBank under accession numbers DW222667 to DW223533. The clustering of ESTs generated 94 contigs (containing 2 or more ESTs) and 385 singletons (containing only 1 EST), yielding 479 unigenes. The redundancy of the library was calculated as 44.8% [(1 − Number of Unigenes/Number of ESTs) × 100%]. Figure 2 shows the distribution of ESTs in unigenes after clustering. Forty-two contigs had more than 2 ESTs, with the largest one containing 77 ESTs.

3.2. Functional Annotation and Classification of C. praecox Unigenes. The 479 unigenes were compared with the nonredundant protein and nucleotide sequences database in NCBI using BLAST. Four hundred five unique sequences, corresponding to 84.6% of all the unigenes, shared significant homology with sequences in the public databases. Of these, 266 were similar to genes of known functions, whereas 139 were similar to putative uncharacterized proteins (Table 2). The remaining 74 unigenes (15.4%) had only very weak or no matches and were considered as novel genes in *C. praecox* flowers.

Table 2: Overview of *C. praecox* flowers ESTs.

Items	Number
Total sequenced cDNA	896
Total number of ESTs analyzed	867
Total reading valid length (bp)	584,201
Average EST length (bp)	673.8
Unigenes	479
Contigs	94
Singletons	385
Redundancy	44.8

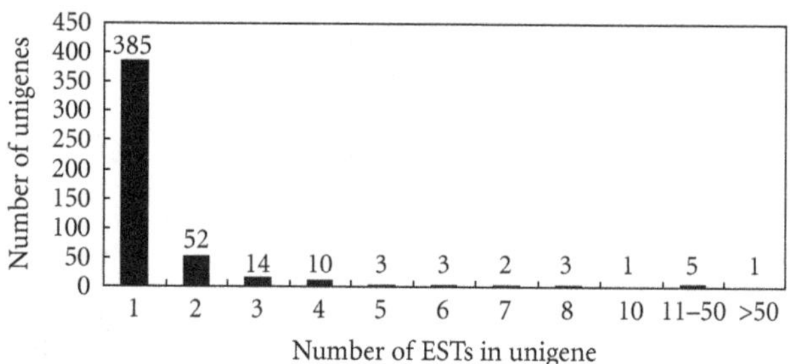

Figure 2: Distribution of *C. praecox* ESTs among unigenes.

Table 3 summarizes the highly expressed genes that contained more than 5 ESTs in one contig. The first and third most abundant ESTs were homologous to lipid transfer protein (LTP); the second most abundant ESTs encoded protein related to the adenine nucleotide translocator and then seed-specific protein, mannose-specific lectin, and LEA III protein (Table 3). These ESTs possibly corresponded to the most abundantly expressed genes in *C. praecox* flowers. Two hundred eighty-one ESTs were found to have more than five copies (Table 3), 385 had single copies of ESTs, and the remaining ESTs contained two to four copies in *C. praecox* flowers.

The database-matched 405 unigenes were found to have BLAST hits with 99 organisms, among which the highest number was from *Arabidopsis thaliana* (32.6%; 132 unigenes), followed by *Oryza sativa* (24.0%; 97 unigenes) and *Nicotiana tabacum* (4.2%; 17 unigenes) (Table 4). The remaining 67 unigenes (16.5%) had BLAST hits with a single organism only. The extensive distribution of the matched organisms may be attributed to the fact that the genome of *C. praecox* significantly differed from the genomes of model plants, as well as the fact that the relative plants' genomes have not yet been widely studied.

The initial annotations were further simplified into plant-specific annotations (plant GO slim; http://amigo geneontology.org/) to obtain additional insights into the putative functions of unigenes. Of the 479 *C. praecox* unigenes, 364 were assigned GO terms in any category (biological, cellular, and molecular). Figures 3, 4, and 5 classify the unigenes according to terms in the biological process ontology, molecular function ontology, and cellular component ontology, respectively.

Table 3: The 15 most abundant ESTs in the *C. praecox* flowers cDNA library.

Gene code	*Length(bp)*	*Putative function*	*E-value*	*Number of ESTs*
Cp1	1319	lipid transfer protein	1.00*E* − 34	77
Cp2	2124	adenine nucleotide translocator	9.00*E* − 80	47
Cp4	1024	lipid transfer protein	2.00*E* − 35	37
Cp6	1144	hypothetical protein	1.00*E* − 34	19
Cp7	710	seed specific protein Bn15D18B	4.00*E* − 19	1
Cp20	896	mannose specific lectin	5.00*E* − 28	17
Cp16	593	putative LEA III protein isoform 2	6.00*E* − 10	10
Cp10	900	glutathione S-transferase GST 22	4.00*E* − 80	8
Cp11	928	palmitoyl-acyl carrier protein thioesterase	7.00*E* − 22	8
Cp14	1531	hypothetical protein	2.00*E* − 79	8
Cp17	717	stearoyl acyl carrier protein desaturase	2.00*E* − 31	7
Cp26	1074	aquaporin	8.00*E* − 97	7
Cp23	696	S28 ribosomal protein	4.00*E* − 20	6
Cp24	574	No hits	NULL	6
Cp27	652	proline-rich protein	1.00*E* − 30	6
Total	14882	15		281

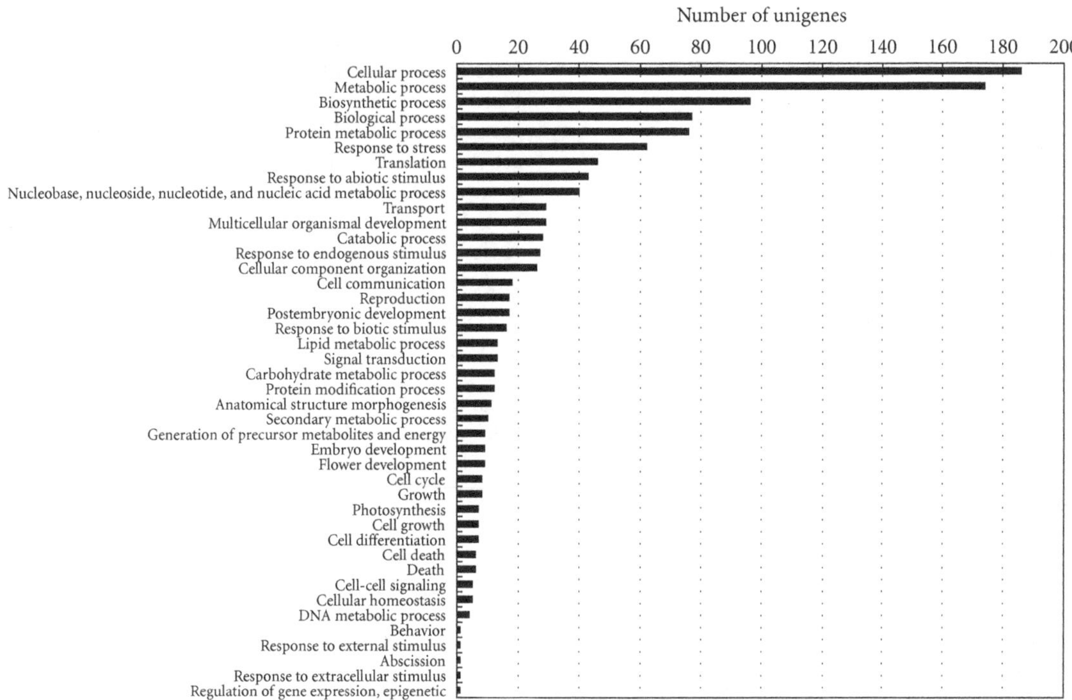

Figure 3: GO classification of the ESTs based on their biological functions in the *C. praecox* flower cDNA library.

3.3. Sequences Related to Stress and Defense. Among the ESTs that matched genes with known or putative functions, approximately 95 unigenes (291 ESTs) involved in stress and defense accounted for 19.8% (95/479) of all unigenes and 33.6% (291/867) of all ESTs. Table 5 shows the nonredundant ESTs that share similarities with genes related to defense and stress response according to GO classifications and previously published data. As expected, the ESTs involved in stress and defense were highly abundant in the library because the *C. praecox* thrives and blossoms in winter, thereby confirming previous reports of related transcripts with higher levels of defense in developing flowers [16–20].

Twelve unigenes (22 ESTs) related to cold stress tolerance were found in the library; these were classified as "response to cold" according to GO terms [e.g., β-amylase, acyl-CoA-binding protein, 3-hydroxyisobutyryl-coenzyme A hydrolase, catalase, low-temperature and salt-responsive protein, glutathione *S*-transferase (GST), membrane channel protein, and abscisic-acid-induced protein] and "cold acclimation" (e.g., fatty acid biosynthesis 1 and WCOR413-like protein). In this class, the most abundant sequences encode GST. Five GSTs encoded by 16 ESTs were identified in the library. GST genes exhibited a diverse range of responses to jasmonates, salicylic acid, ethylene, as well as oxidative

Table 4: Statistics of organism origin of matched-function homologs.

Organism name	No. of unigenes	Percentage (%)[a]
Arabidopsis thaliana	132	32.6%
Oryza sativa (japonica cultivar-group)	97	24.0%
Nicotiana tabacum	17	4.2%
Glycine max	8	2.0%
Lycopersicon esculentum	7	1.7%
Gossypium hirsutum	6	1.5%
Cucumis sativus	5	1.2%
Capsicum annuum	5	1.2%
Pisum sativum	4	1.0%
Solanum tuberosum	4	1.0%
Petunia x hybrida	4	1.0%
Nicotiana attenuata	3	0.7%
Helianthus annuus	3	0.7%
Beta vulgaris	3	0.7%
Calycanthus floridus var. glaucus	3	0.7%
Ipomoea batatas	3	0.7%
Hyacinthus orientalis	3	0.7%
Persea americana	3	0.7%
Ricinus communis	3	0.7%
Lily mottle virus	3	0.7%
Medicago sativa	2	0.5%
Zea mays	2	0.5%
Cicer arietinum	2	0.5%
Triticum aestivum	2	0.5%
Hevea brasiliensis	2	0.5%
Asparagus officinalis	2	0.5%
Panax ginseng	2	0.5%
Malus x domestica	2	0.5%
Elaeis oleifera	2	0.5%
Pinus taeda	2	0.5%
Spinacia oleracea	2	0.5%
The other 67 organisms	1	16.5%

[a] Proportion from unigenes with BLAST hits.

stress in *Arabidopsis* [21], and were induced by heavy metals and hypoxic stress in rice roots [22]. In the current study, however, not enough cold-related unigenes were obtained, as expected. Some unknown functional genes related to cold stress tolerance likely exist in this library.

Another class of genes involved in "response to absence of light," "low light intensity," and "response to red or blue light" according to GO terms was also represented in the library. Eight unigenes (10 ESTs) were identified, including acyl-CoA-binding protein, sadtomato protein, and catalase, among others.

Of the stress- and defense-related unigenes, 39 were possibly related to development. Nine types of LTPs or LTP precursors were encoded by the highest number of ESTs in this study (Table 5). Research has suggested different functions for LTPs in the physiology of plants, such as cutin synthesis, β-oxidation, somatic embryogenesis, pollen development, allergenics, plant signaling, and plant defense [23–25], but the true physiological role of LTPs in *C. praecox* flowers has yet to be determined. Fourteen ESTs encoding two kinds of LEA proteins, groups III and V, were identified. The presence of LEA proteins correlates well with freezing, water deficit, and salt stress [26–28], probably through the prevention of enzyme aggregation [29], and likely plays a similar role in *C. praecox*. Some other development-related unigenes were also found, such as MYB, calmodulin, actin, CCR4-associated factor, ubiquitin-conjugating enzyme E2, and ABC transporter, which are involved in transcription factor activity, signal transduction, cell structure, nucleotide metabolism, protein metabolic process, and transporter activity, respectively.

3.4. Sequences Related to Floral Development. Table 6 shows the 19 unigenes related to floral development. Five of these were homologous to MADS box transcription factor genes. Plant life critically depends on the function of MADS box genes encoding MADS domain transcription factors, which are present to a limited extent in nearly all major eukaryotic groups but constitute a large gene family in land plants [30].

MADS box genes control diverse developmental processes in flowering plants—ranging from root to flower and fruit development—and they especially control the processes of the transition from vegetative to reproductive development and establishment of floral organ identity [31]. The present study has also identified six unigenes related to secondary metabolism that are probably involved in floral color and fragrance.

The role of chilling temperature in dormancy in vegetative buds and induction of flowering has been investigated in many temperate-region species, particularly in *A. thaliana* and *Populus* spp. [32, 33]. The physiological processes of dormancy release and induction of flowering competence rely on longer-term chilling temperature and a period of vernalization, respectively [34, 35]. The current study has identified dormancy and vernalization-related genes; however, only one unigene (Cp82; E-value $= 6.00E^{-15}$) was annotated as a dormancy-related protein in the library The possible reason was only a small-scale sequencing in our study or the processes of dormancy release and induction of flowering competence in *C. praecox* only last a short period.

3.5. Expression Analysis of Cold-Responsive and Floral Development-Related Genes. Real-time RT-PCR was performed for 12 unigenes, including 9 selected from the GO Slim annotation belonging to "response to cold" (Cp88, Cp173, Cp215, Cp274, Cp359, Cp364, Cp375, Cp440, and Cp465), 1 annotated as a dormancy-related protein (Cp82), and 2 without functional annotation (Cp24 and Cp64), to analyze the changes in their expression due to cold stress (Figure 6). The data revealed that Cp82 responded to cold

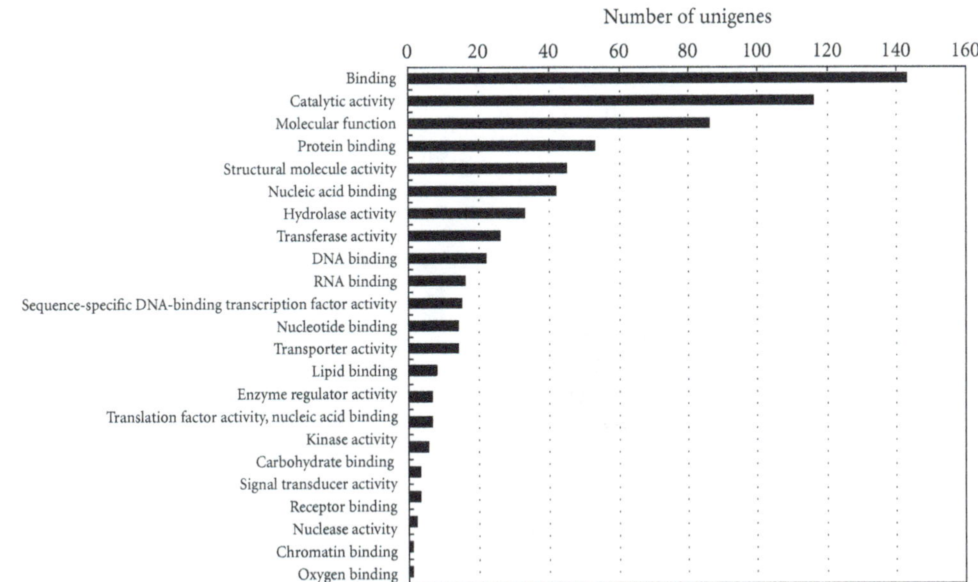

Figure 4: GO classification of the ESTs based on their molecular functions in the *C. praecox* flower cDNA library.

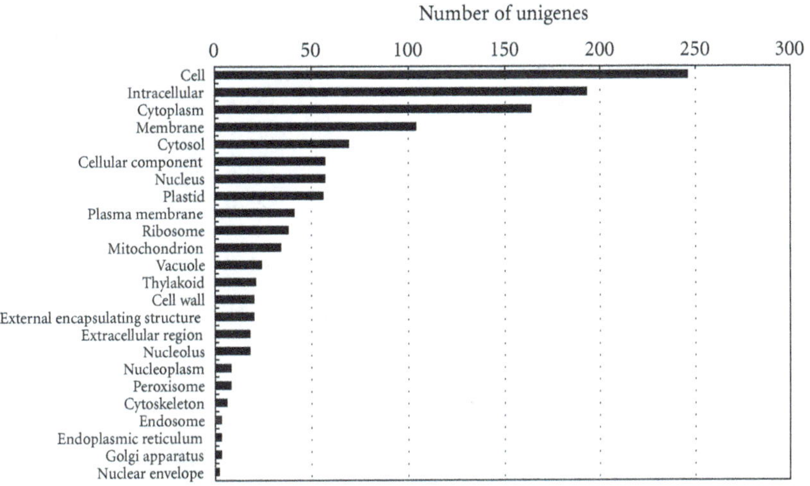

Figure 5: GO classification of the ESTs based on their cellular components in the *C. praecox* flower cDNA library.

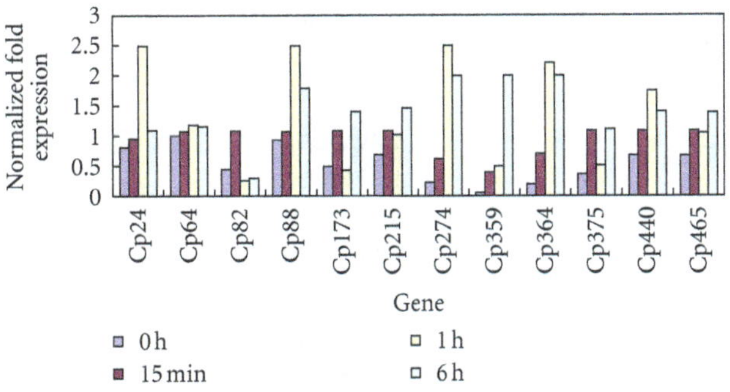

Figure 6: Expression analysis of 12 genes under cold stress (4°C).

stress immediately after treatment and reached peak expression levels as early as 15 min. Cp24 (no annotation), Cp88 (membrane channel protein), Cp274 (GST), Cp364 (cold acclimation protein WCOR413-like protein), and Cp440 (multiple stress-responsive zinc-finger protein) were upregulated after 15 min of treatment and reached their expression peaks at 1 h. Cp173 (low-temperature and salt-responsive protein), Cp215 (abscisic acid-induced protein), Cp359 (1,4-alpha-glucan-maltohydrolase), Cp375 (3-oxoacyl-[acyl-carrier-protein] synthase), and Cp465 (catalase) displayed later responses and reached their peak expression levels at 6 h. Cp64 (no annotation) exhibited minor changes in its transcript level.

Real-time RT-PCR was also applied to validate the expression patterns of 11 genes related to floral development (Cp197, Cp203, Cp268, Cp297, Cp328, Cp330, Cp360, Cp383, Cp423, Cp436, and Cp458), Cp24, Cp64, and Cp82 (Figure 7). The expression patterns of these 14 genes were analyzed in roots, stems, leaves, outer tepals, middle tepals, inner tepals, stamina, and pistils. The results showed

Table 5: Sequences related to stress and defense.

Gene code	Putative function	E-value	Matches no.	Number of ESTs
Cp465	catalase	1.00E − 148	BAC79443.1	1
Cp346	transporter	1.00E − 138	AAP13421.1	2
Cp382	peroxidase	1.00E − 125	AAK52084.1	1
Cp374	GMPase	1.00E − 122	AAT58365.1	1
Cp333	putative aldolase	1.00E − 122	AAM64281.1	1
Cp397	chitinase	1.00E − 105	AAF04454.1	1
Cp452	cysteine synthase	3.00E − 99	BAA05965.1	1
Cp314	tryptophan synthase beta chain	2.00E − 94	AAN15574.1	1
Cp357	Stromal cell-derived factor 2-like protein	2.00E − 93	XP_481233.1	1
Cp191	glutathione S-transferase GST 23	5.00E − 92	AAG34813.1	2
Cp25	dehydroascorbate reductase	2.00E − 89	AAL71857.1	3
Cp438	ras-related protein RAB8-3	9.00E − 89	BAB84324.1	1
Cp459	GTP-binding protein	1.00E − 86	AAN31076.1	1
Cp340	putative 3-Hydroxyisobutyryl-coenzyme A hydrolase	9.00E − 84	AAN41356.1	1
Cp379	cell-autonomous heat shock cognate protein 70	1.00E − 82	gAAN86276.1	1
Cp39	auxin-binding protein	3.00E − 80	AAB51240.1	4
Cp10	glutathione S-transferase GST 22	4.00E − 80	AAG34812.1	8
Cp304	putative 2-Nitropropane dioxygenase	2.00E − 79	AAL34288.1	3
Cp21	glutathione S-transferase GST 22	2.00E − 78	AAG34812.1	3
Cp311	GTP-binding protein	6.00E − 73	AAM12880.1	1
Cp353	patatin-like protein	1.00E − 73	CAB16788.1	2
Cp274	glutathione S-transferase	1.00E − 72	AAF61392.1	1
Cp359	1,4-Alpha-glucan-maltohydrolase	1.00E − 71	CAH60892.1	1
Cp59	BTF3b-like transcription factor	2.00E − 68	AAT67244.1	3
Cp76	NDKB_FLABI Nucleoside diphosphate kinase B	2.00E − 63	P47920	1
Cp166	putative tropinone reductase	6.00E − 59	AAM62552.1	1
Cp313	putative glutathione S-transferase T3	5.00E − 56	AAG16758.1	1
Cp156	protolysis and peptidolysis	6.00E − 56	CAA50022.1	1
Cp120	putative pyruvate dehydrogenase E1 alpha subunit	8.00E − 56	XP_467697.1	1
Cp80	sadtomato protein	7.00E − 55	AAS77347.1	1
Cp440	putative multiple stress-responsive zinc-finger protein	6.00E − 54	BAD35553.1	1
Cp175	epoxide hydrolase	6.00E − 50	AAB02006.1	1
Cp364	cold acclimation protein WCOR413-like protein beta form	4.00E − 49	AAG13394.1	1
Cp471	pathogenesis-related protein 4B	1.00E − 48	eCAA41438.1	1
Cp145	Thioredoxin	4.00E − 44	CAA77847.1	1
Cp406	auxin-responsive family protein	1.00E − 43	NP_565113.1	1
Cp38	basic PR-1 protein precursor	3.00E − 43	AAU20808.1	4
Cp215	putative abscisic acid-induced protein	4.00E − 41	BAD37454.1	1
Cp66	GLRX_RICCO Glutaredoxin	5.00E − 40	sp|$P55143$|	2
Cp235	histone H2A	5.00E − 38	CAA07234.1	1
Cp111	topoisomerase 6 subunit B	5.00E − 37	CAC24690.1	1
Cp232	cysteine proteinase inhibitor-like protein	2.00E − 36	BAB03156.1	2
Cp220	glyoxalase I family protein	2.00E − 35	NP_973579.1	2
Cp214	acyl-CoA-binding protein	3.00E − 35	CAA70200.1	1
Cp27	proline-rich protein	1.00E − 30	AAF78903.1	6
Cp20	lectin	5.00E − 28	AAD45250.1	17
Cp134	metallothionein-like protein	6.00E − 27	CAB52585.1	1
Cp127	Gip1-like protein	2.00E − 24	CAD10106.1	1
Cp296	bZIP transcription factor	2.00E − 24	AAN61914.1	2
Cp375	3-Oxoacyl-[acyl-carrier-protein] synthase	2.00E − 23	AAC78479.1	1
Cp104	metallothionein-like protein type 2	4.00E − 18	CAB77242.1	2
Cp372	allyl alcohol dehydrogenase	5.00E − 17	BAA89423.1	1
Cp36	metallothionein-like protein type 2	1.00E − 13	AAV97748.1	5

Table 5: Continued.

Gene code	Putative function	E-value	Matches no.	Number of ESTs
Cp173	low temperature and salt responsive protein	1.00E − 12	BAC23051.1	1
Cp472	cystatin	2.00E − 07	AAO18638.1	1
Cp88	membrane channel protein	3.00E − 06	CAB83138.1	2
Also related to development				
Cp391	actin	1.00E − 166	AAN40685.1	1
Cp402	xyloglucan endotransglycosylase precursor	1.00E − 110	AAC09388.1	1
Cp26	aquaporin	8.00E − 97	CAE53881.1	7
Cp381	4-Coumarate-CoA ligase-like protein	2.00E − 89	AAP03021.1	1
Cp461	glyceraldehyde-3-phosphate dehydrogenase	1.00E − 86	CAC88118.1	1
Cp291	3-Ketoacyl-CoA thiolase	6.00E − 84	CAA47926.1	1
Cp378	probable ubiquitin-conjugating enzyme E2	2.00E − 83	AAC32141.1	1
Cp400	auxin response factor 4	5.00E − 81	BAD19064.1	1
Cp288	expansin	4.00E − 80	AAO15998.1	2
Cp226	calmodulin	3.00E − 79	AAB68399.1	4
Cp323	putative calmodulin	3.00E − 73	CAC84563.1	2
Cp298	putative adrenal gland protein AD-004	2.00E − 65	XP_477670.1	1
Cp326	UDP-glucose:salicylic acid glucosyltransferase	2.00E − 65	AAF61647.1	1
Cp58	actin-depolymerizing factor 1	1.00E − 63	AAK72617.1	2
Cp167	ABC transporter	4.00E − 60	AAP80385.1	1
Cp286	putative CCR4-associated factor	4.00E − 60	AAN13153.1	1
Cp368	MYB8 protein	1.00E − 60	CAD87008.1	1
Cp324	actin depolymerizing factor	2.00E − 59	AAD23407.1	1
Cp450	TAF9	1.00E − 58	AAR28026.1	1
Cp405	myb family transcription factor-like	2.00E − 52	BAD29385.1	1
Cp77	putative actin-depolymerizing factor 1	4.00E − 48	XP_475079.1	1
Cp121	TAF10	3.00E − 41	AAR28030.1	1
Cp347	putative transcription factor	2.00E − 38	XP_479103.11	2
Cp4	lipid-transfer protein	2.00E − 35	AAS13435.1	37
Cp1	lipid transfer protein	1.00E − 34	CAA63340.1	77
Cp275	lipid transfer protein precursor	4.00E − 30	AAL27855.1	2
Cp9	lipid transfer protein 1 precursor	3.00E − 25	AAT45202.1	1
Cp216	lipid-transfer protein	3.00E − 24	NP_915960.1	4
Cp90	zinc finger homeobox family protein	3.00E − 22	NP_565088.1	1
Cp41	lipid transfer protein	3.00E − 18	BAC77694.1	1
Cp13	lipid transfer protein isoform 4	1.00E − 16	AAO33394.1	1
Cp338	polyprotein	2.00E − 16	CAD92110.1	1
Cp82	dormancy related protein, putative	6.00E − 15	AAG51119.1	1
Cp8	lipid-transfer protein	7.00E − 14	AAS13435.1	1
Cp141	lipid-transfer protein	4.00E − 13	AAS13435.1	1
Cp211	neutral/alkaline invertase 1	2.00E − 12	AAV28809.1	1
Cp451	putative MYB family transcription factor	8.00E − 11	AAD23043.1	1
Cp34	Lea5 protein	3.00E − 10	CAA86851.1	4
Cp16	putative LEA III protein isoform 2	6.00E − 10	CAC39110.1	10

clear differences in their expression. The genes related to floral development, except for Cp203 (LLP-B3 protein) and Cp297 (SRG1-like protein), increased by more than twofold in middle tepals. Cp24 and Cp203 presented an active expression in stamina but showed a very slight accumulation in other tissues. Cp268 (caffeoyl-CoA *O*-methyl-transferase-like protein) and Cp458 (MADS box protein 9) presented a peak in middle tepals but were only slightly or not expressed at all in roots, stems, leaves, and pistils. Cp197 (AGL9.2) was higher in all reproductive organs and presented its peak in middle tepals but was not detected in roots, stems, and leaves. The expression of Cp328 (STYLOSA protein) and that of Cp383 (secondary cell wall-related glycosyltransferase family 47) were not detected in roots; moreover, Cp328

Table 6: Sequences related to floral development.

Gene code	*Putative function*	*E-value*	*Matches no.*	*Number of ESTs*
Cp360	Peroxisomal fatty acid beta-oxidation multifunctional protein AIM1	1.00*E* − 108	XP_464920.1	1
Cp268	caffeoyl-CoA *O*-methyl-transferase-like protein	1.00*E* − 107	CAB80122.1	2
Cp237	MADS box protein	1.00*E* − 106	AAQ83835.1	2
Cp116	MADS box transcription factor AP3-1	2.00*E* − 95	AAF73928.1	1
Cp330	RAC-like G-protein Rac1	1.00*E* − 93	AAD47828.2	1
Cp217	beta-D-glucosidase	2.00*E* − 92	AAQ17461.1	1
Cp413	leucine-rich repeat transmembrane protein kinase	6.00*E* − 86	NP_192248.2	1
Cp408	isoflavone reductase related protein	3.00*E* − 85	AAC24001.1	1
Cp423	putative polyketide synthase	8.00*E* − 76	NP_919053.1	1
Cp335	AGL 6	3.00*E* − 73	AAY25580.1	2
Cp395	isopentenyl pyrophosphate:dimethyllallyl pyrophosphate isomerase	2.00*E* − 70	BAB09611.1	1
Cp297	SRG1-like protein	1.00*E* − 61	CAB81342.1	3
Cp383	secondary cell wall-related glycosyltransferase family 47	3.00*E* − 49	AAX33321.1	1
Cp197	AGL9.2	4.00*E* − 42	AAX15924.1	1
Cp436	allergen-like protein BRSn20	1.00*E* − 37	AAM62935.1	1
Cp407	caffeate *O*-methyltransferase	8.00*E* − 23	AAV36331.1	1
Cp203	LLP-B3 protein	1.00*E* − 22	AAN76546.1	1
Cp328	STYLOSA protein	1.00*E* − 15	CAF18245.1	1
Cp458	MADS-box protein 9	9.00*E* − 15	AAQ72497.1	1

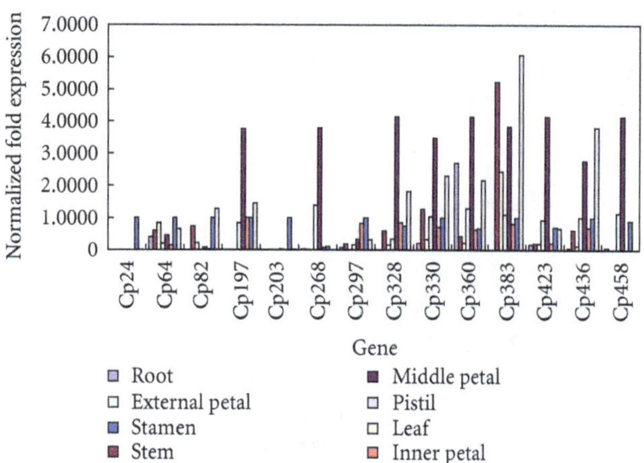

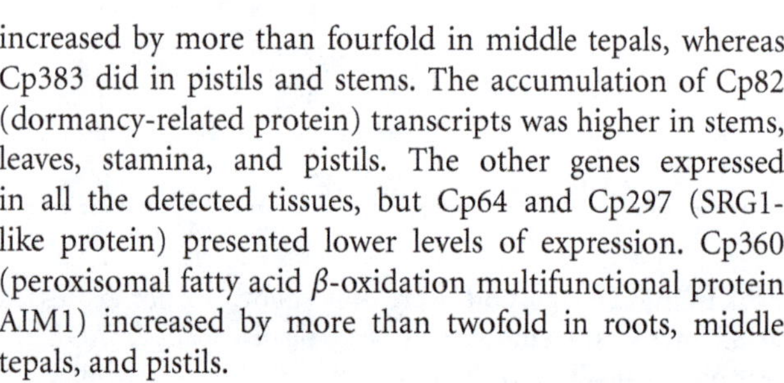

Figure 7: Expression analysis of 14 genes in different tissues of *C. praecox*.

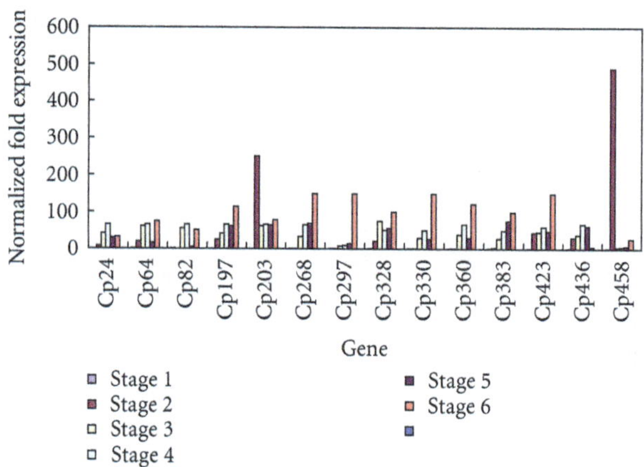

Figure 8: Expression analysis of 14 genes during *C. praecox* flowering.

increased by more than fourfold in middle tepals, whereas Cp383 did in pistils and stems. The accumulation of Cp82 (dormancy-related protein) transcripts was higher in stems, leaves, stamina, and pistils. The other genes expressed in all the detected tissues, but Cp64 and Cp297 (SRG1-like protein) presented lower levels of expression. Cp360 (peroxisomal fatty acid β-oxidation multifunctional protein AIM1) increased by more than twofold in roots, middle tepals, and pistils.

Quantitative real-time PCR methods were used to validate the transcript levels of the 14 genes further during the blooming process in *C. praecox* (Figure 8). The results showed that all these genes were not detected or very slightly expressed in Stage 1 and that Cp64, Cp82, Cp268, Cp297, Cp330, Cp360, and Cp383 had almost no accumulation in Stage 2. The transcript accumulations of Cp203 and Cp458 were sharply elevated to the highest level in Stage 2 but dramatically decreased at the subsequent stages of floral development. The expressions of Cp24, Cp82, and Cp436 presented a peak in Stage 4. The transcript accumulations of the other 9 genes increased during the six developmental stages and reached their peak in Stage 6. The expression of Cp297(SRG1-like protein) was significantly high in Stage 6 but very low in the other stages, and it was associated with flower senescence. The SRG1 gene is reportedly expressed in senescing organs of *Arabidopsis* plants [36].

The current study found few references about bud dormancy in *C. praecox*. The expression pattern of Cp82 was very attractive. However, the results indicate that Cp82 is not certainly related to flower-bud dormancy. The dormancy-related protein CAA93825.1, which matched Cp82 (E-value = $1e^{-15}$), has been reported to play a role in dormancy breaking and in the germination of *Trollius ledebourii* seeds [37]. These data warrant further research into the relativity of Cp82 to dormancy breaking and germination in *C. praecox* seeds.

4. Conclusions

A cDNA library was constructed to generate an EST collection from *C. praecox*, thereby providing a preliminary view into the genomic properties of this species. This collection of high-quality ESTs represents the first EST data set for *C. praecox*. Eight hundred sixty-seven valid EST sequences were generated, and 479 unigenes were assembled, among which 266 unigenes (55.53%) were identified according to their significant similarities with proteins of known functions. The EST sequences have been deposited in GenBank under accession numbers DW222667 to DW223533. Stress response genes and floral development-related genes were also identified. This study evaluated the expression patterns of 23 genes, including 2 novel ones, using real-time RT-PCR. Further investigations in this direction would help in the discovery of promising candidates with a key role in the development of stress tolerance for woody plants to bloom.

Acknowledgments

This work was supported by the National Natural Science Foundation of China (Grant nos. 31070622 and 30872063) and the Fundamental Research Funds for the Central Universities (Grant nos. XDJK2009A004 and XDJK2010C071).

References

[1] K. G. Zhao, M. Q. Zhou, L. Q. Chen, D. Zhang, and G. W. Robert, "Genetic diversity and discrimination of *Chimonanthus praecox* (L.) link germplasm using ISSR and RAPD markers," *HortScience*, vol. 42, no. 5, pp. 1144–1148, 2007.

[2] J. Q. Feng, "New Zealand flower industry—with special reference to wintersweet introduction and commercialization," *Journal of Beijing Forestry University*, vol. 29, supplement 1, pp. 4–8, 2007.

[3] H. Wei, A. L. Dhanaraj, L. J. Rowland, Y. Fu, S. L. Krebs, and R. Arora, "Comparative analysis of expressed sequence tags from cold-acclimated and non-acclimated leaves of *Rhododendron catawbiense* Michx," *Planta*, vol. 221, no. 3, pp. 406–416, 2005.

[4] R. Mahalingam, A. Gomez-Buitrago, N. Eckardt et al., "Characterizing the stress/defense transcriptome of *Arabidopsis*," *Genome Biology*, vol. 4, no. 3, article R20, 2003.

[5] G. Iturriaga, M. A. F. Cushman, and J. C. Cushman, "An EST catalogue from the resurrection plant *Selaginella lepidophylla* reveals abiotic stress-adaptive genes," *Plant Science*, vol. 170, no. 6, pp. 1173–1184, 2006.

[6] J. Zhang, U. P. John, Y. Wang et al., "Targeted mining of drought stress-responsive genes from EST resources in Cleistogenes songorica," *Journal of Plant Physiology*, vol. 168, no. 15, pp. 1844–1851, 2011.

[7] N. O. Ozgenturk, F. Oru, U. Sezerman et al., "Generation and analysis of expressed sequence tags from Olea europaea L," *Comparative and Functional Genomics*, vol. 2010, Article ID 757512, 9 pages, 2010.

[8] S. Channelière, S. Rivière, G. Scalliet et al., "Analysis of gene expression in rose petals using expressed sequence tags," *FEBS Letters*, vol. 515, no. 1–3, pp. 35–38, 2002.

[9] M. S. Vinod, H. M. Sankararamasubramanian, R. Priyanka, G. Ganesan, and A. Parida, "Gene expression analysis of volatile-rich male flowers of dioecious *Pandanus fascicularis* using expressed sequence tags," *Journal of Plant Physiology*, vol. 167, no. 11, pp. 914–919, 2010.

[10] W. C. Tsai, Y. Y. Hsiao, S. H. Lee et al., "Expression analysis of the ESTs derived from the flower buds of *Phalaenopsis equestris*," *Plant Science*, vol. 170, no. 3, pp. 426–432, 2006.

[11] C. L. Wu and N. Z. Hu, "Studies on the flower form and blooming characteristics of the wintersweet," *Acta Horticulturae Sinica*, vol. 22, no. 3, pp. 277–282, 1995.

[12] Q. K. Gao and C. Hu, "Construction of a cDNA library of host recognition kairomone for telenomus theophilae," *Insect Science*, vol. 9, no. 1, pp. 35–39, 2002.

[13] Y. Ogihara, K. Mochida, Y. Nemoto et al., "Correlated clustering and virtual display of gene expression patterns in the wheat life cycle by large-scale statistical analyses of expressed sequence tags," *Plant Journal*, vol. 33, no. 6, pp. 1001–1011, 2003.

[14] F. Sterky, S. Regan, J. Karlsson et al., "Gene discovery in the wood-forming tissues of poplar: analysis of 5,692 expressed sequence tags," *Proceedings of the National Academy of Sciences of the United States of America*, vol. 95, no. 22, pp. 13330–13335, 1998.

[15] S. Carbon, A. Ireland, C. J. Mungall et al., "AmiGO: online access to ontology and annotation data," *Bioinformatics*, vol. 25, no. 2, pp. 288–289, 2009.

[16] P. Fernández, N. Paniego, S. Lew, H. E. Hopp, and R. A. Heinz, "Differential representation of sunflower ESTs in enriched organ-specific cDNA libraries in a small scale sequencing project," *BMC Genomics*, vol. 4, article no. 40, 2003.

[17] T. Lotan, N. Ori, and R. Fluhr, "Pathogenesis-related proteins are developmentally regulated in tobacco flowers," *The Plant cell*, vol. 1, no. 9, pp. 881–887, 1989.

[18] A. D. Neale, J. A. Wahleithner, M. Lund et al., "Chitinase, β-1,3-glucanase, osmotin, and extensin are expressed in tobacco explants during flower formation," *Plant Cell*, vol. 2, no. 7, pp. 673–684, 1990.

[19] Q. Gu, E. E. Kawata, M. J. Morse, H. M. Wu, and A. Y. Cheung, "A flower-specific cDNA encoding a novel thionin in tobacco," *Molecular and General Genetics*, vol. 234, no. 1, pp. 89–96, 1992.

[20] A. H. Atkinson, R. L. Heath, R. J. Simpson, A. E. Clarke, and M. A. Anderson, "Proteinase inhibitors in Nicotiana alata stigmas are derived from a precursor protein which is processed into five homologous inhibitors," *Plant Cell*, vol. 5, no. 2, pp. 203–213, 1993.

[21] U. Wagner, R. Edwards, D. P. Dixon, and F. Mauch, "Probing the diversity of the *Arabidopsis* glutathione S-transferase gene family," *Plant Molecular Biology*, vol. 49, no. 5, pp. 515–532, 2002.

[22] A. Moons, "Osgstu3 and osgtu4, encoding tau class glutathione S-transferases, are heavy metal- and hypoxic stress-induced and differentially salt stress-responsive in rice roots," *FEBS Letters*, vol. 553, no. 3, pp. 427–432, 2003.

[23] A. D. O. Carvalho and V. M. Gomes, "Role of plant lipid transfer proteins in plant cell physiology—a concise review," *Peptides*, vol. 28, no. 5, pp. 1144–1153, 2007.
[24] C. Chen, G. Chen, X. Hao et al., "CaMF2, an anther-specific lipid transfer protein (LTP) gene, affects pollen development in *Capsicum annuum* L.," *Plant Science*, vol. 181, no. 4, pp. 439–448, 2011.
[25] A. Kiełbowicz-Matuk, P. Rey, and T. Rorat, "The organ-dependent abundance of a *Solanum* lipid transfer protein is up-regulated upon osmotic constraints and associated with cold acclimation ability," *Journal of Experimental Botany*, vol. 59, no. 8, pp. 2191–2203, 2008.
[26] C. Ndong, J. Danyluk, K. E. Wilson, T. Pocock, N. P. A. Huner, and F. Sarhan, "Cold-regulated cereal chloroplast late embryogenesis abundant-like proteins. Molecular characterization and functional analyses," *Plant Physiology*, vol. 129, no. 3, pp. 1368–1381, 2002.
[27] K. N. Dramé, D. Clavel, A. Repellin, C. Passaquet, and Y. Zuily-Fodil, "Water deficit induces variation in expression of stress-responsive genes in two peanut (*Arachis hypogaea* L.) cultivars with different tolerance to drought," *Plant Physiology and Biochemistry*, vol. 45, no. 3-4, pp. 236–243, 2007.
[28] S. C. Park, Y. H. Kim, J. C. Jeong et al., "Sweetpotato late embryogenesis abundant 14 (IbLEA14) gene influences lignification and increases osmotic- and salt stress-tolerance of transgenic calli," *Planta*, vol. 233, no. 3, pp. 621–634, 2011.
[29] J. Grelet, A. Benamar, E. Teyssier, M. H. Avelange-Macherel, D. Grunwald, and D. Macherel, "Identification in pea seed mitochondria of a late-embryogenesis abundant protein able to protect enzymes from drying," *Plant Physiology*, vol. 137, no. 1, pp. 157–167, 2005.
[30] L. Gramzow and G. Theissen, "A hitchhiker's guide to the MADS world of plants," *Genome Biology*, vol. 11, no. 6, article no. 214, 2010.
[31] A. Becker and G. Theißen, "The major clades of MADS-box genes and their role in the development and evolution of flowering plants," *Molecular Phylogenetics and Evolution*, vol. 29, no. 3, pp. 464–489, 2003.
[32] W. S. Chao, M. E. Foley, and D. P. Horvath et al., "Signals regulating dormancy in vegetative buds," *International Journal of Plant Developmental Biology*, vol. 1, no. 1, pp. 49–56, 2007.
[33] E. S. Dennis and W. J. Peacock, "Epigenetic regulation of flowering," *Current Opinion in Plant Biology*, vol. 10, no. 5, pp. 520–527, 2007.
[34] P. Chouard, "Vernalization and its relations to dormancy," *Annual Review of Plant Physiology*, vol. 11, pp. 191–238, 1960.
[35] A. Rohde and R. P. Bhalerao, "Plant dormancy in the perennial context," *Trends in Plant Science*, vol. 12, no. 5, pp. 217–223, 2007.
[36] D. Callard, M. Axelos, and L. Mazzolini, "Novel molecular markers for late phases of the growth cycle of arabidopsis thaliana cell-suspension cultures are expressed during organ senescence," *Plant Physiology*, vol. 112, no. 2, pp. 705–715, 1996.
[37] P. C. Bailey, G. W. Lycett, and J. A. Roberts, "A molecular study of dormancy breaking and germination in seeds of Trollius ledebouri," *Plant Molecular Biology*, vol. 32, no. 3, pp. 559–564, 1996.

4

Diversity of Eukaryotic Translational Initiation Factor eIF4E in Protists

Rosemary Jagus,[1] Tsvetan R. Bachvaroff,[2] Bhavesh Joshi,[3] and Allen R. Place[1]

[1] *Institute of Marine and Environmental Technology, University of Maryland Center for Environmental Science, 701 E. Pratt Street, Baltimore, MD 21202, USA*
[2] *Smithsonian Environmental Research Center, 647 Contees Wharf Road, Edgewater, MD 21037, USA*
[3] *BridgePath Scientific, 4841 International Boulevard, Suite 105, Frederick, MD 21703, USA*

Correspondence should be addressed to Rosemary Jagus, jagus@umces.edu

Academic Editor: Thomas Preiss

The greatest diversity of eukaryotic species is within the microbial eukaryotes, the protists, with plants and fungi/metazoa representing just two of the estimated seventy five lineages of eukaryotes. Protists are a diverse group characterized by unusual genome features and a wide range of genome sizes from 8.2 Mb in the apicomplexan parasite *Babesia bovis* to 112,000-220,050 Mb in the dinoflagellate *Prorocentrum micans*. Protists possess numerous cellular, molecular and biochemical traits not observed in "text-book" model organisms. These features challenge some of the concepts and assumptions about the regulation of gene expression in eukaryotes. Like multicellular eukaryotes, many protists encode multiple eIF4Es, but few functional studies have been undertaken except in parasitic species. An earlier phylogenetic analysis of protist eIF4Es indicated that they cannot be grouped within the three classes that describe eIF4E family members from multicellular organisms. Many more protist sequences are now available from which three clades can be recognized that are distinct from the plant/fungi/metazoan classes. Understanding of the protist eIF4Es will be facilitated as more sequences become available particularly for the under-represented opisthokonts and amoebozoa. Similarly, a better understanding of eIF4Es within each clade will develop as more functional studies of protist eIF4Es are completed.

1. Eukaryogenesis and Protein Synthesis

Protein synthesis is an ancient, conserved, complex multienzyme system, involving the participation of hundreds of macromolecules in which the mRNA template is decoded into a protein sequence on the ribosome. The ribosome, a complex and dynamic nucleoprotein machine, provides the platform for amino acid polymerization in all organisms [1, 2]. This process utilizes mRNAs, aminoacyl tRNAs, and a range of protein factors, as well as the inherent peptidyltransferase activity of the ribosome itself. The common origin of protein synthesis in all domains of life is evident in the conservation of tRNA and ribosome structure, as well as some of the additional protein factors. Although the basic molecular mechanisms are conserved across the three domains of life, the Bacteria (eubacteria), Archaea (archaebacteria), and Eukarya (eukaryotes), important divergences have taken place as eukaryotic species have evolved. The origin of the eukaryotic cell is enigmatic. Eukaryotes are thought to have evolved from a fusion of a euryarchaeon with a deep-rooted Gram-positive proteobacteria, the phylum from which mitochondria are derived [3]. It is currently unclear whether the eubacterial fusion partner was distinct from the ancestor of mitochondria or identical to it. This view of the origins of eukaryotes is consistent with the observation that informational genes such as those involved in transcription, translation, and other related processes are most closely related to archaeal genes, whereas operational genes such as those involved in cellular metabolic processes including amino acid biosynthesis, cell envelope, and lipid synthesis are most closely related to eubacterial genes [4]. Such an origin is also consistent with the eukaryotic rooting implied by the presence of an insert within the elongation factor EF-1A that is found in all known eukaryotic and eocytic (crenarchaeal) EF-1A sequences, but lacking in all paralogous EF-G sequences [3].

The mechanisms underlying protein synthesis in all organisms share common features and can be divided into three stages: *initiation, elongation, and termination.* During *initiation*, the ribosome is assembled at the initiation codon in the mRNA with a methionyl initiator tRNA bound in the peptidyl (P) site. During *elongation*, aminoacyl tRNAs enter the acceptor (A) site and the ribosome catalyzes the formation of a peptide bond. After the tRNAs and mRNA are translocated bringing the next codon into the A-site, the elongation process is repeated until a stop codon is encountered. During *termination*, the completed polypeptide is released from the ribosome, after which the ribosomal subunits are dissociated and the mRNA released for reuse. Different sets of protein accessory factors, the translation factors, assist the ribosome at each of these stages. These are referred to as initiation factors, elongation factors, and termination factors, respectively, to reflect the stage at which they are involved. The elongation process and machinery is well conserved from bacteria to eukaryotes, as is termination. However, the mechanisms of the initiation process, including recognition of the correct reading frame, differ, as do the mechanisms by which mRNA is recruited by the ribosome. Genomewide sequencing projects now allow us to assess the components of translational initiation in a wide range of organisms [5, 6].

Our view of protein synthesis is based mainly on information derived from *S. cerevisiae*, *Drosophila*, plant, and mammalian systems, with the translation components identified through sequencing projects. However, these are only narrow windows on the full diversity of extant eukaryotes. The greatest diversity of eukaryotic species is to be found within the protists, with plants and metazoans representing just two of the estimated 75 lineages of eukaryotes [7, 8]. We are only just beginning to uncover the vast diversity of bacterial-sized (pico- and nano-) eukaryotes, first discovered in clone libraries derived by PCR amplification of pooled "environmental" DNAs (culture-independent PCR) [9–11]. Microbial eukaryotes are a diverse group of organisms characterized by many unusual genome features. These features challenge some of the concepts and assumptions about regulation of gene expression in eukaryotes. In this paper, we will focus on a comparison of our current knowledge of the translation initiation factor eIF4E and its family members from protists. We will compare eIF4E in a range of protists and look at translational components in a simplified translation system found in an algal endosymbiont.

The control of gene expression is a complex process. Even after mRNA is transcribed from DNA, mRNAs can undergo many processing and regulatory steps that influence their expression [12]. Gene regulation at the translational level is widespread and significant. The extent of gene regulation at the translational level has been demonstrated during early *Drosophila* embryogenesis on a genomewide basis that was investigated by determining ribosomal density and ribosomal occupancy of over 10,000 transcripts during the first ten hours after egg laying in *Drosophila*. The diversity of the translation profiles indicates multiple mechanisms modulating transcript-specific translation with cluster analyses suggesting that the genes involved in some biological processes are coregulated at the translational level at certain developmental stages [13]. Similarly, protists have been shown to regulate translation over wide range of conditions and physiological changes, with groups like the dinoflagellates showing regulation of translation to be the predominant form of regulation of gene expression.

2. Origin of Eukaryotes

Eubacteria and Archaea show tremendous diversity in their metabolic capabilities but have limited morphological and behavioral diversity; conversely, eukaryotes share similar metabolic machinery but have tremendous morphological and behavioral diversity. Eukaryotes are thought to have evolved from the endosymbiosis of an α-proteobacteria and a phagotropic euryarchaeon approximately 2 billion years ago. The transition from prokaryotes to eukaryotes was the most radical change in cell organization since life began, with a burst of gene transfer, duplication, and the appearance of novel cell structures and processes such as the nucleus, the endomembrane system, actin-based cytoskeleton [14, 15], the spliceosome and splicing, nonsense-mediated decay of mRNA (NMD), and ubiquitin signaling [16, 17]. Although the deep phylogeny of eukaryotes currently should be considered unresolved, Koonin and his colleagues have postulated that the mitochondrial endosymbiont spawned an intron invasion which contributed to the emergence of these principal features of the eukaryotic cell [18–20]. Phagocytosis is thought to be central to the origin of the eukaryotic cell for the acquisition of the bacterial endosymbiont that became the ancestor of the mitochondrion. Findings suggest a hypothetical scenario of eukaryogenesis under which the archaeal ancestor of eukaryotes had no cell wall (like modern *Thermoplasma*) but had an actin-based cytoskeleton that allowed the euryarcheon to produce actin-supported membrane protrusions. These protrusions would enable accidental, occasional engulfment of bacteria, one of which would eventually became the mitochondrion. The acquisition of the endosymbiont triggered eukaryogenesis. From a fused cell with two independent prokaryotic gene expression systems, coordination of cell division developed and gene transfer took place through occasional membrane lysis. Some of eubacterial genes recombined into host chromosomes including group II introns [18]. Group II introns can be found among free-living α-proteobacteria, the ancestors of mitochondria [21]. They evolved specifically from group II introns that invaded the ancestrally intronless eukaryotic genome through the mitochondrial endosymbiont, thereby generating the prediction that group II introns should be found among free-living-proteobacteria, the ancestors of mitochondria [21]. This prediction was borne out supporting the idea that introns could originate from the mitochondrial endosymbiont. The mobility of group II introns in contemporary eubacteria [22] and their prevalence in α-proteobacteria [23] are consistent with such a view. The rapid, coincidental spread of introns following the origin of mitochondria is posited as the selective pressure that forged nucleus-cytosol compartmentalization [18, 20]. The function of the nuclear envelope was to allow mRNA splicing, which is slow, to go to completion so that translation, which is fast, would occur only on mRNA with intact reading frames. The evolutionary relationships of proteins specific to the nuclear envelope and nuclear pore complex reveal that this protein set is a mix of proteins and domains of archaebacterial and eubacterial origins, along with some eukaryotic innovations, suggesting that the nucleus arose in a cell that already contained a mitochondrial endosymbiont [24].

3. Evolution of Translational Initiation and Eukaryogenesis

Eukaryotes inherited from their archaeal ancestor a core of translation initiation factors, which includes eukaryotic initiation factor (eIF)1, eIF1A, eIF2 (all three subunits), eIF2B (α, β, and δ subunits only) subunits), eIF4A, eIF5B, and eIF6 [25–27]. The establishment of the nuclear membrane resulted in the physical separation of transcription and translation and presented early eukaryotes with a different challenge; how to shuttle RNA from the nucleus to the site of protein synthesis in the cytoplasm. In prokaryotes, mRNA is translated as it is being synthesized, whereas in eukaryotes, mRNA is synthesized, and processed in the nucleus, and it is then exported to the cytoplasm. There is also a transition from uncapped and polycistronic mRNAs recognized by the ribosome through the Shine-Dalgarno sequence in the 5′-UTR to capped, polyadenylated, and, in most cases, monocistronic mRNAs and the evolution of the scanning process. The evolution of protein synthesis in the context of eukaryogenesis has been discussed previously by Hernández who proposed that recruitment of mRNAs in early eukaryotes was likely to have been through internal ribosome entry sites (IRESs) based on the functional similarity between IRESs and introns [28]. Although not universal, IRES transacting factors (ITAFs) are required for the proper functioning of most viral and cellular IRESs [29, 30]. ITAFs are predominantly nuclear proteins that also play key roles in pre-mRNA splicing and mRNA transport to the cytoplasm [31, 32]. Furthermore, polypyrimidine tracts, a hallmark of introns, are a common feature of cellular and some viral IRESs [33–35]. Hernández considers that the cellular IRESs are descendants of spliceosomal introns and that some of the ITAFs that existed as components of the splicing machinery (such as the ancestral PTB and hnRNPCs) were later incorporated into the nascent eukaryotic translational process. During this period, 5′-UTRs lacking Shine-Dalgarno motifs that were able to passively recruit the 40S ribosomal subunit would have been positively selected and could, therefore, have become the first examples of an IRES [28].

It also seems possible that capped spliced leader (SL) *trans*-spliced mRNAs may have arisen with eukaryogenesis and represent an early form of 5′ blocked mRNAs. In *trans*-splicing, a short SL exon is spliced from a capped small nuclear RNA and is transferred to pre-mRNA, thereby becoming the 5′-terminal end. The fully functional spliceosome is likely to have existed in the last eukaryote common ancestor, leading to splicing components and pre-mRNA signals that are found throughout eukaryotes and are similar among different eukaryotic lineages. It seems certain that SL *trans*-splicing arose through evolution from *cis*-splicing or *vice versa*. *Trans*-splicing shares the splicing signals and most of the components with *cis*-splicing, indicating a common relationship (reviewed [36]). Considering the similarities between the SL snRNP and the spliceosomal snRNPs, specialized *trans*-splicing SL RNAs could have arisen from a splicing U snRNP in ancestral *cis*-splicing early eukaryote and thus may be an ancient form of 5′-end blocking for emerging eukaryotes. SL *trans*-splicing is now found sporadically across the eukaryotic tree of life in a set of distantly related animal groups including urochordates, nematodes, flatworms, and hydra, as well as in the protist Euglenozoa and dinoflagellates, stimulating the argument that a common evolutionary origin seems unlikely. However, an attractive hypothesis to explain multiple evolutionary origins for the SL genes is that they have derived repeatedly from U-rich small nuclear RNAs (snRNAs) of the Sm-class involved in the nuclear spliceosome machinery [37]. In support of this, phylogenomic studies from *Hydra* indicate that SL genes can evolve rapidly in any organism because constraint on SL exon sequence evolution is low [38]. Furthermore, it has been reported that mammalian cells, which do not have SL *trans*-splicing, can SL *trans*-splice when supplied with the SL RNA of either nematodes or trypanosomes [39]. Duplications of the U1 snRNA gene followed by just a few mutations would be sufficient to lead to the acquisition of *trans*-splicing [39] suggesting that it could have happened in the emerging eukaryote as well as in more recent eukaryotic lines.

The separation of the nucleus from the cytoplasm led to the need for mechanisms to shuttle the transcripts into the cytoplasm and to provide for their protection against degradation. With the exception of eIF5, all the eukaryotic-specific initiation factors that evolved, eIF4E, eIF4G, eIF4B, eIF4H, and eIF3, are involved in the 5′-cap-binding and scanning processes. The 5′-cap structure provides stability from 5′ exonucleases and in extant eukaryotes is recognized by the small ribosomal subunit through the novel eukaryotic initiation factor eIF4E. eIF4E, a translational initiation factor found only in eukaryotes, has a unique alpha/beta fold that is considered to have no homologues outside the eukaryotes, as determined by sequence comparison or structural analyses [25]. Although in extant eukaryotes the main role of eIF4E is in translational initiation through cap recognition, it is possible that the cap structure and eIF4E emerged among the primary adaptive responses to the intron invasion and the need for nucleocytoplasmic RNA export, but initially had no role in translation [40]. For instance, it could have appeared in early eukaryotes either as a mediator of nuclear export of mRNAs, thus enhancing mRNA stability during nuclear export, or as a mediator of cytoplasmic storage of mRNAs. Consistent with this, one of the eIF4E proteins from the primitive eukaryote species *Giardia lamblia* binds only to nuclear noncoding small RNAs and has no function in translation [41]. eIF4E is found within different cytoplasmic bodies involved in such processes as mRNP remodeling, mRNA decay or storage [42–44]. In addition, a fraction of this protein resides in the nucleus where it mediates the export of specific mRNAs to the cytoplasm [44, 45]. Since eIF4E has no ability to interact directly with the ribosome itself, the recruitment of eIF4E-bound mRNAs in emerging eukaryotes was likely to have been IRES-dependent.

4. Diversity of eIF4E Family Members

In eukaryotes, eIF4E is a central component in the initiation and regulation of translation in eukaryotic cells [46–49]. Through its interaction with the 5′-cap structure of mRNA and its translation partner, eIF4G, eIF4E functions to recruit mRNAs to the ribosome [46]. The interaction of eIF4E and eIF4G can be competed out by a family of 4E-binding proteins, the 4E-BPs, which are capable of repressing translation [46]. Three-dimensional structures of eIF4Es bound to cap-analogues resemble "cupped-hands" in which the cap-structure is sandwiched between two conserved Trp residues (W56 and W102 of *H. sapiens* eIF4E) [50–52]. A third conserved Trp residue (W166 of *H. sapiens* eIF4E) recognizes the 7-methyl moiety of the cap-structure. Aromatic residues Trp, Phe, and His show a distinctive pattern across from N- to C-terminus of the conserved core, containing eight similarly spaced tryptophans summarized by W(x2)W(x8–12)W(x17–20)W(x29–31)W(x9–12)W(x17)W(x32–36)W [6]. Multiple eIF4E family members have been identified in a wide range of organisms that includes plants, flies, mammals, frogs, birds, nematodes, fish, and various protists [53–55]. Evolutionarily, it seems that a single early eIF4E gene underwent a series of gene duplications, generating multiple structural classes and in some cases subclasses. Today, eIF4E and its relatives comprise a family of structurally related proteins within a given organism, although not all function as prototypical initiation factors. Sequence similarity is highest in a core region of 160 to 170 amino acid residues identified by evolutionary conservation and functional analyses [6]. Prototypical eIF4E is considered to be eIF4E-1 of mammals, eIF4E and eIF (iso)4E of plants, and eIF4E of *Saccharomyces cerevisiae*. With the exception of eIF4Es from protists, all eIF4Es can be grouped into one of three classes [6].

Class I members from Viridiplantae, Metazoa, and Fungi carry Trp residues equivalent to W43, W46, W56, W73, W102, W113, W130, and W166 of *H. sapiens* eIF4E-1 [6]. Prototypical eIF4Es bind the cap and eIF4G through the motif S/TVE/DE/DFW in which the Trp is W73. Substitution of a nonaromatic amino acid for W73 has been shown to disrupt the ability of eIF4E to interact with eIF4G and 4E-BPs [56, 57]. Substitution of a Gly residue in place of V69 creates an eIF4E variant that still binds 4E-BP1 but has a reduced capacity to interact with both eIF4G and 4E-BP2 [56]. A serine at residue equivalent to S209 in *H. sapiens* eIF4E-1 is the site of phosphorylation. Only Class I eIF4Es are known to function as translation factors. Genes, and cDNAs encoding members of Class I can be identified in species from plants/metazoans/fungi. As judged from completed genomes, many protists also encode Class I-like family members although these have proven hard to characterize and can show extension or compaction relative to prototypical eIF4E family members [6]. Evidence for gene duplication of Class I eIF4E family members can be found in certain plant species, as well as in nematodes, insects, chordates, and some fungi [53–55]. Class I members include the prototypical initiation factor but may also include eIF4Es that recognize alternative cap structures such as IFE-1, -2, and -5 of *Caenorhabditis elegans* [58, 59], or eIF4Es that fulfill regulatory functions such as the vertebrate eIF4E-1Bs [55, 60–62].

Class II members possess W → Y/F/L and W → Y/F substitutions relative to W43 and W56 of *H. sapiens* eIF4E. These substitions are absent from the model ascomycetes *S. cerevisiae* and *Schizosaccharomyces pombe*. Mammalian eIF4E-2 (Class II) binds only to cap and 4E-BPs [54]. They have been shown to regulate specific mRNA recruitment in *Drosophila* [63] and *C. elegans* [64].

Class III members possess a Trp residue equivalent to W43 of *H. sapiens* eIF4E but carry a W → C/Y substitution relative to *H. sapiens* W56. They have been identified primarily in chordates with rare examples in other Coelomata and in Cnidaria [6, 54]. Their biological function has not yet been determined, although mouse eIF4E-3 has been shown to bind both cap and eIF4G [54]. The protist eIF4Es do not fall into any of these three classes and by plant/metazoan/fungal standards appear to be compacted or possess extended sequences between the conserved tryptophans [6].

5. Diversity of Protists and Evolution of Eukaryotic Lineages

The greatest diversity of eukaryotic species is to be found within the protists. Eukaryotes appear to be monophyletic; all extant eukaryotes appear to postdate the acquisition of mitochondria. However, their phylogeny is currently not widely agreed upon. Molecular phylogenetics has the potential to resolve the systematics of eukaryotes. Sequence data continues to accumulate, but with few protists and fewer protist taxa and a distinct bias towards parasites infecting humans (and crop plants). There is increasing availability of multigene data from diverse lineages, although it seems likely that eukaryotic taxonomy will be further complicated by the discovery of ultrasmall eukaryotes. These are scattered across the eukaryotic tree and may include major new supergroups [9, 65]. The root of the eukaryotes remains open to debate, but recent analysis places the eukaryotic root between the monophyletic "unikonts" and "bikonts" [66].

The protists are defined loosely as unicellular eukaryotic organisms that are not plants, animals, or fungi. Eukaryotic features evolved within the protists that thrived for up to a billion years before they gave rise independently to multicellular eukaryotes, the familiar plants, animals, and fungi [67]. Extreme examples of genome sizes, both large and small, can be found among microbial eukaryotes from 8.2 Mb in the apicomplexan *Babesia bovis* to >200,000 Mb in certain dinoflagellates. Roughly forty sequenced genomes are available (depending on classification), some of which are multiple representatives of the same genus, for example, *Plasmodium*, *Leishmania*, and *Trypanosoma*. The last common ancestor of all eukaryotes is believed to have been

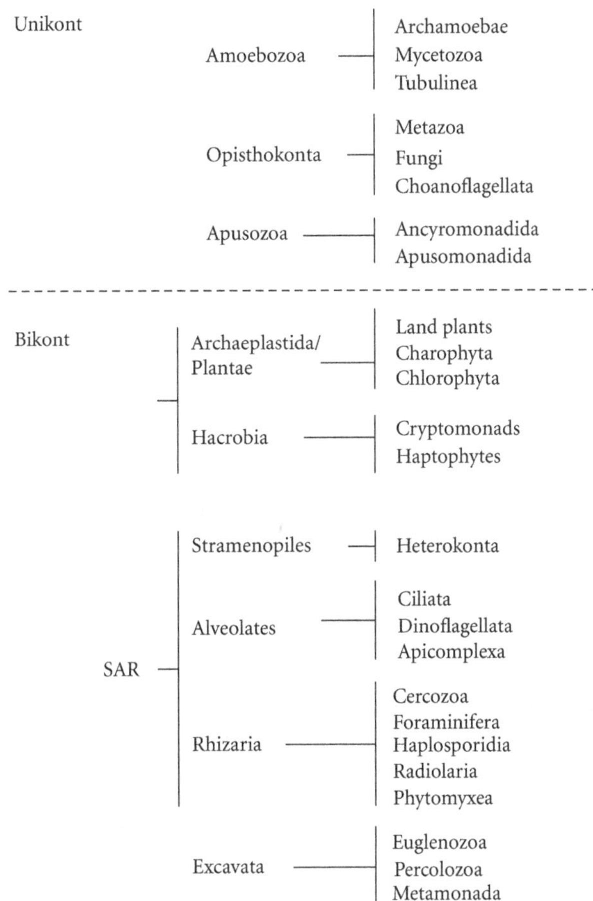

Figure 1: Relationships among major lineages of eukaryotes. Summary tree of eukaryotic relationships based on multigene analyses as outlined by Parfrey et al. [8].

a phagotrophic protist with a nucleus, at least one centriole and cilium, facultatively aerobic mitochondria, sex (meiosis and syngamy) and a dormant cyst with a cell wall of chitin and/or cellulose, and peroxisomes (based on a root along the lineage leading to Euglenozoa). Endosymbiosis led to the spread of plastids. Analyses of multigene genealogies have led to the conclusion that the acquisition of photosynthesis in eukaryotes arose from a primary endosymbiosis between a cyanobacterium and a eukaryotic host. This gave rise to glaucocystophytes (white lineage), red algae (red lineage), and green algae (green lineage, including plants) [7, 68–70]. Plastids spread by secondary endosymbiosis. Other photosynthetic eukaryotes such as cryptomonads, haptophytes, chlorarachniophytes (amoeboflagellate cercozoans), dinoflagellates, diatoms, brown algae, and euglenids are the result of secondary endosymbiosis, tertiary endosymbiosis, and, perhaps, even quaternary endosymbiosis in which a nonphotosynthetic eukaryotic ancestor engulfed a photosynthetic eukaryote [68, 71, 72]. Endosymbiosis resulted in the transfer of hundreds of genes to the host nucleus. Multiple gains and multiple losses of plastids are likely to have occurred, with plastids possibly lost in ciliates and remaining in relict form in apicomplexans [73] and *Perkinsus* [74]. Dinoflagellates have substituted the ancestral plastid several times by tertiary symbioses involving a diverse array of eukaryotes [71, 72].

There is no real consensus on eukaryotic phylogeny currently; part of the problem is that we are still very much in the discovery phase, and another is that some of the divisions are quite ancient. In recent years, eukaryotic taxonomy has shifted towards a new system of six supergroups that aims to portray evolutionary relationships between microbial and macrobial lineages [8, 75–77]. The six supergroups posited are the Amoebozoa, Opisthokonta, Apusozoa, the Archaeplastida/Plantae, SAR (Stramenopiles, Alveolates, and Rhizaria), and the Excavata (Table 1). These break down into two larger groups, those with a single flagellum (unikonts), which may or may not be retained, and those with two flagella (bikonts) (Table 1). A summary tree of eukaryotic relationships based on multigene analyses as outlined by Parfrey et al. [8] is shown in Figure 1.

The Amoebozoa includes a diversity of predominantly amoeboid members such as the tubulinid amoeba, *Amoeba* spp., *Dictyostelium discoideum* (cellular slime mold), and *Entamoeba* spp., which are secondarily amitochondriate. Opisthokonts include the metazoans, fungi, and the choanoflagellates such as *Monosiga brevicollis* that are the sister to the metazoans [78]. This is the best supported

Table 1: Eukaryotic groups and genera used for analysis of eIF4E family members. The six hypothesized supergroups of eukaryotes after Parfrey et al. [8]. Groups (pink) and genera (yellow) from which eIF4E sequences have been used to examine the relationship of protist eIF4E family members are highlighted.

Unranked	Super group	Group	Examples of Genera
Unikont	Amoebozoa	Archamoeba	*Entameoba*
		Mycetozoa	*Dictyostelium*
		Tubulinea	*Amoeba, Acanthamoeba*
Unikont	Opisthokonta	Metazoa	*Drosophila, Homo*
		Fungi	*Saccharomyces*
		Choanoflagellata	*Monosiga*
Unikont	Apusozoa	Ancyromonadida	*Ancyromonas*
		Apusomonadida	*Apusomona*
Bikont	Archaeplastida/Plantae (A/H):		
	Viridiplantae (true plants)	Land plants	*Arabidopsis*
	Hacrobia	Charophyta	*Chara*
		Chlorophyta	*Chlamydomonas, Volvox*
		Haptophytes	*Emiliania*
		Cryptomonads	*Guillardia*
	SAR:		
	Rhizaria	Cercozoa	*Bigelowiellia*
		Foraminifera	*Allogromia*
	Stramenopiles	Haplosporidia	*Bonamia*
	Alveolates	Radiolaria	*Collozoum*
		Phytomyxea	*Polymyxa*
		Heterokonta	*Phytopthora, Thalassiosira, Ectocarpus Phaeodactylum*
		Ciliata	*Tetrahymena, Paramecium*
		Apicomplexa	*Plasmodium, Eimeria, Babesia, Neospora, Theileria, Cryptosporidium, Toxoplasma*
		Dinoflagellata	*Karlodinium, Amoebophyra Amphidinium, Alexandrium*
Bikont	Excavata	Euglenozoa	*Euglena, Trypanosoma*
			Leishmania
		Percolozoa	*Naegleria*
		Metamonada	*Giardia, Trichomonas*

supergroup. The Apusozoa is a supergroup comprising flagellate protozoa, the apusomonads, and ancyromonads. On molecular trees, these two group together, but their relationship to other eukaryotes is uncertain [8]. The supergroup Archaeplastida/Plantae was posited to unite the three lineages with primary plastids: green algae (including land plants), rhodophytes, and glaucophytes with two other lineages, the cryptophytes and haptophytes, both of which have secondary plastids [79]. There is strong support for the SAR supergroup consisting of stramenopiles, alveolates, and plus rhizarians [8]. Within the SAR clade, each of the three members forms distinct lineages [68, 70]. For example, the Rhizaria emerged from molecular data to unite a heterogeneous group of flagellates and amoebae including cercomonads, foraminifera, diverse testate amoebae, and former members of the radiolaria [80] and represents an expansion of the Cercozoa to include foraminifera [81]. The Cercozoa was also recognized from molecular data [82]. Cercozoa and foraminifera appear to share a unique insertion in ubiquitin [83], although there is a paucity of nonmolecular characters uniting the members of this supergroup [8]. Within the alveolates, the Apicomplexa is a large monophyletic group many of which are parasites, including *Plasmodium*, the parasite responsible for malaria. The last supergroup is the Excavata, a supergroup composed predominately of heterotrophic flagellates, and includes many important parasites such as the trypanosomes, *Giardia*, and trichomonads. Within this supergroup, the "euglenozoa," the combination of eugleniids and trypanosomes is a grouping with good support.

6. Unusual Features of Protist eIF4Es

A previous phylogenetic analysis of eIF4E family members from protists indicated that they cannot be grouped with

FIGURE 2: Relationship of selected eIF4E-family members from multiple protist species. Maximum likelihood phylogeny of eIF4E amino acid sequences aligned with T-coffee and trimmed to include only the core region of 453 aligned positions (corresponding to positions 30 to 203 of the human sequence). The tree was constructed using RAxML with the Jones Taylor Thornton gamma distributed model with 100 rapid bootstrap replicates. Bootstrap values above 50% are shown.

the three main classes that describe eIF4E family members from multicellular organisms [6]. At the time of the earlier analysis, very few sequences were available for protists. Many more are now available, though not all in publically available databases. Figure 2 shows a tree describing the overall relationships of selected eIF4E-family members from multiple protists species rooted with *H. sapiens* eIF4E-1. The tree shows maximum likelihood phylogeny of eIF4E amino acid sequences aligned with T-coffee and trimmed to include only the core regions corresponding to amino acids 30 to 203 of the human sequence). The tree was constructed using RAxML with the Jones Taylor Thornton gamma distributed model with 100 rapid bootstrap replicates. Bootstrap values above 50% are shown. Sequences derive predominantly from representatives of SAR mainly heterokonts, ciliates, apicomplexans, dinoflagellates, *Perkinsus*, along with Excavata representatives from Diplomonads (*Giardia*), Euglenozoa (*Trypanosoma*), and Parabasalids (*Trichomonas*).

Three clades stand out and are bracketed with solid lines. All three solid bracketed clades include eIF4Es from dinoflagellates and *Perkinsus* suggesting the possibility of three different classes. The bottom bracket shows a large clade (Clade 1) including eIF4Es from the ciliate, *Tetrahymena thermophila*; *Perkinsus marinus* eIF4E-5, -6, and -7; eIF4E-2a–d sequences from the dinoflagellate, *Karlodinium veneficum*, along with eIF4Es from the dinoflagellates *Amphidinium carterae* and *Amoebophrya*. This clade also includes eIF4Es from the closely related apicomplexans and is the only strong clade with apicomplexans in this tree. Clade 1 also includes "dotted line" clade eIF4E family members from the euglenozoan

Table 2: Summary of protist eIF4E family member characteristics. A selection of Clade 1 and Clade 2 protist eIF4E family members is shown, looking at the residue at positions equivalent to W46, W56, W73, W113 in human eIF4E-1; presence or absence of a Ser residue at the position equivalent to S209 in human eIF4E-1; presence or absence of insertions; the sequence of the sequence of the eIF4G-binding domain. Shading indicates Amoebozoa (Pale yellow); nucleomorph (Gray); haptophyte (Aqua); alveolate, apicomplexan (Yellow); alveolate dinoflagellate/Perkinsus (Pink); excavate (Aquamarine) eIF4Es.

Spp/form	Clade	W46	W56	W73	W113	S209	W46–W56	W73–W102	W113–W130	W130–W166	eIF4GBD
Plant/metazoan/fungi consensus	N/A	W	W	W	W	Y	N	N	N	N	S/TVxxFW
H. sapiens eIF4E-1	N/A	W	W	W	W	Y	N	N	N	N	TVEDFW
A. thaliana	N/A	W	W	W	W	Y	N	N	N	N	TVEDFW
E. histolytica, 180000	1	W	Y	W	W	Y	N	N	N	N	TVENFW
D. discoideum,	2	W	W	W	W	Y	N	N	N	N	SVEDFW
A. castellani eIF4E-1	2	W	W	W	W	Y	N	N	N	U	TVEDFW
G. theta nucleomorph	2	W	W	L	W	N	N	N	N	N	NLEDFL
H. andersonii nucleomorph	2	W	W	W	W	N	N	N	N	N	SIDNFW
C. paramecium nucleomorph	2	W	W	L	W	N	N	N	N	N	DVENFL
E. huxleyi, unk1	1	W	Y	W	W	Y	Ys	Y	N	N	TVEEFW
P. falciparum, unk1	1	W	Y	W	F	Y	N	Y	N	Ys	SVQKFW
N. caninum	1	W	Y	W	F	Y	N	Y	N	Ys	TVQKFW
E. tenella, unk1	1	W	Y	W	F	Y	N	Y	N	Ys	TVQTFW
T. gondii	1	W	Y	W	F	Y	N	Y	N	N	TVQKFW
B. bovis, 548495	1	W	Y	W	F	Y	Ys	N	N	N	SVQSFW
A. tamarense, unc 1	1	W	Y	W	F	Y	Y	Y	N	Y	SVEQFW
K. veneficum 2a	1	W	Y	W	F	Y	N	Y	N	Y	TVQEFW
K. veneficum 2b	1	W	Y	W	F	Y	N	Y	N	Y	TVQEFW
K. veneficum 2c	1	W	Y	W	F	N	N	Y	N	Y	TVQEFW
K. veneficum 2d	1	W	Y	W	F	N	N	Y	N	Y	TVKGFW
P. marinus 5	1	W	W	W	L	N	N	Y	N	Y	TVGEFW
P. marinus 6	1	W	W	W	L	N	N	Y	N	Y	TVGEFW
P. marinus 7	1	W	W	W	L	N	N	Y	N	Y	TVGEFW
L. major, EF4E3	1	Y	F	W	S	Y	N	N	Ys	N	DVESFW
T. brucei, EIF4E3	1	Y	Y	F	T	Y	N	N	Ys	N	DVECFW
N. gruberii, 8859902	1	W	Y	F	W	N	N	N	N	N	DVETFW
G. lamblia eIF4E2	U	F	F	F	K	N	Ys	N	N	N	SLKAFF

excavates, *Leishmania* and *Trypanosoma*, EIF4E3 and 4. The next bracketed clade (Clade 2) includes eIF4E family members from *K. veneficum* (eIF4E-1), *A. carterae* 18399, *P. marinus* eIF4E-8, *Amoebophrya* and the ciliate *T. thermophila* and "dotted line" clade that includes trypanosome sequences *Leishmania* EIF1 and 2. Characteristics of some Clade 1 and Clade 2 eIF4E family members are summarized in Table 2. The top bracketed clade (Clade 3) contains eIF4E family members from *P. marinus*, eIF4E-2, -3, -4, -11, *K. veneficum* eIF4E-1, and *A. carterae* 33977. eIF4Es from ciliates are absent from this top clade, and there is an "orphaned" clade of ciliate sequences. These results suggest gene duplication into three groups prior to divergence of the alveolates with the loss of one copy in *Amoebophrya* and the loss of two copies in apicomplexans. An alternate explanation could be that these copies are not apparent because they are so diverged, or, in the case of *Amoebophrya*, because of poor coverage.

7. eIF4E Family Members in *Giardia lamblia*

Giardia lamblia is an amitochondriate flagellated protozoan parasite that belongs to the diplomonad group (Excavata) that includes both parasitic and free living species [84]. Its genome is compact in structure and content (~11.7 Mb), contains few introns or mitochondrial relics, and has simplified machinery for DNA replication, transcription, RNA processing, and most metabolic pathways [85]. mRNA recruitment in these organisms is unusual in that their transcripts have exceedingly short 5′ untranslated regions

(5′-UTRs), ranging from 0 to 14 nucleotides, and similarly short 3′-UTRs of 10 to 30 nucleotides [86]. Extremely short 5′-UTRs are a highly conserved trait of transcripts from *Trichomonas*, *Entamoeba*, as well as *Giardia*. The precise cap structure in *Giardia* RNAs has not yet been determined, although native *Giardia* mRNAs have blocked 5′-ends and the genome encodes a yeast-like capping apparatus [87]. Furthermore, m^7GpppN-capped mRNA introduced into the cells is expressed well [87, 88]. Eight $m^{2,2,7}$GpppN-capped snRNA species have been identified in *Giardia* [89]. Experimentally, mRNA recruitment occurs efficiently in mRNAs that are capped and in which the first initiation codon is located only 1 nucleotide downstream from the m^7GpppN-cap structure. Recruitment can be decreased when the 5′-UTR between the cap and the initiation codon is lengthened beyond 9 nucleotides [88]. There are two eIF4E family members in *Giardia*, termed eIF4E1 and eIF4E2, which have distinct properties [41]. Of the two, eIF4E2 has been shown to be essential and binds to m^7GTP-Sepharose, suggesting that it functions in protein synthesis. The other, eIF4E1, is not essential and binds only to $m^{2,2,7}$GpppN-Sepharose. eIF4E1 is found concentrated and colocalized with the $m^{2,2,7}$GpppN cap, 16S-like rRNA, and fibrillarin in the nucleolus-like structure in the nucleus [41]. Of the eight conserved tryptophan residues typical of eIF4E Class I sequences, both forms have a Phe residue at the position equivalent to human W56. eIF4E1 has Leu at the position equivalent to human W73, and eIF4E2 has a Phe residue (Table 2). Both forms have poor consensus at the eIF4G binding site with substitutions of W113/Y and W113/I for eIF4E1 and eIF4E2, respectively (numbering as in human eIF4E), eIF4E1 has an insertion between residues 130–166.

8. eIF4E Family Members in Trypanosomatids

Trypanosomatids are a group of kinetoplast protozoa (Excavata/Euglenozoa) distinguished by having only a single flagellum. The haploid genome size in *Leishmania major* is ~36 Mb (haploid). mRNA maturation in trypanosomes differs from the process in most eukaryotes mainly because protein-coding genes are transcribed into polycistronic RNAs in this organism [36, 37, 90]. Transcription of protein coding genes occurs polycistronically, and processing to monocistronic mRNAs occurs through coupled splice leader (SL) *trans*-splicing and polyadenylation (reviewed [36]). The SL *trans*-splicing mechanism was once considered an anomaly of the kinetoplastids, but subsequent identification of *trans*-splicing in dinoflagellates, *Perkinsus*, euglenozoans, and several major invertebrate phyla suggests that this particular form of RNA processing may represent an evolutionarily important aspect of gene expression [36, 37, 91]. There are similarities, particularly in genomic arrangement of SL RNAs, between phyla known to exhibit *trans*-splicing and their mRNAs; however, there is little sequence similarity between the SLs of different organisms. In this RNA-mediated form of *trans*-splicing, a short SL exon is spliced from a capped small nuclear RNA and is transferred to pre-mRNA, thereby becoming the 5′-terminal end and providing an unusual cap structure to mature mRNAs. In *Euglena* (Excavata/Euglenozoa), the SL contribution results in trimethylguanosine, a so-called trimethyl cap, $m^{2,2,7}$GpppG (TMG), in which there are additional methylations to the prototypical monomethyl (m^7GpppN) cap structure found on most eukaryotic mRNAs [92]. In metazoans such as nematodes, where only a percentage of mRNAs are *trans*-spliced, the SL contribution results in a trimethyl cap [93]. In kinetoplastids, all of the mRNAs are *trans*-spliced and the SL contribution results in a highly unique cap structure where additional methylations are apparent. Whereas no more than three modified nucleotides have been described in any metazoan cap structure, the kinetoplastid cap has four consecutive modifiednucleotides (and thus by convention is referred to as a cap-4 structure) [94, 95]. This has been the most highly modified eukaryotic mRNA cap known to date. In trypanosomatids, mRNAs have a common 39-nt long spliced leader sequence at the distal end of the 5′-UTR, which is identical for all mRNAs of a given species. Regulation of gene expression in trypanosomatids is accomplished mainly through posttranscriptional mechanisms such as control of mRNA stability and translation [96–98].

Four eIF4E family members have been characterized from the trypanosomatids *Leishmania major* and *Trypanosoma brucei*, termed EIF4E1, 2, 3, and 4 [99, 100]. All four are expressed in both procyclic and bloodstream forms of the parasites. These four can be broadly classified into two groups (Figure 2). Sequence analysis has identified features that distinguish EIF4E1 and 2 from EIF4E3 and 4 in both *T. brucei* and *L. major*. Similarly, separation of the four eIF4Es into two distinct groups can be made on the basis of localization and function [100]. In *T. brucei*, EIF4E1 and 2 (Group 1, expanded Clade 2) localize both to the nucleus and the cytoplasm and do not seem to be directly involved in translation based on knockdown experiments, although they do perform functions essential for cellular viability [100]. The second group (Group 2, Clade 1) formed by EIF4E3 and 4 is more abundant, is strictly cytoplasmic, is required for translation, and interacts with *T. brucei* eIF4Gs [100].

Group 1 comprises the EIF4E1 and 2 sequences (expanded Clade 2), which are more similar in size to the human and yeast sequences, but show extensions between W102–W113. The function of this extension in Clade 2 eIF4Es in euglenozoans is not known, but the prolines suggest it is solvent exposed and thus could be involved in protein-protein interaction. eIF4E family members from Group 2 (expanded Clade 1), EIF4E3 and 4, share a few unusual features absent from the Group 1 members and distinct from plant, fungi, and metazoan eIF4Es. These include a long N-terminus of more than 150 amino acids which share extensive homology between different orthologues in the EIF4E3 sequences and also contain short segments of limited homology which seem to be conserved between the EIF4E3 and EIF4E4 sequences [100]. Of the eight conserved tryptophan residues typical of eIF4E Class I sequences, most are either conserved in the various trypanosomatid homologues or are replaced by other aromatic residues such as W56Y/F in the Group 2 eIF4Es (human eIF4E numbering) (Table 2) [100]. The only exception is W113, present in the EIF4E1 and EIF4E2 sequences but which is replaced by nonaromatic hydrophilic residues in EIF4E3 and 4. Other substitutions in the trypanosomatid sequences are D104, next to the universally conserved W102/E103, involved in cap binding [50] which is replaced by a histidine in EIF4E2 and 3; V69/E70, part of the eIF4G-binding domain [101], which is missing in EIF42 and EIF4E4 [100].

EIF4E3, the most abundant *Trypanosoma* and *Leishmania* eIF4E family member, is the only confirmed essential homologue in procyclic and bloodstream T. *brucei*. The similarities observed between *T. brucei* EIF4E3 and 4 at the sequence level, their similar subcellular localization, abundance, and their ability to bind to eIF4G partners are consistent with both performing related-roles in translational initiation. Interestingly, *T. brucei* EIF4E1, 2, and 4, but not *T. brucei* EIF4E3, can efficiently bind the m^7G cap. Nevertheless, when compared with EIF4E4 in *L. major*, it binds less efficiently to the trypanosomatid cap4 [99]. Although *T. brucei* EIF4E2 binds to the m^7G Sepharose in a similar manner to *T. brucei* EIF4E1 and 4, *L. major* EIF4E2 does not bind this cap [102], but rather, preferentially binds the methylated cap4 [99]. This difference, plus the existence of unusual insertions in the *L. major* EIF4E2 between W113–W130 that are missing from the *T. brucei* or *T. cruzi* orthologues, implies a divergence in function unique to the *L. major* protein. The earlier prediction [99] that this insertion might be related to the ability of *L. major* eIF4E to bind to the larger cap-4 seems therefore not to be a compelling argument.

9. eIF4E Family Members in Dinoflagellates

Dinoflagellates are alveolate unicellular protists and a sister group to the parasitic apicomplexans such as *Toxoplasma gondii* and *Plasmodium falciparum*. Dinoflagellates are a diversified group that exhibit a wide diversity in size, form, and lifestyle. They also show a wide spread of genome size, from 1500 to 4700 Mb in *Symbiodinium* sp to 112,000 to 220,050 Mb in *Prorocentrum micans* [103]. Ninety percent of all dinoflagellates are marine plankton with the remaining species being benthic, freshwater, or parasitic.

The free-living species are major primary producers, and several are known to produce harmful algal blooms that result in massive fish kills, human and marine mammal intoxications, as well as economic losses in fisheries and tourism. However, scientific interest with dinoflagellates extends beyond their ecological and economic importance. They possess numerous cellular, molecular, and biochemical traits not observed in "text-book" model organisms. It appears that the organization and regulation of genes in dinoflagellates is different from that of typical eukaryotes. DNA is in permanently condensed chromosomes not packaged in nucleosomes and DNA content ranging from 3 to 250 pg per cell (up to almost 60-fold larger than humans) [104]. Within the dinoflagellate genome, there appears to be a high degree of DNA redundancy, with multiple tandem copies (>20 in many cases) of protein coding genes to give complex gene families [103, 105] that are highly and coordinately expressed. Unlike trypanosomes, in which polycistronic mRNAs contain a series of different genes, the examples studied in dinoflagellates consist of tandemly arrayed copies of the same gene [105].

Recent studies find a predominance of posttranscriptional control of gene expression in dinoflagellate gene expression, including circadian controlled processes such as bioluminescence [106], carbohydrate metabolism [107], and the cell cycle [108], as well as a range of stressors [109–114]. The Van Dolah lab, at the NOAA Center for Coastal Environmental Health and Biomolecular Research, has developed an oligonucleotide microarray from 11,937 unique ESTs from the dinoflagellate *Karenia brevis* [115]. Following validation of the microarray, large-scale transcript profiling studies were performed examining diurnally regulated genes and genes involved in the acute stress response. These studies represent the largest transcript profiling experiments in a dinoflagellate species to date and showed only a small percentage of transcripts changing. None of the anticipated genes, under transcriptional control in other eukaryotes (e.g., cell cycle genes, heat shock, etc.), showed changes in mRNA abundance. Consistent with this, a massively parallel signature sequencing (MPSS) analysis of the transcriptome of the dinoflagellate *Alexandrium tamarense* has shown that of a total of 40,029, only 18, 2, and 12 signatures were found exclusively in the nutrient-replete, nitrogen-depleted, and phosphate-depleted cultures, respectively. The presence of bacteria had the most significant impact on the transcriptome, although the changes represented only ~1.0% of the total number of transcribed genes and a total of only ~1.3% signatures were transcriptionally regulated under any condition [116]. Since the levels of many proteins have been well documented to change in a variety of dinoflagellates, these large-scale studies point to translational regulation as a likely regulatory point in dinoflagellate gene expression. Currently, almost nothing is known about translational initiation or its regulation in these organisms.

Dinoflagellates have mRNAs with unique spliced leaders and cap structures: through analysis of sequences representing all major orders of dinoflagellates, nuclear mRNAs from fifteen species were recently found to be *trans*-spliced with the addition of a 22-nt conserved SL [117, 118]. SL *trans*-splicing has not been identified in a ciliate or apicomplexan to date; however, preliminary analysis using the 22-nt dinoflagellate SL revealed the usage of *trans*-splicing in *Perkinsus marinus* and *P. chesapeaki*, phylogenetic intermediates between apicomplexans and dinoflagellates [118]. Recently, SL *trans*-splicing has been identified in *Amoebophrya* sp, a member of the *Syndinales*, a dinoflagellate parasite of dinoflagellates, which represents a basal root of the dinoflagellates [119]. This suggests the SL machinery was present in an early ancestor of dinoflagellates. It is unclear whether all or only a subset of dinoflagellate genes are subject to SL *trans*-splicing, but, given the diversity of the cDNAs found in the full length libraries, a conservative estimate would be that greater than 90% of mRNAs are *trans*-spliced.

The 22-nt sequence found in dinoflagellate SL-RNA is 5′A(T)CCGTAGCCATTTTGGCTCAAG-3′ [118]. The identity of the cap structure for the SL-RNA needs to be verified, but preliminary analysis indicates only a monomethylated

5′ m^7G is present on mRNAs. Based on the SL-RNA sequence and LC-MS analysis, Place has proposed the following novel cap-4 structure for dinoflagellate mRNAs: $m^7GpppA(U)p^{m2'}Cp^{m2'}CpG$ with modifications to A (U) and G still needing to be established (unpublished results). There is no evidence for a trimethylguanosine or 2′O-methyl adenosine.

Dinoflagellates encode unusual eIF4E-family members. Two distinct eIF4E orthologues, eIF4E-1 and -2 have been partially characterized in *K. veneficum* (Jagus and Place, m/s in preparation) (Figure 3). To facilitate comparison of the sequences, the residues conserved in Class I eIF4Es in multicellular organisms are indicated and numbered as in human eIF4E-1: W43, W46, W56, W73, W102, W113, W130, W166 and S209. eIF4E-2 is represented by four distinct but closely related subtypes (eIF4E-2a–d) (Figure 3). Seven contigs encoding eIF4E-1 and 31 contigs encoding eIF4E-2 (approximately equivalent representation by the a–d subtypes) have been identified indicating the eIF4E-2 group is more highly expressed and may represent the dominant isoforms in the cell. A neighbor-joining tree predicts that the dinoflagellate eIF4E-2 is related to eIF4Es from the kinetoplasts *Leishmania* and *Trypanosoma* with 51% bootstrap support. RT-qPCR analysis for eIF4E transcript abundance is consistent with this assertion (Jagus and Place, m/s in preparation). The *K. veneficum* eIF4E sequences are aligned in Figure 3 with prototypical eIF4E-1 from human. Also included are the sequences for additional, as yet uncharacterized eIF4E family members. Additional sequences were uncovered after this paper was initiated and are shown as kv20926 and kv31228 in Figure 2; however, their sequences are not included in Figure 3. Kv20926 groups with *K. veneficum* Clade 1 eIF4E-2 subtypes and kv31228 with *K. veneficum* Clade 2 eIF4E-1. *K. veneficum* eIF4Es show a clear separation into two subclasses, based on an insert of 11 amino acids between W73 and W102 (numbering equivalent to human eIF4E-1) and distribute between three clades. *K. veneficum* eIF4E-1 and eIF4E 2a–d have a Tyr substitution at the position equivalent to human W56, one of the tryptophans involved in cap binding. This is also observed in eIF4Es from the dinoflagellate *Alexandrium tamarense*, but not from *Amphidinium carterae*. In addition, eIF4E-1 has glutamine instead of D/E in the eIFG/4E-BP-binding domain. The eIF4E-2 family members contain extended amino acid stretches between the structural units of the core, between residues equivalent to human W73 to W102, and W130 to W166. In addition, eIF4Es from several alveolate species have a Trp to Phe substitution at W113 [6], a characteristic shared by *K. veneficum* eIF4E-1. It is of interest that the different subtypes of *K. veneficum* eIF4E-2s show marked heterogeneity between W102–W113. The conserved phosphorylation site of eIF4E is only observed in eIF4E-2a and -2b of *K. veneficum*. eIF4E-2a and -2b share the TKS motif at the putative phosphorylation site in which the Lys residue is a sumoylation site in human eIF4E [120, 121]. The sumoylation site at the equivalent of human Lys35 is shared by eIF4E-2b, -2b, -2c, and -2d. eIF4E-1 contains the sumoylation site equivalent to human Lys210. eIF4E-2a, but not eIF4E-1, binds to m^7GTP-Sepharose *in vitro*, although neither interact with TMG. It is not known whether either form interacts with the unique cap-4 of dinoflagellates (Jagus/Place, m/s in preparation). These results are consistent with eIF4E-2a being a functional initiation factor, but not definitive. The *K. veneficum* eIF4E-2s fall into Clade 1 raising the possibility that other eIF4Es of Clade 1 bind to m^7GTP. The eIF4E-1s fall into Clade 2. Unlike *K. veneficum* eIF4E-1, some of the extended Clade 2 members like the *L. major* and *T. bruceii* eIF4E1 and 2 are known to bind m^7GTP but appear not to participate in protein synthesis [100], making it hard to predict function of the *K. veneficum* eIF4E-1s. Three of the *K. veneficum* eIF4Es fall into Clade 3. As with the *K. veneficum* Clade 2 representatives, these do not have the insert between W73 and W102.

10. eIF4E Family Members in *Perkinsus marinus*

Perkinsus marinus is an alveolate with a genome of 86 Mb and is closely related to the dinoflagellates [122]. Like the dinoflagellates, it also exhibits *trans*-splicing. Five different SLs of 21-22 nucleotides (nt) in length have been reported from *P. marinus* [123–125]. Variability at positions 1 and 2 between the different SLs suggests variability of cap structures. Overall these data suggest a complex gene regulatory system both at the level of mRNA generation and of translational control consistent with its complex life style. The *P. marinus* genome encodes eight eIF4E family members along with two very large (>600 amino acid) forms that contain only some of the typical eIF4E signatures. *P. marinus* eIF4E-5, -6, and -7 form a group that aligns most closely with the *K. veneficum* eIF4E-2s in Clade 1, suggesting they will bind m^7GTP caps (Figures 1 and 4 and Table 2). These share the insertions between W73 to W102 and W113 to W133. This group also has TVGEFW at the eIF4G binding domain. In addition, they each have a Trp to Leu substitution at W113. The *L. major* and *T. bruceii* also show a consistent substitution at this position, but to a hydrophilic amino acid. *P. marinus* eIF4E-2, -3, and -4 also form a group in Clade 3 with eIF4Es with two of the *K. veneficum* eIF4Es (Figure 2, Table 2). *P. marinus* eIF4E-8 groups with *K. veneficum* eIF4E-1.

11. Cryptomonads, *Guillardia theta*, and Nucleomorphs

Cryptomonads (Chromalveolata/Cryptophyta) are chimeras of two different eukaryotic cells; a flagellate host and a photosynthetic endosymbiont. These organisms are thought to have arisen by secondary symbiogenesis shortly after the origin of the common ancestor of green plants, red, and glaucophyte algae [126–128]. In the cryptomonad *Guillardia theta*, the flagellate host acquired a chloroplast by engulfing and retaining a red alga. In doing so, the host was able to convert from obligate heterotrophy to an autotrophic way of life [129–131]. In addition to the red algal chloroplast, cryptomonads have retained a vestigial red algal nuclear

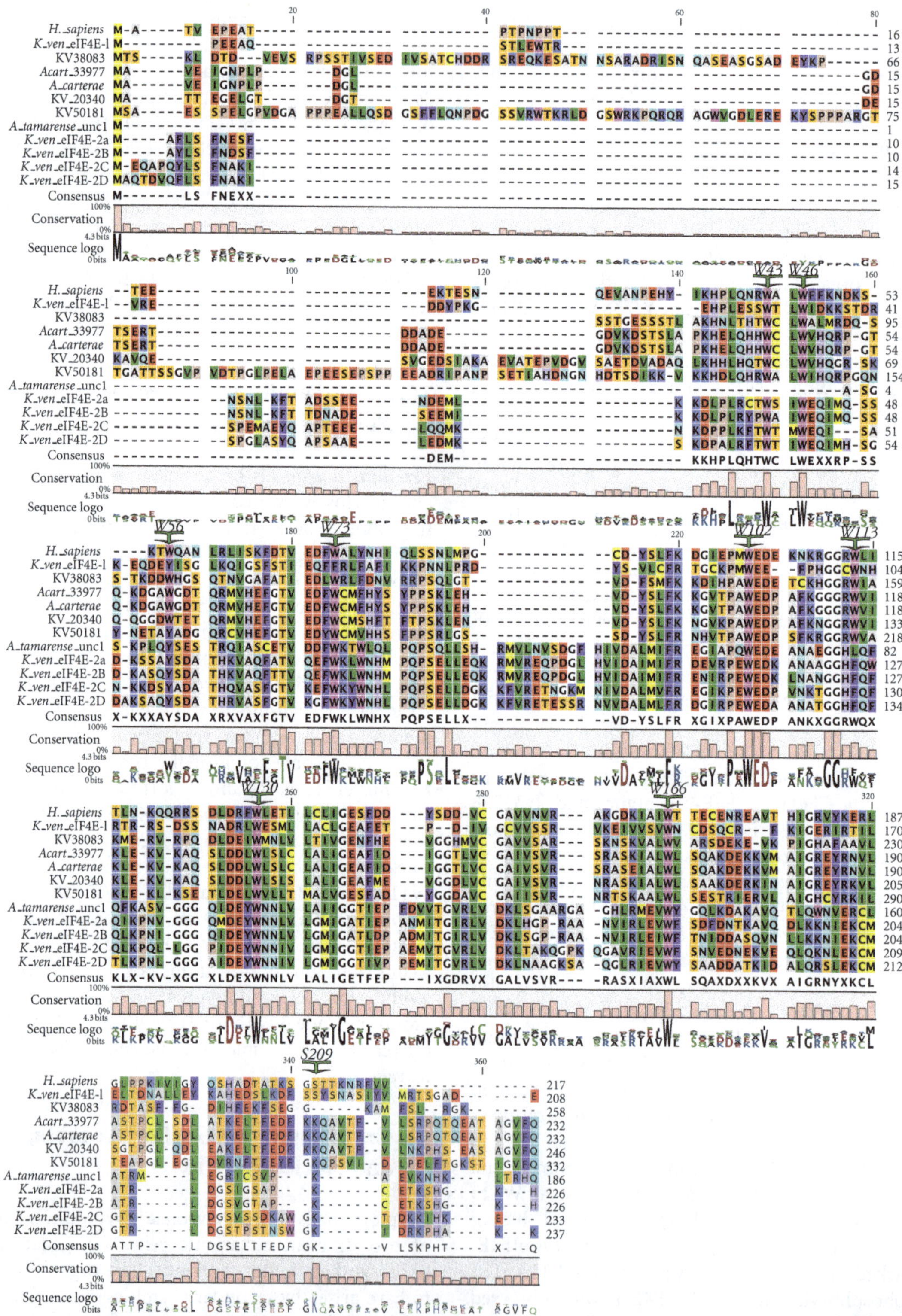

FIGURE 3: Comparison of the sequences of selected eIF4E-family members from *Karlodinium veneficum* and other dinoflagellates. Alignment of the amino acid sequences of selected established eIF4E-family members from *K. veneficum* and other dinoflagellates. Amino acid sequences were aligned with T-coffee using the BLOSUM62MT scoring matrix in CLC Main Workbench. To facilitate comparison of the sequences, the residues conserved in Class I eIF4Es in multicellular organisms are indicated and numbered as in human eIF4E-1: W43, W46, W56, W73, W102, W113, W130, W166, and S209.

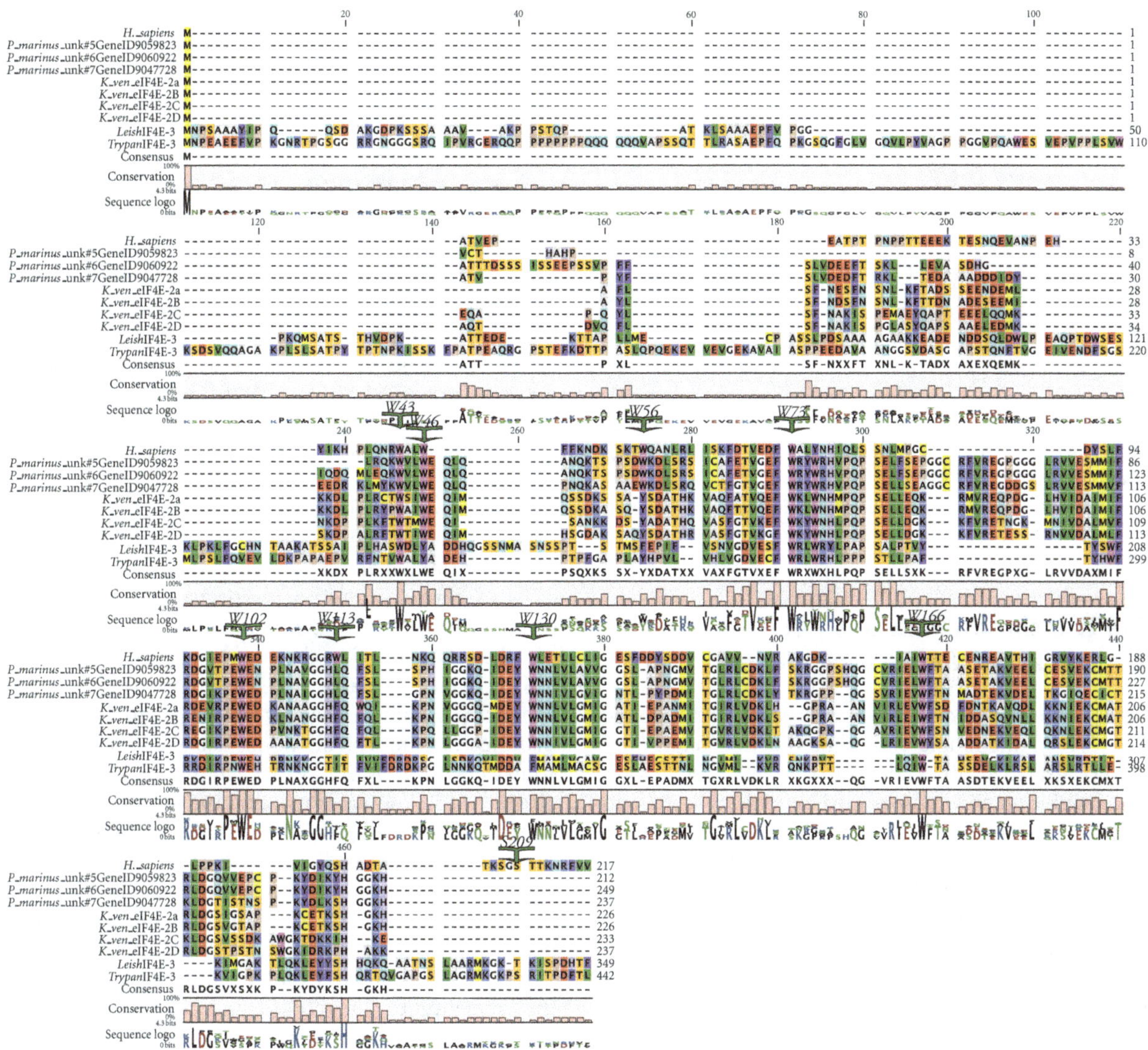

FIGURE 4: Comparison of the sequences of selected eIF4E-family members from *Perkinsus marinus* with related species from *K. veneficum* and trypanosome species. Alignment of the amino acid sequences of selected established eIF4E-family members from *P. marinus* with related species from *K. veneficum*. Amino acid sequences were aligned with T-coffee using the BLOSUM62MT scoring matrix in CLC Main Workbench. To facilitate comparison of the sequences, the residues conserved in Class I eIF4Es in multicellular organisms are indicated and numbered as in human eIF4E-1: W43, W46, W56, W73, W102, W113, W130, W166, and S209.

genome as a minute nucleomorph with three chromosomes [132–134]. The nucleomorph resides in a cell compartment, the periplastid space, that also contains the chloroplast. The cellular organization of *Guillardia theta* is shown in Figure 5.

In the cartoon, former chloroplast genes now inserted in nucleomorph or nuclear chromosomes are indicated in green, and former red algal genes now in the host nucleus are indicated in red. The nucleomorph genome has been sequenced and shown to be 551 kbp with a gene density of 1 gene per 977 bp, encoding 464 putative protein coding genes [133]. This compact genome has infrequent overlapping genes, and short inverted repeats containing rRNA cistrons at its chromosome ends [132, 133, 135]. There is almost a total absence of spliceosomal introns which has facilitated gene annotation. Marked evolutionary compaction [126–128] has eliminated almost all the nucleomorph genes for metabolic functions, but left a few hundred housekeeping genes, and 30 genes encoding chloroplast-located proteins [133]. The housekeeping genes are limited to nuclear maintenance and transport, translation, protein degradation and folding, and microtubule/centrosome functions [133, 135]. More than 20% of the housekeeping genes encode components of the translational machinery. The nucleomorph and its periplastid space can be viewed as providing a minimum eukaryotic expression system for a small number of nucleomorph-encoded chloroplast proteins. The endosymbiont has been reduced to an organelle, equivalent to a "complex plastid." The relict, enslaved red alga is referred to here as the endosymbiont for convenience, although strictly speaking it should be considered an organelle.

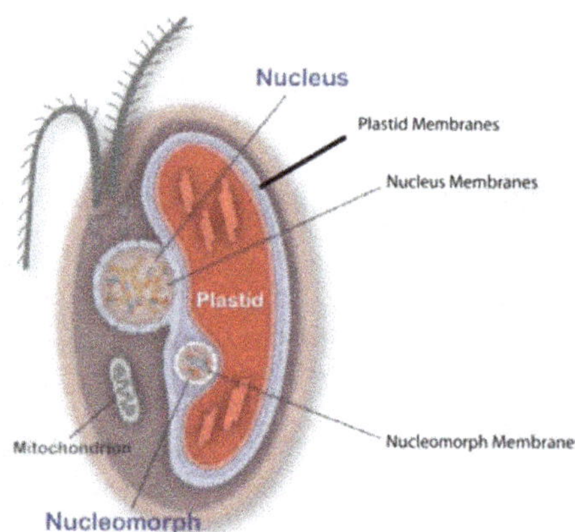

FIGURE 5: Cellular organization of *Guillardia theta*. Former chloroplast genes now inserted in nucleomorph or nuclear chromosomes are indicated in green, and former red algal genes now in the host nucleus are indicated in red. The four endosymbiont membranes are clearly represented.

12. The Translation Machinery of the *Guillardia theta* Nucleomorph

The endosymbiont encodes its own rRNA and 65 ribosomal proteins [133]. Functioning endosymbiont ribosomes have been demonstrated in the endosymbiont cytoplasm [136]. The endosymbiont has its own mRNAs with 5′-caps and poly(A) tails, elongation, and release factors, but only a subset of translational initiation factors. The nucleomorph encodes eIF1, eIF1A, eIF4A, eIF2 (all subunits, although the alpha subunit is truncated), eIF4E (truncated), eIF5B, eIF6, and poly(A) binding protein. It does not appear to encode any of the subunits of eIF2B, the factor that promotes guanine nucleotide exchange on eIF2. Furthermore, several initiation factors thought to be essential for eukaryotic initiation have not been identified; the nucleomorph does not encode eIF4B, eIF5, or the scaffold proteins eIF3 (any subunit) or eIF4G. All of these initiation factors have been shown to be essential in yeast (reviewed [137]). The nucleomorph is also without the eIF4E regulatory proteins, the 4E-BPs. Since the genome of the *G. theta* nucleomorph has been so severely compacted, it is hypothesized that the genes encoding complex cellular functions, such as protein synthesis, are limited to the minimal set needed to accomplish the function. Beyond the reduction in the number of initiation factors, several of the translational initiation factors encoded are truncated compared to their counterparts in nonprotist eukaryotes. This system can be considered to represent a natural experiment in deletion analysis and may tell us much about structure/function relationships in initiation factors, in addition to deepening our knowledge of this branch of the eukaryotic Tree of Life.

The factors eIF1, eIF1A, eIF2, eIF2B, eIF3, eIF4A, eIF4E, eIF4G, and eIF5 are all essential in yeast. eIF5B is not essential, although its deletion produces a severe slow growth phenotype [138]. The possibility that the lack of eIF2B, eIF3, eIF4G, and eIF5 in the *G. theta* endosymbiont reflects a primitive condition is unlikely since the deeply rooted, free-living red alga, *Cyanidioschyzon merolae*, encodes eIF4G, eIF5, and all the subunits of eIF2B and eIF3 [139]. *C. merolae* is considered to have the smallest genome of any free-living photosynthetic organism and molecular analyses support the primitiveness of this alga [140]. However, like the *Guillardia theta* nucleomorph, *C. merolae* does not appear to encode eIF4B, suggesting that eIF4B is a later evolutionary development [139]. Consistent with this, eIF4B is not essential in yeast, although its disruption results in a slow growth and cold-sensitive phenotype [141]. The endosymbiont has either evolved a minimal system of initiation through compaction of the genome, has made mechanistic adjustments to overcome factor deficiencies, or uses host factors. Use of host factors would require transport across the outer two membranes into the periplastidial compartment PPC and across all four membranes into the stroma [142].

The predicted eIF4E sequence of the *G. theta* nucleomorph is compacted, lacking extended amino-terminal and carboxy-terminal domains relative to the core of prototypical eIF4E (Figure 6) [6]. Although comparable forms from yeast, produced from deletion mutants, are still able to support life, they show considerably slower growth rates [143, 144]. This is likely to reflect a role of the N-terminal domain in enhancing stability. Scrutiny of the alignment also shows that the nucleomorph eIF4E has Leu at amino acid positions equivalent to V69 and W73 in human eIF4E-1. In human eIF4E-1, it is known that mutation to give a nonaromatic amino acid at position W73 disrupts the interaction with the adaptor protein, eIF4G, as does mutation of V69 to G [56, 57]. It is therefore unclear whether the nucleomorph eIF4E has the capacity to bind to eIF4G or indeed whether it needs to. It is possible that the nucleomorph eIF4E interacts with eIF4G imported from the host cytoplasm, although the sequence of the eIF4G-binding domain makes this unlikely. Alternatively, mRNA recruitment via an alternate interaction may be occurring. Interestingly, eIF4E sequences are available from additional nucleomorphs, those of another cryptophyte *Cryptomonas paramecium* and the heterokont *Haplogloia andersonii*. Both of these are truncated at the N-terminus, and both show substitutions in essential amino acids in the eIF4G binding domain.

13. Entamoeba and Mimivirus

Mimivirus is a double-stranded DNA virus isolated from amoebae [145]. It was first isolated from the water of a cooling tower in Bradford, England, during a study following a pneumonia outbreak in 1992 [146, 147]. Its name is derived from "mimicking microbe" because of the bacterium-like appearance of the particle and its Gram$^+$ staining. It has a cycle of viral transmission and replication that is typical of many dsDNA viruses. The study of mimivirus grown in *Acanthamoeba polyphaga* reveals a mature particle with the characteristic morphology of an icosahedral capsid with a diameter of at least 400 nm. At the beginning of the life cycle, the virus enters the amoeba and the viral genome is released. After expression of viral proteins and replication of the genome, the virus DNA is packaged into capsids and viral

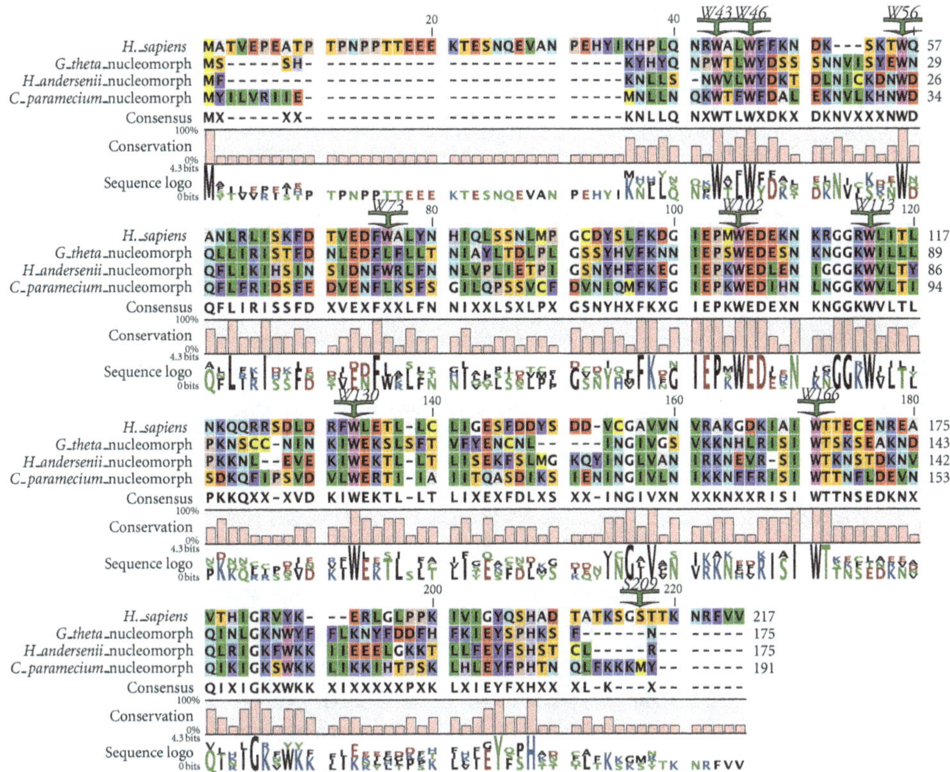

FIGURE 6: Comparison of the sequences of selected nucleomorph eIF4Es. Alignment of the amino acid sequences of the eIF4E from the nucleomorphs of *Guillardia theta*, *Haplogloia andersonii*, and *Cryptomonas paramecium*. Amino acid sequences were aligned with T-coffee using the BLOSUM62MT scoring matrix in CLC Main Workbench. To facilitate comparison of the sequences, the residues conserved in Class I eIF4Es in multicellular organisms are indicated and numbered as in human eIF4E-1: W43, W46, W56, W73, W102, W113, W130, W166, and S209.

particles are released from the amoeba [148]. Mimivirus has the largest known viral genome, 1.18 megabase pairs, and predicted to contain 1,262 genes, a very complex life cycle at the molecular level [146]. It encodes an unprecedented number of components of the transcriptional, translational, and replication machinery, many of which have not previously been described in viruses.

Although the mimivirus genome has more components resembling cellular genes than any other virus, it is still dependent on its host cell for the synthesis of proteins. Currently, the strategies by which mimivirus appropriates the host translation machinery have not been uncovered. Mimivirus exhibits many features that distinguish it from other nucleocytoplasmic large DNA viruses (NCLDVs). The most unexpected is the presence of numerous genes encoding central protein-translation components, encoding 10 proteins central to the translation apparatus: four aminoacy tRNA synthetases, eIF4E, ORF L496, eIF1A, eIF4A, eEF-1, and peptide chain release factor eRF1 [149, 150]. In addition, mimivirus encodes its own mRNA capping enzyme, and its own RNA cap guanine-N2 methyltransferase [151, 152]. Interestingly, mimivirus does not encode the mimic of the α-subunit of eIF2, found in many NCLDVs, that functions as a substrate to protect endogenous eIF2 from phosphorylation by an infection-activated kinase PKR. Finding these components of the translation apparatus in mimivirus calls into question the prevailing view that viruses rely entirely on the host translation machinery for protein synthesis [153]. Although the molecular mechanisms of its replicative cycle are yet to be uncovered, the detailed genome analysis has provided useful information on what viral genes may be involved in DNA replication and DNA repair, transcription, and protein folding, virion morphogenesis, and intracellular transport and suggests a complex life cycle.

The atypical eIF4E-family member of mimivirus is shown in Figure 7 aligned with the amino acid sequences of eIF4E-family members from *Acanthamoeba*. Mimivirus eIF4E has F49, W109, and E110, in positions equivalent to W56, W102, and E103 of human eIF4E-1, predicting that it should function in cap binding. However, mimivirus eIF4E, like the many protist eIF4E-family members, has extended stretches of amino acids between structural units of the core tryptophans. The positions of these stretches in mimivirus eIF4E resemble the extensions found in eIF4E-family members from Alveolata and Stramenopiles. However, the stretch of amino acids between residues equivalent to W102 and W166 of mouse/human eIF4E-1 are considerably longer in mimivirus eIF4E than those found in *P. falciparum* or other known stramenopile/alveolate eIF4E family members. Mimivirus eIF4E also differs from other eIF4E-family members in that it lacks a Trp residue equivalent to W73 of mouse

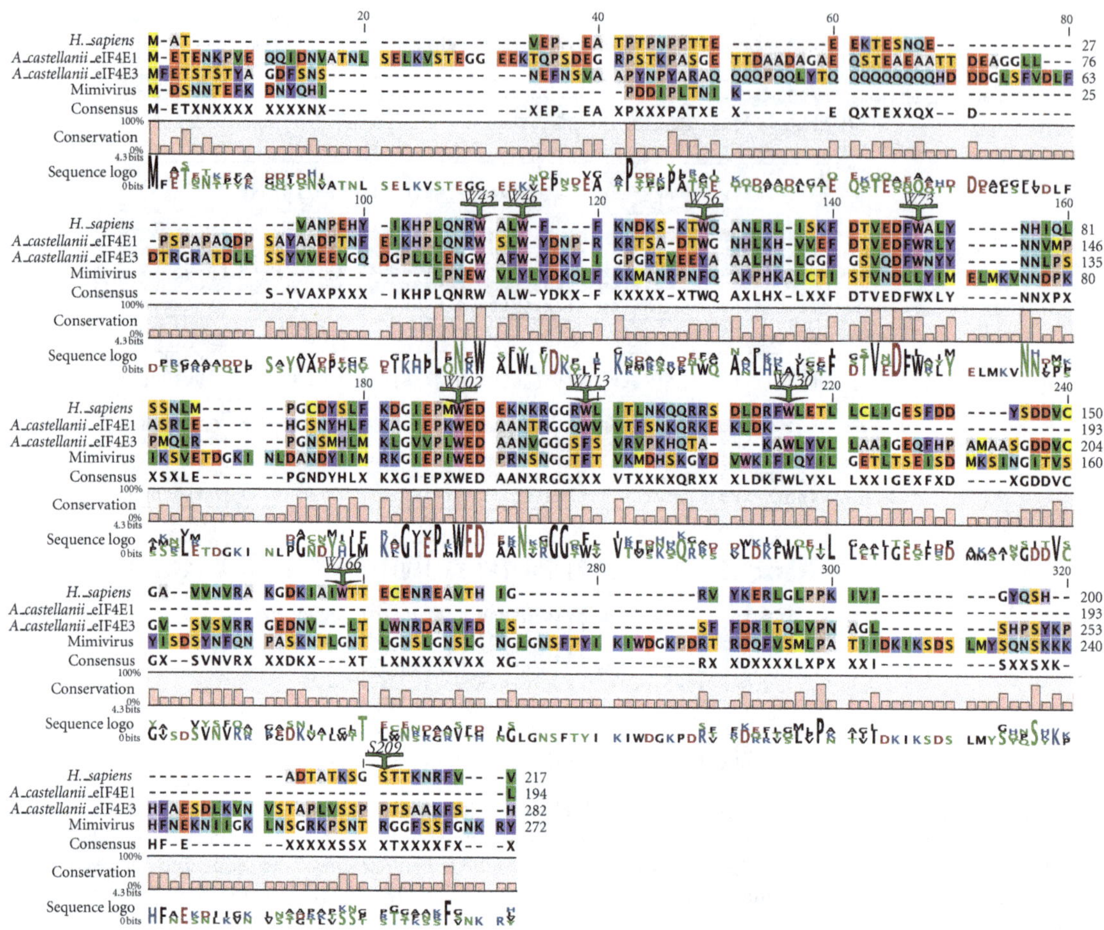

FIGURE 7: Comparison of the sequence of mimivirus eIF4E with those of an *Acanthamoeba* species. Alignment of the amino acid sequences of mimivirus eIF4E with eIF4Es from its host *Acanthamoeba castellani*. Amino acid sequences were aligned with T-coffee using the BLOSUM62MT scoring matrix in CLC Main Workbench. *A. castellani* sequences were derived from the Protist EST Program (PEP) in advance of scientific publication and acceptance by GenBank at http://amoebidia.bcm.umontreal.ca/public/pepdb/agrm.php. This site is no longer publically available. To facilitate comparison of the sequences, the residues conserved in Class I eIF4Es in multicellular organisms are indicated and numbered as in human eIF4E-1: W43, W46, W56, W73, W102, W113, W130, W166, and S209.

eIF4E-1 suggesting that the protein may not interact with eIF4G or 4E-BPs.

The host *A. castellanii* expresses at least five eIF4E-family members. None of the *A. castellanii* eIF4E family members shows extended stretches of amino acids in positions similar to those found in mimivirus eIF4E. Furthermore, *A. castellanii* eIF4E-family members possess conserved residues equivalent to V69 and W73 of human eIF4E-1 important for interaction with eIF4G and 4E-BPs, unlike mimivirus eIF4E. As a consequence of these differences in significant residues, it seems unlikely that mimivirus eIF4E has been acquired from the *Acanthamoeba* host. The sequence of mimivirus eIF4E predicts that it is likely to bind to 5′-cap structures but may not interact with eIF4G, suggesting that it could function as an inhibitor of cap-dependent translation. However, the mimivirus genome encodes genes for mRNA capping enzymes [151, 152], as do related NCLD viruses, suggesting that mimivirus mRNAs are capped and that the virus requires cap-dependent translation of its mRNAs. Since mimivirus can use both *A. polyphaga* and human as hosts, it will be of use to consider the role of its eIF4E in the context of mRNA recruitment in both environments.

14. Overview of Protist eIF4Es

Like multicellular eukaryotes, many protists encode multiple eIF4E family members. However, these do not fall into the eIF4E classes found in plants/metazoans/fungi. Of the eight conserved tryptophan residues typical of eIF4E Class I sequences, most are either conserved in protist eIF4E family members or are replaced by other aromatic residues. In many bikont protists, extensions are found between the conserved aromatic amino acids which vary with clade and phylogenetic grouping. Figure 2 shows the relationships of the protist eIF4Es and suggests that they fall into three clades. eIF4Es from dinoflagellates/*Perkinsus* and heterokonts can be found in all three clades. eIF4Es from ciliates and the parasitic dinoflagellate *Amoebophrya* are present in only two clades. Unfortunately, at the current time, there are many more eIF4E sequences available for alveolates and excavates than for other protist groups, particularly the opisthokonts and amoebozoa. Furthermore, there is insufficient data on the functional characteristics of the eIF4Es in each of these clades to allow for any confident classification at this stage. Nevertheless, it is known that the *Leishmania* and *Trypanosoma* eIF4Es, EIF4E3 and 4 function as initiation factors and that the dinoflagellate eIF4E-2s from *K. veneficum* bind cap structures suggesting that this clade contains eIF4E family members that function as initiation factors. Table 2 shows the characteristics of some of the members from "Clades 1" and "2." As genome sequencing projects are completed, it is expected that the number of protist eIF4E family members available for scrutiny will increase dramatically in the near future. A wider representation of taxa will allow a more complete understanding of the relationships between these eIF4Es, as will a much needed expansion of functional studies particularly in the non-parasitic representatives.

Acknowledgments

This was supported by MCB no. 0626678, to A. R. Place and MCB no. 0134013 to R. Jagus. The authors are grateful to Drs. Terry Gaasterland, John Gill, Senjie Lin, Yu-Hui Rogers, and Huang Zhang, co-PIs with A. R. Place of the NSF Microbial Genome Sequencing Program Grant no. EF-0626678, "Dinoflagellate full-length cDNA sequencing," which made possible searching of the *K. veneficum* cDNA database. The authors would like to thank Charles Delwiche (supported by NSF DEB 0629624) for allowing them to use unpublished sequences for *Amoebophyra*. In addition, thanks are extended to Dr. Joseph Pitula, University of Maryland Eastern Shore, for useful discussions on the *P. marinus* eIF4Es and splice leaders and to Jorge Rodriguez, University of Maryland Eastern Shore, for initial alignments of the *P. marinus* eIF4Es. This paper represents contribution no. #12-234 from IMET and #4660 from UMCES.

References

[1] A. Ben-Shem, N. GarreaudeLoubresse, S. Melnikov, L. Jenner, G. Yusupova, and M. Yusupov, "The structure of the eukaryotic ribosome at 3.0 Å resolution," *Science*, vol. 334, pp. 1524–1529, 2011.

[2] A. Korostelev, D. N. Ermolenko, and H. F. Noller, "Structural dynamics of the ribosome," *Current Opinion in Chemical Biology*, vol. 12, no. 6, pp. 674–683, 2008.

[3] M. C. Rivera and J. A. Lake, "Evidence that eukaryotes and eocyte prokaryotes are immediate relatives," *Science*, vol. 257, no. 5066, pp. 74–76, 1992.

[4] M. C. Rivera, "Genomic analyses and the origin of the eukaryotes," *Chemistry and Biodiversity*, vol. 4, no. 11, pp. 2631–2638, 2007.

[5] G. Hernández and P. Vazquez-Pianzola, "Functional diversity of the eukaryotic translation initiation factors belonging to eIF4 families," *Mechanisms of Development*, vol. 122, no. 7-8, pp. 865–876, 2005.

[6] B. Joshi, K. Lee, D. L. Maeder, and R. Jagus, "Phylogenetic analysis of eIF4E-family members," *BMC Evolutionary Biology*, vol. 5, article 48, 2005.

[7] C. L. McGrath and L. A. Katz, "Genome diversity in microbial eukaryotes," *Trends in Ecology and Evolution*, vol. 19, no. 1, pp. 32–38, 2004.

[8] L. W. Parfrey, J. Grant, Y. I. Tekle et al., "Broadly sampled multigene analyses yield a well-resolved eukaryotic tree of life," *Systematic Biology*, vol. 59, no. 5, pp. 518–533, 2010.

[9] D. Moreira and P. López-García, "The molecular ecology of microbial eukaryotes unveils a hidden world," *Trends in Microbiology*, vol. 10, no. 1, pp. 31–38, 2002.

[10] J. D. Bangs, P. F. Crain, T. Hashizume, J. A. McCloskey, and J. C. Boothroyd, "Mass spectrometry of mRNA cap 4 from trypanosomatids reveals two novel nucleosides," *The Journal of Biological Chemistry*, vol. 267, no. 14, pp. 9805–9815, 1992.

[11] W. Marande, P. López-García, and D. Moreira, "Eukaryotic diversity and phylogeny using small- and large-subunit ribosomal RNA genes from environmental samples," *Environmental Microbiology*, vol. 11, no. 12, pp. 3179–3188, 2009.

[12] J. D. Keene, "Minireview: global regulation and dynamics of ribonucleic acid," *Endocrinology*, vol. 151, no. 4, pp. 1391–1397, 2010.

[13] X. Qin, S. Ahn, T. P. Speed, and G. M. Rubin, "Global analyses of mRNA translational control during early Drosophila embryogenesis," *Genome Biology*, vol. 8, no. 4, article R63, 2007.

[14] J. B. Dacks, A. A. Peden, and M. C. Field, "Evolution of specificity in the eukaryotic endomembrane system," *International Journal of Biochemistry and Cell Biology*, vol. 41, no. 2, pp. 330–340, 2009.

[15] M. C. Field and J. B. Dacks, "First and last ancestors: reconstructing evolution of the endomembrane system with ESCRTs, vesicle coat proteins, and nuclear pore complexes," *Current Opinion in Cell Biology*, vol. 21, no. 1, pp. 4–13, 2009.

[16] B. L. Semler and M. L. Waterman, "IRES-mediated pathways to polysomes: nuclear versus cytoplasmic routes," *Trends in Microbiology*, vol. 16, no. 1, pp. 1–5, 2008.

[17] A. Schröder-Lorenz and L. Rensing, "Circadian changes in protein-synthesis rate and protein phosphorylation in cell-free extracts of *Gonyaulax polyedra*," *Planta*, vol. 170, no. 1, pp. 7–13, 1987.

[18] E. V. Koonin, "The origin of introns and their role in eukaryogenesis: a compromise solution to the introns-early versus introns-late debate?" *Biology Direct*, vol. 1, article 22, 2006.

[19] E. V. Koonin, "The origin and early evolution of eukaryotes in the light of phylogenomics," *Genome Biology*, vol. 11, no. 5, article 209, 2010.

[20] W. Martin and E. V. Koonin, "Introns and the origin of nucleus-cytosol compartmentalization," *Nature*, vol. 440, no. 7080, pp. 41–45, 2006.

[21] T. Cavalier-Smith, "Intron phylogeny: a new hypothesis," *Trends in Genetics*, vol. 7, no. 5, pp. 145–148, 1991.

[22] B. Cousineau, S. Lawrence, D. Smith, and M. Belfort, "Retrotransposition of a bacterial group II intron," *Nature*, vol. 404, no. 6781, pp. 1018–1021, 2000.

[23] A. M. Lambowitz and S. Zimmerly, "Mobile group II introns," *Annual Review of Genetics*, vol. 38, pp. 1–35, 2004.

[24] B. J. Mans, V. Anantharaman, L. Aravind, and E. V. Koonin, "Comparative genomics, evolution and origins of the nuclear envelope and nuclear pore complex," *Cell Cycle*, vol. 3, no. 12, pp. 1612–1637, 2004.

[25] L. Aravind and E. V. Koonin, "Eukaryote-specific domains in translation initiation factors: implications for translation regulation and evolution of the translation system," *Genome Research*, vol. 10, no. 8, pp. 1172–1184, 2000.

[26] D. Benelli and P. Londei, "Translation initiation in Archaea: conserved and domain-specific features," *Biochemical Society Transactions*, vol. 39, no. 1, pp. 89–93, 2011.
[27] P. Londei, "Evolution of translational initiation: new insights from the archaea," *FEMS Microbiology Reviews*, vol. 29, no. 2, pp. 185–200, 2005.
[28] G. Hernández, "Was the initiation of translation in early eukaryotes IRES-driven?" *Trends in Biochemical Sciences*, vol. 33, no. 2, pp. 58–64, 2008.
[29] K. Sawicka, M. Bushell, K. A. Spriggs, and A. E. Willis, "Polypyrimidine-tract-binding protein: a multifunctional RNA-binding protein," *Biochemical Society Transactions*, vol. 36, no. 4, pp. 641–647, 2008.
[30] S. M. Lewis and M. Holcik, "For IRES *trans*-acting factors, it is all about location," *Oncogene*, vol. 27, no. 8, pp. 1033–1035, 2008.
[31] B. L. Semler and M. L. Waterman, "IRES-mediated pathways to polysomes: nuclear versus cytoplasmic routes," *Trends in Microbiology*, vol. 16, no. 1, pp. 1–5, 2008.
[32] A. A. Komar and M. Hatzoglou, "Cellular IRES-mediated translation: the war of ITAFs in pathophysiological states," *Cell Cycle*, vol. 10, no. 2, pp. 229–240, 2011.
[33] O. Elroy-Stein and W. C. Merrick, "Translation initiation via cellular internal ribosome entry sites," in *Translational Controlin Biology and Medicine*, M. B. Mathews, N. Sonenberg, and J. W. B. Hershey, Eds., pp. 155–172, Cold Spring Harbor Laboratory Press, Cold Spring Harbor, NY, USA, 2007.
[34] J. A. Doudna and P. Sarnow, "Translation initiation by viral internal ribosome entry sites," in *Translational Control in Biology and Medicine*, M. B. Mathews, N. Sonenberg, and J. W. B. Hershey, Eds., pp. 129–154, Cold Spring Harbor Laboratory Press, Cold Spring Harbor, NY, USA, 2007.
[35] A. Pacheco and E. Martinez-Salas, "Insights into the biology of IRES elements through riboproteomic approaches," *Journal of Biomedicine and Biotechnology*, vol. 2010, Article ID 458927, 12 pages, 2010.
[36] E. L. Lasda and T. Blumenthal, "*Trans*-splicing," *Wiley Interdisciplinary Reviews*, vol. 2, pp. 417–34, 2011.
[37] K. E. M. Hastings, "SL *trans*-splicing: easy come or easy go?" *Trends in Genetics*, vol. 21, no. 4, pp. 240–247, 2005.
[38] R. Derelle, T. Momose, M. Manuel, C. Da Silva, P. Wincker, and E. Houliston, "Convergent origins and rapid evolution of spliced leader *trans*-splicing in Metazoa: insights from the Ctenophora and Hydrozoa," *RNA*, vol. 16, no. 4, pp. 696–707, 2010.
[39] J. P. Bruzik and T. Maniatis, "Spliced leader RNAs from lower eukaryotes are *trans*-spliced in mammalian cells," *Nature*, vol. 360, no. 6405, pp. 692–695, 1992.
[40] G. Hernández, "On the origin of the cap-dependent initiation of translation in eukaryotes," *Trends in Biochemical Sciences*, vol. 34, no. 4, pp. 166–175, 2009.
[41] L. Li and C. C. Wang, "Identification in the ancient protist *Giardia lamblia* of two eukaryotic translation initiation factor 4E homologues with distinctive functions," *Eukaryotic Cell*, vol. 4, no. 5, pp. 948–959, 2005.
[42] M. A. Andrei, D. Ingelfinger, R. Heintzmann, T. Achsel, R. Rivera-Pomar, and R. Lührmann, "A role for eIF4E and eIF4E-transporter in targeting mRNPs to mammalian processing bodies," *RNA*, vol. 11, no. 5, pp. 717–727, 2005.
[43] N. P. Hoyle, L. M. Castelli, S. G. Campbell, L. E. A. Holmes, and M. P. Ashe, "Stress-dependent relocalization of translationally primed mRNPs to cytoplasmic granules that are kinetically and spatially distinct from P-bodies," *Journal of Cell Biology*, vol. 179, no. 1, pp. 65–74, 2007.
[44] L. Rong, M. Livingstone, R. Sukarieh et al., "Control of eIF4E cellular localization by eIF4E-binding proteins, 4E-BPs," *RNA*, vol. 14, no. 7, pp. 1318–1327, 2008.
[45] I. G. Goodfellow and L. O. Roberts, "Eukaryotic initiation factor 4E," *International Journal of Biochemistry and Cell Biology*, vol. 40, no. 12, pp. 2675–2680, 2008.
[46] A. C. Gingras, B. Raught, and N. Sonenberg, "eIF4 initiation factors: effectors of mRNA recruitment to ribosomes and regulators of translation," *Annual Review of Biochemistry*, vol. 68, pp. 913–963, 1999.
[47] T. von der Haar, J. D. Gross, G. Wagner, and J. E. G. McCarthy, "The mRNA cap-binding protein eIF4E in post-transcriptional gene expression," *Nature Structural and Molecular Biology*, vol. 11, no. 6, pp. 503–511, 2004.
[48] N. Sonenberg, "eIF4E, the mRNA cap-binding protein: from basic discovery to translational research," *Biochemistry and Cell Biology*, vol. 86, no. 2, pp. 178–183, 2008.
[49] I. Topisirovic, Y. V. Svitkin, N. Sonenberg, and A. J. Shatkin, "Cap and cap-binding proteins in the control of gene expression," *Wiley Interdisciplinary Reviews*, pp. 277–98, 2011.
[50] J. Marcotrigiano, A. C. Gingras, N. Sonenberg, and S. K. Burley, "Co-crystal structure of the messenger RNA5′ cap-binding protein (eIF4E) bound to 7-methyl-GDP," *Cell*, vol. 89, no. 6, pp. 951–961, 1997.
[51] H. Matsuo et al., "Structure of translation factor eIF4E bound to m^{7}GDP and interaction with 4E-binding protein," *Natural Structural Biology*, vol. 4, pp. 717–24.
[52] A. Niedzwiecka, J. Marcotrigiano, J. Stepinski et al., "Biophysical studies of eIF4E cap-binding protein: recognition of mRNA 5′ cap structure and synthetic fragments of eIF4G and 4E-BP1 proteins," *Journal of Molecular Biology*, vol. 319, no. 3, pp. 615–635, 2002.
[53] G. Hernández, M. Altmann, J. M. Sierra et al., "Functional analysis of seven genes encoding eight translation initiation factor 4E (eIF4E) isoforms in *Drosophila*," *Mechanisms of Development*, vol. 122, no. 4, pp. 529–543.
[54] B. Joshi, A. Cameron, and R. Jagus, "Characterization of mammalian eIF4E-family members," *European Journal of Biochemistry*, vol. 271, no. 11, pp. 2189–2203, 2004.
[55] J. Robalino, B. Joshi, S. C. Fahrenkrug, and R. Jagus, "Two zebrafish eIF4E family members are differentially expressed and functionally divergent," *The Journal of Biological Chemistry*, vol. 279, no. 11, pp. 10532–10541, 2004.
[56] M. Ptushkina, T. Von der Haar, S. Vasilescu, R. Frank, R. Birkenhäger, and J. E. G. McCarthy, "Cooperative modulation by eIF4G of eIF4E-binding to the mRNA5′ cap in yeast involves a site partially shared by p20," *The EMBO Journal*, vol. 17, no. 16, pp. 4798–4808, 1998.
[57] S. Pyronnet, H. Imataka, A. C. Gingras, R. Fukunaga, T. Hunter, and N. Sonenberg, "Human eukaryotic translation initiation factor 4G (eIF4G) recruits Mnk1 to phosphorylate eIF4E," *The EMBO Journal*, vol. 18, no. 1, pp. 270–279, 1999.
[58] M. Jankowska-Anyszka, B. J. Lamphear, E. J. Aamodt et al., "Multiple isoforms of eukaryotic protein synthesis initiation factor 4E in *Caenorhabditis elegans* can distinguish between mono- and trimethylated mRNA cap structures," *The Journal of Biological Chemistry*, vol. 273, no. 17, pp. 10538–10541, 1998.
[59] B. D. Keiper, B. J. Lamphear, A. M. Deshpande et al., "Functional characterization of five eIF4E isoforms in *Caenorhabditis elegans*," *The Journal of Biological Chemistry*, vol. 275, no. 14, pp. 10590–10596, 2000.

[60] N. Minshall, M. H. Reiter, D. Weil, and N. Standart, "CPEB interacts with an ovary-specific eIF4E and 4E-T in early *Xenopus oocytes*," *The Journal of Biological Chemistry*, vol. 282, no. 52, pp. 37389–37401, 2007.

[61] N. Standart and N. Minshall, "Translational control in early development: CPEB, P-bodies and germinal granules," *Biochemical Society Transactions*, vol. 36, no. 4, pp. 671–676, 2008.

[62] A. V. Evsikov and C. Marín de Evsikova, "Evolutionary origin and phylogenetic analysis of the novel oocyte-specific eukaryotic translation initiation factor 4E in Tetrapoda," *Development Genes and Evolution*, vol. 219, no. 2, pp. 111–118, 2009.

[63] P. F. Cho, C. Gamberi, Y. Cho-Park, I. B. Cho-Park, P. Lasko, and N. Sonenberg, "Cap-dependent translational inhibition establishes two opposing morphogen gradients in *Drosophila* embryos," *Current Biology*, vol. 16, no. 20, pp. 2035–2041, 2006.

[64] T. D. Dinkova, B. D. Keiper, N. L. Korneeva, E. J. Aamodt, and R. E. Rhoads, "Translation of a small subset of *Caenorhabditis elegans* mRNAs is dependent on a specific eukaryotic translation initiation factor 4E isoform," *Molecular and Cellular Biology*, vol. 25, no. 1, pp. 100–113, 2005.

[65] S. L. Baldauf, "The deep roots of eukaryotes," *Science*, vol. 300, no. 5626, pp. 1703–1706, 2003.

[66] R. Derelle and B. F. Lang, "Rooting the eukaryotic tree with mitochondrial and bacterial proteins," *Molecular Biology and Evolution*, vol. 29, no. 4, pp. 1277–89, 2012.

[67] E. J. Javaux, A. H. Knoll, and M. R. Walter, "Morphological and ecological complexity in early eukaryotic ecosystems," *Nature*, vol. 412, no. 6842, pp. 66–69, 2001.

[68] P. J. Keeling, "Chromalveolates and the evolution of plastids by secondary endosymbiosis," *Journal of Eukaryotic Microbiology*, vol. 56, no. 1, pp. 1–8, 2009.

[69] P. J. Keeling, "Diversity and evolutionary history of plastids and their hosts," *American Journal of Botany*, vol. 91, no. 10, pp. 1481–1493, 2004.

[70] P. J. Keeling, "The endosymbiotic origin, diversification and fate of plastids," *Philosophical Transactions of the Royal Society B*, vol. 365, no. 1541, pp. 729–748, 2010.

[71] S. Y. Hwan, J. D. Hackett, F. M. Van Dolah, T. Nosenko, K. L. Lidie, and D. Bhattacharya, "Tertiary endosymbiosis driven genome evolution in dinoflagellate algae," *Molecular Biology and Evolution*, vol. 22, no. 5, pp. 1299–1308, 2005.

[72] C. F. Delwiche, "Tracing the thread of plastid diversity through the tapestry of life," *American Naturalist*, vol. 154, no. 4, pp. S164–S177, 1999.

[73] S. Sato, "The apicomplexan plastid and its evolution," *Cellular and Molecular Life Sciences*, vol. 68, no. 8, pp. 1285–1296, 2011.

[74] J. A. Fernández Robledo et al., "The search for the missing link: a relic plastid in *Perkinsus*?" *International Journal for Parasitology*, vol. 41, pp. 1217–1229, 2011.

[75] P. J. Keeling et al., "The tree of eukaryotes," *Trends in Ecology and Evolution*, vol. 20, pp. 670–676, 2005.

[76] S. M. Adl et al., "Diversity, nomenclature, and taxonomy of protists," *Systematics Biology*, pp. 684–689.

[77] C. E. Lane and J. M. Archibald, "The eukaryotic tree of life: endosymbiosis takes its TOL," *Trends in Ecology and Evolution*, vol. 23, no. 5, pp. 268–275, 2008.

[78] N. King, "The unicellular ancestry of animal development," *Developmental Cell*, vol. 7, no. 3, pp. 313–325, 2004.

[79] T. Cavalier-Smith, "Eukaryote kingdoms: seven or nine?" *BioSystems*, vol. 14, no. 3-4, pp. 461–481, 1981.

[80] T. Cavalier-Smith, "The phagotrophic origin of eukaryotes and phylogenetic classification on protozoa," *International Journal of Systematic and Evolutionary Microbiology*, vol. 52, no. 2, pp. 297–354, 2002.

[81] T. Cavalier-Smith, "A revised six-kingdom system of life," *Biological Reviews of the Cambridge Philosophical Society*, vol. 73, no. 3, pp. 203–266, 1998.

[82] T. Cavalier-Smith and E. E. Y. Chao, "Phylogeny and classification of phylum Cercozoa (Protozoa)," *Protist*, vol. 154, no. 3-4, pp. 341–358, 2003.

[83] J. M. Archibald and P. J. Keeling, "Actin and ubiquitin protein sequences support a cercozoan/foraminiferan ancestry for the plasmodiophorid plant pathogens," *Journal of Eukaryotic Microbiology*, vol. 51, no. 1, pp. 113–118, 2004.

[84] R. D. Adam, "Biology of *Giardia lamblia*," *Clinical Microbiology Reviews*, vol. 14, no. 3, pp. 447–475, 2001.

[85] J. Jerlström-Hultqvist, O. Franzén, J. Ankarklev et al., "Genome analysis and comparative genomics of a *Giardia intestinalis* assemblage E isolate," *BMC Genomics*, vol. 11, no. 1, article 543, 2010.

[86] R. D. Adam, "The *Giardia lamblia* genome," *International Journal for Parasitology*, vol. 30, no. 4, pp. 475–484, 2000.

[87] S. Hausmann, M. A. Altura, M. Witmer, S. M. Singer, H. G. Elmendorf, and S. Shuman, "Yeast-like mRNA capping apparatus in *Giardia lamblia*," *The Journal of Biological Chemistry*, vol. 280, no. 13, pp. 12077–12086, 2005.

[88] L. Li and C. C. Wang, "Capped mRNA with a single nucleotide leader is optimally translated in a primitive eukaryote, *Giardia lamblia*," *The Journal of Biological Chemistry*, vol. 279, no. 15, pp. 14656–14664, 2004.

[89] X. Niu, T. Hartshorne, X. Y. He, and N. Agabian, "Characterization of putative small nuclear RNAs from *Giardia lamblia*," *Molecular and Biochemical Parasitology*, vol. 66, no. 1, pp. 49–57, 1994.

[90] X. H. Liang, A. Haritan, S. Uliel, and S. Michaeli, "*Trans* and *cis* splicing in trypanosomatids: mechanism, factors, and regulation," *Eukaryotic Cell*, vol. 2, no. 5, pp. 830–840, 2003.

[91] V. Douris, M. J. Telford, and M. Averof, "Evidence for multiple independent origins of *trans*-splicing in Metazoa," *Molecular Biology and Evolution*, vol. 27, no. 3, pp. 684–693, 2010.

[92] M. Keller, L. H. Tessier, R. L. Chan, J. H. Weil, and P. Imbault, "In *Euglena*, spliced-leader RNA (SL-RNA) and 5S rRNA genes are tandemly repeated," *Nucleic Acids Research*, vol. 20, no. 7, pp. 1711–1715, 1992.

[93] T. Blumenthal, "*Trans*-splicing and polycistronic transcription in *Caenorhabditis elegans*," *Trends in Genetics*, vol. 11, no. 4, pp. 132–136, 1995.

[94] J. D. Bangs, P. F. Crain, T. Hashizume, J. A. McCloskey, and J. C. Boothroyd, "Mass spectrometry of mRNA cap 4 from trypanosomatids reveals two novel nucleosides," *The Journal of Biological Chemistry*, vol. 267, no. 14, pp. 9805–9815, 1992.

[95] K. L. Perry, K. P. Watkins, and N. Agabian, "Trypanosome mRNAs have unusual 'cap 4' structures acquired by addition of a spliced leader," *Proceedings of the National Academy of Sciences of the United States of America*, vol. 84, no. 23, pp. 8190–8194, 1987.

[96] C. E. Clayton, "Life without transcriptional control? From fly to man and back again," *The EMBO Journal*, vol. 21, no. 8, pp. 1881–1888, 2002.

[97] S. Haile and B. Papadopoulou, "Developmental regulation of gene expression in trypanosomatid parasitic protozoa," *Current Opinion in Microbiology*, vol. 10, no. 6, pp. 569–577, 2007.

[98] R. Queiroz, C. Benz, K. Fellenberg, J. D. Hoheisel, and C. Clayton, "Transcriptome analysis of differentiating trypanosomes reveals the existence of multiple post-transcriptional regulons," *BMC Genomics*, vol. 10, article 1471, p. 495, 2009.

[99] Y. Yoffe, J. Zuberek, A. Lerer et al., "Binding specificities and potential roles of isoforms of eukaryotic initiation factor 4E in *Leishmania*," *Eukaryotic Cell*, vol. 5, no. 12, pp. 1969–1979, 2006.

[100] E. R. Freire, R. Dhalia, D. M. N. Moura et al., "The four trypanosomatid eIF4E homologues fall into two separate groups, with distinct features in primary sequence and biological properties," *Molecular and Biochemical Parasitology*, vol. 176, no. 1, pp. 25–36, 2011.

[101] J. Marcotrigiano, A. C. Gingras, N. Sonenberg, and S. K. Burley, "Cap-dependent translation initiation in eukaryotes is regulated by a molecular mimic of elF4G," *Molecular Cell*, vol. 3, no. 6, pp. 707–716, 1999.

[102] R. Dhalia, C. R. S. Reis, E. R. Freire et al., "Translation initiation in *Leishmania major*: characterisation of multiple eIF4F subunit homologues," *Molecular and Biochemical Parasitology*, vol. 140, no. 1, pp. 23–41, 2005.

[103] Y. Hou and S. Lin, "Distinct gene number-genome size relationships for eukaryotes and non-eukaryotes: gene content estimation for dinoflagellate genomes," *PLoS ONE*, vol. 4, no. 9, Article ID e6978, 2009.

[104] S. Moreno Díaz de la Espina, E. Alverca, A. Cuadrado, and S. Franca, "Organization of the genome and gene expression in a nuclear environment lacking histones and nucleosomes: the amazing dinoflagellates," *European Journal of Cell Biology*, vol. 84, no. 2-3, pp. 137–149, 2005.

[105] T. R. Bachvaroff and A. R. Place, "From stop to start: tandem gene arrangement, copy number and *Trans*-splicing sites in the dinoflagellate *Amphidinium carterae*," *PLoS ONE*, vol. 3, no. 8, Article ID e2929, 2008.

[106] M. Mittag, D. H. Lee, and J. W. Hastings, "Circadian expression of the luciferin-binding protein correlates with the binding of a protein to the 3′ untranslated region of its mRNA," *Proceedings of the National Academy of Sciences of the United States of America*, vol. 91, no. 12, pp. 5257–5261, 1994.

[107] T. Fagan, D. Morse, and J. W. Hastings, "Circadian synthesis of a nuclear-encoded chloroplast glyceraldehyde-3-phosphate dehydrogenase in the dinoflagellate *Gonyaulax polyedra* is translationally controlled," *Biochemistry*, vol. 38, no. 24, pp. 7689–7695, 1999.

[108] S. A. Brunelle and F. M. Van Dolah, "Post-transcriptional regulation of S-Phase genes in the dinoflagellate, *Karenia brevis*," *Journal of Eukaryotic Microbiology*, vol. 58, no. 4, pp. 373–382, 2011.

[109] A. Schröder-Lorenz and L. Rensing, "Circadian changes in protein-synthesis rate and protein phosphorylation in cell-free extracts of *Gonyaulax polyedra*," *Planta*, vol. 170, no. 1, pp. 7–13, 1987.

[110] M. R. Ten Lohuis and D. J. Miller, "Light-regulated transcription of genes encoding peridinin chlorophyll a proteins and the major intrinsic light-harvesting complex proteins in the dinoflagellate *Amphidinium carterae* hulburt (Dinophycae): changes in cytosine methylation accompany photoadaptation," *Plant Physiology*, vol. 117, no. 1, pp. 189–196, 1998.

[111] C. Rossini, W. Taylor, T. Fagan, and J. W. Hastings, "Lifetimes of mRNAs for clock-regulated proteins in a dinoflagellate," *Chronobiology International*, vol. 20, no. 6, pp. 963–976, 2003.

[112] F. W. F. Lee, D. Morse, and S. C. L. Lo, "Identification of two plastid proteins in the dinoflagellate *Alexandrium affine* that are substantially down-regulated by nitrogen-depletion," *Journal of Proteome Research*, vol. 8, no. 11, pp. 5080–5082, 2009.

[113] H. Akimoto, T. Kinumi, and Y. Ohmiya, "Circadian rhythm of a TCA cycle enzyme is apparently regulated at the translational level in the dinoflagellate *Lingulodinium polyedrum*," *Journal of Biological Rhythms*, vol. 20, no. 6, pp. 479–489, 2005.

[114] D. Morse, P. M. Milos, E. Roux, and J. W. Hastings, "Circadian regulation of bioluminescence in *Gonyaulax* involves translational control," *Proceedings of the National Academy of Sciences of the United States of America*, vol. 86, no. 1, pp. 172–176, 1989.

[115] F. M. Van Dolah, K. B. Lidie, J. S. Morey et al., "Microarray analysis of diurnal- and circadian-regulated genes in the Florida red-tide dinoflagellate *Karenia brevis* (Dinophyceae)," *Journal of Phycology*, vol. 43, no. 4, pp. 741–752, 2007.

[116] A. Moustafa, A. N. Evans, D. M. Kulis et al., "Transcriptome profiling of a toxic dinoflagellate reveals a gene-rich protist and a potential impact on gene expression due to bacterial presence," *PLoS ONE*, vol. 5, no. 3, Article ID e9688, 2010.

[117] K. B. Lidie and F. M. van Dolah, "Spliced leader RNA-mediated *trans*-splicing in a dinoflagellate, *Karenia brevis*," *Journal of Eukaryotic Microbiology*, vol. 54, no. 5, pp. 427–435, 2007.

[118] H. Zhang, Y. Hou, L. Miranda et al., "Spliced leader RNA *trans*-splicing in dinoflagellates," *Proceedings of the National Academy of Sciences of the United States of America*, vol. 104, no. 11, pp. 4618–4623, 2007.

[119] T. R. Bachvaroff, A. R. Place, and D. W. Coats, "Expressed sequence tags from *Amoebophrya sp.* infecting *Karlodinium veneficum*: comparing host and parasite sequences," *Journal of Eukaryotic Microbiology*, vol. 56, no. 6, pp. 531–541, 2009.

[120] X. Xu, J. Vatsyayan, C. Gao, C. J. Bakkenist, and J. Hu, "Sumoylation of eIF4E activates mRNA translation," *EMBO Reports*, vol. 11, no. 4, pp. 299–304, 2010.

[121] X. Xu, J. Vatsyayan, C. Gao, C. J. Bakkenist, and J. Hu, "HDAC2 promotes eIF4E sumoylation and activates mRNA translation gene specifically," *The Journal of Biological Chemistry*, vol. 285, no. 24, pp. 18139–18143, 2010.

[122] T. R. Bachvaroff, S. M. Handy, A. R. Place, and C. F. Delwiche, "Alveolate phylogeny inferred using concatenated ribosomal proteins," *Journal of Eukaryotic Microbiology*, vol. 58, no. 3, pp. 223–233, 2011.

[123] J. A. F. Robledo, P. Courville, M. F. M. Cellier, and G. R. Vasta, "Gene organization and expression of the divalent cation transporter Nramp in the protistan parasite *Perkinsus marinus*," *Journal of Parasitology*, vol. 90, no. 5, pp. 1004–1014, 2004.

[124] J. L. Hearne and J. S. Pitula, "Identification of two spliced leader RNA transcripts from *Perkinsus marinus*," *Journal of Eukaryotic Microbiology*, vol. 58, no. 3, pp. 266–268, 2011.

[125] H. Zhang, D. A. Campbell, N. R. Sturm, C. F. Dungan, and S. Lin, "Spliced leader RNAs, mitochondrial gene frameshifts and multi-protein phylogeny expand support for the genus *Perkinsus* as a unique group of alveolates," *PLoS ONE*, vol. 6, no. 5, Article ID e19933, 2011.

[126] T. Cavalier-Smith, "Principles of protein and lipid targeting in secondary symbiogenesis: euglenoid, dinoflagellate, and sporozoan plastid origins and the eukaryote family tree," *Journal of Eukaryotic Microbiology*, vol. 46, no. 4, pp. 347–366, 1999.

[127] T. Cavalier-Smith and M. J. Beaton, "The skeletal function of non-genic nuclear DNA: new evidence from ancient cell chimaeras," *Genetica*, vol. 106, no. 1-2, pp. 3–13, 1999.

[128] T. Cavalier-Smith, "Membrane heredity and early chloroplast evolution," *Trends in Plant Science*, vol. 5, no. 4, pp. 174–182, 2000.

[129] S. E. Douglas, C. A. Murphy, D. F. Spencer, and M. W. Grayt, "Cryptomonad algae are evolutionary chimaeras of two phylogenetically distinct unicellular eukaryotes," *Nature*, vol. 350, no. 6314, pp. 148–151, 1991.

[130] S. E. Douglas, "Eukaryote-eukaryote endosymbioses: insights from studies of a cryptomonad alga," *BioSystems*, vol. 28, no. 1–3, pp. 57–68, 1992.

[131] U. G. Maier, C. J. B. Hofmann, S. Eschbach, J. Wolters, and G. L. Igloi, "Demonstration of nucleomorph-encoded eukaryotic small subunit ribosomal RNA in cryptomonads," *Molecular and General Genetics*, vol. 230, no. 1-2, pp. 155–160, 1991.

[132] T. Cavalier-Smith, "Nucleomorphs: enslaved algal nuclei," *Current Opinion in Microbiology*, vol. 5, no. 6, pp. 612–619, 2002.

[133] S. Douglas, S. Zauner, M. Fraunholz et al., "The highly reduced genome of an enslaved algal nucleus," *Nature*, vol. 410, no. 6832, pp. 1091–1096, 2001.

[134] U. G. Maier, S. E. Douglas, and T. Cavalier-Smith, "The nucleomorph genomes of cryptophytes and chlorarachniophytes," *Protist*, vol. 151, no. 2, pp. 103–109, 2000.

[135] S. Zauner, M. Fraunholz, J. Wastl et al., "Chloroplast protein and centrosomal genes, a tRNA intron, and odd telomeres in an unusually compact eukaryotic genome, the cryptomonad nucleomorph," *Proceedings of the National Academy of Sciences of the United States of America*, vol. 97, no. 1, pp. 200–205, 2000.

[136] G. I. McFadden, P. R. Gilson, and S. E. Douglas, "The photosynthetic endosymbiont in cryptomonad cells produces both chloroplast and cytoplasmic-type ribosomes," *Journal of Cell Science*, vol. 107, no. 2, pp. 649–657, 1994.

[137] L. D. Kapp and J. R. Lorsch, "The molecular mechanics of eukaryotic translation," *Annual Review of Biochemistry*, vol. 73, pp. 657–704, 2004.

[138] W. L. Zoll, L. E. Horton, A. A. Komar, J. O. Hensold, and W. C. Merrick, "Characterization of mammalian eIF2A and identification of the yeast homolog," *The Journal of Biological Chemistry*, vol. 277, no. 40, pp. 37079–37087, 2002.

[139] M. Matsuzaki, O. Misumi, T. Shin-I et al., "Genome sequence of the ultrasmall unicellular red alga *Cyanidioschyzon merolae* 10D," *Nature*, vol. 428, no. 6983, pp. 653–657, 2004.

[140] N. Ohta, N. Sato, H. Nozaki, and T. Kuroiwa, "Analysis of the cluster of ribosomal protein genes in the plastid genome of a unicellular red alga *Cyanidioschyzon merolae*: translocation of the *str* cluster as an early event in the rhodophyte-chromophyte lineage of plastid evolution," *Journal of Molecular Evolution*, vol. 45, no. 6, pp. 688–695, 1997.

[141] M. Altmann, P. P. Muller, B. Wittmer, F. Ruchti, S. Lanker, and H. Trachsel, "A *Saccharomyces cerevisiae* homologue of mammalian translation initiation factor 4B contributes to RNA helicase activity," *The EMBO Journal*, vol. 12, no. 10, pp. 3997–4003, 1993.

[142] S. B. Gould, "Ariadne's thread: guiding a protein across five membranes in cryptophytes," *Journal of Phycology*, vol. 44, no. 1, pp. 23–26, 2008.

[143] B. Joshi, J. Robalino, E. J. Schott, and R. Jagus, "Yeast "knock-out-and-rescue" system for identification of eIF4E-family members possessing eIF4E-activity," *BioTechniques*, vol. 33, no. 2, pp. 392–401, 2002.

[144] S. Vasilescu, M. Ptushkina, B. Linz, P. P. Müller, and J. E. G. McCarthy, "Mutants of eukaryotic initiation factor eIF-4E with altered mRNA cap binding specificity reprogram mRNA selection by ribosomes in *Saccharomyces cerevisiae*," *The Journal of Biological Chemistry*, vol. 271, no. 12, pp. 7030–7037, 1996.

[145] B. La Scola, T. J. Marrie, J. P. Auffray, and D. Raoult, "Mimivirus in pneumonia patients," *Emerging Infectious Diseases*, vol. 11, no. 3, pp. 449–452, 2005.

[146] D. Raoult, S. Audic, C. Robert et al., "The 1.2-megabase genome sequence of Mimivirus," *Science*, vol. 306, no. 5700, pp. 1344–1350, 2004.

[147] D. Raoult, B. La Scola, and R. Birtles, "The discovery and characterization of mimivirus, the largest known virus and putative pneumonia agent," *Clinical Infectious Diseases*, vol. 45, no. 1, pp. 95–102, 2007.

[148] M. Suzan-Monti, B. La Scola, and D. Raoult, "Genomic and evolutionary aspects of Mimivirus," *Virus Research*, vol. 117, no. 1, pp. 145–155, 2006.

[149] J. M. Claverie and C. Abergel, "Mimivirus and its virophage," *Annual Review of Genetics*, vol. 43, pp. 49–66, 2009.

[150] J. M. Claverie, C. Abergel, and H. Ogata, "Mimivirus," *Current Topics in Microbiology and Immunology*, vol. 328, pp. 89–121, 2009.

[151] D. Benarroch, P. Smith, and S. Shuman, "Characterization of a trifunctional mimivirus mRNA capping enzyme and crystal structure of the RNA triphosphatase domain," *Structure*, vol. 16, no. 4, pp. 501–512, 2008.

[152] D. Benarroch, Z. R. Qiu, B. Schwer, and S. Shuman, "Characterization of a mimivirus RNA cap guanine-N2 methyltransferase," *RNA*, vol. 15, no. 4, pp. 666–674, 2009.

[153] J. M. Claverie and C. Abergel, "Mimivirus: the emerging paradox of quasi-autonomous viruses," *Trends in Genetics*, vol. 26, no. 10, pp. 431–437, 2010.

5

The Bic-C Family of Developmental Translational Regulators

Chiara Gamberi and Paul Lasko

Department of Biology, McGill University, 3649 Promenade Sir William Osler, Montréal, QC, Canada H3G 0B1

Correspondence should be addressed to Chiara Gamberi, chiara.gamberi@mcgill.ca

Academic Editor: Greco Hernández

Regulation of mRNA translation is especially important during cellular and developmental processes. Many evolutionarily conserved proteins act in the context of multiprotein complexes and modulate protein translation both at the spatial and the temporal levels. Among these, Bicaudal C constitutes a family of RNA binding proteins whose founding member was first identified in *Drosophila* and contains orthologs in vertebrates. We discuss recent advances towards understanding the functions of these proteins in the context of the cellular and developmental biology of many model organisms and their connection to human disease.

1. Introduction

Translational regulation of mRNA distributed asymmetrically in the early *Drosophila* embryo underlies pattern formation and germ cell specification. Furthermore, expression of certain proteins occurs only at definite stages of development. Exquisite, often partially redundant mechanisms of control ensure the coordination of the spatial and temporal expression of proteins with morphogenetic potential. These mechanisms have been reviewed recently [6]. Here we will discuss the case of one of such translational regulators, Bicaudal C (Bic-C), which is evolutionarily conserved, and for which there is recent accumulating functional evidence from both invertebrate and vertebrate model organisms suggesting that Bic-C is a fundamental regulator of cellular processes and an outstanding example of the fascinating complexity of the developmental mechanisms.

2. Materials and Methods

The sequences shown in this paper are listed in Table 1, and they were recovered by running BLAST [7] with the *Drosophila* sequence and the NCBI sequence database, using the Homologene feature at the NCBI. The sequences for the different *Drosophila* species were retrieved from FlyBase [8]. Sequences were aligned with Clustal W [1, 2].

3. Results and Discussion

3.1. Bic-C . The *Bic-C* gene was originally identified during a *Drosophila* screen for maternal genes affecting embryonic polarity [9]. In fact, adult females bearing *Bic-C* mutations in one of their second chromosomes produce embryos exhibiting anterior-posterior defects of severity ranging from anterior defects, to the development of bicaudal embryos composed of as few as four segments arranged as two, mirror-image posterior ends, to embryos that fail to cellularize [3]. This pleiotropy indicates that Bic-C participates in (or influences) many different pathways.

Early work demonstrated that Bic-C is required during oogenesis to establish anterior-posterior polarity in the oocyte [3, 5, 9, 10]. It encodes a 905-amino-acid (aa) RNA binding protein containing two canonical and three non-canonical KH RNA binding domains (KH2, 4 and KH 1, 3, 5, resp., aa 56–524) [3, 11, 12], a C-terminal Sterile Alpha Motif domain (SAM domain, aa 805–868, Prosite) [13], and a region rich in serine and glycine (aa 598–693). In the Bic-C protein, both the region containing the KH domains and the full-length, recombinant protein possess affinity for RNA [14, 15] with the full-length protein exhibiting more selective binding of synthetic probes *in vitro*. RNA binding is likely important to Bic-C function in fruit flies, as a spontaneous mutation (G296R) that affects the third KH domain, decreases RNA affinity *in vitro*, and exhibits

TABLE 1: Sequences used in this study.

Sequences	Species
Bic-C	
Gene Bank ID	
gi|24584539	*D. melanogaster* B isoform
gi|158300058	*A. gambiae*
gi|13994223	*M. musculus*
gi|109509376	*R. norvegicus*
gi|122937472	*H. sapiens*
gi|114631037	*P. troglodytes*
gi|73953060	*C. familiaris*
gi|194679417	*B. taurus*
gi|292623098	*D. rerio*
gi|212646112	*C. elegans*
gi|118092391	*G. gallus*
FlyBase ID	
FBpp0080362	*D. melanogaster* B isoform
FBpp0080363	*D. melanogaster* D isoform
FBpp0080361	*D. melanogaster* A isoform
FBpp0118127	*D. ananassae*
FBpp0143734	*D. erecta*
FBpp0144300	*D. grimshawi*
FBpp0166588	*D. mojavensis*
FBpp0179414	*D. persimilis*
FBpp0287937	*D. pseudobscura*
FBpp0200128	*D. sechellia*
FBpp0222439	*D. simulans*
FBpp0232468	*D. virilis*
FBpp0253912	*D. willistoni*
FBpp0266309	*D. yakuba*
Not3/5	
Gene Bank ID	
gi|39945962	*Magnaportae oryzae*
gi|85075997	*Neurospora crassa*
gi|19115701	*S. pombe*
gi|19921660	*D. melanogaster*
gi|158299738	*A. gambiae*
gi|22122717	*M. musculus*
gi|34854462	*R. norvegicus*
gi|7657387	*H. sapiens*
gi|114678945	*P. troglodytes*
gi|73946891	*C. familiaris*
gi|119911200	*B. taurus*
gi|53933228	*D. rerio*
gi|133901756	*C. elegans*
gi|238481292	*A. thaliana*
gi|115454389	*O. sativa japonica*
FlyBase ID	
FBpp0085398	*D. melanogaster*
FBpp0125948	*D. ananassae*
FBpp0129398	*D. erecta*
FBpp0147530	*D. grimshawi*
FBpp0160933	*D. mojavensis*
FBpp01852	*D. persimilis*
FBpp0288020	*D. pseudobscura*

TABLE 1: Continued.

Sequences	Species
FBpp0197981	*D. sechellia*
FBpp0208756	*D. simulans*
Bpp0227498	*D. virilis*
FBpp0243918	*D. willistoni*
FBpp0264455	*D. yakuba*

a strong phenotype *in vivo* [3]. However, this mutation may be affecting more than RNA binding of the whole protein, for example, by perturbing secondary structure in its neighbourhood, as it may be the case for a similar mutation occurring in another KH domain [12]. If this were the case, the severity of the phenotype may be due to the combination of lack of RNA interaction and other defective pathways under Bic-C control in the wild type. The region containing the KH domains in two Bic-C orthologs shows conserved RNA binding capability in the mouse Biccl [16] and, surprisingly, not in the *C. elegans* GLD-3 [12].

SAM domains are ancient modules present in most species that are commonly engaged in mediating protein-protein interaction [13, 17] and can multimerize [18, 19]. Multimerization of RNA binding proteins and RNA is most likely the basis for building RNP particles and a target of regulation. Interestingly, the SAM domain of the human BICC1 can form polymers *in vitro* [20] and some KH domains can mediate interactions between proteins [21, 22]. This is also the case for the C. elegans GLD-3 that interacts with the GLD-2 polymerase via its first KH domain [23] therefore it is likely that Bic-C is part of multiprotein complexes such as cellular RNPs. Certain SAM domains have also been implicated in RNA binding, as the case of *Drosophila* Smaug and *S. cerevisiae* Vts1 [24]. Interestingly, among all the *Drosophila* SAM domains, Bic-C contains the one most similar to Smaug's, which includes the critical residues for RNA interaction [25], suggesting the possibility that it may contribute to the Bic-C RNA binding capacity in the cell [17]. Studies of the vertebrate Bic-C homologs, whose targets are largely unknown, have suggested that presence of the SAM domain may mediate association with the P-bodies [26, 27]. Another interesting possibility is that the putative RNA binding and protein-protein interaction capabilities of the SAM domain may be regulated, possibly via posttranslational modifications. In this scenario protein modification in this domain may change the specificity and/or affinity of Bic-C for RNA to switch between protein and RNA binding activities in certain cellular or developmental contexts. Interestingly, a tyrosine residue in position 822 that can be phosphorylated in other SAM domains to regulate their activity is also conserved [28] (Figure 1).

3.2. Evolutionary Conservation of the Bic-C Protein. Bic-C is found in all the sequenced *Drosophila* species and its homologs are virtually identical to each other, except for regions of

(a)

```
D. melanogaster-PB   MLSCASFNKLMYPSAADVAKPP--------------MVGLEV-EAGSIGSLSSLHALPST   45
D. melanogaster-PA   MLSCASFNKLMYPSAADVAKPP--------------MVGLEV-EAGSIGSLSSLHALPST   45
D. melanogaster-PD   ------------------------------------------------------------
D. sechellia         MLSCASFNKLMYPSAADVAKPP--------------MVGLDV-EAGSIGSLSSLQALPST   45
D. simulans          MLSCASFNKLMYPSAADVAKPP--------------MVGLEV-EAGSIGSLSSLQALPST   45
D. erecta            MLSCASFNKLMYPSAADVAKPP--------------MVGLDV-EAGSIGSLSSLQALPST   45
D. yakuba            MLSCASFNKLMYPSAADVAKPP--------------MVGLEV-EGGSIGSLSSLQALPST   45
D. ananassae         MLSCASFNKLIYPTAADVAKSP--------------VEGVGVGLAGSIGSLASLQTLTSV   46
D. willistoni        MLSCASFNKLLYPTATGGSVTP--------------VTSGKSPLLGSLANLP-LATGPAG   45
D. pseudobscura      MLSCASFNKLIYPSAADVSAVASGKSSTVAVDGGVGADSGGGVNPVGIGSLASLQALPSG   60
D. mojavensis        MLSCASFNKLIYPTAAEVTAMA--------------------SGKTTPVG-ANLNSLPLP   39
D. virilis           MLSCAPFNKLIYPTAADISAMA--------------------SAKATPVA-VGLTTLSLP   39
D. grimshawi         MLSCAPFNKLIYPTTADVPAMS--------------------NGKGTPLAPVGLSTLSLP   40

                                                                       ←
D. melanogaster-PB   TSVG--SGAPSETQSEISSVDSDWSDIRAIAMKLGVQNPDDLHTERFKVDRQKLEQLIKA   103
D. melanogaster-PA   TSVG--SGAPSETQSEISSVDSDWSDIRAIAMKLGVQNPDDLHTERFKVDRQKLEQLIKA   103
D. melanogaster-PD   ------------------------------------------------------------
D. sechellia         TSVG--SGAPSETQSEISSVDSDWSDIRAIAMKLGVQNPDDLHTERFKVDRQKLEQLIKA   103
D. simulans          TSVG--SGAPSETQSEISSVDSDWSDIRAIAMKLGVQNPDDLHTERFKVDRQKLEQLIKA   103
D. erecta            TSVG--SGAPSETQSEISSVDSDWSDIRAIAMKLGVQNPDDLHTERFKVDRQKLEQLIKA   103
D. yakuba            TSVG--SGAPSETQSEISSVDSDWSDIRAIAMKLGVQNPDDLHTERFKVDRQKLEQLIRA   103
D. ananassae         TSMG--SGAPSETQSEISSVDSDWSDIRAIAMKLGVKNPDDLHTERFKVDRQKLEQLIKA   104
D. willistoni        NPVGGGSGAPSETQSEISSVDSDWSDIRAIALKLGVQNPDELHTERFKVDRQKLEQFITA   105
D. pseudobscura      TLAG--SGAPSETQSEISSVDSDWSDIRAIALKLGVQNPDDLHTERFKVDRQKLEQLIKA   118
D. mojavensis        R-------APSETQSEISSVDSDWSDIRAIALKLGVQNVDDLHTERFKVDRQKLEQLLKA   92
D. virilis           TGPAGASGAPSETQSEISSVDSDWSDIRAIALKLGVQNPDDLHTERFKVDRQKLERLIMA   99
D. grimshawi         TGP----GGGSGAPSEISSVDSDWSDIRAIALKLGVQNPDDLHTERFKVDRQKLEQLIKA   96

                                           KH-like
D. melanogaster-PB   ESSIEGMNGAEYFFHDIMNTTDTYVSWPCRLKIGAKSKKDPHVRIVGKVDQVQRAKERIL   163
D. melanogaster-PA   ESSIEGMNGAEYFFHDIMNTTDTYVSWPCRLKIGAKSKKDPHVRIVGKVDQVQRAKERIL   163
D. melanogaster-PD   -----------------MNTTDTYVSWPCRLKIGAKSKKDPHVRIVGKVDQVQRAKERIL   43
D. sechellia         ESSIEGMNGAEYFFHDIMNTTDTYVSWPCRLKIGAKSKKDPHVRIVGKVDQVQRAKERIL   163
D. simulans          ESSIEGMNGAEYFFHDIMNTTDTYVSWPCRLKIGAKSKKDPHVRIVGKVDQVQRAKERIL   163
D. erecta            ESSIEGMNGAEYFFHDIMNTTDTYVSWPCRLKIGAKSKKDPHVRIVGKVDQVQRAKERIL   163
D. yakuba            ESSIEGMNGAEYFFHDIMNTTDTYVSWPCRLKIGAKSKKDPHVRIVGKVDQVQRAKDRIL   163
D. ananassae         DSAIEGMNGAEYFFDDIMNTTDTYVSWPCRLKIGAKSKKDPHVRIVGKVEQVQRAKERIL   164
D. willistoni        DSAIEGMNGAEYFFNDIMNTTDTYVSWPCRLKIGAKSKKDPHVRIVGKVDEVSRAKERIL   165
D. pseudobscura      DSAIEGMNGAEYFFHDIMNTTDTYVSWPCRLKIGAKSKKDPHVRIVGKVEQVQRAKERIL   178
D. mojavensis        DSAIEGMNGAEYFFDNIMSTTDTYVSWPCRLKIGAKSKKDPHVRIVGKVDQVQRAKDHIL   152
D. virilis           DSAIEGMNGAEYFFDDIMNTTDTYVSWPCRLKIGAKSKKDPHVRIVGKVEQVQRAKDHIL   159
D. grimshawi         DSAIEGMNGAEYFFDDIMNTTDTYVSWPCRLKIGAKSKKDPHVRIVGKMEQVQRAKDHIL   156
                                      *.*****************************:::*.***::**

                     ------------>< <-------------KH---------------------------
D. melanogaster-PB   SSLDSRGTRVIMKMDVSYTDHSYIIGRGGNNIKRIMDDTHTHIHFPDSNRSNPTEKSNQV   223
D. melanogaster-PA   SSLDSRGTRVIMKMDVSYTDHSYIIGRGGNNIKRIMDDTHTHIHFPDSNRSNPTEKSNQV   223
D. melanogaster-PD   SSLDSRGTRVIMKMDVSYTDHSYIIGRGGNNIKRIMDDTHTHIHFPDSNRSNPTEKSNQV   103
D. sechellia         SSLDSRGTRVIMKMDVSYTDHSYIIGRGGNNIKRIMDDTHTHIHFPDSNRSNPTEKSNQV   223
D. simulans          SSLDSRGTRVIMKMDVSYTDHSYIIGRGGNNIKRIMDDTHTHIHFPDSNRSNPTEKSNQV   223
D. erecta            SSLDSRGTRVIMKMDVSYTDHSYIIGRGGNNIKRIMDDTHTHIHFPDSNRSNPTEKSNQV   223
D. yakuba            SSLDSRGTRVIMKMDVSYTDHSYIIGRGGNNIKRIMDDTHTHIHFPDSNRSNPTEKSNQV   223
D. ananassae         SSLDSRGTRVIMKMDVSYTDHSYIIGRGGNNIKRIMDDTHTHIHFPDSNRSNPTEKSNQV   224
D. willistoni        GSLDSRGTRVIMKMDVSYTDHSYIIGRGGNNIKRIMDDTHTHIHFPDSNRSNPTEKSNQV   225
D. pseudobscura      SSLDSRGTRVIMKMDVSYTDHSYIIGRGGNNIKRIMDDTHTHIHFPDSNRSNPTEKSNQV   238
D. mojavensis        GSLDSRGTRVIMKMDVSYTDHSYIIGRGGNNIKRIMDDTHTHIHFPDSNRSNPTEKSNQV   212
D. virilis           GSLDSRGTRVIMKMDVSYTDHSYIIGRGGNNIKRIMDDTHTHIHFPDSNRSNPTEKSNQV   219
D. grimshawi         SSLDSRGTRVIMKMDVSYTDHSYIIGRGGNNIKRIMDDTNTHIHFPDSNRSNATEKSNQV   216
                     .**************************************:************.*******

                     ------------------>< <---------KH-like--------------------
D. melanogaster-PB   SLCGSLEGVERARALVRLSTPLLISFEMPVMGPNKPQPDHETPYIKMIETKFNVQVIFST   283
D. melanogaster-PA   SLCGSLEGVERARALVRLSTPLLISFEMPVMGPNKPQPDHETPYIKMIETKFNVQVIFST   283
D. melanogaster-PD   SLCGSLEGVERARALVRLSTPLLISFEMPVMGPNKPQPDHETPYIKMIETKFNVQVIFST   163
D. sechellia         SLCGSLEGVERARALVRLSTPLLISFEMPVMGPNKPQPDHETPYIKMIETKFNVQVIFST   283
D. simulans          SLCGSLEGVERARALVRLSTPLLISFEMPVMGPNKPQPDHETPYIKMIETKFNVQVIFST   283
D. erecta            SLCGSLEGVERARALVRLSTPLLISFEMPVMGPNKPQPDHETPYIKMIETKFNVQVIFST   283
D. yakuba            SLCGSLEGVERARALVRLSTPLLISFEMPVMGPNKPQPDHETPYIKMIETKFNVQVIFST   283
D. ananassae         SLCGSLEGVERARALVRLSTPLLISFEMPVMGPSKPQPDHETPYIKMIESKFNVQVIFST   284
```

FIGURE 1: Continued.

```
D. willistoni          SLCGSLEGVERARALVRLSTPLLISFEMPVMGPNKPQPDHETPYIKMIESKFNVQVIFST   285
D. pseudobscura        SLCGSLEGVERARALVRLSTPLLISFEMPVMGPSKQQPDHDTPYIKMIESKFNVQVIFST   298
D. mojavensis          SLCGSLEGVERARALVRLSTPLLISFEMPVMGPGKTQPDHETPYIKMIESKFNVQVIFSS   272
D. virilis             SLCGSLDGVERARALVRLSTPLLISFEMPVMGPGKPQPDHETPYIKMIESKFNVQVIFSS   279
D. grimshawi           SLCGTLEGVEHARALVRLSTPLLISFEMPVMGPGKPQPDHETPYIKMIESKFNVQVIFSS   276
                       ****:*:***:**********************.* ****:********:*********:

D. melanogaster-PB     RPKLHTSLVLVKGSEKESAQVRDATQLLINFACESIASQILVNVQMEISPQHHEIVKGKN   343
D. melanogaster-PA     RPKLHTSLVLVKGSEKESAQVRDATQLLINFACESIASQILVNVQMEISPQHHEIVKGKN   343
D. melanogaster-PD     RPKLHTSLVLVKGSEKESAQVRDATQLLINFACESIASQILVNVQMEISPQHHEIVKGKN   223
D. sechellia           RPKLHTSLVLVKGSEKESAQVRDATQLLINFACESIASQILVNVQMEISPQHHEIVKGKN   343
D. simulans            RPKLHTSLVLVKGSEKESAQVRDATQLLINFACESIASQILVNVQMEISPQHHEIVKGKN   343
D. erecta              RPKLHTSLVLVKGSEKESGQVRDATQLLINFACESIASQILVNVQMEISPQHHEIVKGKN   343
D. yakuba              RPKLHTSLVLVKGSEKESSQVRDATQLLINFACESIASQILVNVQMEISPQHHEIVKGKN   343
D. ananassae           RPKLHTSLVLVKGSERESAQVRDATQLLINFACESIASQILVNVQMEISPQHHEIVKGKN   344
D. willistoni          RPKLHTSLVLVKGSEKESAQVRDATQLLINFACESIASQILVNVQMEISPQHHEIVKGKN   345
D. pseudobscura        RPKLHTSLVLVKGSEKESAQVRDATQLLINFACESIASQILVNVQMEISPQHHEIVKGKN   358
D. mojavensis          RPKLHTSLVLVKGSEKESAQVRDATQLLINFAFESIASQILVNVQMEISPQHHEIVKGKN   332
D. virilis             RPKLHTSLVLVKGSEKESAQVRDATQLLINFAFESIASQILVNVQMEISPQHHEIVKGKN   339
D. grimshawi           RPKLHTSLVLVKGSEKESAQVRDATQLLINFAFESIASQILVNVQMEISPQHHEVVKGKN   336

                       ***************:**.************* **********************:*****
                                                    KH
D. melanogaster-PB     NVNLLSIMERTQTKIIFPDLSDMNVKPLKKSQVTISGRIDDVYLARQQLLGNLPVALIFD   403
D. melanogaster-PA     NVNLLSIMERTQTKIIFPDLSDMNVKPLKKSQVTISGRIDDVYLARQQLLGNLPVALIFD   403
D. melanogaster-PD     NVNLLSIMERTQTKIIFPDLSDMNVKPLKKSQVTISGRIDDVYLARQQLLGNLPVALIFD   283
D. sechellia           NVNLLSIMERTQTKIIFPDLSDMNVKPLKKSQVTISGRIDDVYLARQQLLGNLPVALIFD   403
D. simulans            NVNLLSIMERTQTKIIFPDLSDMNVKPLKKSQVTISGRIDDVYLARQQLLGNLPVALIFD   403
D. erecta              NVNLLSIMERTQTKIIFPDLSDMNVKPLKKSQVTISGRIDDVYLARQQLLGNLPVALIFD   403
D. yakuba              NVNLLSIMERTQTKIIFPDLSDMNVKPLKKSQVTISGRIDDVYLARQQLLGNLPVALIFD   403
D. ananassae           NVNLLSIMERTQTKIIFPDLSDMNVKPLKKSQVTISGRIDNVYLARQQLLGNLPVALIFD   404
D. willistoni          NVNLLSIMERTQTKIIFPDLSDMNVKPLKKSQVTISGRIDDVYKARQQLLGNLPVALIFD   405
D. pseudobscura        NVNLLSIMERTQTKIIFPDLSDMNVKPLKKSQVTISGRIDDVYKARQQLLGNLPVALIFD   418
D. mojavensis          NVNLLSIMERTQTKIIFPDLSDMNVKPLKKSQVTISGRIDDVYRARQQLLGNMPVALIFD   392
D. virilis             NVNLLSIMDRTQTKIIFPDLTDMNVKPLKKSQVTISGRIDDVYRARQQLLGNMPVALIFD   399
D. grimshawi           NVNLLSIMDRTQTKIIFPDLTDINVKPLKKSQVTISGRIDDVYKARQQLLGNMPVALIFD   396
                       ********:***********:*:*****************:** *********:*******
                                                                 KH-like
D. melanogaster-PB     FPDNHNDASEIMSLNTKYGVYITLRQKQRQSTLAIVVKGVEKFIDKIYEARQEILRLATP   463
D. melanogaster-PA     FPDNHNDASEIMSLNTKYGVYITLRQKQRQSTLAIVVKGVEKFIDKIYEARQEILRLATP   463
D. melanogaster-PD     FPDNHNDASEIMSLNTKYGVYITLRQKQRQSTLAIVVKGVEKFIDKIYEARQEILRLATP   343
D. sechellia           FPDNHNDASEIMSLNTKYGVYITLRQKQRQSTLAIVVKGVEKFIDKIYEARQEILRLATP   463
D. simulans            FPDNHNDASEIMSLNTKYGVYITLRQKQRQSTLAIVVKGVEKFIDKIYEARQEILRLATP   463
D. erecta              FPDNHNDASEIMSLNTKYGVYITLRQKQRQSTLAIVVKGVEKFIDKIYEARQEILRLATP   463
D. yakuba              FPDNHNDASEIMSLNTKYGVYITLRQKQRQSTLAIVVKGVEKFIDKIYEARQEILRLATP   463
D. ananassae           FPDNQNDASEIMGLNTKYGVYITLRQKQRQSTLAIVVKGVEKFIDKIYEARQEILHLATP   464
D. willistoni          FPDNQNDASDIMSLNTKYGVLITLRQKQRQSTLAIVIKGLEKFIDKIYEARQEILCLSSP   465
D. pseudobscura        FPDNQNDASEIMSLNTKYGVYITLRQKQRQSTLAIVVKGVEKFIDKIYEARQEILRLATP   478
D. mojavensis          FPDNQTDASEIMGLNIKHGVYITLRQKQRQSTLAIVIKGIEKFIDKIYEARQEILHLTTP   452
D. virilis             FPDNQTDASEIMGLNLKYGVYITLRQKQRQSTLAIVIKGIEKFIDKIYEARQEILHLTTP   459
D. grimshawi           FPDNQTDASDIMGLNAKYGVYITLRQKQRQSTLAIVIKGIEKFIDKIYEARQEILQLSTP   456
                       ****:.***:**.** *:** ****************:**:**************** *::*

D. melanogaster-PB     FVKPEIPDYYFMPKDKDLNLAYRTQLTALLAGYVDSPKTP-SLLPPSLAGQLTPYANN--   520
D. melanogaster-PA     FVKPEIPDYYFMPKDKDLNLAYRTQLTALLAGYVDSPKTP-SLLPPSLAGQLTPYANN--   520
D. melanogaster-PD     FVKPEIPDYYFMPKDKDLNLAYRTQLTALLAGYVDSPKTP-SLLPPSLAGQLTPYANN--   400
D. sechellia           FVKPEIPDYYFMPKDKDLNLAYRTQLTALLAGYVDSPKTP-SLLPPALAGQLTPYANN--   520
D. simulans            FVKPEIPDYYFMPKDKDLNLAYRTQLTALLAGYVDSPKTP-SLLPPALAGQLTPYANN--   520
D. erecta              FVKPEIPEYYFMPKDKDLNLAYRTQLTALLAGYVDSPKTP-SLLPPALTGQLTPYANN--   520
D. yakuba              FVKPEIPDYYFMPKDKDLNLAYRTQLTALLAGYVDSPKTP-SLLPPALTGQLTPYANN--   520
D. ananassae           SVKPEIPELYFMPKDKDLSLAYRTQLTALLAGYVDSPKTP-SLLPPALAGQLTPYANN--   521
D. willistoni          AIQPVIPDHYFMPKDKDLNLAYRTQLTALLGGYSDNLKSPPGLLPPGLSNQLTPYANN--   523
D. pseudobscura        AIKPEVPDHYFMPKDKDLNLAYRTQLTALLAGYVDSPKTP-SLLPPALAGQLTPYANN--   535
D. mojavensis          AIKPDIPDYYFMPKDSDVNLAYRSQLTALLAGYPDSPKTP-SLLPPTMGGQLTPYDNNGK   511
D. virilis             VIKPEIPDHYYMPKDKDVSLAYRSQLTALLAGYPDSPKTP-SLLPPTMGGQLTPYGN--K   516
D. grimshawi           ALRPEIPEHYYMPKDKAVNAAYRAQLTALLAGYPDSPKTP-SLLPPIIA-QLAAYGNK-S   513
                        ::* :*: *:****. :. ***:******.** *. *:* .****  :   **:.* *

D. melanogaster-PB     NHLLLNANG---------LATPTGVCAPTQKYMQLHN-SFQQAQ---------------   554
D. melanogaster-PA     NHLLLNANG---------LATPTGVCAPTQKYMQLHN-SFQQAQ---------------   554
D. melanogaster-PD     NHLLLNANG---------LATPTGVCAPTQKYMQLHN-SFQQAQ---------------   434
```

Figure 1: Continued.

```
D. sechellia           NHLLLNANG---------LATPTGVCAPTQKYMQLHN-SFQQTQ----------------   554
D. simulans            NHLLLNANG---------LATPTGVCAPTQKYMQLHN-SFQQTQ----------------   554
D. erecta              NHLLLNANG---------LATPTGVCAPTQKYMQLHN-SFQQTQ----------------   554
D. yakuba              NHLLLNANG---------LATPTGVCAPTQKYMQLHN-SFQQTQ----------------   554
D. ananassae           NHLLLNANGGV---AVGGLATPTGVCAPTQKYMQLHNSAFQQGQ----------------   562
D. willistoni          NHLLLNANASVNGSGGGGLSTPTGICAPTQKYMQMHN-NFQQAQ----------------   566
D. pseudobscura        NHLLLNANAAVG-----GLATPTGICAPTQKYMQLHNSAFQHQQ----------------   574
D. mojavensis          GHMLLGA---------AGLATPTGICAPTQKYMQLHNNNYQPRPLSAIN-----------   551
D. virilis             AHMLLAANVG------VGLTTPTGICAPTQKYMQLHNSSYQPRQVSTMNNLSNCSNNNSS   570
D. grimshawi           HNVLLGNSVG------VGLATPTGICAPTQKYMQLHNSNYQPR-----------------   550
                        ::**             *:****:*********:**  :*

D. melanogaster-PB     ----------------NRSMVAG-----------GQSNNGNYLQVPG------AVAPP--   579
D. melanogaster-PA     ----------------NRSMVAG-----------GQSNNGNYLQVPG------AVAPP--   579
D. melanogaster-PD     ----------------NRSMVAG-----------GQSNNGNYLQVPG------AVAPP--   459
D. sechellia           ----------------NRSMVAG-----------GQNNNGNYLQVPG------AVAP---   578
D. simulans            ----------------NRSMVAG-----------GQNNNGNYLQVPG------AVAP---   578
D. erecta              ----------------GRSMVAG-----------GQSSNGNYLQVPG------AVAP---   578
D. yakuba              ----------------GRSMVAG-----------GQSSNCNYLQVPG------AVAP---   578
D. ananassae           ----------------VGTVQAGR--------PLGVNHNNNYLQVPGGL---GGVAGNG-   594
D. willistoni          ----------------AQQQQQQQQHVQQVAPRQSVVANNNYLQVPGS---KPPLNVG--   605
D. pseudobscura        ----------------LQGQGQVQGP-GQGRPGPVPNHNNNYLQVPGTANAGAGVGAGAG   617
D. mojavensis          NNNNNSSSNNNNNTTTTSNNISNNNNNNNNNINN---NNNNYLQVPGAGLLKPPANLPPT   608
D. virilis             NNNNNNNSNNNNCSSNNNNNISNNSNNINNNNNNNISNNNNYLQVPGSGLLKPPPAPMPS   630
D. grimshawi           --------------QPLAPVAGTGNG-------------------TGTGSVATAPA----   573
                                                                    .*

D. melanogaster-PB     ------------LKPPTVSPRNSCSQNTSGYQSFSSSTTSLEQSYPPYAQLPGTVSSTSS   627
D. melanogaster-PA     ------------LKPPTVSPRNSCSQNTSGYQSFSSSTTSLEQSYPPYAQLPGTVSSTSS   627
D. melanogaster-PD     ------------LKPPTVSPRNSCSQNTSGYQSFSSSTTSLEQSYPPYAQLPGTVSSTSS   507
D. sechellia           ------------LKPPTVSPRNSCSQNTSGYQSFSSSTTSLEQSYPPYAQLPGTVSSTSS   626
D. simulans            ------------LKPPTVSPRNSCSQNTSGYQSFSSSTTSLEQSYPPYAQLPGTVSSTSS   626
D. erecta              ------------LKPPTVSPRNSCSQNTSGYQSFSSSTTSLEQSYPPYAQLPGTVSSTSS   626
D. yakuba              ------------LKPPTVSPRNSCSQNTSGYQSFSSSTTSLEQSYPPYAQLPGTVSSTSS   626
D. ananassae           ---------QLKPLPMNVSPRNSCSQNTSGYQSFSSSTTSLEQSYPPYAQLQGAVSSTSS   645
D. willistoni          ------------SNTVNVSPRNSCSQNTSGYQSFSSSTTSLEQSYPPYAQLQATVSSTSS   653
D. pseudobscura        MLKPPPPPSSGVGGGMNVSPRNSCSQNTSGYQSFSSSTTSLEQSYPPYAQLQGAVSSTSS   677
D. mojavensis          ISVTG---------SINLSPRNSCSQNTSGYQSFSSSTTSLEQSYPPYGQVQTTVSSTSS   659
D. virilis             TNVGPP-----PTVGVNLSPRNSCSQNTSGYQSFSSSTTSLEQSYPPYAQVQAAVSSTSS   685
D. grimshawi           --------------AVQLSPRNSCSQNTSGYQSFSSSTTSLEQSYPPFAQVQTVVSSTSS   619
                                        :*****************************:.*:  .******

D. melanogaster-PB     STAG------SQNRAHYSPDSTYGSEGGGV-GGGGGGGARLGRRLSDGVLLGLSN-----   675
D. melanogaster-PA     STAG------SQNRAHYSPDSTYGSEGGGV-GGGGGGGARLGRRLSDGVLLGLSN-----   675
D. melanogaster-PD     STAG------SQNRAHYSPDSTYGSEGGGV-GGGGGGGARLGRRLSDGVLLGLSN-----   555
D. sechellia           STAG------SQNRAHYSPDSTYGSEGGGV-GGGGGGGARLGRRLSDGVLLGLGN-----   674
D. simulans            STAG------SQNRAHYSPDSTYGSEGGGV-GGGGGGGARLGRRLSDGVLLGLGN-----   674
D. erecta              STAG------SQNRAHYSPDSTYGSEGGGV-GGGGGGGARLGRRLSDGVLLGLGN-----   674
D. yakuba              STAG------SQSRAHYSPDSTYGSEGGGV-GGGGGGGARLGRRLSDGVLLGLGN-----   674
D. ananassae           S-AG------CANRAHYSPDSTYGSEAGSVPGGGGGGGARLGRRLSDGVLLGLGN-----   693
D. willistoni          S-SS------CANRAHYSPDSTYSSEGGGG-GLGMGASARLGRRLSDGVLLGLSNAAGGV   705
D. pseudobscura        T--G------CGSRAHYSPDSTYSSEAGSI-----GGAARLGRRLSDGVLLGLGN-----   719
D. mojavensis          S-SG-------ANRAHYSPDSTYNSEVGGSIVG----AARLGRRLSDGVLLGLSN-----   702
D. virilis             SSAG-------ANRAHYSPDSTYSSEAGSIAGA----AARLGRRLSDGVLLGLGN-----   729
D. grimshawi           SSGGGAGGLGCASRSHYSPDSTYSSEAGSIAG-----AARLGRRLSDGVLLGLGS-----   669
                       :  .        .*:*********.** *.        .***************..

D. melanogaster-PB     -SNGGGGNSGG-AHLLPGSAESYRSLHYDLGG--------------------NKHS-GHR   712
D. melanogaster-PA     -SNGGGGNSGG-AHLLPGSAESYRSLHYDLGG--------------------NKHS-GHR   712
D. melanogaster-PD     -SNGGGGNSGG-AHLLPGSAESYRSLHYDLGG--------------------NKHS-GHR   592
D. sechellia           -SSGGGGNSGGGAHLLPGSAESYRSLHYDLGG--------------------NKHS-GHR   712
D. simulans            -SSGGGGNSGGGAHLLPGSAESYRSLHYDLGG--------------------NKHS-GHR   712
D. erecta              -SSGGGGNAGGGAHLLPGSAESYRSLHYDLGG--------------------NKHS-SHR   712
D. yakuba              -SSGGGANSGGGAHLLPGSAESYRSLHYDLGG--------------------NKHS-SHR   712
D. ananassae           -GSSGG------APLLPGSAESYRSLHYDLTGS-------GSISGSGTGAAGNKHTNIHR   739
D. willistoni          GGSMGGAGGGGGAHLLPGSAESYRSLHYDLAG---------------------NGQLTHR   744
D. pseudobscura        --SGGGG-----AHLLPGSAESYRSLHYDLAGG-------GG------GAKHHHQHATHR   759
D. mojavensis          ---ANNGINSGGAHLLPGSAESYRNLHYDLAAVAGKQQQHQQQQQQHQQQQQQQQQQQQR   759
D. virilis             ---ATG----GGAHLLPGSAESYRNLHYDLAA---------------------QQQQQQR   761
D. grimshawi           ---ATT----GGAHLLPGSAESYRNLHYEHQQ----QQQQQQHQQQHHHQQQQQQQQQQR   718
                                   * **********.***:                          :   :*
```

Figure 1: Continued.

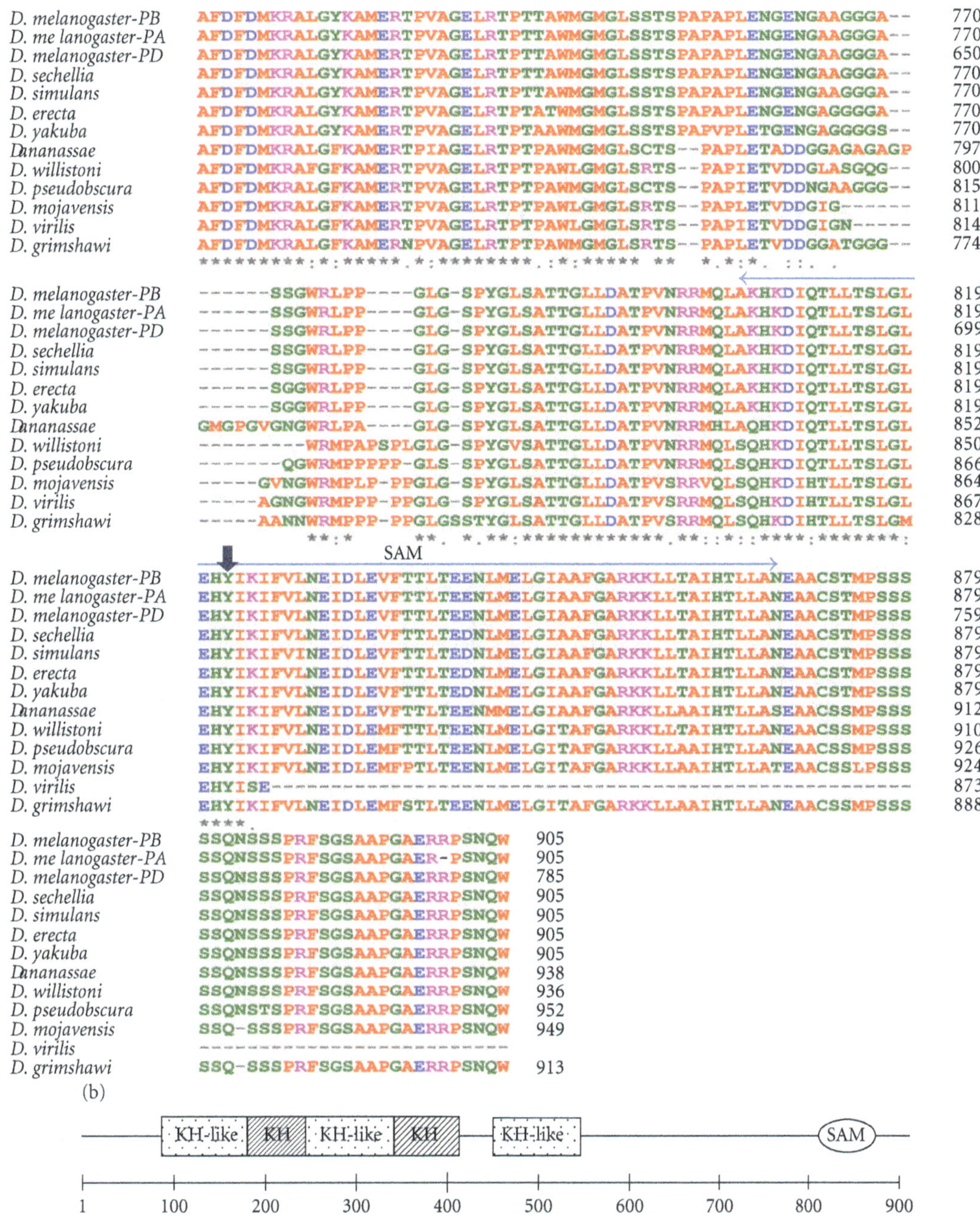

FIGURE 1: (a) Alignment of Bic-C sequences from 11 *Drosophila* species. Clustal W [1, 2] was used to align sequences extracted from FlyBase. Amino acid (aa) color coding is from Clustal W: red, small aliphatic, hydrophobic, and aromatics; blue, acidic; magenta, basic; green, hydroxyl, sulphydryl, amine, and glycine; grey, unusual aa. Symbols for aa conservation are from Clustal W: (asterisk $*$): positions with a single, fully conserved residue. (Colon :): conservation between groups of strongly similar properties scoring >0.5 in the Gonnet PAM 250 matrix. (Period .): conservation between groups of weakly similar properties scoring ≤0.5 in the Gonnet PAM 250 matrix. All three *D. melanogaster* Bic-C isoforms are shown (*PA, PB, PD*). The two canonical (KH) and three noncanonical (KH-like) KH RNA-binding modules are indicated (arrows, top). Domain assignment is as in [3] except for the fourth KH-related motif and the SAM domains, that are labelled according to the Pfam database [4]. A conserved, potentially phosphorylated, tyrosine is also indicated (arrowhead, top). Divergence occurs in regions of low complexity in the encoding DNA. Relative to the numbering of the *Drosophila* sequence: insertion at 555, variable length of the serine stretches around aa 623, and between aa 647–658 in the serine-glycine rich region. Further, after aa 715 there seems to be insertions of glutamine stretches of various lengths in *D. mojavensis, D. virilise*, and *D. grimshawi*. Finally, *D. ananassae* shows a short insertion at aa 770. The *D. virilise*, sequence results truncated. A TBLASTn search with the C-terminal region of Bic-C from *D. melanogaster* reveals many ESTs with similarity to the *D. melanogaster* sequence, suggesting a possible misannotation (not shown). Another region of possible sequencing misannotation in the *D. virilis* and the *D. mojavensis* Bic-C is italicized and not in bold type. Note that the *Bic-C* gene in *D. melanogaster* has nine mapped introns [5], and there is the possibility that the sequence was misannotated with this respect. (b) Block structure of the *D. melanogaster* Bic-C highlighting the protein motifs described in the text.

low complexity, where there are stretches of adjacent identical amino acids whose number varies in different species, the possible result of evolutionary mechanisms acting on triplet repeats or of stuttering sequencing polymerases (Figure 1).

An alignment of Bic-C orthologs from different animals reveals extensive sequence conservation from aa 83 to 268, (referring to the *Drosophila* sequence). Between aa 269 and 303, the vertebrate proteins lack the acidic residues present in the two Dipterans (*D. melanogaster and Anopheles gambiae*) while the basic residues between aa 281 and 286 are conserved (Figure 2).

Similarly, between aa 417 and 423 the acidic residues are exchanged with a basic (K) or a neutral (G) residues, while the adjacent phenylalanine 424 is changed conservatively into a tyrosine, suggesting that the overall protein folding may be preserved and that the electrostatic environment may be different between the insect and the vertebrate proteins. Since this region contains possible KH-domain-like modules, this may influence their ability to interact with RNA by contributing positive charges that might help retain or stabilize the interaction with RNA. At aa 458, vertebrate sequences diverge from those of *Drosophila*, *Anopheles*, and *Caenorhabditis elegans.* These sequences show blocks of conservation (aa 712–737 and 815–863) interspersed with regions of divergence and one insertion of 38 residues at aa 778. The SAM domain is one such block of conservation, with its phosphorylatable tyrosine [28] that is invariant in all the sequences analysed and the identity (or conservative substitution) of most of the amino acids that contribute to create an environment conducive to RNA binding in the case of Smaug [24].

3.3. Bic-C and Translational Regulation. Evidence that Bic-C was involved in control of mRNA translation came first from studies in *Drosophila* where it was observed that Oskar, a well-studied morphogen, was upregulated in ovaries from Bic-C mutated females [14]. The identification of other mRNA targets coimmunoprecipitated with Bic-C yielded the *Bic-C* mRNA itself and several mRNAs encoding factors involved in the Wnt pathway, vesicular trafficking, and organization of the actin cytoskeleton [15]. Bic-C interacts directly with the Not3/5 subunit of the CCR4 deadenylase complex, and it is believed that, when bound to its target RNA, it is able to recruit the deadenylase. This shifts the cellular balance between polyadenylation and deadenylation towards the latter, impairing translation [15]. Since Not3/5 is also evolutionarily conserved, it is discussed below in the perspective of its contribution to the Bic-C complexes.

The other invertebrate family member for which there is substantial functional information is the *C. elegans* GLD-3. GLD-3 is involved in germline development and embryogenesis by regulating the time of expression of developmental factors [23, 29, 30]. GLD-3, via its first KH domain, interacts with GLD-2, a noncanonical polyA polymerase devoid of an RNA interaction domain of its own [23, 30]. Although it was expected that GLD-3 may tether GLD-2 to the RNA, a recent structural study could not find any RNA binding activity for the GLD-3 KH region [12]; therefore further studies are needed to elucidate how GLD-3 participates to *C. elegans* development.

In the *Drosophila* ovary Bic-C is present in cytoplasmic granules enriched for Trailer Hitch (Tral) and Me31B [31, 32], two proteins marking sponge bodies, ovarian organelles related to the repression of mRNA translation [33–35]. Mouse and *Xenopus* Bicc1 in cultured cells are also found within subcellular structures associated with mRNA silencing, the processing granules (P granules, [26, 27, 36]), strongly suggesting that the members of the Bic-C protein family may share a conserved function in translational control. For example, P bodies may destabilize mRNAs via the action of decapping enzymes such as Dcp1 in many tissues undergoing rapid mRNA turnover, while certain yeast mRNAs can be reversibly associated with P-bodies [37]. Further, in metazoans, deadenylation is often the rate-limiting, first step of mRNA decay [38]. While in the kidney, high turnover of certain mRNA may be instrumental to rapidly adapt organ function to the environmental changes, in tissues with a strong "anabolic" activity such as the ovary it would not be surprising to find that some maternal mRNAs are silenced and stored in cellular compartments refractory to translation during oogenesis, to be deployed later in the early embryo. Consistent with the possibility that Bic-C may not function by destabilizing its mRNA targets, no global changes in *Bic-C* mRNA stability were observed in the *Drosophila* ovary, neither by quantitative RT-PCR of ovarian total mRNA nor by *in situ* hybridization (Bic-C negatively regulates its own mRNA) [15]. While there seems to be a mild effect on stability of the polycystic kidney disease 2 (*Pkd2*) mRNA in the kidneys of the *Bicc1*$^{-/-}$ KO mice, in this case, no direct association of this mRNA with the Bicc1 protein was formally demonstrated [27]. It is also possible that only a fraction of the cellular Bic-C pool is involved in destabilization and degradation of mRNA targets, possibly constituting a distinct compartment. This scenario would have escaped detection via traditional biochemical methods because they cannot preserve the integrity of the tissues analyzed. Until more regulatory targets for the Bic-C family members will be identified, validated, and characterized functionally, this current puzzle will remain unanswered.

3.4. Not3/5: An Evolutionarily Conserved Bic-C Partner Affecting mRNA Translation. Not3 is one of the subunits of the CCR4-NOT deadenylase, which is the predominant deadenylase, at least in the yeast *S. cerevisiae* [39–41]. Other subunits include CCR4, CAF1, NOT1-5 [40–44]. In *Drosophila* homologous genes are present for each of these subunits, with the exception of NOT3 and NOT5, for which there is only one gene displaying homology to both proteins [45]. Interestingly, Not3/5 does not contain any known protein domain, as identified via Prosite [46].

Drosophila Not3/5 proteins are virtually identical in 12 species, the differences being concentrated in areas of low-sequence complexity (Figure 3). A BLAST search [7] reveals that besides insects and vertebrates, there are Not3/5 orthologs, in *fungi* (*S. cerevisiae*, *Schizosaccharomyces pombe*, as well as the mushrooms *Laccaria bicolor*, *Coprinopsis*

```
Drosophila_melanogaster    MLSCASFNKLMYPSAADVAKPPMVGLEVEAGSIGSLSSLHALPSTTSVGS 50
Anopheles_gambiae          --------------------------PVRCKMMASCSSFN---KHIFLNG 21
Mus_musculus               -----------------------------------MASQSEPGYLAAAQS 15
Rattus_norvegicus          --------------------------------------------------
Homo_sapiens               ------------------------------------MAAQGEPGYLAAQS 14
Pan_troglodytes            --------------------------------------------------
Canis_familiaris           --------------------------------------------------
Bos_taurus                 --------------------------------------------------
Danio_rerio                ----------------------------------------MAEPLSFMHH 10
Caenorhabditis_elegans     --------------------------------------------------

Drosophila_melanogaster    GAPSETQSEISSVDSDWSDIRAIAMKLGVQNPDDLHTERFKVDRQKLEQL 100
Anopheles_gambiae          GPPSETTSEISSVESDWGDLRLIAAQLGVANPDDLHVERFKVDRQKLEDM 71
Mus_musculus               DPGSNSERSTDSPVAGSEDDL--VAAAPLLHSPEWSEERFRVDRKKLEAM 63
Rattus_norvegicus          --------------------------------------------------
Homo_sapiens               DPGSNSERSTDSPVPGSEDDL--VAGA-TLHSPEWSEERFRVDRKKLEAM 61
Pan_troglodytes            ------------------------------MTPERCEQ------------ 8
Canis_familiaris           --------------------------------------------------
Bos_taurus                 --------------------------------------------------
Danio_rerio                DHGSSSERSDDSPSAVSEDDSSGHCGHISPPDPDWTEERFRVDRKKLETM 60
Caenorhabditis_elegans     ---------------------MLREDTVIQLPDGRFEQKIQVDRRKLESM 29

                                             KH-like
Drosophila_melanogaster    IK--AESSIEGMNGAEYFFHDIMNTTDTYVSWPCRLKIGAKSKKDPHVRI 148
Anopheles_gambiae          IK--VETYSEGMNSAEEFFTNIMKETTTYVSWPCRLKIGAKTKKDPHIRI 119
Mus_musculus               LQ--AAAEGKGRS-GEDFFQKIMEETNTQIAWPSKLKIGAKSKKDPHIKV 110
Rattus_norvegicus          ----------------------MEETNTQIAWPSKLKIGAKSKKDPHIKV 28
Homo_sapiens               LQ--AAAEGKGRS-GEDFFQKIMEETNTQIAWPSKLKIGAKSKKDPHIKV 108
Pan_troglodytes            -------------------------------------------MDPHIKV 15
Canis_familiaris           --------------------------------------------------
Bos_taurus                 ----------------------MEETNTQIAWPSKLKIGAKSKKDPHIKV 28
Danio_rerio                LL--AANEGR-IN-GDDFFQKVMDETNTQIAWPSKLKIGAKSKKDPHIKV 106
Caenorhabditis_elegans     ITGRIDNTSHQLPTAESFFANVMSYSNAEVIWPSQLKIGAKTKKDPYVKV 79

Drosophila_melanogaster    VGKVDQVQRAKERILSSLDSRGTRVIMKMDVSYTDHSYIIGRGGNNIKRI 198
Anopheles_gambiae          VGKMADVLRAKDKVMARLDSRGSRVIMKMDVSYTDHSFIIGRGGNNIKKI 169
Mus_musculus               SGKKEDVKEAKEMIMSVLDTKSNRVTLKMDVSHTEHSHVIGKGGNNIKKV 160
Rattus_norvegicus          SGKKEAVKEAKEMIMAVLDTKSNRVTLKMDVSHTEHSHVIGKGGNNIKKV 78
Homo_sapiens               SGKKEDVKEAKEMIMSVLDTKSNRVTLKMDVSHTEHSHVIGKGGNNIKKV 158
Pan_troglodytes            SGKKEDVKEAKEMIMSVLDTKSNRVTLKMDVSHTEHSHVIGKGGNNIKKV 65
Canis_familiaris           ----------------------------MDVSHTEHSHVIGKGGNNIKKV 22
Bos_taurus                 SGKKEDVKEAKEMIMSVLDTKSNRVTLKMDVSHTEHSHVIGKGGNNIKKV 78
Danio_rerio                SGKRDDVREAKEKIMSVLDTKSHRVTLKMDVSHTEHSHVIGKGGHNIKRV 156
Caenorhabditis_elegans     IGSIEQIESARTLVLNSLQIKKERVSLKMELHHSLHSHIIGKGGRGIQKV 129
                                                       *:: :: **.:**:**..*:::
                                           KH
Drosophila_melanogaster    MDDTHTHIHFPDSNRSNPTEKSNQ-VSLCGSLEGVERARALVRLSTPLLI 247
Anopheles_gambiae          MEETATHIHFPDSNRSNPTEKSNQ-VSMCGSIEGVERARSLVRNSTPLLI 218
Mus_musculus               MEDTGCHIHFPDSNRNNQAEKSNQ-VSIAGQPAGVESARARIRELLPLVL 209
Rattus_norvegicus          MEDTGCHIHFPDSNRNNQVEKSNQ-VSIAGQPAGVESARARIRELLPLVL 127
Homo_sapiens               MEETGCHIHFPDSNRNNQAEKSNQ-VSIAGQPAGVESARVRIRELLPLVL 207
Pan_troglodytes            MEETGCHIHFPDSNRNNQAEKSNQVVSIAGQPAGVESARVRIRELLPLVL 115
Canis_familiaris           MEETGCHIHFPDSNRNNQAEKSNQ-VSIAGQPAGVESARVRIRELLPLVL 71
Bos_taurus                 MEETGCHIHFPDSNRNNQAEKSNQ-VSIAGQPAGVESARVRIRELLPLVL 127
Danio_rerio                MEETGCHIHFPDSNRHSQAEKSNQ-VSIAGQLTGVEAARVKIRELLPLVL 205
Caenorhabditis_elegans     MKMTSCHIHFPDSNKYSDSNKSDQ-VSISGTPVNVFEALKHLRSMCPLTV 178
                           *. *  ********: .  :**:* **:.*   .*  *   :*   ** :
                                             KH-like
Drosophila_melanogaster    SFEMPVMGPNKPQPDHETPYIKMIETKFNVQVIFSTRPKLHTSLVLVKGS 297
Anopheles_gambiae          SFELPILAPGKTPPDNDTPYVKEIEAEYGVQVIFSTRPKLHSSLVLVKGS 268
Mus_musculus               MFELPIAGILQPVPDPNTPSIQHISQTYSVSVSFKQRSRMYGATVTVRGS 259
Rattus_norvegicus          MFELPIAGILQPVPDPNTPSIQHISQTYSVSVSFKQRSRMYGATVIVRGS 177
Homo_sapiens               MFELPIAGILQPVPDPNSPSIQHISQTYNISVSFKQRSRMYGATVIVRGS 257
Pan_troglodytes            MFELPIAGILQPVPDPNSPSIQHISQTYNISVSFKQRSRMYGATVIVRGS 165
Canis_familiaris           MFELPIAGILQPVPDPNSPSIQHISQTYNISVSFKQRSRMYGATVIVRGS 121
Bos_taurus                 MFELPIAGILQPVPDPNSPSIQHISQMYNISVSFKQRSRMYGATVIVRGS 177
Danio_rerio                MFEC--SGVVQLV-DCSSPVVQHISHTYNVSISFRPPSRLYGNTAIVRAN 252
Caenorhabditis_elegans     YMKLPWYNPGQPD-------LRPLMSQMDLDVSVEQN--IYSLAIKMTGS 219
                            ::       :         :: :    .:.: .     ::     : ..
```

FIGURE 2: Continued.

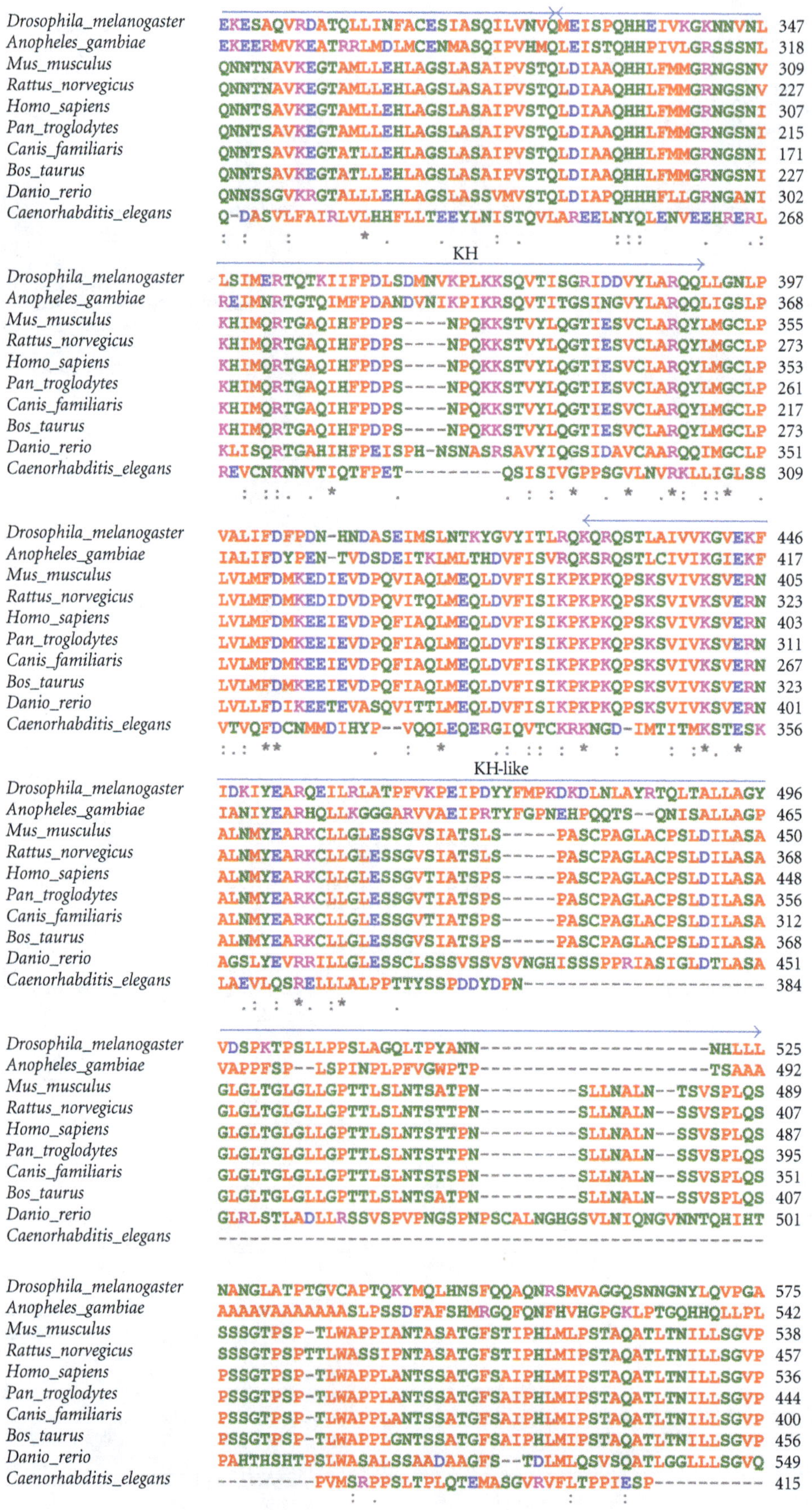

```
Drosophila_melanogaster   EKESAQVRDATQLLINFACESIASQILVNVQMEISPQHHEIVKGKNNVNL 347
Anopheles_gambiae         EKEERMVKEATRRLMDLMCENMASQIPVHMQLEISTQHHPIVLGRSSSNL 318
Mus_musculus              QNNTNAVKEGTAMLLEHLAGSLASAIPVSTQLDIAAQHHLFMMGRNGSNV 309
Rattus_norvegicus         QNNTNAVKEGTAMLLEHLAGSLASAIPVSTQLDIAAQHHLFMMGRNGSNV 227
Homo_sapiens              QNNTSAVKEGTAMLLEHLAGSLASAIPVSTQLDIAAQHHLFMMGRNGSNI 307
Pan_troglodytes           QNNTSAVKEGTAMLLEHLAGSLASAIPVSTQLDIAAQHHLFMMGRNGSNI 215
Canis_familiaris          QNNTSAVKEGTATLLEHLAGSLASAIPVSTQLDIAAQHHLFMMGRNGSNI 171
Bos_taurus                QNNTSAVKEGTATLLEHLAGSLASAIPVSTQLDIAAQHHLFMMGRNGSNI 227
Danio_rerio               QNNSSGVKRGTALLLEHLAGSLASSVMVSTQLDIAPQHHHFLLGRNGANI 302
Caenorhabditis_elegans    Q-DASVLFAIRLVLHHFLLTEEYLNISTQVLAREELNYQLENVEEHRERL 268
                          : :   :      * .    .   : .           :::     .  .:
                                              KH
Drosophila_melanogaster   LSIMERTQTKIIFPDLSDMNVKPLKKSQVTISGRIDDVYLARQQLLGNLP 397
Anopheles_gambiae         REIMNRTGTQIMFPDANDVNIKPIKRSQVTITGSINGVYLARQQLIGSLP 368
Mus_musculus              KHIMQRTGAQIHFPDPS----NPQKKSTVYLQGTIESVCLARQYLMGCLP 355
Rattus_norvegicus         KHIMQRTGAQIHFPDPS----NPQKKSTVYLQGTIESVCLARQYLMGCLP 273
Homo_sapiens              KHIMQRTGAQIHFPDPS----NPQKKSTVYLQGTIESVCLARQYLMGCLP 353
Pan_troglodytes           KHIMQRTGAQIHFPDPS----NPQKKSTVYLQGTIESVCLARQYLMGCLP 261
Canis_familiaris          KHIMQRTGAQIHFPDPS----NPQKKSTVYLQGTIESVCLARQYLMGCLP 217
Bos_taurus                KHIMQRTGAQIHFPDPS----NPQKKSTVYLQGTIESVCLARQYLMGCLP 273
Danio_rerio               KLISQRTGAHIHFPEISPH-NSNASRSAVYIQGSIDAVCAARQQIMGCLP 351
Caenorhabditis_elegans    REVCNKNNVTIQTFPET---------QSISIVGPPSGVLNVRKLLIGLSS 309
                            : ::. . *      .          . : : *  . *  .*: ::*  .

Drosophila_melanogaster   VALIFDFPDN-HNDASEIMSLNTKYGVYITLRQKQRQSTLAIVVKGVEKF 446
Anopheles_gambiae         IALIFDYPEN-TVDSDEITKLMLTHDVFISVRQKSRQSTLCIVIKGIEKF 417
Mus_musculus              LVLMFDMKEDIEVDPQVIAQLMEQLDVFISIKPKPKQPSKSVIVKSVERN 405
Rattus_norvegicus         LVLMFDMKEDIDVDPQVITQLMEQLDVFISIKPKPKQPSKSVIVKSVERN 323
Homo_sapiens              LVLMFDMKEEIEVDPQFIAQLMEQLDVFISIKPKPKQPSKSVIVKSVERN 403
Pan_troglodytes           LVLMFDMKEEIEVDPQFIAQLMEQLDVFISIKPKPKQPSKSVIVKSVERN 311
Canis_familiaris          LVLMFDMKEEIEVDPQFIAQLMEQLDVFISIKPKPKQPSKSVIVKSVERN 267
Bos_taurus                LVLMFDMKEEIEVDPQFIAQLMEQLDVFISIKPKPKQPSKSVIVKSVERN 323
Danio_rerio               LVLLFDIKEETEVASQVITTLMEQLDVFISIKPKPKQPSKSVIVKSVERN 401
Caenorhabditis_elegans    VTVQFDCNMMDIHYP--VQQLEQERGIQVTCKRKNGD-IMTITMKSTESK 356
                          :.: **         .  :  *     .: :: : *  :     : :*. *
                                                 KH-like
Drosophila_melanogaster   IDKIYEARQEILRLATPFVKPEIPDYYFMPKDKDLNLAYRTQLTALLAGY 496
Anopheles_gambiae         IANIYEARHQLLKGGGARVVAEIPRTYFGPNEHPQQTS--QNISALLAGP 465
Mus_musculus              ALNMYEARKCLLGLESSGVSIATSLS-----PASCPAGLACPSLDILASA 450
Rattus_norvegicus         ALNMYEARKCLLGLESSGVSIATSLS-----PASCPAGLACPSLDILASA 368
Homo_sapiens              ALNMYEARKCLLGLESSGVTIATSPS-----PASCPAGLACPSLDILASA 448
Pan_troglodytes           ALNMYEARKCLLGLESSGVTIATSPS-----PASCPAGLACPSLDILASA 356
Canis_familiaris          ALNMYEARKCLLGLESSGVTIATSPS-----PASCPAGLACPSLDILASA 312
Bos_taurus                ALNMYEARKCLLGLESSGVSIATSPS-----PASCPAGLACPSLDILASA 368
Danio_rerio               AGSLYEVRRILLGLESSCLSSSVSSVSVNGHISSSPPRIASIGLDTLASA 451
Caenorhabditis_elegans    LAEVLQSRELLLALPPTTYSSPDDYDPN---------------------- 384
                             .: : *. :*     .

Drosophila_melanogaster   VDSPKTPSLLPPSLAGQLTPYANN---------------------NHLLL 525
Anopheles_gambiae         VAPPFSP--LSPINPLPFVGWPTP---------------------TSAAA 492
Mus_musculus              GLGLTGLGLLGPTTLSLNTSATPN---------SLLNALN--TSVSPLQS 489
Rattus_norvegicus         GLGLTGLGLLGPTTLSLNTSTTPN---------SLLNALN--SSVSPLQS 407
Homo_sapiens              GLGLTGLGLLGPTTLSLNTSTTPN---------SLLNALN--SSVSPLQS 487
Pan_troglodytes           GLGLTGLGLLGPTTLSLNTSTTPN---------SLLNALN--SSVSPLQS 395
Canis_familiaris          GLGLTGLGLLGPTTLSLNTSTSPN---------SLLNALN--SSVSPLQS 351
Bos_taurus                GLGLTGLGLLGPTTLSLNTSATPN---------SLLNALN--SSVSPLQS 407
Danio_rerio               GLRLSTLADLLRSSVSPVPNGSPNPSCALNGHGSVLNIQNGVNNTQHIHT 501
Caenorhabditis_elegans    --------------------------------------------------

Drosophila_melanogaster   NANGLATPTGVCAPTQKYMQLHNSFQQAQNRSMVAGGQSNNGNYLQVPGA 575
Anopheles_gambiae         AAAAVAAAAAAAASLPSSDFAFSHMRGQFQNFHVHGPGKLPTGQHHQLLPL 542
Mus_musculus              SSSGTPSP-TLWAPPIANTASATGFSTIPHLMLPSTAQATLTNILLSGVP 538
Rattus_norvegicus         SSSGTPSPTTLWASSIPNTASATGFSTIPHLMIPSTAQATLTNILLSGVP 457
Homo_sapiens              PSSGTPSP-TLWAPPLANTSSATGFSAIPHLMIPSTAQATLTNILLSGVP 536
Pan_troglodytes           PSSGTPSP-TLWAPPLANTSSATGFSAIPHLMIPSTAQATLTNILLSGVP 444
Canis_familiaris          PSSGTPSP-TLWAPPLANTSSATGFSAIPHLMIPSTAQATLTNILLSGVP 400
Bos_taurus                PSSGTPSP-TLWAPPLGNTSSATGFSAIPHLMIPSTAQATLTNILLSGVP 456
Danio_rerio               PAHTHSHTPSLWASALSSAADAAGFS--TDLMLQSVSQATLGGLLLSGVQ 549
Caenorhabditis_elegans    ---------PVMSRPPSLTPLQTEMASGVRVFLTPPIESP---------- 415
                                   : .                 :       :
```

FIGURE 2: Continued.

```
Drosophila_melanogaster   VAPPLKPPTVSPRN--------------------------------SCSQ 593
Anopheles_gambiae         SLPPGLERTVPGGSS-----------------AGKMNHLSSPHLLLTVSQ 575
Mus_musculus              TYGHT-APSPPPGLTPVDVHINSMQTEGKNISASINGHVQPANMKYGPLS 587
Rattus_norvegicus         TYGHT-APSPPPGLTPVDVHINSMQTEGKNISASINGHVQPPNMKYGPLS 506
Homo_sapiens              TYGHT-APSPPPGLTPVDVHINSMQTEGKKISAALNGHAQSPDIKYGAIS 585
Pan_troglodytes           TYGHT-APSPPPGLTPVDVHINSMQTEGKKISAALNGHAQSPNIKYGAIS 493
Canis_familiaris          TYGHT-APSPPPGLTPVDVHINTMQTEGKKISASLNGHAQSPNIKYGAIP 449
Bos_taurus                TYGHT-APSPPPGLTPVDVHINTMQAEGKKISAALNGHTQSPSLKYGAIS 505
Danio_rerio               SQAHTHTPSLPPGLAPIHKTVS---------AEHLNGHLASS--VYSRIS 588
Caenorhabditis_elegans    --------------------------------------------------

Drosophila_melanogaster   N-------TSGYQSFSSSTTSLEQSYPPYAQLPGTVSSTSSSTAGSQNRA 636
Anopheles_gambiae         NSSHNDIHSSGYQSLNCSSNSLDQQFQSNSSASGSVSQVSSNSLLNNSPD 625
Mus_musculus              TSSLGEKVLSSNHGDPSMQTAGPEQASPKSNSVEGCNDAFVEVGMPRSPS 637
Rattus_norvegicus         TSSLGEKVLSSNHGDPSMQTAGPEQASPKSNSVEGCNDAFVEVGMPRSPS 556
Homo_sapiens              TSSLGEKVLSANHGDPSIQTSGSEQTSPKSSPTEGCNDAFVEVGMPRSPS 635
Pan_troglodytes           TSSLGEKVLSANHGDPSIQTSGSEQTSPKSSPTEGCNDAFVEVGMPRSPS 543
Canis_familiaris          TSSLGEKVLSGNHGDPSRQTTGPEQASPKSNPTEGCNDAFVEVGMPRSPS 499
Bos_taurus                TSSLGEKVLSANHGDPSRQTAGSEQTSPKSNPTEGCNDAFVEVGMPRSPS 555
Danio_rerio               SVSL-----NSAHCDTAQEGIGHTQSEAKS--TDEGSDTFVEVGMPRSPS 631
Caenorhabditis_elegans    ------------------KSPDPEDSPLAASILKGAKDISKNSDIWKKKS 447
                                                    .    :      ..   .    ..

Drosophila_melanogaster   HYSP--DSTYGSEGGGVGGGGGGGARLGRRLSDGVLLGLSNSNGGGGNSG 684
Anopheles_gambiae         HQSPGAAGTSGLNRCRLSVCTPESPHYQSELEQRTPLAFEQKVG-----V 670
Mus_musculus              HSGNAGDLKQMLGASKVSCAKRQTVELLQGTKNSHLHGTDRLLSDPELSA 687
Rattus_norvegicus         HSGNAGDLKQMLGPSKVSCAKRQTVELLQGTKNSHLHSTDRLLSDTELSA 606
Homo_sapiens              HSGNAGDLKQMMCPSKVSCAKRQTVELLQGTKNSHLHSTDRLLSDPELSA 685
Pan_troglodytes           HSGNAGDLKQMMCPSKVSCAKRQTVELLQGTKNSHLHSTDRLLSDPELSA 593
Canis_familiaris          HSGNAGDLKQMMGPSKVSCAKRQTVELLQGTKNSHLHSTDRLLSDPELST 549
Bos_taurus                HSGNAGDLKQMMGPSKVACAKRQTVELLQGTKNSHLHSTDRLLSDPELSA 605
Danio_rerio               HSANGSELKQMLASCTVSPGKRQTVELLQRTKNTLLH-VECVLAD----S 676
Caenorhabditis_elegans    KADRG--------------------------------------------- 452
                          :

Drosophila_melanogaster   GAHLLPGSAESYRSLHYDLGGNK--------------HSGHRAFDFDMKR 720
Anopheles_gambiae         VRRCLPVHLKRLTVLGNHLQSS---------------LADTFLFNLDPRV 705
Mus_musculus              TESPLADKKAPGSERAAERAAAAQQKSERARLASQPTYVHMQAFDYEQKK 737
Rattus_norvegicus         TESPLADKKAPGSERAAERAAAAQQNSERARLASQPPYVHMQAFDYEQKK 656
Homo_sapiens              TESPLADKKAPGSERAAERAAAAQQNSERAHLAPRSSYVNMQAFDYEQKK 735
Pan_troglodytes           TESPLADKKAPGSERAAERAAAAQQNSERAHLAPRSSYVNMQAFDYEQKK 643
Canis_familiaris          TESPLADKKAPGSERAAERAAAAQQNSERARLAPRPSYVNMQAFDYEQKK 599
Bos_taurus                AESPLADKKAPGSERAAER--AAQQNNERARLAPRPSYVNMQAFDYEQKK 653
Danio_rerio               DDNPMTDKRAPGSERAAER-----------RLAP-----HMQAFDYEKKK 710
Caenorhabditis_elegans    -----------------------------------------------EML 455

Drosophila_melanogaster   ALGYKAMERTPVAGELRTPTTAWMGMGLSSTSPAP--------------- 755
Anopheles_gambiae         VAGYKAMHMSPQQGEIRTPTLSWQGLGLSQSSPAPLE------------- 742
Mus_musculus              LLATKAMLKKPVVTEVRTPTNTWSGLGFSKSMPAETIKELRRANHVSYKP 787
Rattus_norvegicus         LLATKAMLKKPVVTEVRTPTNTWSGLGFSKSMPAETIKELRRANHVSYKP 706
Homo_sapiens              LLATKAMLKKPVVTEVRTPTNTWSGLGFSKSMPAETIKELRRANHVSYKP 785
Pan_troglodytes           LLATKAMLKKPVVTEVRTPTNTWSGLGFSKSMPAETIKELRRANHVSYKP 693
Canis_familiaris          LLATKAMLKKPVVTEVRTPTNTWSGLGFSKSMPAETIKELRRANHVSYKP 649
Bos_taurus                LLATKAMLKKPVVTEVRTPTNTWSGLGFSKSMPAETIKELRRANHVSYKP 703
Danio_rerio               LLATKAMLKKPVVTEIRTPTNTWSGLGFSKSMPAESIKELRRAHHVPYKP 760
Caenorhabditis_elegans    IKATQAIFDDSVLSSPRYPTDLWSGYGFSSSLPADLLKGMMDLSTNEPST 505
                            .  :*:    .    . * **  * * *:*.: **

Drosophila_melanogaster   ------------APLENGENGAAGGGASSGWRLPP--------------- 778
Anopheles_gambiae         ----ACDLSWANTSSSSSTGGGRDGGGGSGCANTS--------------- 773
Mus_musculus              TMTTAYEGSSLSLSRSSSREHLASGSESDNWRDRN------GIGPMGHSE 831
Rattus_norvegicus         TMTTAYEGSSLSLSRSSSREHLASGSESDNWRDRN------GIGPMGHSE 750
Homo_sapiens              TMTTTYEGSSMSLSRSNSREHLGGGSESDNWRDRN------GIGPGSHSE 829
Pan_troglodytes           TMTTTYEGSSMSLSRSNSREHLGSGSESDNWRDRN------GIGPGSHSE 737
Canis_familiaris          TMTTTFEGSSMSLSRSNSREHLGSGSESDNWRDRN------GIGPPSPSE 693
Bos_taurus                TMTTTFEGSSMSLSRSNSREHLGSGSESDNWRDRN------GIGPASHGE 747
Danio_rerio               SMGTTYEDSHLSMSHSGIQEGLINDTKSDNWGDLNGNVNINGNGPSGNSE 810
Caenorhabditis_elegans    NGPPMMNHSQRGLCSVREEDEELSDFSASSTNYGMS-------------- 541
                                                  ..   ...
```

FIGURE 2: Continued.

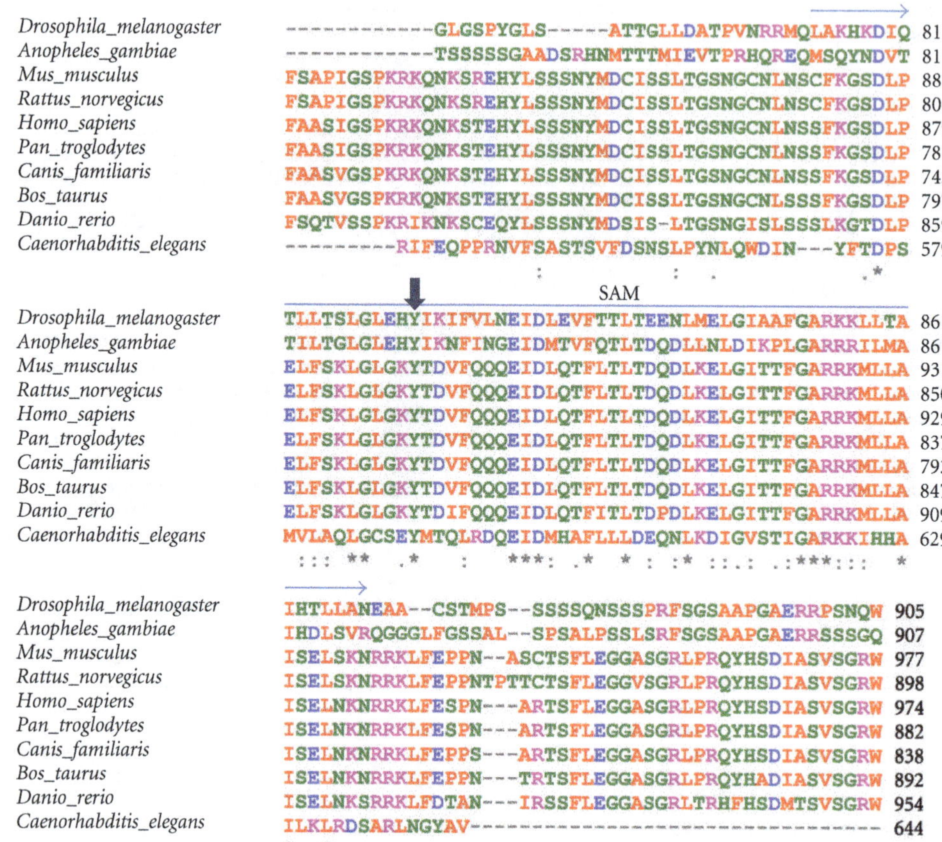

Figure 2: Bic-C orthologs. Clustal W [1, 2] was used to align sequences extracted from the NCBI sequence database. As in Figure 1, the two canonical (KH) and three noncanonical (KH-like) KH RNA-binding modules are indicated (arrows, top). Domain assignment is as in [3] except for the fourth KH-related motif and the SAM domains, that are labelled according to the Pfam database [4]. A conserved, potentially phosphorylated, tyrosine is also indicated (arrowhead, top). Amino acid (aa) color-coding is from Clustal W: red, small aliphatic, hydrophobic and aromatics; blue, acidic; magenta, basic; green, hydroxyl, sulphydryl, amine, and glycine; grey, unusual aa. Symbols for aa conservation are from Clustal W: (asterisk $*$): positions with a single, fully conserved residue. (Colon :): conservation between groups of strongly similar properties-scoring >0.5 in the Gonnet PAM 250 matrix. (Period .): conservation between groups of weakly similar properties-scoring ≤0.5 in the Gonnet PAM 250 matrix. Highlighted yellow: residues that contribute to RNA binding in the Smaug protein. Grey highlight denotes mild (versus strong) basic charges. Light blue highlights a charged aa in a conserved position, but an opposite electrical charge. The *Gallus gallus* genome also contains a predicted sequence with extensive homology to Bic-C (Table 1) and with a long extension at the N terminal end. Since there is no experimental evidence of the true starting methionine we did not include it in this alignment.

cinerea, and *Schizophyllum commune*), vascular plants (e.g., *Arabidopsis thaliana*, *Oryza sativa*, *Vitis vinifera*, and *Ustilago maydis*), and mosses (*Physcomitrella patens*). An alignment of complete sequences is shown in Figure 3. The conservation is highest at the N- and C-terminus of the protein (aa 1–238 and 680–844, with reference to the *Drosophila* sequence) where all the family members show extensive identity. Between aa 330 and 679 the sequences diverge with the orthologues from the two insects (*D. melanogaster* and *Anopheles gambiae*), the *fungi*, the higher eukaryotes, and the plants being more similar with each other than with members of a different group. Notably, the vertebrate sequences, with the exception of zebrafish that contains various small deletions, have blocks of almost complete identity in this region (Figure 4). The partial divergence in the central region of Not3/5 is likely due to the fact that the *Drosophila* gene is homologous to both the NOT3 and NOT5 genes and likely plays the functional roles of both yeast proteins, [45] a seemingly unique feature of *Drosophila [47]*. Not3/5 was recovered in a two-hybrid screen for proteins interacting with *Drosophila* Bic-C, and multiple pieces of evidence support the existence of this interaction *in vivo*: there is genetic interaction between *Bic-C* and *twin*, the *Drosophila* gene encoding for CCR4; other subunits of the CCR4-NOT complex can be coimmunoprecipitated with Bic-C from ovary extracts and the Bic-C target mRNAs that were tested were found with longer polyA tails in *Bic-C* mutants [15]. Although one study of vertebrate models could not detect differences in polyadenylation in a presumptive Bic-C target [27], due to the high homology of the Bic-C and NOT orthologs it is possible that Bic-C from other species can interact with NOT homologs and, possibly, other subunits of the deadenylase complex. These may contribute to the interaction only in the context of the assembled complex and

```
D. melanogaster    MAATRKLQGEIDRCLKKVAEGVETFEDIWKKVHNATNTNQKQKHLQEKYEADLKKEIKKL 60
D. sechellia       MAATRKLQGEIDRCLKKVAEGVETFEDIWKKVHNATNTNQKQKHLQEKYEADLKKEIKKL 60
D. simulans        MAATRKLQGEIDRCLKKVAEGVETFEDIWKKVHNATNTNQKQKHLQEKYEADLKKEIKKL 60
D. erecta          MAATRKLQGEIDRCLKKVAEGVETFEDIWKKVHNATNTNQKQKHLQEKYEADLKKEIKKL 60
D. yakuba          MAATRKLQGEIDRCLKKVAEGVETFEDIWKKVHNATNTNQKQKHLQEKYEADLKKEIKKL 60
D. persimilis      MAATRKLQGEIDRCLKKVAEGVETFEDIWKKVHNATNTNQKQKHLQEKYEADLKKEIKKL 60
D. pseudobscura    MAATRKLQGEIDRCLKKVAEGVETFEDIWKKVHNATNTNQKQKHLQEKYEADLKKEIKKL 60
D. grimshawi       MAATRKLQGEIDRCLKKVGEGVETFEDIWKKVHNATNTNQKQKHLQEKYEADLKKEIKKL 60
D. virilis         MAATRKLQGEIDRCLKKVGEGVETFEDIWKKVHNATNTNQKQKHLQEKYEADLKKEIKKL 60
D. mojavensis      MAATRKLQGEIDRCLKKVGEGVETFEDIWKKVHNATNTNQKQKHLQEKYEADLKKEIKKL 60
D. willistoni      MAATRKLQGEIDRCLKKVAEGVETFEDIWKKVHNATNTNQKQKHLQEKYEADLKKEIKKL 60
D. ananassae       MAATRKLQGEIDRCLKKVAEGVETFEDIWKKVHNATNTNQKQKHLQEKYEADLKKEIKKL 60
                   ******************.*****************************************

D. melanogaster    QRLRDQIKSWIASAEIKDKSSLLENRRLIET---QMERFKVVERETKTKAYSKEGLGAAQ 117
D. sechellia       QRLRDQIKSWIASAEIKDKSSLLENRRLIET---QMERFKVVERETKTKAYSKEGLGAAQ 117
D. simulans        QRLRDQIKSWIASAEIKDKSSLLENRRLIET---QMERFKVVERETKTKAYSKEGLGAAQ 117
D. erecta          QRLRDQIKSWIASAEIKDKSSLLENRRLIET---QMERFKVVERETKTKAYSKEGLGAAQ 117
D. yakuba          QRLRDQIKSWIASAEIKDKSSLLENRRLIET---QMERFKVVERETKTKAYSKEGLGAAQ 117
D. persimilis      QRLRDQIKSWIASAEIKDKSSLLENRRLIET---QMERFKVVERETKTKAYSKEGLGAAQ 117
D. pseudobscura    QRLRDQIKSWIASAEIKDKSSLLENRRLIET---QMERFKVVERETKTKAYSKEGLGAAQ 117
D. grimshawi       QRLRDQIKSWIASAEIKDKSALLENRRLIET---QMERFKVVERETKTKAYSKEGLGAAQ 117
D. virilis         QRLRDQIKSWIASAEIKDKSALLENRRLIETASCQMERFKVVERETKTKAYSKEGLGAAQ 120
D. mojavensis      QRLRDQIKSWIASAEIKDKSALLENRRLIET---QMERFKVVERETKTKAYSKEGLGAAQ 117
D. willistoni      QRLRDQIKSWIASAEIKDKSALLENRRLIET---QMERFKVVERETKTKAYSKEGLGAAQ 117
D. ananassae       QRLRDQIKSWIASAEIKDKSALLENRRLIET---QMERFKVVERETKTKAYSKEGLGAAQ 117
                   ********************:**********   **************************

D. melanogaster    KMDPAQRIKDDARNWLTSSISSLQIQIDQYESEIESLLAGKKKRLDRDKQERMDDLRGKL 177
D. sechellia       KMDPAQRIKDDARNWLTSSISSLQIQIDQYESEIESLLAGKKKRLDRDKQERMDDLRGKL 177
D. simulans        KMDPAQRIKDDARNWLTSSISSLQIQIDQYESEIESLLAGKKKRLDRDKQERMDDLRGKL 177
D. erecta          KMDPAQRIKDDARNWLTSSISSLQIQIDQYESEIESLLAGKKKRLDRDKQERMDDLRGKL 177
D. yakuba          KMDPAQRIKDDARNWLTSSISSLQIQIDQYESEIESLLAGKKKRLDRDKQERMDDLRGKL 177
D. persimilis      KMDPAQRIKDDARNWLTSSISSLQIQIDQYESEIESLLAGKKKRLDRDKQERMDDLRAKL 177
D. pseudobscura    KMDPAQRIKDDARNWLTSSISSLQIQIDQYESEIESLLAGKKKRLDRDKQERMDDLRAKL 177
D. grimshawi       KMDPAQRIKDHARNWLTNSISALQIQIDQYESEIESLLAGKKKRLDRDKQERMDDLRSKL 177
D. virilis         KMDPAQRIKDHARNWLTGSISTLQIQIDQYESEIESLLAGKKKRVDRDKQERMDDLRSKL 180
D. mojavensis      KMDPAQRIKDHARNWLTGSISTLQIQIDQYESEIESLLAGKKKRLDRDKQERMDDLRSKL 177
D. willistoni      KMDPAQRIKDDARNWLTSSISSLQIQIDQYESEIESLLAGKKKRLDRDKQERMDDLRSKL 177
D. ananassae       KMDPAQRIKDDARNWLTSSISSLQIQIDQYESEIESLLAGKKKRLDRDKQERMDDLRGKL 177
                   **********.******.***:**********************:************.**

D. melanogaster    DRHKFHITKLETLLRLLDNDGVEAEQVNKIKDDVEYYIDSSQEPDFEENEFIYDDIIGLD 237
D. sechellia       DRHKFHITKLETLLRLLDNDGVEAEQVNKIKDDVEYYIDSSQEPDFEENEFIYDDIIGLD 237
D. simulans        DRHKFHITKLETLLRLLDNDGVEAEQVNKIKDDVEYYIDSSQEPDFEENEFIYDDIIGLD 237
D. erecta          DRHKFHITKLETLLRLLDNDGVEAEQVNKIKDDVEYYIDSSQEPDFEENEFIYDDIIGLD 237
D. yakuba          DRHKFHITKLETLLRLLDNDGVEAEQVNKIKDDVEYYIDSSQEPDFEENEFIYDDIIGLD 237
D. persimilis      DRHKFHITKLETLLRLLDNDGVEADQVNKIKDDVEYYIDSSQEPDFEENEFIYDDIIGLD 237
D. pseudobscura    DRHKFHITKLETLLRLLDNDGVEADQVNKIKDDVEYYIDSSQEPDFEENEFIYDDIIGLD 237
D. grimshawi       DRHKFHITKLETLLRLLDNDGVEAEQVNKIKDDVEYYIDSSQEPDFEENEFIYDDIIGLD 237
D. virilis         DRHKFHITKLETLLRLLDNDGVEADQVNKIKDDVEYYIDSSQEPDFEENEFIYDDIIGLD 240
D. mojavensis      DRHKFHITKLETLLRLLDNDGVEADQVNKIKDDVEYYIDSSQEPDFEENEFIYDDIIGLD 237
D. willistoni      DRHKFHITKLETLLRLLDNDGVEAEQVNKIKDDVEYYIDSSQEPDFEENEFIYDDIIGLD 237
D. ananassae       DRHKFHISKLETLLRLLDNDGVEAEQVNKIKDDVEYYIDSSQDPDFEENEFIYDDIIGLD 237
                   *******:****************:*****************:*****************

D. melanogaster    EVELSGTATTDSNNSNETSGSPSSVTSGGSPSQSPVTVQQILNTSSQ------------G 285
D. sechellia       EVELSGTATTDSNNSNETSGSPSSVTSGGSPSQSPVTVQQILNTSSQ------------G 285
D. simulans        EVELSGTATTDSNNSNETSGSPSSVTSGGSPSQSPVTVQQILNTSSQ------------G 285
D. erecta          EVELSGTATTDSNNSNETSGSPSSVTSGGSPSQSPVTVQQILNTSSQ------------G 285
D. yakuba          EVELSGTATTDSNNSNETSGSPSSVTSGGSPSQSPVTVQQILNASSQ------------G 285
D. persimilis      EVELSGTATTDSNNSNETSGSPSSVTSGGSPSQSPVTVQQVLPASVQ------------A 285
D. pseudobscura    EVELSGTATTDSNNSNETSGSPSSVTSGGSPSQSPVTVQQVLPASVQ------------A 285
D. grimshawi       EVELSGTATTDSNNSNETSGSPSSVTSGGSPSQSPVTVQQILPSSSS-----------SG 286
D. virilis         EVELSGTATTDSNNSNETSGSPSSVTSGGSPSQSPVTVQQVLPSSSTQPQ---SAMAGSS 297
D. mojavensis      EVELSGTATTDSNNSNETSGSPSSVTSGGSPSQSPVTVQQVLPSSSSSAQQQQTSTAGSS 297
D. willistoni      EVELSGTATTDSNNSNETSGSPSSVTSGGSPSQSPVTVQQVLPSGASSGG-------GSS 290
D. ananassae       EVELSGTATTDSNNSNETSGSPSSVTSGGSPSQSPVTVQQVLPPSMP-----------VA 286
                   ****************************************:*  ..            .
```

Figure 3: Continued.

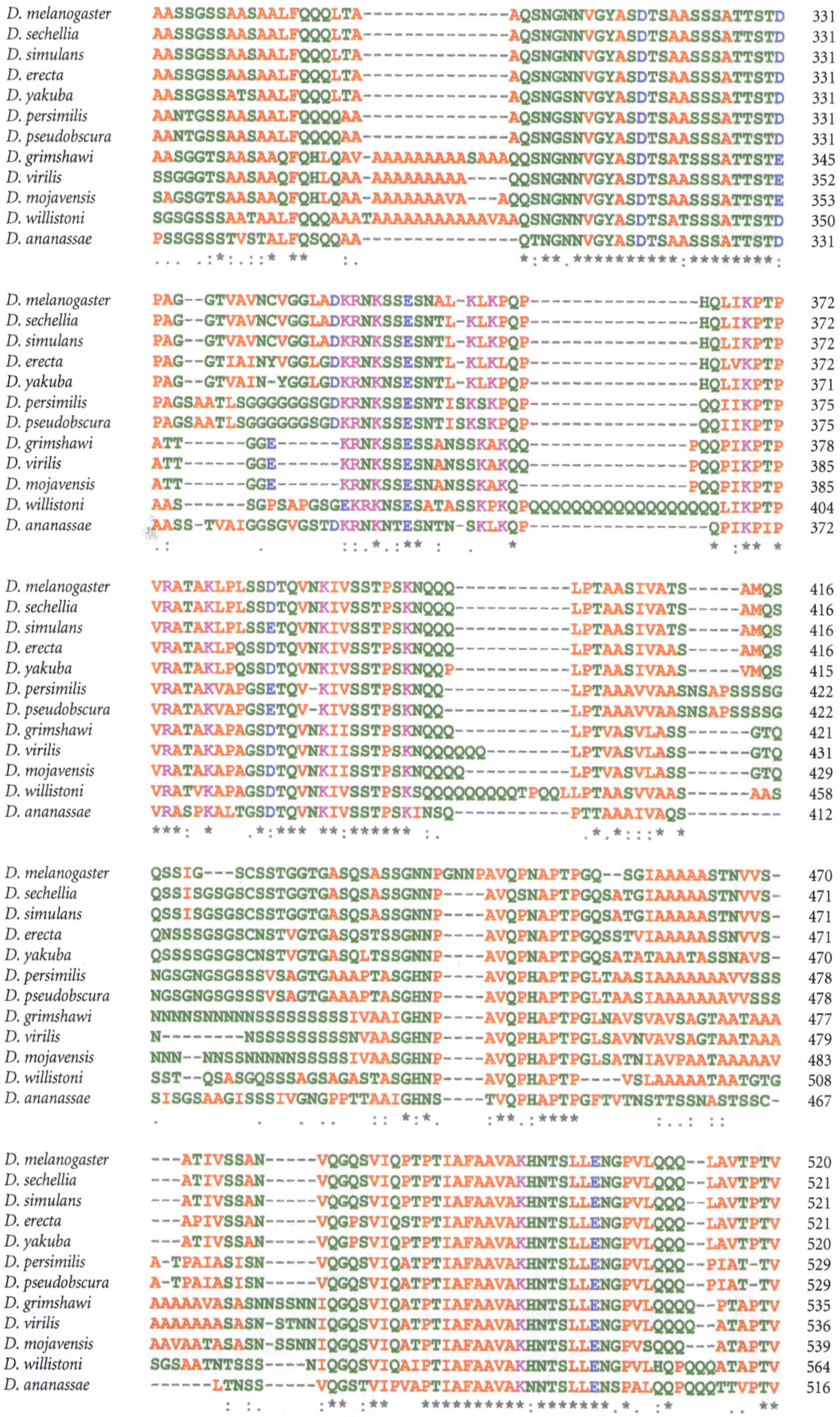

```
D. melanogaster   AASSGSSAASAALFQQQLTA--------------AQSNGNNVGYASDTSAASSSATTSTD   331
D. sechellia      AASSGSSAASAALFQQQLTA--------------AQSNGNNVGYASDTSAASSSATTSTD   331
D. simulans       AASSGSSAASAALFQQQLTA--------------AQSNGNNVGYASDTSAASSSATTSTD   331
D. erecta         AASSGSSAASAALFQQQLTA--------------AQSNGNNVGYASDTSAASSSATTSTD   331
D. yakuba         AASSGSSATSAALFQQQLTA--------------AQSNGSNVGYASDTSAASSSATTSTD   331
D. persimilis     AANTGSSAASAALFQQQQAA--------------AQSNGSNVGYASDTSAASSSATTSTD   331
D. pseudobscura   AANTGSSAASAALFQQQQAA--------------AQSNGSNVGYASDTSAASSSATTSTD   331
D. grimshawi      AASGGTSAASAAQFQHLQAV-AAAAAAAAASAAAQQSNGNNVGYASDTSATSSSATTSTE   345
D. virilis        SSGGGTSAASAAQFQHLQAA-AAAAAAAAA----QQSNGNNVGYASDTSAASSSATTSTE   352
D. mojavensis     SAGSGTSAASAAQFQHLQAA-AAAAAAAVA---AQQSNGNNVGYASDTSAASSSATTSTE   353
D. willistoni     SGSGSSSAATAALFQQQAAATAAAAAAAAAAAVAAQSNGNNVGYASDTSATSSSATTSTD   350
D. ananassae      PSSGSSSTVSTALFQSQQAA---------------QTNGNNVGYASDTSAASSSATTSTD   331
                  ... .:*:.::* **   :.               *:**.**********:********:

D. melanogaster   PAG--GTVAVNCVGGLADKRNKSSESNAL-KLKPQP----------------HQLIKPTP   372
D. sechellia      PAG--GTVAVNCVGGLADKRNKSSESNTL-KLKPQP----------------HQLIKPTP   372
D. simulans       PAG--GTVAVNCVGGLADKRNKSSESNTL-KLKPQP----------------HQLIKPTP   372
D. erecta         PAG--GTIAINYVGGLGDKRNKSSESNTL-KLKLQP----------------HQLVKPTP   372
D. yakuba         PAG--GTVAIN-YGGLGDKRNKNSESNTL-KLKPQP----------------HQLIKPTP   371
D. persimilis     PAGSAATLSGGGGGGSGDKRNKSSESNTISKSKPQP----------------QQIIKPTP   375
D. pseudobscura   PAGSAATLSGGGGGGSGDKRNKSSESNTISKSKPQP----------------QQIIKPTP   375
D. grimshawi      ATT------GGE------KRNKSSESSANSSKAKQQ---------------PQQPIKPTP   378
D. virilis        ATT------GGE------KRNKSSESNANSSKAKQQ---------------PQQPIKPTP   385
D. mojavensis     ATT------GGE------KRNKSSESNANSSKAKQ----------------PQQPIKPTP   385
D. willistoni     AAS------SGPSAPGSGEKRKNSESATASSKPKQPQQQQQQQQQQQQQQQQQQLIKPTP   404
D. ananassae      AASS-TVAIGGSGVGSTDKRNKNTESNTN-SKLKQP-----------------QPIKPIP   372
                  .:        .       ::.*.:** :  .   *                  * :** *

D. melanogaster   VRATAKLPLSSDTQVNKIVSSTPSKNQQQ-----------LPTAASIVATS-----AMQS   416
D. sechellia      VRATAKLPLSSDTQVNKIVSSTPSKNQQQ-----------LPTAASIVATS-----AMQS   416
D. simulans       VRATAKLPLSSETQVNKIVSSTPSKNQQQ-----------LPTAASIVATS-----AMQS   416
D. erecta         VRATAKLPQSSDTQVNKIVSSTPSKNQQQ-----------LPTAASIVAAS-----AMQS   416
D. yakuba         VRATAKLPQSSDTQVNKIVSSTPSKNQQP-----------LPTAASIVAAS-----VMQS   415
D. persimilis     VRATAKVAPGSETQV-KIVSSTPSKNQQ------------LPTAAAVVAASNSAPSSSSG   422
D. pseudobscura   VRATAKVAPGSETQV-KIVSSTPSKNQQ------------LPTAAAVVAASNSAPSSSSG   422
D. grimshawi      VRATAKAPAGSDTQVNKIISSTPSKNQQQ-----------LPTVASVLASS------GTQ   421
D. virilis        VRATAKAPAGSDTQVNKIISSTPSKNQQQQQQ--------LPTVASVLASS------GTQ   431
D. mojavensis     VRATAKAPAGSDTQVNKIISSTPSKNQQQQ----------LPTVASVLASS------GTQ   429
D. willistoni     VRATVKAPAGSDTQVNKIVSSTPSKSQQQQQQQQQTPQQLLPTAASVVAAS------AAS   458
D. ananassae      VRASPKALTGSDTQVNKIVSSTPSKINSQ-----------PTTAAAIVAQS---------   412
                  ***: *   .*:*** **:****** :.             .*.*:::* *

D. melanogaster   QSSIG---SCSSTGGTGASQSASSGNNPGNNPAVQPNAPTPGQ--SGIAAAAASTNVVS-   470
D. sechellia      QSSISGSGSCSSTGGTGASQSASSGNNP----AVQSNAPTPGQSATGIAAAAASTNVVS-   471
D. simulans       QSSISGSGSCSSTGGTGASQSASSGNNP----AVQPNAPTPGQSATGIAAAAASTNVVS-   471
D. erecta         QNSSSGSGSCNSTVGTGASQSTSSGNNP----AVQPNAPTPGQSSTVIAAAAASSNVVS-   471
D. yakuba         QSSSSGSGSCNSTVGTGASQLTSSGNNP----AVQPNAPTPGQSATATAAATASSNAVS-   470
D. persimilis     NGSGNGSGSSSVSAGTGAAAPTASGHNP----AVQPHAPTPGLTAASIAAAAAAAVVSSS   478
D. pseudobscura   NGSGNGSGSSSVSAGTGAAAPTASGHNP----AVQPHAPTPGLTAASIAAAAAAAVVSSS   478
D. grimshawi      NNNNSNNNNNSSSSSSSSSIVAAIGHNP----AVQPHAPTPGLNAVSVAVSAGTAATAAA   477
D. virilis        N--------NSSSSSSSSSNVAASGHNP----AVQPHAPTPGLSAVNVAVSAGTAATAAA   479
D. mojavensis     NNN--NNSSNNNNNSSSSSIVAASGHNP----AVQPHAPTPGLSATNIAVPAATAAAAAV   483
D. willistoni     SST--QSASGQSSSAGSAGASTASGHNP----AVQPHAPTP----VSLAAAAATAATGTG   508
D. ananassae      SISGSAAGISSSIVGNGPPTTAAIGHNS----TVQPHAPTPGFTVTNSTTSSNASTSSC-   467
                  .         .   . ..   :: *:*.    :**.:****       :..: ::

D. melanogaster   ---ATIVSSAN-----VQGQSVIQPTPTIAFAAVAKHNTSLLENGPVLQQQ--LAVTPTV   520
D. sechellia      ---ATIVSSAN-----VQGQSVIQPTPTIAFAAVAKHNTSLLENGPVLQQQ--LAVTPTV   521
D. simulans       ---ATIVSSAN-----VQGQSVIQPTPTIAFAAVAKHNTSLLENGPVLQQQ--LAVTPTV   521
D. erecta         ---APIVSSAN-----VQGPSVIQSTPTIAFAAVAKHNTSLLENGPVLQQQ--LAVTPTV   521
D. yakuba         ---ATIVSSAN-----VQGPSVIQPTPTIAFAAVAKHNTSLLENGPVLQQQ--LAVTPTV   520
D. persimilis     A-TPAIASISN-----VQGQSVIQATPTIAFAAVAKHNTSLLENGPVLQQQ--PIAT-TV   529
D. pseudobscura   A-TPAIASISN-----VQGQSVIQATPTIAFAAVAKHNTSLLENGPVLQQQ--PIAT-TV   529
D. grimshawi      AAAAAVASASNNSSNNIQGQSVIQATPTIAFAAVAKHNTSLLENGPVLQQQQ--PTAPTV   535
D. virilis        AAAAAAASASN-STNNIQGQSVIQATPTIAFAAVAKHNTSLLENGPVLQQQQ--ATAPTV   536
D. mojavensis     AAVAATASASN-SSNNIQGQSVIQATPTIAFAAVAKHNTSLLENGPVSQQQ---ATAPTV   539
D. willistoni     SGSAATNTSSS----NIQGQSVIQAIPTIAFAAVAKHNTSLLENGPVLHQPQQQATAPTV   564
D. ananassae      ------LTNSS-----VQGSTVIPVAPTIAFAAVAKNNTSLLENSPALQQPQQQTTVPTV   516
                         : :.     :** :**   ***********:*******.*. :*     .. **
```

FIGURE 3: Continued.

```
D. melanogaster   AAIVGAGTQAQQKHVPP--------LSNLQTNSPHIQNGLPVSD---------STNDNSC   563
D. sechellia      AAIVGAGTQAQQKHVPP--------LSNLQTNSPHIQNGLPVSD---------SNNDNSC   564
D. simulans       AAIVGAGTQAQQKHVPP--------LSNLQTNSPHIQNGLPVSD---------SNNDNSC   564
D. erecta         AAIVGAGTQAQQKHVQP--------LSNLQTNSPHIQNGLPVSD---------SNNDTSC   564
D. yakuba         AAIVGSGTQAQQKHVPP--------LSNLQTNSPHIQNGLPVSD---------SNNDTSC   563
D. persimilis     AAIVGAGAQTQQPQHQQQQ---PAQLSNLQTNSSHLQNGLPVSV---------SSSDSNS   577
D. pseudobscura   AAIVGAGAQTQQPQHQQQQ---PAQLSNLQTNSSHLQNGLPVSV---------SSSDSNS   577
D. grimshawi      AAIVSASAQAQQQQQQQQQ--QAAQLSNLQTNSSHLQNGLPVSLSSSSSGSSNSSSNNNS   593
D. virilis        AAIVSASAQAQQQQQQQ----QAAQLSNLQTNSSHMQNGLPVSLSSSSSGSSNSSTNNNS   592
D. mojavensis     AAIVSANAQAQQQQQQQQ---QAAQLSNLQTNSSHMQNGLPASLSSSSSGSSNSSSNNNS   596
D. willistoni     ASIVGAGAQSQQQQQQQQQQQQATQLSNLQTNSSHIQNGLPVSIGSSGS----SSENSNS   620
D. ananassae      AAIVGAGPQTLQQQQQPQ----AAQLSNLQTNSSHIQNGLSVSD---------GNSDNNN   563
                  *:**.:..*: * :           ********.*:****..*          .. :..

D. melanogaster   NVVDTISLKTMAQDAINRSAIDPNSLNQQ-----QTSSIDLRQPQSQK------------   606
D. sechellia      NVVDTISLKTMAQDAINRSAIDPNSLNQQ-----QTSSIDLRQPQSQK------------   607
D. simulans       NVVDTISLKTMAQDAINRSAIDPNSLNQQ-----QTSSIDLRQPQSQK------------   607
D. erecta         -TVDTISLKTMAQDAINRSAIDTISLNQQ-----QTSSIDLRQQQSQK------------   606
D. yakuba         NVVDTISLKTMAQDAINRSAIDTNSLTQQ-----HTSSIDLRQQQSQK------------   606
D. persimilis     NVVDAISLKTMAQEAINRSAIDVNSLSQQPQTQQQQSNIDTRQQQSQQ------------   625
D. pseudobscura   NVVDAISLKTMAQEAINRSAIDVNSLSQQPQTQQQQSNIDTRQQQSQQ------------   625
D. grimshawi      NAVDAISLKTIAQEAINRSVIEPNSLNQQ-----QASNIDTRQQQQQQQQQQQQQQQQQQ   648
D. virilis        NAVDAISLKTMAQEAINRSVIEPNSLNQQ-----QASNIDARQQQQQQQQQQQQQQQQPQ   647
D. mojavensis     NAVDAISLKTMAQEAINRSVIEPNSLNQQ-----QASNIDTRQQQQQQQQQQQAQQQAAQ   651
D. willistoni     NAVESISLKTMAQEAINRSVIDNTSLNPQ-----QQQ--QSQQQQAASNIDTRQQQQTTQ   673
D. ananassae      SIIDAISLKSMAQEAINRSAIDTTTLIQQ-----QPNNMDTRQSQSQQPSSQQ-------   611
                    :::****::**:*****.*:   :*  *     :  .  :  :* *   .

D. melanogaster   -------------SLLQ-HFNSETNTNQQQ-LTSQQQ---------QQLQNNSLAATTGS   642
D. sechellia      -------------SLLQ-HFNSETNTNQQQ-LTSQQQ---------QQLQNNSLAATTGS   643
D. simulans       -------------SLLQ-HFNSETNTNQQQQLTSQQQ---------QQLQNNSLAATTGS   644
D. erecta         -------------SLLQ-HFNSETNTNQQQ-LTSQQQ---------QQIQNSSLATTAGS   642
D. yakuba         -------------SLLQ-HFNSETNTNQQQ-LTSQQ-----------QLQNNSLSAVAVS   640
D. persimilis     -------------SLLQQHFSTENTANQPQQVASQQQ---------QNAAAAAAAAAAAA   663
D. pseudobscura   -------------SLLQQHFSTENTANQPQQVASQQQ---------QNAAAAAAAAAAAA   663
D. grimshawi      QQQAAQQQAAAQQSLLQQHFSTETTANQQQQQQ-------------LQQQATAAAAQQQQ   695
D. virilis        QQQAAP------QSLLQQHFSTETTANQQQQQQQQQQQQQQQQLQQQQQQATAAAAAQQQ   701
D. mojavensis     Q------------SLLQQHFSTETTAKQQQQQQLQ-----------QQQAATAAAAAAAQ   688
D. willistoni     Q------------TLLQQHFSTETTANQQQQQQ-------------AAAAAAAAAAAQQQ   708
D. ananassae      -------------SLMQQHFNTENTANQLQQHQVPPQLQN------VGASGTLISSVVAV   652
                               :*:* **.:*..::* *                        ::

D. melanogaster   -------------------------NNGTSTGSGLMNVANA-TGQPISGNAK--------   668
D. sechellia      -------------------------NNGPSTGSGLMNVANA-TGQPISGNAK--------   669
D. simulans       -------------------------NNGPSTGSGLMNVANA-TGQPISGNAK--------   670
D. erecta         -------------------------NNGPSTGSGLINVSNA-TGQPISGNAK--------   668
D. yakuba         -------------------------NNGPSTGSGLMNVANA-TGQPISGNVK--------   666
D. persimilis     AAA-------AAMGTIAAAAHAAAASNGSTTGSGLMNLANA-AGQPTSGSAK--------   707
D. pseudobscura   AAA-------AAMGTIAAAAHAAAASNGSTTGSGLMNLANA-AGQPTSGSAK--------   707
D. grimshawi      QQN----AAAAAAAVAAIGTIVSATSNGPTAGSGLMNLANA-AGQPTSGSGG--------   742
D. virilis        QQQQQQQNAAAAAAVAAIGTIVAAASNGPTAGSGLMNLANA-AGQPSTGNGGVVVAANAK   760
D. mojavensis     QHN------------------------GPTAGSGLMNLANA-AGQPTAVSGGVVVTANAK   723
D. willistoni     QQQ-----QQNAAAAAAVAAIAAASSNGPTAGSGLMNLANATAGQPPSGGNS-------K   756
D. ananassae      THG-------------------ATASNGPTAG--LINVTNA-AGQPTSSNTK--------   682
                                             *.::*  *:*::** :*** :  .

D. melanogaster   ---------------THACQPQQMATTEAHIPTLLGVTPLGPTPLQKEHQMQFQMMEAAY   713
D. sechellia      ---------------THTCQPQQMATTEAHIPTLLGVTPLGPTPLQKEHQMQFQMMEAAY   714
D. simulans       ---------------THTCQPQQMATTEAHIPTLLGVTPLGPTPLQKEHQMQFQMMEAAY   715
D. erecta         ---------------THTCQPQQMATTEAHIPTLLGVTPLGPTPLQKEHQMQFQMMEAAY   713
D. yakuba         ---------------THTCQPQQMATTEAHIPTLLGVTPLGPTPLQKEHQMQFQMMEAAY   711
D. persimilis     ---------------THTSQ-QQMATTEAHIPTLLGVTPLGPTPLQKEHQVQFQMMEAAY   751
D. pseudobscura   ---------------THTSQ-QQMATTEAHIPTLLGVTPLGPTPLQKEHQVQFQMMEAAY   751
D. grimshawi      ---------------QQQQQQQQIATTEAHIPPLLGVTPLGPTPLQKEHQLQFQMMEAAY   787
D. virilis        THTAQQQQQQQQQQQQQQQQQQQIATTEAHIPPLLGVTPLGPTPLQKEHQLQFQMMEAAY   820
D. mojavensis     THTAQQQQQQQQ---QQQQQQQQMATTEAHIPPLLGVTPLGPTPLQKEHQLQFQMMEAAY   780
D. willistoni     TPQQQQQQQQQQ---QLQQQQQQQQQQEAHIPPLLGVTPLGPLPLQKEHQMQFQMMEAAY   813
D. ananassae      ----------------PHSHQQQLATTEAHIPTLLGVTPLGPTPLQKEHQVQFQMMEAAY   726
                                    : **   *****.********* *******:*********
```

Figure 3: Continued.

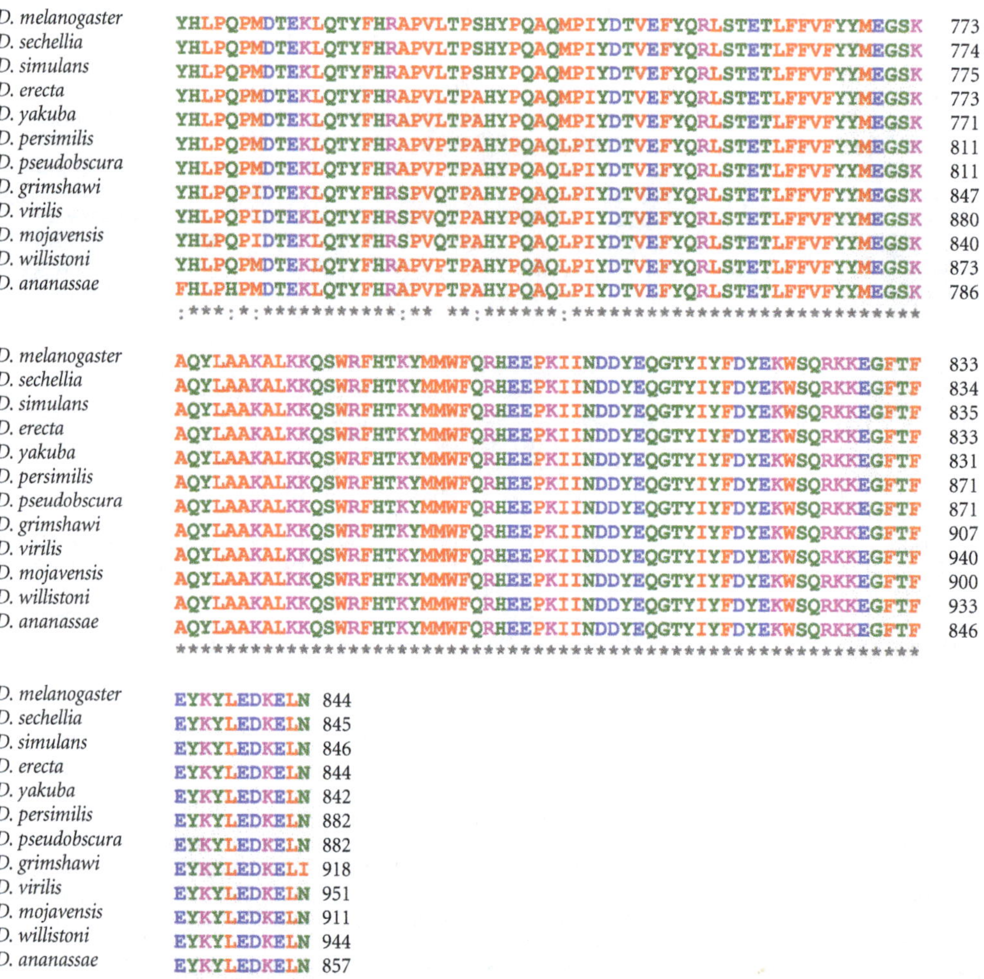

FIGURE 3: Not3/5 homologs from 12 *Drosophila* species. Clustal W [1, 2] was used to align sequences extracted from FlyBase. Amino acid (aa) color-coding is from Clustal W: red, small aliphatic, hydrophobic, and aromatics; blue, acidic; magenta, basic; green, hydroxyl, sulphydryl, amine, and glycine; grey, unusual aa. Symbols for aa conservation are from Clustal W: (asterisk $*$): positions with a single, fully conserved residue. (Colon :): conservation between groups of strongly similar properties-scoring >0.5 in the Gonnet PAM 250 matrix. (Period .): conservation between groups of weakly similar properties-scoring ≤0.5 in the Gonnet PAM 250 matrix.

may have therefore escaped detection in the *Drosophila* two-hybrid screen. Coimmunoprecipitation studies from tissue extracts and the precise mapping of the interaction domains on both proteins will be required to resolve this issue.

3.5. Multiple Bic-C Isoforms. Drosophila Bic-C has three predicted mRNA isoforms, RA, RB, and RD, that encode two identical (RA and RB) and one shorter (RD) proteins lacking the first 120 aa (Figure 1). These mRNA isoforms are expressed at different times during development (FlyBase): *Bic-C-RA* is expressed in the early embryo (0–6 hrs old) and in the adult female (i.e., most likely in the ovary), and *Bic-C-RB* is found mostly in late embryogenesis (7–22 hrs old). This is also consistent with our earlier report of multiple protein isoforms [14]. During the larval phases *Bic-C* is undetectable, and during pupation *Bic-C* expression is resumed, with its *RD* isoform being the most abundant and remaining prominent in adult males (FlyBase). The presence of two distinct mRNAs encoding the same amino acid sequence at definite developmental stages also suggests the possibility that they may be subjected to distinct regulation(s) in different tissues or at different developmental times and that the Bic-C activity may be required in specific time windows. This is consistent with a report that *Bic-C* function is especially needed at embryonic day (*E*) 18.5 during mouse development [27].

Interestingly, the mouse *Bicc1* gene and human *BICC1* also produce two distinct mRNAs by alternative splicing, which differ for the presence of exon 21 [26, 48] although no further functional information is known to date, so it is difficult to speculate if the presence of multiple Bic-C isoforms has conserved functional roles.

```
Magnaportae oryzae     -MAARKLQQEVDKCFKKVAEGVAEFESIYE--KIEQSSNISQK-----EK   42
N. crassa              -MAARKLAQEVDKCFKKVAEGVQEFEAIYE--KIEQSNNPAQK-----DK   42
S. pombe               -MIA----------FYLHLEKIAIFDEVYE--KLSASNSVSQK-----EK   32
D. melanogaster        MAATRKLQGEIDRCLKKVAEGVETFEDIWK--KVHNATNTNQKQKHLQEK   48
A. gambiae             QNVSSVFAGEIDRCLKKVTEGVETFEDIWQ--KVHNATNSNQK---VCEK   45
M. musculus            MADKRKLQGEIDRCLKKVSEGVEQFEDIWQ--KLHNAANANQK-----EK   43
R. norvegicus          MADKRKLQGEIDRCLKKVSEGVEQFEDIWQ--KLHNAANANQK-----EK   43
H. sapiens             MADKRKLQGEIDRCLKKVSEGVEQFEDIWQ--KLHNAANANQK-----EK   43
P. troglodytes         MADKRKLQGEIDRCLKKVSEGVEQFEDIWQ--KLHNAANANQK-----EK   43
C. familiaris          MADKRKLQGEIDRCLKKVSEGVEQFEDIWQ--KLHNAANANQK-----EK   43
B. taurus              MADKRKLQGEIDRCLKKVSEGVEQFEDIWQ--KLHNAANANQK-----EK   43
D. rerio               MADKRKLQGEIDRCLKKVAEGVEQFEDIWK--KLHNAANANQK-----EK   43
C. elegans             MAEKRKLLAEIDKCFKKIDEGVELFEETME--KMHEANSDNQR-----DK   43
A. thaliana            MGASRKLQGEIDRVLKKVQEGVDVFDSIWNKWNVYDTDNVNQK-----EK   45
O. sativa japonica     MGASRKLQGEIDRVLKKVQEGVDVFDSIWN--KVYDTENANQK-----EK   43
                                   :   * : *:  :  :: : . *:    :*
```

```
Magnaportae oryzae     YEDQLKREIKKLQRLRDQIKTWAASNDIKDK-------APLLENRRKIET   85
N. crassa              LEDNLKREIKKLQRLRDQIKTWAASNDIKDK-------APLLEHRRLIET   85
S. pombe               LEGDLKTQIKKLQRLRDQIKTWASSNDIKDK-------KALLENRRLIEA   75
D. melanogaster        YEADLKKEIKKLQRLRDQIKSWIASAEIKDK-------SSLLENRRLIET   91
A. gambiae             YEADLKKEIKKLQRLRDQIKSWIASGEIKDK-------SALLENRRLIET   88
M. musculus            YEADLKKEIKKLQRLRDQIKTWVASNEIKDK-------RQLIENRKLIET   86
R. norvegicus          YEADLKKEIKKLQRLRDQIKTWVASNEIKDK-------RQLIENRKLIET   86
H. sapiens             YEADLKKEIKKLQRLRDQIKTWVASNEIKDK-------RQLIDNRKLIET   86
P. troglodytes         YEADLKKEIKKLQRLRDQIKTWVASNEIKDK-------RQLIDNRKLIET   86
C. familiaris          YEADLKKEIKKLQRLRDQIKTWVASNEIKDK-------RQLIDNRKLIET   86
B. taurus              YEADLKKEIKKLQRLRDQIKTWVASNEIKDK-------RQLIDNRKLIET   86
D. rerio               YEADLKKEIKKLQRLRDQIKTWVASNEIKDK-------RQLVENRKLIET   86
C. elegans             YQDDLKKEIKKLQRLRDQVKNWQNASEIKDK-------DKLNSYRKLIEQ   86
A. thaliana            FEADLKKEIKKLQRYRDQIKTWIQSSEIKDKKVSASYEQSLVDARKLIEK   95
O. sativa japonica     FEADLKKEIKKLQRYRDQIKTWIQSSEIKDK-------KALMDARKQIER   86
                     : :** :****** ***:*.*   : :****          * . *: **
```

```
Magnaportae oryzae     QMERFKAVEKAMKTKAYSKEGLSAAAKLDPKEQAKAEASEFLGNMIDTLE   135
N. crassa              QMEKFKAVEKAMKTKAYSKEGLSAAAKLDPKEQAKLEAGEFLSQMVDELE   135
S. pombe               KMEEFKAVEREMKIKAFSKEGLSIASKLDPKEKEKQDTIQWISNAVEELE   125
D. melanogaster        QMERFKVVERETKTKAYSKEGLGAAQKMDPAQRIKDDARNWLTSSISSLQ   141
A. gambiae             QMERFKVVERETKTKAYSKEGLGAAQKMDPAQREKEEISTWLTSSITSLQ   138
M. musculus            QMERFKVVERETKTKAYSKEGLGLAQKVDPAQKEKEEVGQWLTNTIDTLN   136
R. norvegicus          QMERFKVVERETKTKAYSKEGLGLAQKVDPAQKEKEEVGQWLTNTIDTLN   136
H. sapiens             QMERFKVVERETKTKAYSKEGLGLAQKVDPAQKEKEEVGQWLTNTIDTLN   136
P. troglodytes         QMERFKVVERETKTKAYSKEGQGLAQKVDPAQKEKEEVGQWLTNTIDTLN   136
C. familiaris          QMERFKVVERETKTKAYSKEGLGLAQKVDPAQKEKEEVGQWLTNTIDTLN   136
B. taurus              QMERFKVVERETKTKAYSKEGLGLAQKVDPAQKEKEEVGQWLTNTIDTLN   136
D. rerio               QMERFKVVERETKTKAYSKEGLGLAQKVDPAQKEKEETEQWLTNTIDTLN   136
C. elegans             RMEQFKDVERENKTKPHSKLGLSAEEKLDPKEKEKAETMDWIQHQIRSLN   136
A. thaliana            EMERFKICEKETKTKAFSKEGLGQQPKTDPKEKAKSETRDWLNNVVSELE   145
O. sativa japonica     EMERFKVCEKETKTKAFSKEGLGQQPKTDPKEKAKAETRDWLNNVVSDLE   136
                       .**.**   *:   * *..** *  .    * ** :: * :    ::   :  *:
```

```
Magnaportae oryzae     LQIEALEAEAEQIQATV-----KKGKIQGA----KAERMANIEQIIERHK   176
N. crassa              QQIETLEAESESIQATM-----KRGKGHGA----KADRISEIERIIERHK   176
S. pombe               RQAELIEAEAESLKATF-----KRGKKDLS----KLSHLSELESRIERHK   166
D. melanogaster        IQIDQYESEIESLLAG------KKKRLDRD----KQERMDDLRGKLDRHK   181
A. gambiae             IQIDQFECEVESLLAG------KKKKLDKD----KQDKMDELKGKLERHK   178
M. musculus            MQVDQFESEVESLSVQT-----RKKKGDKD----KQDRIEGLKRHIEKHR   177
R. norvegicus          MQVDQFESEVESLSVQT-----RKKKGDKD----KQDRIEGLKRHIEKHR   177
H. sapiens             MQVDQFESEVESLSVQT-----RKKKGDKD----KQDRIEGLKRHIEKHR   177
P. troglodytes         MQVDQFESEVESLSVQT-----RKKKGDKD----KQDRIEGLKRHIEKHR   177
C. familiaris          MQVDQFESEVESLSVQT-----RKKKGDKDQ---KQDRIEGLKRHIEKHR   178
B. taurus              MQVDQFESEVESLSVQT-----RKKKGDKD----KQDRIEGLKRHIEKHR   177
D. rerio               MQVDQFESEVESLSVQT-----RKKKGDKE----KQDRIEELKRLIERHR   177
C. elegans             EEVDRTEMQLESLSNTDTGKGKRGKKEDAKTKNEREKRVEGLKHHLERIN   186
A. thaliana            SQIDSFEAELEGLSVKK----------GKT----RPPRLTHLETSITRHK   181
O. sativa japonica     NQIDNFEAEVEGLSIKK----------GKQ----RPPRLVHLEKSITRHK   172
                        : :  * : * :                      :  ::  :.  : : .
```

Figure 4: Continued.

```
Magnaportae oryzae     WHQGKLELIRRSLENGGVDTEQVTD-IEENIRYYVSDGMQDDFMDDD-TL 224
N. crassa              WHQGKLELIRRSLENGGVETEQVNE-LEESIRYYVTDGMNEDFMDDE-GI 224
S. pombe               WHQDKLELIMRRLENSQISPEAVND-IQEDIMYYVECSQSEDFAEDE-NL 214
D. melanogaster        FHITKLETLLRLLDNDGVEAEQVNK-IKDDVEYYIDSSQEPDFEENE-FI 229
A. gambiae             FHVTKLETLLRMLDNDGVEVEQIKK-IKEDVEYYIDSSQEPDFEENE-YI 226
M. musculus            YHVRMLETILRMLDNDSILVDAIRK-IKDDVEYYVDSSQDPDFEENE-FL 225
R. norvegicus          YHVRMLETILRMLDNDSILVDAIRK-IKDDVEYYVDSSQDPDFEENE-FL 225
H. sapiens             YHVRMLETILRMLDNDSILVDAIRK-IKDDVEYYVDSSQDPDFEENE-FL 225
P. troglodytes         YHVRMLETILRMLDNDSILVDAIRK-IKDDVEYYVDSSQDPDFEENE-FL 225
C. familiaris          YHVRMLETILRMLDNDSILVDAIRK-IKDDVEYYVDSSQDPDFEENE-FL 226
B. taurus              YHVRMLETILRMLDNDSILVDAIRK-IKDDVEYYVDSSQDPDFEENE-FL 225
D. rerio               YHIRMLETILRMLDNDSIQVDAIHK-IKDDVEYYIDSSQDPDFEENE-FL 225
C. elegans             FHIEKLEICMRMISNESLNAKMVLETLKEPIETYVEMMNEEDSEEADNYD 236
A. thaliana            DHIIKLELILRLLDNDELSPEQVND-VKDFLDDYVERNQDDFDEFSDVDE 230
O. sativa japonica     AHIKKLESILRLLDNDELSPEQVND-VKDFLDDYVERNQEDFDEFSDVEE 221
                        *   **   * :.*  :   . : . ::: :  *:     .      :

Magnaportae oryzae     YDDLALGEEEDAYGMNQDNDKGSSQDAQSVHEDSLEDTRPTPPAPVAKPR 274
N. crassa              YDDLNLEEEEDAYGMNVDNDKGSSQDAQSIQDEPEPEPKPAS-VPATKQR 273
S. pombe               YDELNLDEASASY-----DAERSGRSSSSSHSPSPSASSSSSSSENLLQDK 259
D. melanogaster        YDDIIGLDEVELSGTATTDSNNSNETSGSPSSVTSGGSPSQSPVTVQQIL 279
A. gambiae             YDDIIGLDDVEISGNFFVFRNNSNETAGSPSSLISGTSPAQSPVLN---Y 273
M. musculus            YDDLD-LEDIPQALVATSPPSHSHMEDEIFNQSSSTPTSTTSSSPIPPSP 274
R. norvegicus          YDDLD-LEDIPQALVATSPPSHSHMEDEIFNQSSSTPTSTTSSSPIPPSP 274
H. sapiens             YDDLD-LEDIPQALVATSPPSHSHMEDEIFNQSSSTPTSTTSSSPIPPSP 274
P. troglodytes         YDDLD-LEDIPQALVATSPPSHSHMEDEIFNQSSSTPTSTTSSSPIPPSP 274
C. familiaris          YDDLD-LEDIPQALVATSPPSHSHMEDEIFNQSSSTPTSTTSSSPIPPSP 275
B. taurus              YDDLD-LEDIPQALVATSPPSHSHMEDEIFNQSSSTPTSTTSSSPIPPSP 274
D. rerio               YDDLD-LEDIP---------------------TSNGTGTGASIGLLGSSP 253
C. elegans             PDDAYDELNLEKLCQQIGGVNVASVDDEHRENGHELGIDTAESGAVSGSR 286
A. thaliana            LYSTLPLDEVEGLEDLVTAGP--LVKGTP-------LSMKSSLAASASQV 271
O. sativa japonica     LYSTLPMEKVEALEDMVSLAPSSLVKGVASVSTTAVLSTKSSVATSPTQA 271
                         .

Magnaportae oryzae     AAAVEATVAAGRRPSTQMKSPLPTLATLHT-PLPTISNGSSSSAGMKPAP 323
N. crassa              TPADTVAASSIRRSSAQLKSPLPTLATVHNNTMPSISNTPASNVSMKPAS 323
S. pombe               AEAEEKVSADASVQDIAEKESLDADKELATNDQEDDEEENQAETQKDGAI 309
D. melanogaster        NTSSQGAASSGSSAASAALFQQQLTAAQSNGNNVGYASDTSAASSSATTS 329
A. gambiae             SASTLHNHSSDLSADNNNLNEKR---SKSEGTKITVTKTTRMLPRRYPPC 320
M. musculus            ANCTTENSEDDKKRGRSTDSEVSQSPAKNGSKPVHSNQHPQSPAVPPTYP 324
R. norvegicus          ANCTTENSEDDKKRGRSTDSEVSQSPAKNGSKPVHSNQHPQSPAVPPTYP 324
H. sapiens             ANCTTENSEDDKKRGRSTDSEVSQSPAKNGSKPVHSNQHPQSPAVPPTYP 324
P. troglodytes         ANCTTENSEDDKKRGRSTDSEVSQSPAKNGSKPVHSNQHPQSPAVPPTYP 324
C. familiaris          ANCTTENSEDDKKRGRSTDSEVSQSPAKNGSKPVHSNQHPQSPALPPSYP 325
B. taurus              ANCTTENSEDDKKRGRSTDSEVSQSPAKNGSKPVHSSQHPQSPAVPPSYP 324
D. rerio               GHGTLTGGILNLVQGQS---------ALQGS--------TQVPVSPVGTA 286
C. elegans             HTSG-ENGQPPSPAGRRIVPLSMPSPHAVTPELKRLASKDSNVDRPRTPP 335
A. thaliana            RSISLP--THHQEKTEDTSLPDSSAEMVPKTPPPKNGAG--LHSAPSTPA 317
O. sativa japonica     TVSAAPSLSVSQDQAEETASQESNPESAPQTPPSKVGSQPSVPVVPTTIS 321

Magnaportae oryzae     APTRPAGEGLKYAS------------------------------------ 337
N. crassa              LPTRPA-EGLKYAS------------------------------------ 336
S. pombe               SNNENMQSEVQTTNP----------------------------------- 324
D. melanogaster        TDPAGGTVAVNCVGGLADKRNKSSESNALKLKPQPHQLIKPTPVRATAKL 379
A. gambiae             WCYRSRPTVYRSSGPLLLPLQNNIPVSIFEWKRERERKKMRTLCVHMKEI 370
M. musculus            SGPPPTTSALSSTPGNNGASTPAAPTSALGPKASPAP------------- 361
R. norvegicus          SGPPPATSALSSTPGNNGASTPAAPPSALGPKASPAP------------- 361
H. sapiens             SGPPPAASALSTTPGNNGVPAPAAPPSALGPKASPAP------------- 361
P. troglodytes         SGPPPAASALSTTPGNNGVPA----------------------------- 345
C. familiaris          PGPPPATSALSTTPGNNGASTPAAPTSALGPKASPAP------------- 362
B. taurus              PGPPPAASALSATPGSNGAPAAAAPASALGAKASPAP------------- 361
D. rerio               PGGGTGESGLGGNGSSSGVSG----------------------------- 307
C. elegans             VTPASAAPPPPGIPYNSVAAG----------------------------- 356
A. thaliana            GGRPSLNVPAGNVSN---------TSVTLSTSIPTQTSIESMG------- 351
O. sativa japonica     TSTAAVSVSAETISSPVRPIVPTTTAAVLPASVTARSAPENIP------- 364
```

Figure 4: Continued.

```
Magnaportae oryzae   --------------------------------------------------
N. crassa            --------------------------------------------------
S. pombe             --------------------------------------------------
D. melanogaster      PLSSDTQVNKIVSSTPSKNQQ-QLPTAASIVATSAMQSQSSIGSCSSTGG   428
A. gambiae           ALLLSTGYWSCVALMDSFFSLSFLLENGSILQPSTPTTGAGASSASSTSG   420
M. musculus          ---------SHNSGTPAPYAQAVAPPNASGPSNAQPRPPSAQPSGGSGGG   402
R. norvegicus        ---------SHNSGTPAPYAQAVAPPNASGPSNAQPRPPSAQPSGGSGGG   402
H. sapiens           ---------SHNSGTPAPYAQAVAPPAPSGPSTTQPRPPSVQPSGGGGGG   402
P. troglodytes       --------------------------------------------------
C. familiaris        ---------SHSSGTPAPYAQAVAPPAPSGSSTTQPRPPSVQPG------   397
B. taurus            ---------SHSAGTPAPYAQAVAPPAPSGPPSAQPRPPSAQPGAGSGGG   402
D. rerio             ---------------------------GVGTNVAPARPPS----------   320
C. elegans           ------------------------------RSTTTPVPSTP---------   367
A. thaliana          ---------SLSPVAA------KEEDATTLPSRKPPSSVADTPL-RGIGR   385
O. sativa japonica   ---------AVTSAPANSSSTLKDDDNMSFPSRRSSPAVTEIGLGRGITR   405

Magnaportae oryzae   ---------------------AAAAAAASDKNNVGIAPLPPPPGA-----   361
N. crassa            ---------------------AAAAAAASDKSGVGIAPLPPPPTT-----   360
S. pombe             ---------------------SASTSAVTNITKPTLIQNPSTPLS-----   348
D. melanogaster      TGASQSASSGNNP-GNNPAVQPNAPTPGQSGIAAAAASTNVVSAT-----   472
A. gambiae           PLQTQAPNSSNIPPGQNSMLLHNALSSASSTESNNHVMSTSSAST-----   465
M. musculus          SGGSSSN---SNSGTGGGAGKQNGATSYSSVVADSPAEVTLSSSG-----   444
R. norvegicus        SGGSSSN---SNSGTGGGAGKQNGATSYSSVVADSPAEVALSSSG-----   444
H. sapiens           SGGGGSSS-SSNSSAGGGAGKQNGATSYSSVVADSPAEVALSSSG-----   446
P. troglodytes       --------------------------RYSSVVADSPAEVALSSSG-----   364
C. familiaris        ------------------AGKQNGATSYSSVVADSPAEVALSSSG-----   424
B. taurus            GNSGG----------GGGAGKQNGATSYSSVVADSPAEAALSSTG-----   437
D. rerio             ------------------GLKQNGATSYSAVVADNTPDSSLSSAS-----   347
C. elegans           -------------------ISANSPAPSLAQAAPIAAASPVFPPA-----   393
A. thaliana          VGIPNQPQPSQPPSPIPANGSRISATSAAEVAKRNIMGVESNVQP-----   430
O. sativa japonica   -GLTSQGLGSAPISIGPVSGN-GSVSALTDLSKRNMLNTDERINSGGISQ   453

Magnaportae oryzae   -APVSTISPQAKASAANSPIVMAAQPA-----------------------   387
N. crassa            -NSSLPASQHVKTSAANSPSVATVQP------------------------   385
S. pombe             -VSNSKVASPETPNATHTAPKVEMRYA-----------------------   374
D. melanogaster      -IVSS-ANVQGQSVIQPTPTIAFAAVAKHNTSLLENGPVLQQQLAVTPTV   520
A. gambiae           -ISSSGANVINNCVSPSNSAVITAFSSNFGFSLCPLFPVFVFVVVLT-TL   513
M. musculus          -GSSASSQALGPTSGPHNPAPSTSKES-----------------------   470
R. norvegicus        -GSSASSQALGPTSGPHNPAPSTLKES-----------------------   470
H. sapiens           -GNNASSQALGPPSGPHNPPPSTSKEP-----------------------   472
P. troglodytes       -GNNASSQGLGPPSGPHNPPPRTSKEP-----------------------   390
C. familiaris        -GSGASSQALGPPSGPHNPPPSTSKEP-----------------------   450
B. taurus            -GSSTGSQALGPPPGPHNPPPSTAKEP-----------------------   463
D. rerio             -QSQNS----------HSSSSSSSTNQ-----------------------   363
C. elegans           -AAAASKPVLAQSVSEMPQKKESITST-----------------------   419
A. thaliana          -LTSPLSKMVLPP-TAKGNDGTASDSNPGDVAASIG-RAFSPSIVSGSQW   477
O. sativa japonica   QLISPLGNKAQPQQVLRTTDTISSDSSNTNESTVLGGRIFSPPVVSGVQW   503

Magnaportae oryzae   ----VSAASQPQTQPPATAASPVKIENAKPASSRSTGKAPATSNASASES   433
N. crassa            ----VAQERIVNAVLPAVGGS---VTNTPVPS-----KTEPAKNVSSRDK   423
S. pombe             ----SAAAAAAAALAKESPSHHYIMQQVRPETP----NSPRLNSTVIQSK   416
D. melanogaster      AAIVGAGTQAQQKHVPPLSNLQTNSPHIQNGLPVSDSTNDNSCNVVDTIS   570
A. gambiae           SPTSSPTFTPYTHPKHNDAVLCNTCVCVCVLAHVNDSLMLFPCSFSCSLV   563
M. musculus          ----STAAPSGAGNVASGSGNNSGGPSLLVPLPVNPPSSPTPSFSEAKAA   516
R. norvegicus        ----STAAPSGAGSVASGSGNNSGGPSLLVPLPVNPPSSPTPSFSEAKAA   516
H. sapiens           ----SAAAPTGAGGVAPGSGNNSGGPSLLVPLPVNPPSSPTPSFSDAKAA   518
P. troglodytes       ----SAAAPTGAGGVAPXSRNNSRRPNLLVPLPVNPPSSPTPSFSDAKAA   436
C. familiaris        ----SAAAPAGAGGVAPGSGNNTGGPSLLVPLPVNPPSSPTPSFSEAKAA   496
B. taurus            ----SATAPVGAGGVAPGSGNNAGGPSLLVPLPVNPPSSPTPSFNEAKAA   509
D. rerio             ----TLDN---------------GPSLLSSITL-PPSSPSPAFTDSTPG   392
C. elegans           ---TSRGSAAAPATTTTTTTTTTSSEPAEVPLVVQQTVSETFVNGVDSPA   466
A. thaliana          RP----GSPFQSQNETVRGRTEIAPDQREKFLQRLQQVQQGHGNLLGIPS   523
O. sativa japonica   RPQNTAGLQNQSEAGQFCGRPEISADQREKYLQRLQQVQQ-QGSLLNVSH   552
```

Figure 4: Continued.

```
Magnaportae oryzae     SEAAGKASSSKSRKGALGEASNQSSA------------------------   459
N. crassa              ASAPVPAATATTSKATPEPEAVKTQP------------------------   449
S. pombe               WDSLGHTASPKMQTQPVR--SVSQSS------------------------   440
D. melanogaster        LKTMAQDAINRSAIDPNSLNQQQTSSIDL---------------------   599
A. gambiae             HLGVAQEAG----PVPSNQTPQPQSGGGG---------------------   588
M. musculus            G-TLLNGPPQFS-TTPEIKAPEPLSS------------------------   540
R. norvegicus          G-TLLNGPPQFS-TTPEIKAPEPLSS------------------------   540
H. sapiens             G-ALLNGPPQFS-TAPEIKAPEPLSS------------------------   542
P. troglodytes         G-ALLNGPPQFS-TAPEIKAPEPLSS------------------------   460
C. familiaris          G-ALLNGPPQFS-TAPEIKAPEPLSS------------------------   520
B. taurus              G-SLLNGPPQFS-AAPEIKAPEPLSS------------------------   533
D. rerio               GGSLLNGPHSYTPNTEAIKAPEPPSS------------------------   418
C. elegans             AATRLTQQERQQ----QLQQQHHHQS------------------------   488
A. thaliana            LSGGNEKQFSSQQQNPLLQQSSSISPHGSLGIGVQAPGFNVMSSASLQQQ   573
O. sativa japonica     ITGISQKQFPSQQPNPLLQQFNSQSSS-------------ISSQAGIG--   587
                                                .

Magnaportae oryzae     ------------TSSHTNGVTNGVKSK-----------AKGKSGKGQPQ-   485
N. crassa              ------------QVPQTNGATNGIKP------------------------   463
S. pombe               ------------ATTETN-----VKP------------------------   449
D. melanogaster        -----RQPQSQKSLLQHFNSETNTNQQQLTSQQQQQLQNNSLAATTGSN-   643
A. gambiae             -----GGGAGQSLMVDASGVPAGAN----AGNNLLPTSSATAAITNGPN-   628
M. musculus            ----------LKSMAERAAISSGIED-----------PVPTLHLTDRD--   567
R. norvegicus          ----------LKSMAERAAISSGIED-----------PVPTLHLTDRD--   567
H. sapiens             ----------LKSMAERAAISSGIED-----------PVPTLHLTERD--   569
P. troglodytes         ----------LKSMAERAAISSGIED-----------PVPTLHLTERD--   487
C. familiaris          ----------LKSMAERAAISSGIED-----------PVPTLHLTERD--   547
B. taurus              ----------LKSMAERAAISSGIED-----------PVPTLHLTERD--   560
D. rerio               ----------LKAMAERAALGLALDG-----------EIPSLHLTDRDS-   446
C. elegans             -----------TIIPTTPTTTTTSSS-----------MLGGMMSTDDPA-   515
A. thaliana            SNAMSQQLGQQPSVADVDHVRNDDQSQ---QNLPDDSASIAASKAIQSED   620
O. sativa japonica     -------LG-QVQVPESGHTKSEEQQQSFAEDVSVESVATAGANKHMSED   629
                                               .

Magnaportae oryzae     -------------VQAQEEPAEEESIYHLPA------------------S   504
N. crassa              ----------------IEEVEEEESIYHLPA------------------S   479
S. pombe               -----------------TKEENADVPVSSPD------------------Y   464
D. melanogaster        -------------NGTSTGSGLMNVANATGQPISGNAKTHACQPQQMATT   680
A. gambiae             -------------TIINTNSSISNAANVNSAGGGGGGGPGGMKP-SAGTH   664
M. musculus            --------------IILSSTSAPP-TSSQPP---------------LQLS   587
R. norvegicus          --------------IILSSTSAPP-TSSQPP---------------LQLS   587
H. sapiens             --------------IILSSTSAPP-ASAQPP---------------LQLS   589
P. troglodytes         --------------IILSSTSAPP-ASAQPP---------------LQLS   507
C. familiaris          --------------IILSSTSAPP-ASAQPP---------------LQLS   567
B. taurus              --------------ILLSSTSAPP-ASAQPP---------------LQLS   580
D. rerio               -------------LELFSGSSAPPGPTTAPQ---------------PAVS   468
C. elegans             -------------AALQAALNMAAASQQAVA---------------TGPK   537
A. thaliana            DSKVLFDTPSGMPSYMLDPVQVSSGPDFSPGQPIQPGQSSSSLGVIGRRS   670
O. sativa japonica     DTKIPFSNPS---ASITEGTQLSRDPDLPAGQPLQPGMSSSGVGVIGRRS   676

Magnaportae oryzae     LQDLVDSYEMSK--KRPAQANSTSTLRAMSHSQANLPDLTDAEAPSSYQP   552
N. crassa              LQDLVESYEVTK--KCPASVDALATQRMHAVAVANKPSALDTELPRPYYP   527
S. pombe               LKDLVNALNTSKE-QHKGAIDKEKLTEALNISCVYVPDATDAAKPQYYIP   513
D. melanogaster        EAHIPTLLGVTPLGPTPLQKEHQMQFQMMEAAYYHLPQPMDTEKLQTYFH   730
A. gambiae             EACIPPLLGVAPLGTSKLQKEHQIQFQLMEAAYYHLPTPSDSERLRPYLQ   714
M. musculus            EVNIPLSLGVCPLGPVSLTKEQLYQQAMEEAAWHHMPHPSDSERIRQYLP   637
R. norvegicus          EVNIPLSLGVCPLGPVSLTKEQLYQQAMEEAAWHHMPHPSDSERIRQYLP   637
H. sapiens             EVNIPLSLGVCPLGPVPLTKEQLYQQAMEEAAWHHMPHPSDSERIRQYLP   639
P. troglodytes         EVNIPLSLGVCPLGPVPLTKEQLYQQAMEEAAWHHMPHPSDSERIRQYLP   557
C. familiaris          EVNIPLSLGVCPLGPVPLTKEQLYQQAMEEAAWHHMPHPSDSERIRT-FP   616
B. taurus              EVNIPLSLGVCPLGPVPLTKEQLYQQAMEEAAWHHMPHPSDSERIRQYLP   630
D. rerio               EVSLPPSLGACPLGPTPLTKEQLYQQAMQEAAWTHMPHPSDSERIRQYLM   518
C. elegans             RAHIPAWLGASPLGRTSMTQEFDGQLAALELACAKATFPLDSEKPRNYLS   587
A. thaliana            NSELGAIGDPSAVG---PMHDQMHNLQMLEAAFYKRPQPSDSERPRPYSP   717
O. sativa japonica     VSDLGAIGDNLSVASASTSHDLLYNLQMLEAAFHRLPQPKDSERVKNYIP   726
                          :                :          :    .   *:
```

Figure 4: Continued.

```
Magnaportae oryzae  EVRVQSSSEYPQELLPIFSDVRLYNRLD-----TDTLFYIFYYKQGTYQQ  597
N. crassa           DVRYHTHNQFPQEPLAIFEDPRLYQRID-----PDTLFYVFYYKQGTYQQ  572
S. pombe            KDPYPVPHYYPQQPLPLFDSSEMTELVD-----PDTLFYMFYYRPGTYQQ  558
D. melanogaster     RAPVLTPSHYPQAQMPIYDTVEFYQRLS-----TETLFFVFYYMEGSKAQ  775
A. gambiae          RQPVQTPPHYPQQQLPHSETVEFFQRLS-----PETLFFVFYYMEGTKAQ  759
M. musculus         RNPCPTPPYHHQMPPPHSDTVEFYQRLS-----TETLFFIFYYLEGTKAQ  682
R. norvegicus       RNPCPTPPYHHQMPPPHSDTVEFYQRLS-----TETLFFIFYYLEGTKAQ  682
H. sapiens          RNPCPTPPYHHQMPPPHSDTVEFYQRLS-----TETLFFIFYYLEGTKAQ  684
P. troglodytes      RNPCPTPPYHHQMPPPHSDTVEFYQRLS-----TETLFFIFYYLEGTKAQ  602
C. familiaris       RNPCPTPPYHHQMPPPHSDTVEFYQRLS-----TETLFFIFYYLEGTKAQ  661
B. taurus           RNPCPTPPYHHQMPPPHSDTVEFYQRLS-----TETLFFIFYYLEGTKAQ  675
D. rerio            RNPCPTPPFHHQMPPHHSDSIEFYQRLS-----TETLFFIFYYLEGTKAQ  563
C. elegans          KVSFPVPSWYGQTAPNTSDSLEYYLRLA-----PDTLFFIFYYMEGTRAQ  632
A. thaliana         RNPAITPQTFPQTQAPIINNPLLWERLGSDAYGTDTLFFAFYYQQNSYQQ  767
O. sativa japonica  KHPAVTPASFPQIQAPVVSNPAFWERMGGDSLSTDLLFFAFYYQQNTYQQ  776
                             . *      .       :      .: **: ***  .:  *

Magnaportae oryzae  YLAAKALKEQSWRFHKQYQTWFQRHEEPKNITEEFE--------------  633
N. crassa           YLAAKALKDQSWRFHKQYQTWFQRHEEPKSITEEFE--------------  608
S. pombe            YIAGQELKKQSWRFHKKYTTWFQRHEEPKMITDEFE--------------  594
D. melanogaster     YLAAKALKKQSWRFHTKYMMWFQRHEEPKIINDDYE--------------  811
A. gambiae          YLAAKALKKQSWRFHTKYMMWFQRHEEPKVINEEYE--------------  795
M. musculus         YLAAKALKKQSWRFHTKYMMWFQRHEEPKTITDEFE--------------  718
R. norvegicus       YLAAKALKKQSWRFHTKYMMWFQRHEEPKTITDEFE--------------  718
H. sapiens          YLAAKALKKQSWRFHTKYMMWFQRHEEPKTITDEFE--------------  720
P. troglodytes      YLAAKALKKQSWRFHTKYMMWFQRHEEPKTITDEFEQIPDHLLVHSLTAF  652
C. familiaris       YLAAKALKKQSWRFHTKYMMWFQRHEEPKTITDEFE--------------  697
B. taurus           YLAAKALKKQSWRFHTKYMMWFQRHEEPKTITDEFE--------------  711
D. rerio            YLSAKALKKQSWRFHTKYMMWFQRHEEPKTITDEFE--------------  599
C. elegans          LLAAKALKKLSWRFHTKYLTWFQRHEEPKQITDDYE--------------  668
A. thaliana         YLAAKELKKQSWRYHRKFNTWFQRHKEPKIATDEYE--------------  803
O. sativa japonica  FLSARELKKQSWRFHRKYNTWFQRHVEPQVTTDEYE--------------  812
                     ::.: **. ***:* ::  ***** **:  .:::*

Magnaportae oryzae  ---QGTYRFFDYEST--------WMNRRKADFKFAYKFLEDEV-----  665
N. crassa           ---QGTYRFFDYEST--------WMNRRKADFKFTYKFLEDEV-----  640
S. pombe            ---SGSYRYFDFEGD--------WVQRKKADFRFTYQYLEDDDDWTR-  630
D. melanogaster     ---QGTYIYFDYEK---------WSQRKKEGFTFEYKYLEDKELN---  844
A. gambiae          ---QGTYIYFDYEK---------WGQRKKEGFTFEYKYLEDRDLN---  828
M. musculus         ---QGTYIYFDYEK---------WGQRKKEGFTFEYRYLEDRDLQ---  751
R. norvegicus       ---QGTYIYFDYEK---------WGQRKKEGFTFEYRYLEDRDLQ---  751
H. sapiens          ---QGTYIYFDYEK---------WGQRKKEGFTFEYRYLEDRDLQ---  753
P. troglodytes      SPGQGTYIYFDYEK---------WGQRKKEGFTFEYRYLEDRDLQ---  688
C. familiaris       ---QGTYIYFDYEK---------WGQRKKEGFTFEYRYLEDRDLQ---  730
B. taurus           ---QGTYIYFDYEK---------WGQRKKEGFTFEYRYLEDRDLQ---  744
D. rerio            ---QGTYIYFDYEK---------WGQRKKEGFTFEYRYLEDRDLQ---  632
C. elegans          ---QGTYVYFDFEK---------WSQRKKESFTFEYKFLEDKEFD---  701
A. thaliana         ---QGAYVYFDFQTPKDENQEGGWCQRIKNEFTFEYSYLEDELVV---  845
O. sativa japonica  ---RGSYVYFDFHVIDDGTG-SGWCQRIKNDFTFEYNFLEDELSVQTN  856
                        *:* :**:.          * :* *  * * * :***
```

Figure 4: Not3/5 orthologs. Clustal W [1, 2] was used to align sequences extracted from the NCBI database. Amino acid (aa) color-coding is from Clustal W: red, small aliphatic, hydrophobic, and aromatics; blue, acidic; magenta, basic; green, hydroxyl, sulphydryl, amine, and glycine; grey, unusual aa. Symbols for aa conservation are from Clustal W: (asterisk $*$): positions with a single, fully conserved residue. (Colon :): conservation between groups of strongly similar properties-scoring >0.5 in the Gonnet PAM 250 matrix. (Period .): conservation between groups of weakly similar properties-scoring ≤0.5 in the Gonnet PAM 250 matrix.

3.6. Bic-C and Polycystic Kidney Disease. In humans, two polycystic kidney disease (PKD) forms are caused by mutations in the *PKD1* and *PKD2* genes (autosomal, dominant [49–55]) or in PKHD1 (autosomal recessive, [52–55]). The link between Bic-C malfunction and PKD is compelling: two mouse models developing polycystic kidneys harbor mutations of the *Bicc1* gene [56]; Bic-C inactivation in *Xenopus* induces cystic kidneys [27, 57]; recently, a zebrafish model of PKD was validated that inhibits the Bicc1 function [58]. Finally, human studies on patients with renal disorders identified two mutations associated with the *BICC1* gene: one affecting the first KH domain and the other affecting the SAM domain [48], proving the relevance of the *Bic-C* animal models for understanding the etiology of this incurable disease.

In 3D cultures of mouse IMCD cells, depleting Bicc1 disrupts cadherin-mediated cell adhesion, normal epithelial polarization, proliferation, and apoptosis that prevent tubulomorphogenesis *in vitro* [59]. Interestingly, aspects of the *Drosophila* phenotype also affect cell migration and may influence cell-cell interaction and polarization. For example, migration of the follicle cells (FCs) in the ovary is defective in Bic-C mutant [3], resulting in eggs that remain open at the anterior end. This defect may occur because of inefficient communication between germ line and somatic cells, although to date we do not know the molecular pathway underlying this phenomenon (for an alternate possibility, see also Section 3.7).

In a recent paper [27] Tran and colleagues report that in a novel *Biccl*$^{-/-}$ mutant mice and in *Xenopus* depleted for Biccl the *Pkd2* mRNA and its cognate protein are downregulated (29 and 54%, resp.), while both *Pkd1* and *Pdhd1* levels are unaffected. In the mouse these effects are clearest specifically at stage E18.5. The regulation appears to be mediated via a cellular microRNA, *miR-17* [27] that is also amplified in certain cancers [60]. Here Biccl may relieve the *miR-17*-mediated repression via a mechanism that does not involve regulation of the polyadenylation state of at least the mRNAs tested and may mildly impact mRNA stability [27]. The fact that the Biccl protein may bind multiple mRNAs and that it may be involved in the possible antagonistic regulation of the *miR-17* complexes, also assembled on multiple mRNAs, reinforces the view that the Bic-C orthologs are central to the regulation of many cellular processes and that many more aspects of their function await elucidation.

3.7. Other Bic-C Functions. Another hint to Bic-C function comes again from *Drosophila*, where the *Bic-C* mutants exhibit disrupted pattern of the cortical filamentous actin in the growing oocyte and abnormal actin-containing structures in the ooplasm that trap both the dorsal fate determinant Gurken [61–63] and other proteins that would normally be secreted [31, 32]. This function requires Trailer hitch, a protein originally identified in a screen for mutants for axial polarity that may regulate expression of endoplasmic reticulum (ER) exit site components on the ER surface. A malfunctioning secretory pathway could affect communication between the oocyte and the overlying FC and may affect their migration. Since many mRNAs involved in vesicular trafficking and/or organization of the actin cytoskeleton were also recovered in Bic-C immunoprecipitates [15], it is possible that their posttranscriptional control may contribute to the observed Bic-C defects. Lastly, and not mutually exclusive, the altered actin dynamics exhibited by the *Bic-C* and *Tral* mutants must also add to the observed inhibition of the normal dumping of nurse cell contents into the nascent oocyte during late oogenesis.

4. Concluding Remarks

Bic-C is an ancient protein conserved from *Drosophila* to man. Its mutation induces a pleiotropic phenotype. In fruit flies the Bic-C protein binds to RNAs involved in establishing the embryonic polarity, the Wnt pathway, actin dynamics and results in many observed defects, including abnormal development. In the vertebrates the better characterized aspect of lack of Bic-C function is the induction of cystic kidneys and the alteration of cell proliferation and three dimensional organization; however, defects in pancreatic and liver function and heterotaxia (i.e., randomization of the left-right symmetry) of the visceral organs have also been observed [26, 27]. Further, effects on the Wnt pathway have also been reported in human patients with renal displasia [48], as well as in mice and frogs [26]. Biccl is also expressed in the nervous system [58] which suggests that there may be novel aspects of its function ready to be discovered and that Bic-C homologs may be involved in fundamental, evolutionarily conserved mechanisms of determination of polarity, from establishment of the body axes to planar cell polarity.

The experimental evidence so far also suggests that Bic-C function may also be required at specific times of development in many species. Since Bic-C is a negative regulator of translation, we can expect at least part of the mutant phenotypes to be linked with inappropriate spatial and/or temporal regulation of gene expression. Further, Bic-C has multiple mRNA targets, and it exists in multiple isoforms in many organisms. At least in the case of one of the Bic-C interacting partners, the CCR4 deadenylase, it is proposed that multiple forms of this complex exist in higher vertebrates [47], as there are documented isoforms for a few of the complex subunits. Therefore, it is possible that the Bic-C-CCR4-dependent regulation acts via and is regulated by combinatorial mechanisms, with variant complexes having partially redundant function. This could also explain why all the individual molecular effects/phenotypes described for Bic-C tend to be mild and why years of concerted experimental efforts have yielded only a few proven targets for this gene, since many of the real targets would presumably not have been highly enriched compared to the controls.

References

[1] M. A. Larkin, G. Blackshields, N. P. Brown et al., "Clustal W and Clustal X version 2.0," *Bioinformatics*, vol. 23, no. 21, pp. 2947–2948, 2007.

[2] M. Goujon, H. McWilliam, W. Li et al., "A new bioinformatics analysis tools framework at EMBL-EBI," *Nucleic Acids Research*, vol. 38, pp. W695–W699, 2010.

[3] M. Mahone, E. E. Saffman, and P. F. Lasko, "Localized Bicaudal-C RNA encodes a protein containing a KH domain, the RNA binding motif of FMR1," *The Embo Journal*, vol. 14, no. 9, pp. 2043–2055, 1995.

[4] R. D. Finn, J. Mistry, J. Tate et al., "The Pfam protein families database," *Nucleic Acids Research*, vol. 38, supplement 1, pp. D211–D222, 2010.

[5] M. Ashburner, P. Thompson, J. Roote et al., "The genetics of a small autosomal region of *Drosophila* melanogaster containing the structural gene for alcohol dehydrogenase—VII. Characterization of the region around the snail and cactus loci," *Genetics*, vol. 126, no. 3, pp. 679–694, 1990.

[6] P. Lasko, "Posttranscriptional regulation in *Drosophila* oocytes and early embryos," *Wiley Interdisciplinary Reviews: RNA*, vol. 2, no. 3, pp. 408–416, 2011.

[7] S. F. Altschul, T. L. Madden, A. A. Schäffer et al., "Gapped BLAST and PSI-BLAST: a new generation of protein database search programs," *Nucleic Acids Research*, vol. 25, no. 17, pp. 3389–3402, 1997.

[8] P. McQuilton, S. E. St Pierre, and J. Thurmond, "FlyBase 101—the basics of navigating flyBase," *Nucleic Acids Research*, vol. 40, no. 1, pp. D706–D714, 2012.

[9] J. Mohler and E. F. Wieschaus, "Dominant maternal-effect mutations of *Drosophila* melanogaster causing the production of double-abdomen embryos," *Genetics*, vol. 112, no. 4, pp. 803–822, 1986.

[10] T. Schupbach and E. Wieschaus, "Germline autonomy of maternal-effect mutations altering the embryonic body pattern of *Drosophila*," *Developmental Biology*, vol. 113, no. 2, pp. 443–448, 1986.

[11] N. V. Grishin, "KH domain: one motif, two folds," *Nucleic Acids Research*, vol. 29, no. 3, pp. 638–643, 2001.

[12] K. Nakel, S. A. Hartung, F. Bonneau, C. R. Eckmann, and E. Conti, "Four KH domains of the C. elegans Bicaudal-C ortholog GLD-3 form a globular structural platform," *RNA*, vol. 16, no. 11, pp. 2058–2067, 2010.

[13] C. A. Kim and J. U. Bowie, "SAM domains: uniform structure, diversity of function," *Trends in Biochemical Sciences*, vol. 28, no. 12, pp. 625–628, 2003.

[14] E. E. Saffman, S. Styhler, K. Rother, W. Li, S. Richard, and P. Lasko, "Premature translation of *oskar* in oocytes lacking the RNA-binding protein Bicaudal-C," *Molecular and Cellular Biology*, vol. 18, no. 8, pp. 4855–4862, 1998.

[15] J. Chicoine, P. Benoit, C. Gamberi, M. Paliouras, M. Simonelig, and P. Lasko, "Bicaudal-C Recruits CCR4-NOT deadenylase to target mRNAs and regulates Oogenesis, Cytoskeletal organization, and its own expression," *Developmental Cell*, vol. 13, no. 5, pp. 691–704, 2007.

[16] D. J. Bouvrette, S. J. Price, and E. C. Bryda, "K homology domains of the mouse polycystic kidney disease-related protein, Bicaudal-C (Bicc1), mediate RNA binding *in vitro*," *Nephron Experimental Nephrology*, vol. 108, no. 1, pp. e27–e34, 2008.

[17] B. M. Lunde, C. Moore, and G. Varani, "RNA-binding proteins: modular design for efficient function," *Nature Reviews*, vol. 8, no. 6, pp. 479–490, 2007.

[18] A. J. Peterson, M. Kyba, D. Bornemann, K. Morgan, H. W. Brock, and J. Simon, "A domain shared by the Polycomb group proteins Scm and ph mediates heterotypic and homotypic interactions," *Molecular and Cellular Biology*, vol. 17, no. 11, pp. 6683–6692, 1997.

[19] F. Qiao and J. U. Bowie, "The many faces of SAM," *Science's STKE*, vol. 2005, no. 286, article re7, 2005.

[20] M. J. Knight, C. Leettola, M. Gingery, H. Li, and J. U. Bowie, "A human sterile alpha motif domain polymerizome," *Protein Science*, vol. 20, no. 10, pp. 1697–1706, 2011.

[21] T. Chen, B. B. Damaj, C. Herrera, P. Lasko, and S. Richard, "Self-association of the single-KH-domain family members Sam68, GRP33, GLD-1, and Qk1: role of the KH domain," *Molecular and Cellular Biology*, vol. 17, no. 10, pp. 5707–5718, 1997.

[22] M. Di Fruscio, T. Chen, S. Bonyadi, P. Lasko, and S. Richard, "The identification of two *Drosophila* K homology domain proteins: KEP1 and SAM are members of the Sam68 family of GSG domain proteins," *Journal of Biological Chemistry*, vol. 273, no. 46, pp. 30122–30130, 1998.

[23] C. R. Eckmann, S. L. Crittenden, N. Suh, and J. Kimble, "GLD-3 and control of the mitosis/meiosis decision in the germline of *Caenorhabditis elegans*," *Genetics*, vol. 168, no. 1, pp. 147–160, 2004.

[24] T. Aviv, Z. Lin, S. Lau, L. M. Rendl, F. Sicheri, and C. A. Smibert, "The RNA-binding SAM domain of Smaug defines a new family of post-transcriptional regulators," *Nature Structural Biology*, vol. 10, no. 8, pp. 614–621, 2003.

[25] C. Gamberi, O. Johnstone, and P. Lasko, "*Drosophila* RNA binding proteins," *International Review of Cytology*, vol. 248, pp. 43–139, 2006.

[26] C. Maisonneuve, I. Guilleret, P. Vick et al., "Bicaudal C, a novel regulator of Dvl signaling abutting RNA-processing bodies, controls cilia orientation and leftward flow," *Development*, vol. 136, no. 17, pp. 3019–3030, 2009.

[27] U. Tran, L. Zakin, A. Schweickert et al., "The RNA-binding protein bicaudal C regulates *polycystin* 2 in the kidney by antagonizing miR-17 activity," *Development*, vol. 137, no. 7, pp. 1107–1116, 2010.

[28] J. Schultz, C. P. Ponting, K. Hofmann, and P. Bork, "SAM as a protein interaction domain involved in developmental regulation," *Protein Science*, vol. 6, no. 1, pp. 249–253, 1997.

[29] C. R. Eckmann, B. Kraemer, M. Wickens, and J. Kimble, "GLD-3, a bicaudal-C homolog that inhibits FBF to control germline sex determination in *C. elegans*," *Developmental Cell*, vol. 3, no. 5, pp. 697–710, 2002.

[30] L. Wang, C. R. Eckmann, L. C. Kadyk, M. Wickens, and J. Kimble, "A regulatory cytoplasmic poly(A) polymerase in *Caenorhabditis elegans*," *Nature*, vol. 419, no. 6904, pp. 312–316, 2002.

[31] J. M. Kugler, J. Chicoine, and P. Lasko, "Bicaudal-C associates with a trailer hitch/Me31B complex and is required for efficient gurken secretion," *Developmental Biology*, vol. 328, no. 1, pp. 160–172, 2009.

[32] M. J. Snee and P. M. Macdonald, "Bicaudal C and trailer hitch have similar roles in *gurken* mRNA localization and cytoskeletal organization," *Developmental Biology*, vol. 328, no. 2, pp. 434–444, 2009.

[33] M. K. Jaglarz, M. Kloc, W. Jankowska, B. Szymanska, and S. M. Bilinski, "Nuage morphogenesis becomes more complex: two translocation pathways and two forms of nuage coexist in *Drosophila* germline syncytia," *Cell and Tissue Research*, vol. 344, no. 1, pp. 169–181, 2011.

[34] M. J. Snee and P. M. Macdonald, "Dynamic organization and plasticity of sponge bodies," *Developmental Dynamics*, vol. 238, no. 4, pp. 918–930, 2009.

[35] M. Wilsch-Bräuninger, H. Schwarz, and C. Nüsslein-Volhard, "A sponge-like structure involved in the association and transport of maternal products during *Drosophila* oogenesis," *Journal of Cell Biology*, vol. 139, no. 3, pp. 817–829, 1997.

[36] M. Olszewska, J. J. Bujarski, and M. Kurpisz, "P-bodies and their functions during mRNA cell cycle: mini-review," *Cell Biochemistry and Function*, vol. 30, no. 3, pp. 177–182, 2012.

[37] R. Parker and U. Sheth, "P bodies and the control of mRNA translation and degradation," *Molecular Cell*, vol. 25, no. 5, pp. 635–646, 2007.

[38] C. Y. Chen and A. B. Shyu, "AU-rich elements: characterization and importance in mRNA degradation," *Trends in Biochemical Sciences*, vol. 20, no. 11, pp. 465–470, 1995.

[39] M. Tucker, R. R. Staples, M. A. Valencia-Sanchez, D. Muhlrad, and R. Parker, "Ccr4p is the catalytic subunit of a Ccr4p/Pop2p/Notp mRNA deadenylase complex in Saccharomyces cerevisiae," *The Embo Journal*, vol. 21, no. 6, pp. 1427–1436, 2002.

[40] M. Tucker, M. A. Valencia-Sanchez, R. R. Staples, J. Chen, C. L. Denis, and R. Parker, "The transcription factor associated Ccr4 and Caf1 proteins are components of the major cytoplasmic mRNA deadenylase in *Saccharomyces cerevisiae*," *Cell*, vol. 104, no. 3, pp. 377–386, 2001.

[41] J. Chen, Y. C. Chiang, and C. L. Denis, "CCR4, a 3′-5′ poly(A) RNA and ssDNA exonuclease, is the catalytic component of the cytoplasmic deadenylase," *The Embo Journal*, vol. 21, no. 6, pp. 1414–1426, 2002.

[42] S. Thore, F. Mauxion, B. Séraphin, and D. Suck, "X-ray structure and activity of the yeast Pop2 protein: A nuclease subunit of the mRNA deadenylase complex," *EMBO Reports*, vol. 4, no. 12, pp. 1150–1155, 2003.

[43] M. C. Daugeron, F. Mauxion, and B. Séraphin, "The yeast POP2 gene encodes a nuclease involved in mRNA deadenylation," *Nucleic Acids Research*, vol. 29, no. 12, pp. 2448–2455, 2001.

[44] J. Chen, J. Rappsilber, Y. C. Chiang, P. Russell, M. Mann, and C. L. Denis, "Purification and characterization of the 1.0 MDa CCR4-NOT complex identifies two novel components of the complex," *Journal of Molecular Biology*, vol. 314, no. 4, pp. 683–694, 2001.

[45] U. Oberholzer and M. A. Collart, "Characterization of NOT5 that encodes a new component of the Not protein complex," *Gene*, vol. 207, no. 1, pp. 61–69, 1998.

[46] C. J. A. Sigrist, L. Cerutti, E. de Castro et al., "PROSITE, a protein domain database for functional characterization and annotation," *Nucleic Acids Research*, vol. 38, pp. D161–D166, 2010.

[47] C. Temme, S. Zaessinger, S. Meyer, M. Simonelig, and E. Wahle, "A complex containing the CCR4 and CAF1 proteins is involved in mRNA deadenylation in *Drosophila*," *The Embo Journal*, vol. 23, no. 14, pp. 2862–2871, 2004.

[48] M. R. Kraus, S. Clauin, Y. Pfister et al., "Two mutations in human BICC1 resulting in Wnt pathway hyperactivity associated with cystic renal dysplasia," *Human Mutation*, vol. 33, no. 1, pp. 86–90, 2012.

[49] T. C. Burn, T. D. Connors, W. R. Dackowski et al., "Analysis of the genomic sequence for the autosomal dominant polycystic kidney disease (PKD1) gene predicts the presence of a leucine-rich repeat: the American PKD1 consortium (APKD1 Consortium)," *Human Molecular Genetics*, vol. 4, no. 4, pp. 575–582, 1995.

[50] J. Hughes, C. J. Ward, B. Peral et al., "*The polycystic kidney disease* 1 (PKD1) gene encodes a novel protein with multiple cell recognition domains," *Nature Genetics*, vol. 10, no. 2, pp. 151–160, 1995.

[51] T. Mochizuki, G. Wu, T. Hayashi et al., "PKD2, a gene for polycystic kidney disease that encodes an integral membrane protein," *Science*, vol. 272, no. 5266, pp. 1339–1342, 1996.

[52] F. Hildebrandt, E. Otto, C. Rensing et al., "A novel gene encoding an SH3 domain protein is mutated in nephronophthisis type 1," *Nature Genetics*, vol. 17, no. 2, pp. 149–153, 1997.

[53] L. F. Onuchic, L. Furu, Y. Nagasawa et al., "*PKHD1*, the polycystic kidney and hepatic disease 1 gene, encodes a novel large protein containing multiple immunoglobulin-like plexin-transcription-factor domains and parallel beta-helix 1 repeats," *American Journal of Human Genetics*, vol. 70, no. 5, pp. 1305–1317, 2002.

[54] C. J. Ward, M. C. Hogan, S. Rossetti et al., "The gene mutated in autosomal recessive polycystic kidney disease encodes a large, receptor-like protein," *Nature Genetics*, vol. 30, no. 3, pp. 259–269, 2002.

[55] H. Xiong, Y. Chen, Y. Yi et al., "A novel gene encoding a TIG multiple domain protein is a positional candidate for autosomal recessive polycystic kidney disease," *Genomics*, vol. 80, no. 1, pp. 96–104, 2002.

[56] C. Cogswell, S. J. Price, X. Hou, L. M. Guay-Woodford, L. Flaherty, and E. C. Bryda, "Positional cloning of jcpk/bpk locus of the mouse," *Mammalian Genome*, vol. 14, no. 4, pp. 242–249, 2003.

[57] U. Tran, L. M. Pickney, B. D. Özpolat, and O. Wessely, "Xenopus bicaudal-C is required for the differentiation of the amphibian pronephros," *Developmental Biology*, vol. 307, no. 1, pp. 152–164, 2007.

[58] D. J. Bouvrette, V. Sittaramane, J. R. Heidel, A. Chandrasekhar, and E. C. Bryda, "Knockdown of bicaudal C in zebrafish (*Danio rerio*) causes cystic kidneys: a nonmammalian model of polycystic kidney disease," *Comparative Medicine*, vol. 60, no. 2, pp. 96–106, 2010.

[59] Y. Fu, I. Kim, P. Lian et al., "Loss of Bicc1 impairs tubulomorphogenesis of cultured IMCD cells by disrupting E-cadherin-based cell-cell adhesion," *European Journal of Cell Biology*, vol. 89, no. 6, pp. 428–436, 2010.

[60] M. Jovanovic and M. O. Hengartner, "miRNAs and apoptosis: RNAs to die for," *Oncogene*, vol. 25, no. 46, pp. 6176–6187, 2006.

[61] F. S. Neuman-Silberberg and T. Schupbach, "The *Drosophila* dorsoventral patterning gene gurken produces a dorsally localized RNA and encodes a TGFα-like protein," *Cell*, vol. 75, no. 1, pp. 165–174, 1993.

[62] R. P. Ray and T. Schüpbach, "Intercellular signaling and the polarization of body axes during *Drosophila* oogenesis," *Genes & Development*, vol. 10, no. 14, pp. 1711–1723, 1996.

[63] J. E. Wilhelm, M. Buszczak, and S. Sayles, "Efficient protein trafficking requires trailer hitch, a component of a ribonucleoprotein complex localized to the ER in *Drosophila*," *Developmental Cell*, vol. 9, no. 5, pp. 675–685, 2005.

Expression and Functional Analysis of Storage Protein 2 in the Silkworm, *Bombyx mori*

Wei Yu,[1,2] Meihui Wang,[1,2] Hanming Zhang,[1,2] Yanping Quan,[1,2] and Yaozhou Zhang[1,2]

[1] *Institute of Biochemistry, College of Life Sciences, Zhejiang Sci-Tech University, Hangzhou, Zhejiang 310018, China*
[2] *Zhejiang Provincial Key Laboratory of Silkworm Bioreactor and Biomedicine, Hangzhou, Zhejiang 310018, China*

Correspondence should be addressed to Wei Yu; yuwei7198@yahoo.com.cn

Academic Editor: Margarita Hadzopoulou-Cladaras

Storage protein 2 (SP2) not only is an important source of energy for the growth and development of silkworm but also has inhibitory effects on cell apoptosis. Endothelial cell (EC) apoptosis is an important contributing factor in the development of atherosclerosis; therefore, study of the antiapoptotic activity of SP2 on ECs provides information related to the treatment of atherosclerosis and other cardiovascular diseases. In this study, the *sp2* gene was cloned and expressed in *Escherichia coli* to produce a 6xHis-tagged fusion protein, which was then used to generate a polyclonal antibody. Western blot results revealed that SP2 levels were higher in the pupal stage and hemolymph of fifth-instar larvae but low in the egg and adult stages. Subcellular localization results showed that SP2 is located mainly on the cell membrane. In addition, a Bac-to-Bac system was used to construct a recombinant baculovirus for SP2 expression. The purified SP2 was then added to a culture medium for human umbilical vein ECs (HUVECs), which were exposed to staurosporine. A cell viability assay demonstrated that SP2 could significantly enhance the viability of HUVEC. Furthermore, both ELISA and flow cytometry results indicated that SP2 has anti-apoptotic effects on staurosporine-induced HUVEC apoptosis.

1. Introduction

The vascular endothelium provides a cellular interface between the circulating blood and the vascular smooth muscle of the blood vessel walls. It also has an important role in maintaining the balance of the endovascular environment [1]. Many factors, such as peroxide, ox-low density lipoprotein (LDL), angiotensin I, and tumor-necrosis-factor-(TNF-) α can induce the production of reactive oxygen species (ROS) by NADPH oxidase in endothelium and vascular smooth muscle [2], which results in oxidative stress to the endothelial cells (ECs) and, thus, the induction of cell apoptosis. EC dysfunction is a trigger factor for the development of atherosclerosis and other cardiovascular diseases [3]. Many studies have shown that EC apoptosis is one of several atherogenic factors [4, 5].

Additional studies have also shown that hemolymph from the silkworm (*Bombyx mori*) can inhibit insect and mammalian cell apoptosis that is induced by viruses and several chemical inducers, such as staurosporine, camptothecin, and actinomycin D [6–8]. As one such antiapoptotic component in silkworm hemolymph, the 30 K protein has been studied widely [6, 7]. A recent study showed that another protein in silkworm hemolymph, storage protein 2 (SP2), can also inhibit staurosporine-induced HeLa cell apoptosis and ROS generation [8]. Owing to the influence of juvenile hormone and ecdysone, SP2 is synthesized by the fat body of feeding larvae and released into the hemolymph [9]. At the end of the feeding period, it is selectively reabsorbed by the fat body cells [10]. SP2 is presumed to be used as a store of the amino acids required for the development of adult tissues [11].

In the current study, we cloned and expressed the gene encoding silkworm SP2 (*sp2*) both in a prokaryotic expression system and a silkworm baculovirus system. The expressed protein in *Escherichia coli* was used to generate polyclonal antibodies. The distribution of SP2 in different developmental stages and different tissues was detected by Western blot. In addition, the expressed protein in the silkworm baculovirus system was used to study the anti-apoptotic effects of SP2 on human umbilical vein ECs

(HUVECs) induced by peroxidation. This work will lay a foundation for the development and utilization of protein drugs from economically important insects for treatment of vascular diseases.

2. Materials and Methods

2.1. Strains, Cell Lines, Animals, and Reagents. *Escherichia coli* strains TG1 and BL21 (DE3) were grown at 37°C in LB medium. The pET-28a(+) expression vector and pFastBac HTB vector were conserved in our laboratory. The silkworm-derived cell line BmN was maintained at 27°C in TC-100 medium (Gibco, USA) supplemented with 10% fetal bovine serum (FBS). *B. mori* nuclear polyhedrosis virus (bacmid, conserved by our lab) was propagated in BmN cells. HUVECs were purchased from the American Type Culture Collection (Manassas, VA, USA) and cultured in Dulbecco's modified Eagle medium (Gibco, USA) supplemented with 10% FBS (Gibco, USA) in a humidified incubator at 37°C in an atmosphere of 5% CO_2. Fifth-instar silkworm larvae (Jingsong × Haoyue, Showa) were reared on fresh mulberry leaves at 25°C. Male New Zealand rabbits were purchased from the Animal Research Center of Hangzhou Normal University. A DNA gel purification kit and Cy3-labeled goat anti-rabbit IgG were purchased from Promega (Madison, USA). HRP-labeled goat anti-rabbit IgG was purchased from Dingguo Biotechnology (Beijing, China). Cellfectin II Reagent and a Ni-NTA Purification System were sourced from Invitrogen (Carlsbad, USA); a Cell Death Detection ELISA kit was sourced from Roche Diagnostics (Castle Hill, Australia); the Annexin V-FITC/PI apoptosis detection kit was purchased from Invitrogen (Carlsbad, USA); staurosporine was sourced from the Beyotime Company (Shanghai, China). Other reagents were locally purchased products of analytical grade.

2.2. Bioinformatics Analyses. Similarity analyses for the nucleotide and protein sequences were carried out using GenBank BLASTn and BLASTp algorithms. The hydrophobicity analysis and the isoelectric point prediction were conducted on the ExPASy website (http://www.expasy.org/). Simulation for the protein tertiary structure was generated by using the SWISS-MODEL program (http://swissmodel.expasy.org/).

2.3. Preparation of Polyclonal Antibodies. The open reading frame (ORF) of *sp2* was amplified by PCR using the cDNA library of silkworm pupa constructed by our laboratory as a template. The forward and reverse primers were as follows: P1, 5′-CGCGGATCCATGAAGTCTGTCTTAATT-3′; and P2, 5′-CCGCTCGAGTTAATTTTTTGGAACAAC-3′, which contained *Bam*HI and *Xho*I restriction sites, respectively. The *sp2* gene was subcloned into the expression vector pET-28a(+). *Escherichia coli* BL21 (DE3) was transformed with the recombinant plasmid and cultured in LB media, that included 50 μg/mL kanamycin, at 37°C until OD_{600} reached approximately 0.5. Recombinant protein expression was induced with 1 mmol/L IPTG for 4 h. The His-tagged fusion protein was extracted from the bacteria and purified by Ni^{2+}-affinity chromatography. The purified protein was subsequently concentrated and desalted by dialysis. The protein content was analyzed following the method of Bradford and then used to immunize the male New Zealand rabbits to prepare the polyclonal antibodies. An ELISA was utilized to determine the polyclonal antibody titer, and the specificity of the polyclonal antibody was detected by Western blot analysis.

2.4. Western Blot Assay for the Distribution of SP2. Western blot analysis was used to evaluate the distribution of SP2 in different tissues of the fifth-instar larvae and other developmental stages of the silkworm. Tissues from fifth-instar larvae (head, epidermis, hemolymph, fat body, silk gland, trachea, the Malpighian Tubule, and midgut) were isolated and pulverized into powder in liquid nitrogen. The powdered tissues were then resuspended in a lysis buffer (50 mM Tris-Cl, pH 8.0; 0.15 M NaCl; 5 mM EDTA; 0.5% NP-40; 1 mM dithiothreitol; 5 mg/mL sodium deoxycholate; 100 mg/L PMSF; 5 μg/mL aprotinin, Sigma), and the mixture was incubated on ice for 30 min. The homogenates were centrifuged at 12,000 ×g for 15 minutes at 4°C. The supernatants containing total proteins were equalized by assaying the protein content (following the method of Bradford). Samples were verified by running on a 12% SDS-PAGE and electrophoretically transferred onto polyvinylidene difluoride (PVDF) membranes. The membranes were blocked with 3% skim milk in phosphate-buffered saline (PBS) at 4°C overnight, and rabbit anti-SP2 antibody was added and the solution incubated at room temperature for 2 h. After washing with PBST (PBS + 0.05% Tween-20), the bound antibodies were detected by anti-rabbit IgG followed by a DAB detection system.

2.5. Subcellular Localization of SP2 in BmN Cells. BmN cells were cultured overnight on a glass coverslip that could be used specifically for confocal microscopy. After removing the culture medium, the cells were washed three times for 5 min in PBS and fixed in 4% polyformaldehyde in PBS (pH 7.4) at room temperature for 15 min, followed by permeabilization with 0.2% Triton X-100/PBS for 10 min. The fixed cells were preblocked in 3% BSA in PBS at 4°C for 4 h, followed by incubation with the anti-SP2 polyclonal antibody (diluted 1 : 1000 in blocking buffer) at room temperature for 2 h; cells were incubated simultaneously with control serum, which was obtained from the rabbits before immunization with the antigen. After three 10 min PBST washes, cells were incubated with Cy3-labeled goat anti-rabbit IgG (diluted 1 : 1000, Promega) at 37°C for 2 h and washed twice in PBST. The cells were then incubated with 4-6-diamidino-2-phenylindole (1 g/mL in PBS) at 37°C for 30 min. After the cells were washed once with PBST, they were photographed on a Nikon Eclipse TE 2000-E Confocal Microscope (Nikon, Japan). Images were analyzed using EZ-C1 software (Nikon, Japan).

2.6. Expression of SP2 in a Bac-to-Bac System. The *sp2* gene was inserted into the MCS of the transfer plasmid pFastBac-HTB between *Bam*H I and *Xho* I sites and transformed into

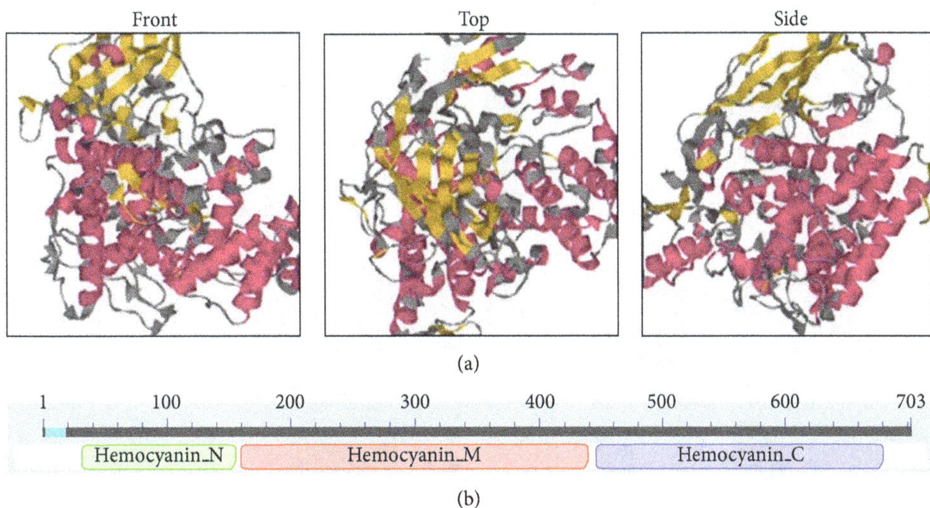

FIGURE 1: The tertiary structure prediction (a) and three conserved domains (b) of silkworm SP2 protein. (a) Coloring is defined by the dss rasmol script, to show predicted secondary structure (red: helix; yellow: strand; gray: other); (b) Three conservative structural domains of SP2: hemocyanin_N domain (amino acids 33–157), hemocyanin_M domain (amino acids 161–442), and hemocyanin_C domain (amino acids 446–680).

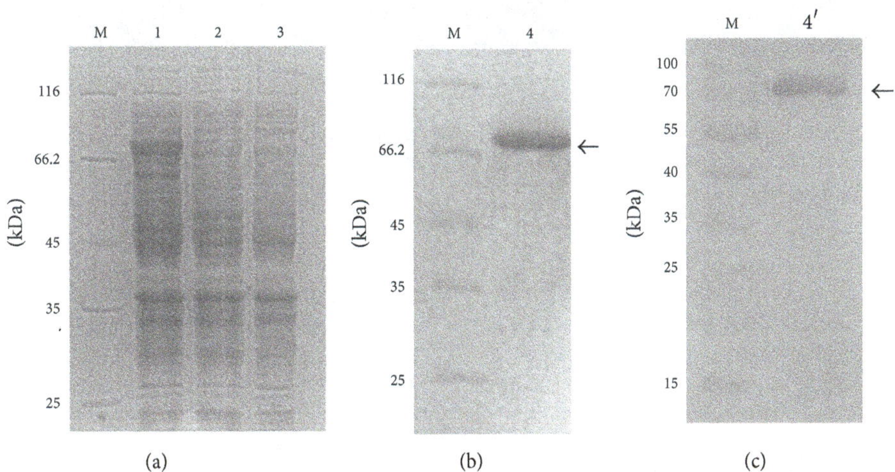

FIGURE 2: Expression (a), purification (b), and Western blot analysis (c) of the SP2 protein expressed in *Escherichia coli*. M: protein molecular weight marker; 1: recombinant bacteria after inducing; 2: recombinant bacteria before inducing; 3: empty vector after inducing; 4: purified of the fusion protein; 4′: Western blot analysis.

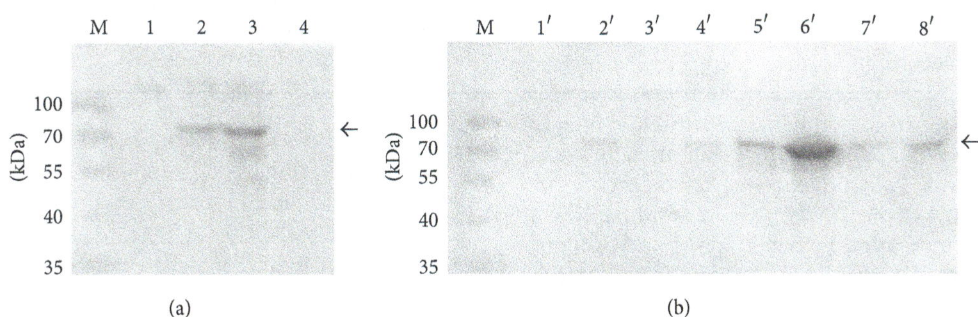

FIGURE 3: The expression level of *sp2* in the developmental stages of the silkworm *Bombyx mori* (a) and in different tissues of fifth-instar *B. mori* larva (b). 1: egg; 2: larva; 3: pupa; 4: adult; 1′: midgut; 2′: the Malpighian tubule; 3′: trachea; 4′: silk gland; 5′: fat body; 6′: hemolymph; 7′: epidermis; 8′: head. The arrow indicates the SP2 protein.

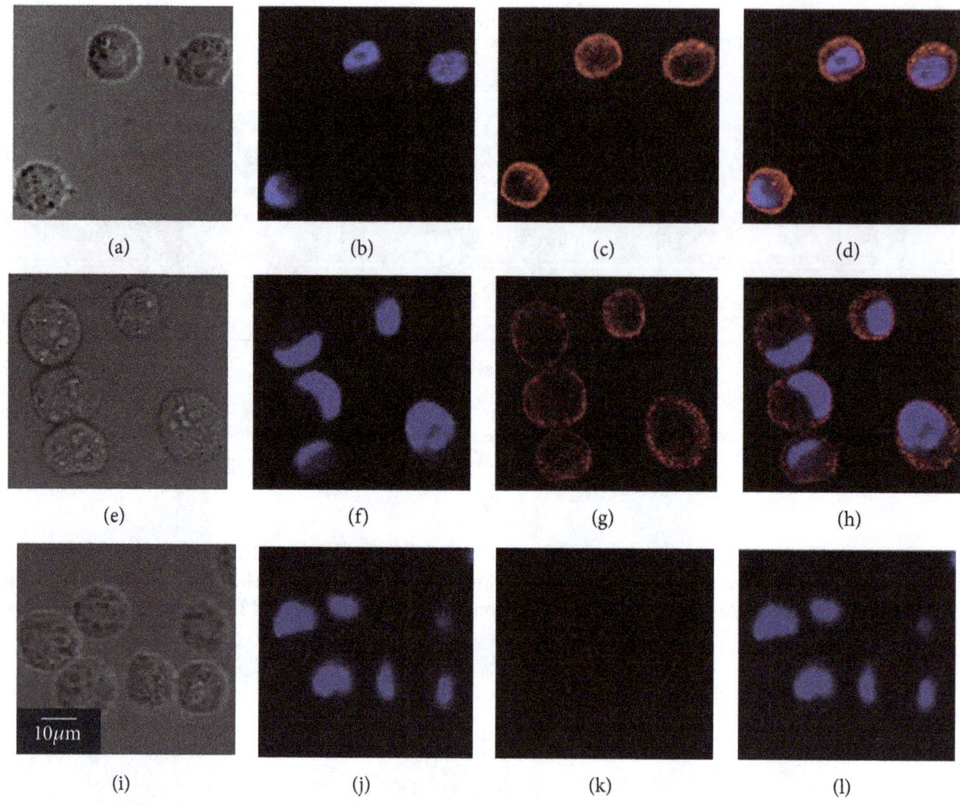

FIGURE 4: Subcellular localization of SP2 detected using Cy3-labeled goat anti-rabbit IgG and DAPI. ((a), (e), and (i)) Cells at the transmission light; ((b), (f), and (j)) nucleolus dyed with DAPI; ((c), (g), and (k)) SP2 dyed with Cy-3; ((d), (h), and (l)) merged images; ((i), (j), (k), and (l)) negative control group, using control rabbit serum.

DH10Bac cells. The *sp2* gene was then transferred into a wild type bacmid (wtbacmid) DNA by homologous recombination to construct the recombinant baculovirus bacmid-*sp2*. After white-blue plaque selection, the positive colonies were selected and analyzed by PCR with M13 universal primers, *sp2* forward and reverse primers. The recombinant bacmid was transfected into BmN cells for amplification. The third-generation virus (MOI = 10 pfu/cell) was further used to infect BmN cells (1×10^6 cells/flask) for subsequent protein expression. After cultivation for 48 h, the cells were harvested by phosphate buffer solution (PBS, pH 7.4) washes, pulsed sonication and centrifugation at 12,000 rpm for 10 min. The supernatant was subjected to Ni^{2+}-affinity chromatography according to the manufacturer's instructions (Invitrogen). SDS-PAGE and Western blot were used to detect the purified fusion protein. The wtbacmid infected BmN cells were used as a control. After purification by affinity chromatography, the SP2 protein was concentrated, and imidazole was removed by dialysis against 25 mM HEPES [4-(2-hydroxyethyl) piperazine-1-ethanesulfonic acid), Sigma]; pH 7.5, 100 mM NaCl. The protein content was analyzed by following the method of Bradford.

2.7. Induction of HUVEC Apoptosis. HUVECs were isolated by 0.25% trypsinization and cultured in Dulbecco's modified Eagle medium (Gibco) supplemented with 10% FBS and maintained in a humidified atmosphere with 5% CO_2 at 37°C air until 70–80% confluence was reached. Cells (from the experimental group) were pretreated for 24 h with culture medium containing purified SP2 protein (at a final concentration of 0.5, 1, and 2 μg/mL). Negative and normal controls did not receive any intervention. After washing twice with Hank's balanced salt solution (Gibco), the experimental and negative groups were both exposed to 2 mL STS (1 μM) for 2 h to induce apoptosis, whereas the normal group was treated with 2 mL Hank's balanced salt solution. The experimental group was then washed twice with Hank's balanced salt solution and cultured in culture medium in the presence of purified SP2 protein (at a final concentration of 0.5, 1, and 2 μg/mL). HUVEC viability and apoptosis were evaluated after 12 h.

2.8. Cell Viability Analysis Using an MTT Assay. Cell viability was estimated using MTT (3-[4,5-dimethylthiazol-2-yl]-2,5-diphenyl tetrazolium bromide; Sigma) assay. Cells were seeded into 96-well plates at densities of 1×10^3 cells/well and incubated overnight. Following the treatments as indicated above, 10 μL MTT (5 mg/mL) was added to each well and incubated for 4 h. The insoluble formazan crystals were dissolved in 200 μL/well dimethylsulfoxide, and absorbance was measured at 490 nm to calculate the relative cell viability ratio. Each experimental condition was repeated at least three times.

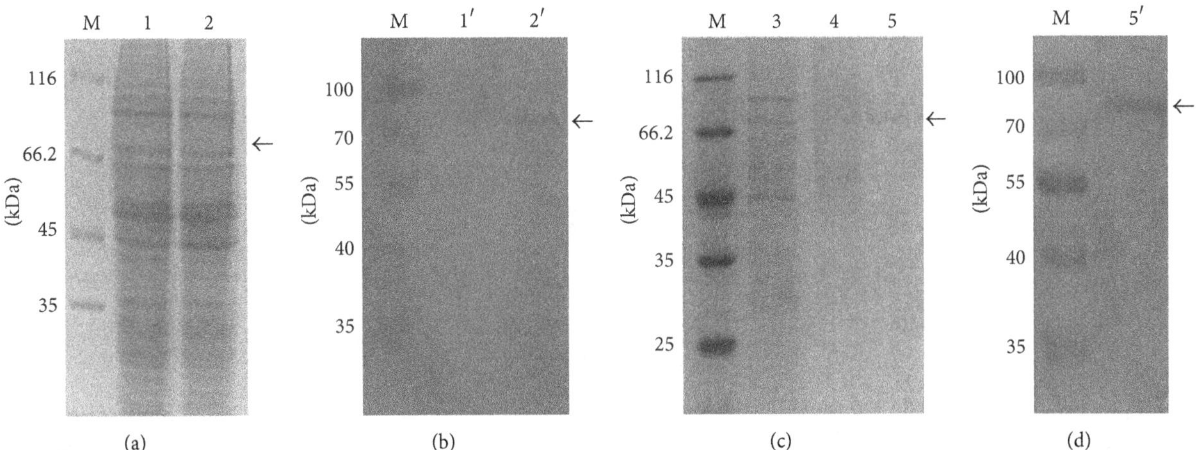

FIGURE 5: SDS-PAGE and Western blot analysis of SP2 protein expressed in BmN cells. (a) SDS-PAGE analysis of BmN cells lysates; (b) Western blot analysis of BmN cells lysates; (c) SDS-PAGE analysis of gradient purified SP2 protein; (d) Western blot analysis of purified SP2 protein; M: protein molecular weight marker; 1, 1′: total protein of BmN cells infected by wtbacmid; 2, 2′: total protein of the BmN cells infected by vBm-*sp2*; 3: protein that discharges out from the column; 4: eluted protein with 20 mM imidazole; 5: eluted protein with 250 mM imidazole; 5′: Western blot analysis of the purified fusion protein.

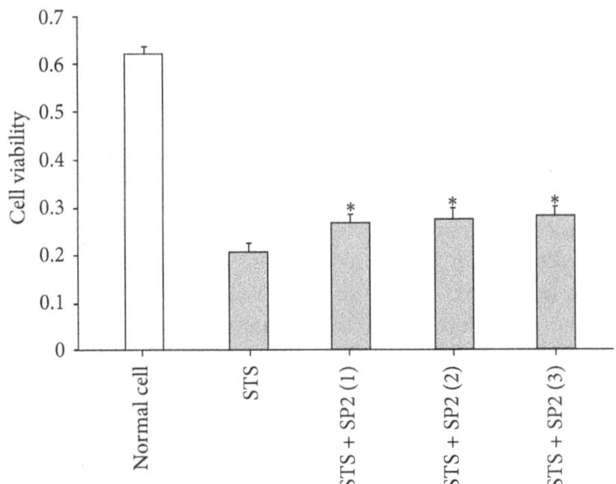

FIGURE 6: The effect of SP2 on the viability of HUVEC. The vertical axis is the 490 nm absorbance; STS is the negative control; STS + SP2 (1–3) are the treatment groups, in which the final concentrations of the SP2 sample were 0.5 μg/mL, 1 μg/mL, and 2 μg/mL, respectively. Data are mean ± SEM, $n = 3$. *Indicates values significantly different from negative control using Student's t-test; $P < 0.05$.

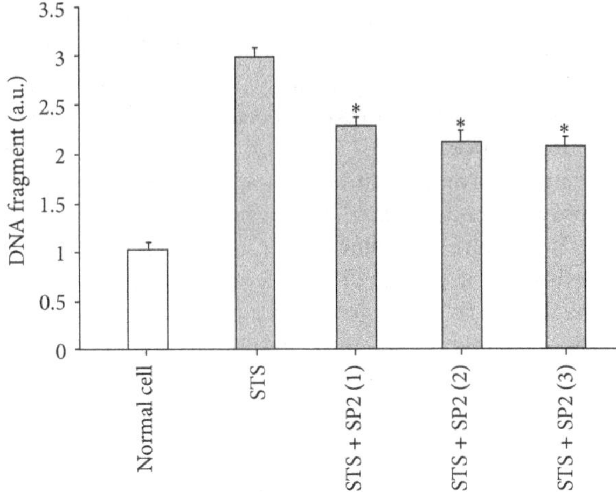

FIGURE 7: The effects of SP2 on DNA fragmentation caused by STS-induced apoptosis of HUVEC. The vertical axis is the DNA fragment unit; STS is the negative control; STS + SP2 (1–3) are the treatment groups, in which the final concentrations of SP2 sample were 0.5 μg/mL, 1 μg/mL, and 2 μg/mL, respectively. Values are expressed as mean ± SEM ($n = 3$). *Indicates values significantly different from negative control using Student's t-test; $P < 0.05$.

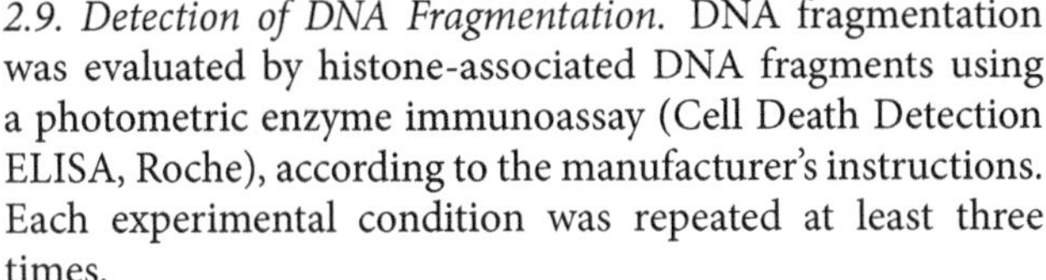

2.9. Detection of DNA Fragmentation. DNA fragmentation was evaluated by histone-associated DNA fragments using a photometric enzyme immunoassay (Cell Death Detection ELISA, Roche), according to the manufacturer's instructions. Each experimental condition was repeated at least three times.

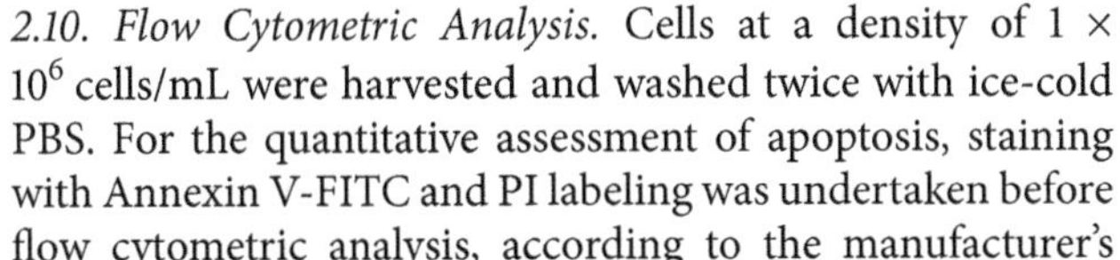

2.10. Flow Cytometric Analysis. Cells at a density of 1×10^6 cells/mL were harvested and washed twice with ice-cold PBS. For the quantitative assessment of apoptosis, staining with Annexin V-FITC and PI labeling was undertaken before flow cytometric analysis, according to the manufacturer's recommendations (Annexin V-FITC/PI apoptosis detection kit, Invitrogen). FACS analysis was accomplished using a FACSCalibur (Becton Dickinson, San Jose, CA, USA).

2.11. Statistical Analysis. The data were expressed as mean ± SEM and analyzed by the Student's t-test and the Newman-Keuls test. Differences with a value of $P < 0.05$ were considered statistically significant.

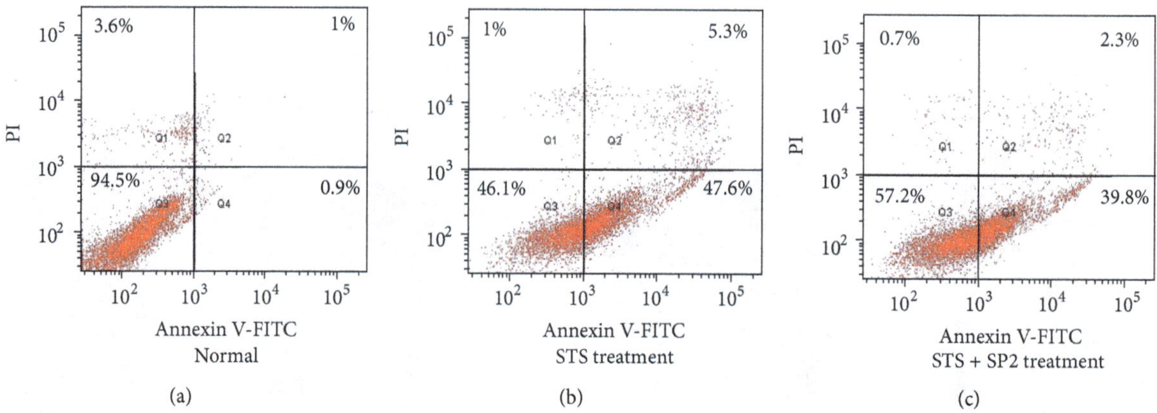

FIGURE 8: Detection of the antiapoptotic effect of SP2 by FCM. The flow cytometry profile represents Annexin V-FITC staining on x-axis and PI on y-axis. The number represents the percentage of early apoptotic cells in each condition (lower right quadrant). The cells in the upper right quadrant indicate Annexin-positive/PI-positive, late apoptotic cells. (a) Normal cell group; (b) STS treatment group; (c) STS + SP2 (1 μg/mL) treatment group; damaged (Annexin V^-/PI^+), apoptotic (Annexin V^+/PI^-), and vital (Annexin V^-/PI^-) cells are shown in the Q1, Q2/Q4, and Q3 regions, respectively.

3. Results

3.1. Bioinformatics Analysis. Sequence analysis revealed that the ORF of *sp2* (accession number DQ443358) was 2112 bp in length and encoded a protein of 703 amino acids, with a predicted molecular mass of 83.45 kD and a theoretical isoelectric point of 5.7. Through BLASTP comparison, three conservative structural domains of the hemocyanin superfamily were discovered: hemocyanin_N domain, hemocyanin_M domain, and hemocyanin_C domain. Hemocyanin can be any of a group of copper-containing respiratory proteins that serve an oxygen-carrying function in the blood of some arthropods and most mollusks [12]. The tertiary structure of silkworm SP2 was predicted using SWISS-MODEL (Figure 1).

3.2. Preparation of Silkworm SP2 Polyclonal Antibodies. The ORF for *sp2* was subcloned into the prokaryotic expression vector pET-28a (+). The His-tagged fusion protein was expressed in *E. coli* BL21 (DE3) (Figure 2(a)) and purified by Ni^{2+}-affinity chromatography. The purified His-SP2 fusion protein was successfully detected by SDS-PAGE (Figure 2(b)). The predicted molecular weight of the fusion protein, including a 3.56-kD His-tag, was 87 kD, which is concordant with the calculated value. The concentration of purified SP2 protein was determined by use of the Bradford assay to be 1.5 mg/mL. With the affinity-purified proteins, anti-SP2 polyclonal antibodies were generated by immunizing a male New Zealand rabbit. The titer of the polyclonal antibody, as determined by ELISA, was 1 : 6400. Western blot analysis indicated that the antibody reacted specifically with purified His-SP2 fusion protein (Figure 2(c)).

3.3. Expression Analysis of SP2 in Silkworm. To determine the distribution of SP2 in different tissues of the fifth-instar larva and other development stages of the silkworm, Western blot analysis was performed on protein extracts. The results showed that the level of SP2 was very high in the pupal and larval stages but extremely low in the egg and adult stages (Figure 3(a)). In the fifth instar larvae, the SP2 level was highest in the hemolymph and fat body and lowest in the trachea and midgut (Figure 3(b)).

3.4. Subcellular Localization of SP2. The treated cells were examined under a Nikon ECLIPSE TE2000-E Confocal Microscope, and images were analyzed using EZ-C1 software. DAPI-stained nuclei fluoresce red when stimulated with 353 nm light, and Cy3-labeled goat anti-rabbit IgG fluoresces red when stimulated with 550 nm light. Our results indicated that SP2 is mainly located in the cell membrane and only partly in the cytoplasm (Figure 4).

3.5. Expression and Purification of SP2 in a Bac-to-Bac System. To study the function of SP2 protein, recombinant bacmid-*sp2* was used to infect BmN cells for expression of the His-tagged protein, and wtbacmid was used as a control (Figure 5(a)). Owing to the low expression level of natural SP2 protein in BmN cells, the results of Western blot did not show the specific band in the BmN cells lysates infected by wtbacmid (Figure 5(b)). The BmN cell lysates were subjected to Ni^{2+}-affinity chromatography. The concentration of purified SP2 protein was determined by use of the Bradford assay to be 0.108 mg/mL. The results of SDS-PAGE and Western blot analysis indicated that the purified fusion protein SP2 had a molecular mass of approximately 87 kD (Figures 5(c) and 5(d)).

3.6. STS-Induced HUVEC Apoptosis and Apoptosis Assay. STS triggers within the cell both morphological changes and internucleosomal DNA fragmentation, which represents apoptosis [13]. An increasing concentration of STS was applied to examine the effects on HUVECs apoptosis. Apoptotic cells accompanying DNA fragmentation were analyzed at 24 h after STS treatment using ELISA. A low concentration

(0.1 μM) of STS did not affect apoptosis, whereas a high concentration (2 μM) induced necrosis, and 1 μM triggered DNA fragmentation efficiently compared with the positive control (camptothecin) provided by the kit. Based on these data, the following experiments were examined using 1 μM STS.

Cell viability was ameliorated by addition of the purified recombinant SP2 protein in a dose-dependent manner (Figure 6), indicating that SP2 can enhance apoptotic HUVEC viability. To investigate the inhibitory effect of SP2 protein on STS-induced HUVEC apoptosis, an ELISA was used to detect the DNA fragmentation. As shown in Figure 7, HUVEC apoptosis as measured by DNA fragmentation was significantly attenuated by recombinant SP2 protein in a dose-related manner. To quantify the apoptotic HUVECs, an Annexin V-FITC/PI apoptosis detection kit was used and assessed by flow cytometry. Apoptotic cells were identified by fluorescein (FITC) conjugated to the human anticoagulant-Annexin V, which was able to bind to phosphatidyl serine (PS) in the outer leaflet of the membrane of apoptotic cells. To distinguish cells that had lost membrane integrity, propidium iodide (PI) was added before analysis. As shown in Figure 8, the apoptosis rate of untreated HUVECs (normal control) was 1.9%, whereas the percentage of apoptotic HUVECs (negative control) that had been treated with 1μM STS was 52.9%. Compared with the negative control, the apoptosis rate of STS + SP2 (1 μg/mL)-treated HUVECs decreased to 42.1%. These results indicate that SP2 has anti-apoptotic effects on HUVEC apoptosis induced by STS.

4. Discussion

As the major regulator of vascular homeostasis, the endothelium exerts several vasoprotective effects, such as vasodilation, suppression of smooth muscle cell growth, and inhibition of inflammatory responses [14]. EC apoptosis has an important role in the diseased state of atherosclerosis [15]. A large body of evidence has suggested that endothelial dysfunction is caused by superoxide and other ROS [16, 17]. Supplementation with antioxidant vitamins C and E can restore endothelial function and, therefore, retard the progression of atherosclerosis [18]. Other factors, such as vascular endothelial growth factor (VEGF), low concentration of nitric oxide (NO) in serum, calcium antagonists, prostacyclin, and estrogen, can also inhibit EC apoptosis [19].

Many clinical trials have been done to investigate ways of treating diseases associated with apoptosis with anti-apoptotic genes and proteins [20, 21]. Silkworm hemolymph and its 30 K protein have been reported to exhibit anti-apoptotic activity in various mammalian and insect cell systems [7] and could have therapeutic potential in diseases related to apoptosis. In the current study, another potential anti-apoptotic protein in silkworm hemolymph (SP2) was studied. As a member of the hemocyanin superfamily, the SP2 protein has multiple biological functions [8–11]. Hemocyanins are large copper-containing proteins that transport oxygen in the hemolymph of many arthropod and mollusk species [22]. Most hexamerins are considered to be storage proteins, providing energy, serving as carrier protein or possibly involved in the humoral immune response. Therefore, they are considered to be an important immune molecule in arthropod [23]. Owing to the important role of hexamerins during arthropod development, we speculated that silkworm SP2 protein might also have an important role during cell apoptosis. To explore its functions, the SP2 protein was expressed and purified both in a prokaryotic expression system and silkworm baculovirus system. The purified SP2 prokaryotic expression product was used to generate monoclonal antibodies to study the distribution and location of SP2 in silkworm, whereas the purified expression product was used to study its antiapoptotic function.

Our analysis of the level of SP2 in different tissues of the fifth-instar larva and other developmental stages of the silkworm showed that SP2 levels were highest in the hemolymph and fat body and lowest in the egg and adult stages. These results suggest that *sp2* gene activity is affected by juvenile and molting hormones [24]. From the analysis of the subcellular localization of SP2, we found that, in the BmN cell line, SP2 was localized to both the cell membrane and cytoplasm but was found primarily in the cell membrane. Because SP2 is a secreted protein and has a signal peptide predicted by bioinformation (data not shown), our results suggest that it is translocated across the cell membrane guided by the signal peptide.

To study the anti-apoptotic activity of SP2 on vascular EC apoptosis, we constructed an STS-induced HUVEC apoptosis model. The STS treatment was able to induce generation of ROS, which results in oxidative stress to the cells. STS at a concentration of 1 μM exerted prominent apoptotic effects in cultured HUVEC. An MTT assay, DNA fragmentation detection, and flow cytometric analysis showed that the purified recombinant SP2 protein could significantly enhance the viability of HUVEC and inhibit HUVEC apoptosis induced by STS. These findings provide new insights into the prevention of endothelial dysfunction and could ultimately provide therapies for atherosclerosis.

Acknowledgments

This work was supported by the National High-Tech R&D Program (863 Program) (no. 2011AA100603) and Natural Science Foundation of Zhejiang Province (no. Y3090339, no. Y3110187).

References

[1] J. R. Vane, E. E. Anggard, and R. M. Botting, "Regulatory functions of the vascular endothelium," *The New England Journal of Medicine*, vol. 323, no. 1, pp. 27–36, 1990.

[2] R. Ray and A. M. Shah, "NADPH oxidase and endothelial cell function," *Clinical Science*, vol. 109, no. 3, pp. 217–226, 2005.

[3] N. Werner and G. Nickenig, "Endothelial progenitor cells in health and atherosclerotic disease," *Annals of Medicine*, vol. 39, no. 2, pp. 82–90, 2007.

[4] S. Dimmeler, C. Hermann, and A. M. Zeiher, "Apoptosis of endothelial cells. Contribution to the pathophysiology of

atherosclerosis?" *European Cytokine Network*, vol. 9, no. 4, pp. 697–698, 1998.

[5] J. C. Choy, D. J. Granville, D. W. C. Hunt, and B. M. McManus, "Endothelial cell apoptosis: biochemical characteristics and potential implications for atherosclerosis," *Journal of Molecular and Cellular Cardiology*, vol. 33, no. 9, pp. 1673–1690, 2001.

[6] S. C. Shin, J. R. Won, and H. P. Tai, "Inhibition of human cell apoptosis by silkworm hemolymph," *Biotechnology Progress*, vol. 18, no. 4, pp. 874–878, 2002.

[7] E. J. Kim, H. J. Park, and T. H. Park, "Inhibition of apoptosis by recombinant 30K protein originating from silkworm hemolymph," *Biochemical and Biophysical Research Communications*, vol. 308, no. 3, pp. 523–528, 2003.

[8] W. J. Rhee, E. H. Lee, J. H. Park, J. E. Lee, and T. H. Park, "Inhibition of HeLa cell apoptosis by storage-protein 2," *Biotechnology Progress*, vol. 23, no. 6, pp. 1441–1446, 2007.

[9] T. Fujii, H. Sakurai, S. Izumi, and S. Tomino, "Structure of the gene for the arylphorin-type storage protein SP2 of *Bombyx mori.*," *The Journal of Biological Chemistry*, vol. 264, no. 19, pp. 11020–11025, 1989.

[10] D. B. Roberts and H. W. Brock, "The major serum proteins of Dipteran larvae," *Experientia*, vol. 37, no. 2, pp. 103–110, 1981.

[11] K. Scheller, H. P. Zimmermann, and C. E. Sekeris, "Calliphorin, a protein involved in cuticle formation of the blowfly," *Zeitschrift für Naturforschung C*, vol. 35, pp. 387–389, 1980.

[12] T. Burmester and T. Hankeln, "The respiratory proteins of insects," *Journal of Insect Physiology*, vol. 53, no. 4, pp. 285–294, 2007.

[13] R. Bertrand, E. Solary, P. O'Connor, K. W. Kohn, and Y. Pommier, "Induction of a common pathway of apoptosis by staurosporine," *Experimental Cell Research*, vol. 211, no. 2, pp. 314–321, 1994.

[14] J. Davignon and P. Ganz, "Role of endothelial dysfunction in atherosclerosis," *Circulation*, vol. 109, no. 23, pp. 27–32, 2004.

[15] S. Verma and T. J. Anderson, "Fundamentals of endothelial function for the clinical cardiologist," *Circulation*, vol. 105, no. 5, pp. 546–549, 2002.

[16] E. S. Biegelsen and J. Loscalzo, "Endothelial function and atherosclerosis," *Coronary Artery Disease*, vol. 10, no. 4, pp. 241–256, 1999.

[17] H. Cai and D. G. Harrison, "Endothelial dysfunction in cardiovascular diseases: the role of oxidant stress," *Circulation Research*, vol. 87, no. 10, pp. 840–844, 2000.

[18] M. M. Engler, M. B. Engler, M. J. Malloy et al., "Antioxidant vitamins C and E improve endothelial function in children with hyperlipidemia: endothelial assessment of risk from lipids in youth trial," *Circulation*, vol. 108, no. 9, pp. 1059–1063, 2003.

[19] E. L. Schiffrin, "A critical review of the role of endothelial factors in the pathogenesis of hypertension," *Journal of Cardiovascular Pharmacology*, vol. 38, supplement 2, pp. S3–S6, 2002.

[20] P. Straten and M. H. Andersen, "The anti-apoptotic members of the Bcl-2 family are attractive tumor-associated antigens," *Oncotarget*, vol. 1, no. 4, pp. 239–245, 2010.

[21] J. Kountouras, C. Zavos, and D. Chatzopoulos, "Apoptotic and anti-angiogenic strategies in liver and gastrointestinal malignancies," *Journal of Surgical Oncology*, vol. 90, no. 4, pp. 249–259, 2005.

[22] H. Decker, N. Hellmann, E. Jaenicke, B. Lieb, U. Meissner, and J. Markl, "Recent progress in hemocyanin research," *Integrative and Comparative Biology*, vol. 47, no. 4, pp. 631–644, 2007.

[23] T. Burmester, "Origin and evolution of arthropod hemocyanins and related proteins," *Journal of Comparative Physiology B*, vol. 172, no. 2, pp. 95–107, 2002.

[24] T. S. Dhadialla, G. R. Carlson, and D. P. Le, "New insecticides with ecdysteroidal and juvenile hormone activity," *Annual Review of Entomology*, vol. 43, pp. 545–569, 1998.

7

Genomewide Analysis of Carotenoid Cleavage Dioxygenases in Unicellular and Filamentous Cyanobacteria

Hongli Cui,[1,2] Yinchu Wang,[1,2] and Song Qin[1]

[1] *The Coastal Zone Bio-Resource Laboratory, Yantai Institute of Coastal Zone Research, Chinese Academy of Sciences, Yantai 264003, China*
[2] *Yantai Institute of Coastal Zone Research, Graduate University of the Chinese Academy of Sciences, Beijing 100049, China*

Correspondence should be addressed to Song Qin, sqin@yic.ac.cn

Academic Editor: Diego Pasini

Carotenoid cleavage dioxygenases (CCDs) are a group of enzymes that catalyze the oxidative cleavage steps from carotenoids to various carotenoid cleavage products. Some *ccd* genes have been identified and encoded enzymes functionally characterized in many higher plants, but little in cyanobacteria. We performed a comparative analysis of *ccd* sequences and explored their distribution, classification, phylogeny, evolution, and structure among 37 cyanobacteria. Totally 61 putative *ccd* sequences were identified, which are abundant in *Acaryochloris marina* MBIC 11017, filamentous N_2-fixing cyanobacteria, and unicellular cyanobacterial *Cyanothece*. According to phylogenetic trees of 16S *rDNA* and CCD, *nced* and *ccd8* genes occur later than the divergence of *ccd7*, *apco*, and *ccd1*. All CCD enzymes share conserved basic structure domains constituted by a single loop formed with seven β-strands and one helix. In this paper, a general framework of sequence-function-evolution connection for the *ccd* has been revealed, which may provide new insight for functional investigation.

1. Introduction

Cyanobacteria, also known as blue-green algae and blue-green bacteria, are among the earliest branching groups on earth, dating back 2.5–3.5 billion years, based on the fossil evidence [1]. They may be unicellular or filamentous and can be found in almost every conceivable environment, such as marine and freshwater habitats, soil, rocks, and plants [2, 3]. With the capacity of oxygenic photosynthesis similar to the process found in higher plants, cyanobacteria constitute a group of species diverse not only in ecological habitat, but also in genome size and the number of gene, indicating the significance of comparative genome research. The genome size varied from 1.6 Mb (*Prochlorococcus* sp. MIT9301) to 9.0 Mb (*Nostoc punctiforme* PCC 73102), and the number of gene ranged from 1,756 (*Prochlorococcus marinus* MED4) to 8,462 (*Acaryochloris marina* MBIC11017) [4–6].

A lot of information on the evolutionary history of cyanobacteria has strongly supported an underlying meaning of comparative genome research. Three major clades are observed in cyanobacteria phylogenetic tree (Figure 1). The unicellular cyanobacteria (*Prochlorococcus* and *Synechococcus*) from ocean form the first monophyletic group (BS: 98%). They maintain the smallest genome size and account for significant biomass and primary production of marine biosphere [7]. Two *Synechococcus elongatus* PCC (6301 and 7942) are found at the base of this monophyletic group. Three thermophilic cyanobacteria (*Synechococcus* sp. JA-2-3B′a (2-13), *Thermosynechococcus elongatus* BP-1, and *Synechococcus* sp. JA-3-3Ab) and a thylakoid-lacking marine cyanobacterium (*Gloeobacter* sp. PCC 7421) [8] together compose the second monophyletic group (BS: 60%). In the last monophyletic group (BS: 64%), the filamentous N_2-fixing cyanobacteria (*Nostoc* sp. PCC 7120, *A. variabilis* ATCC 29413, *N. punctiforme* ATCC 29133, *A. platensis* NIES-39, and *T. erythraeum* IMS101) build the sister group (BS: 75%) with freshwater unicellular cyanobacteria (*Cyanothece* sp. ATCC 51142, 8801, 7424, bloom forming *Synechocystis* sp. PCC 6803, and toxic bloom *Microcystis aeruginosa* NIES-843). An animal-cyanobacterial symbiont (*Acaryochloris marina* MBIC11017) and *Cyanothece* sp. PCC 7425 build the basal branch of this monophyletic group (see Figure 1

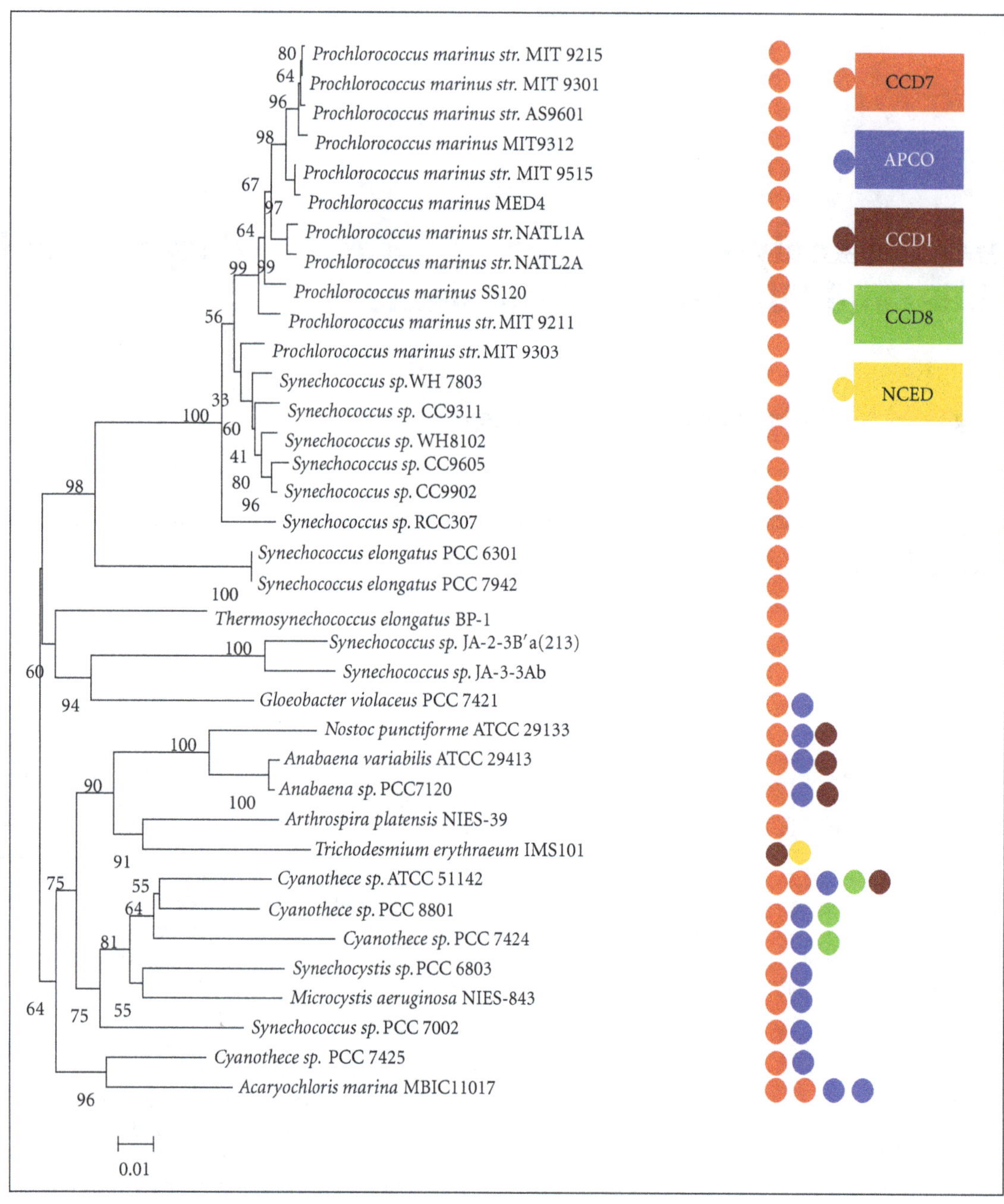

Figure 1: Phylogenetic tree of the sequenced cyanobacterial strains and the distribution of *ccd* sequences genes. Left: 37 fully sequenced cyanobacteria strains were performed based on 16 s *rRNA* as described in the methods section. An identical topology was obtained with two different methods (ML, NJ) and different models applied (for ML, generalized time reversible (GTR); for NJ, Kimura 2-parameter). *Numbers* at the node indicated bootstrap values (%) for 1,000 replicates. Right: the distribution of *ccd* genes across 37 cyanobacterial strains. Circular boxes represented the different *ccd* sequences that were labeled in the distinct color. The numbers of *ccd* sequences were presented by the numbers of circular boxes.

and Supplementary Material Table 1 available online at doi: 10.1155/2012/164690).

The first identified gene encoding carotenoid cleavage dioxygenase was the *Vp14* maize gene which was required for the formation of the abscisic acid (ABA) precursor xanthoxin, representing the rate-limiting step in ABA biosynthesis [9, 10]. Subsequently, homology-based analysis of the existence in all taxa was carried out *in silico*, which allowed the elucidation of the biosynthesis of several carotenoid-derived compounds, such as retinal in animals [11, 12] and fungi [13], the pigments bixin [14], saffron [15], and neurosporaxanthin [16]. On the basis of the identified substrates and presumed mechanism of catalysis, these enzymes are referred to as carotenoid cleavage dioxygenases (CCDs).

Carotenoid cleavage dioxygenases (CCDs) are a group of enzymes that catalyze the oxidative cleavage steps from

TABLE 1: Carotenoid cleavage dioxygenases protein sequences from distinct organisms in this paper.

Name	Species	Accession number	Type
NCED	*Arabidopsis thaliana*	NP_189064.1	*Streptophyta*
NCED	*Solanum lycopersicum*	CAB10168.1	*Streptophyta*
NCED	*Phaeodactylum tricornutum* CCAP 1055/1	XP_002177588.1	*Heterokontophyta*
CCD1	*Zea mays*	AAZ22348.1	*Streptophyta*
CCD4	*Arabidopsis thaliana*	sp|O49675.1	*Streptophyta*
CCD7	*Arabidopsis thaliana*	AEC10494.1	*Streptophyta*
CCD7	*Solanum lycopersicum*	ACY39882.1	*Streptophyta*
CCD8	*Arabidopsis thaliana*	NP_195007.2	*Streptophyta*
RPE	*Ostreococcus tauri*	XP_003080965.1	*Chlorophyta*
RPE	*Ajellomyces dermatitidis* ATCC 18188	EGE78156.1	Fungi
RPE	*Pyrenophora tritici-repentis* Pt-1C-BFP	EDU50284.1	Fungi
CO	*Mycobacterium vanbaalenii* PYR-1	YP_951059.1	Bacteria
PP	*Chlamydomonas reinhardtii*	EDP06596.1	*Chlorophyta*
LSD	*Ostreococcus tauri*	CAL53034.1	*Chlorophyta*
APCO	*Ostreococcus tauri*	CAL50095.1	*Chlorophyta*

carotenoids to various carotenoid cleavage products (apocarotenoids). These apocarotenoids function as signalling molecules with diverse functions including the ubiquitous chromophore retinal, plant hormone abscisic acid, and strigolactones [17, 18]. Other apocarotenoids with unknown functions in plants with high economic value are bixin in *Bixa orellana* and saffron in *Crocus sativus* [19]. These enzymes are present in all taxa [17, 20] and exhibit a high degree of regio- and stereospecificity for certain double bond positions as opposed to their frequent promiscuity towards substrates [21]. CCD enzymes have been widely studied in higher plants. In *Arabidopsis thaliana*, the CCD family includes nine members forming the basis for CCD classification (NCED, CCD1, CCD4, CCD7, and CCD8) in plants [22, 23]. The members of CCD family, *Cmccd4a*, *Cmccd4b*, *Cmnced3a*, and *Cmnced3b* have, been identified and functionally investigated in *Chrysanthemum morifolium* [24]. Among CCD family, NCED subfamily involves in abscisic acid (ABA) biosynthesis [10, 17, 23]. The majority of CCD has been shown to reside in plastids, whereas the only exception is CCD1 and its orthologous enzymes in higher plants, which act in the cytoplast to generate C13 and C14 apocarotenoids. Known pathways of carotenoid cleavage leading to various apocarotenoids have been discussed [17, 20, 21, 25, 26]. According to endosymbiotic theory, chloroplasts in plants and eukaryotic algae evolved from cyanobacterial ancestors *via* endosymbiosis.

It is reasonable to speculate that the CCD enzymes are widely separated in cyanobacteria as well. At present, many *ccd* genes have been cloned and functionally characterized *in vitro* from some sequenced cyanobacteria strains, including *Synechocystis* sp. PCC 6803 [27] and *Nostoc* sp. PCC 7120 [28–30]. With the completion of genome sequencing of several cyanobacterial species, modifications and supplements are needed.

Recently, 37 genomes of unicellular and filamentous cyanobacteria have been available, which facilitate cyanobacterial systemic analysis for metacaspases family [31], serine/threonine protein kinases [32], restriction modification system [33], and carotenoids biosynthesis [34]. These genome-sequencing projects undoubtedly bring great convenience to the searching for novel *ccd* genes by bioinformatic tools. In this study, five characterized *ccd* genes from *Arabidopsis thaliana* and *Zea mays* were selected as queries to search for cyanobacterial *ccd* genes. A BLASTP-plus-phylogeny reconstruction approach was employed to analyze CCD protein sequences in cyanobacteria, emphasizing an overall view on their distribution, classifications, phylogeny, evolution, and structure. Better understanding of cyanobacterial CCD enzymes may provide deeper insights into the evolution and the functional investigation of CCD enzymes in all organisms.

2. Materials and Methods

2.1. Computational Search for Novel ccd Genes. 37 species of cyanobacteria, including *Prochlorococcus*, *Synechococcus*, *Synechocystis*, *Gloeobacter*, *Cyanothece*, *Microcystis*, *Trichodesmium*, *Acaryochloris*, *Anabaena*, and *Nostoc* (http://genome.kazusa.or.jp/cyanobase/), were used in this analysis. The *ccd* genes from higher plants *Arabidopsis thaliana*, *Tomato*, and *Zea may* (Table 1) were obtained from National Center for Biotechnology Information (NCBI) and were used to construct a query protein set. Each protein in this query data set was used to search for the potential novel sequences in all cyanobacterial species from whole genome sequences available, by using the basic local alignment search tool (BLASTP) with E-value $< 1E-10$ [35–37]. Proteins found by this method that fit the criteria for genuine CCD enzymes were added to the query set for another round of BLASTP searches. The searches were iterated until convergence.

2.2. Multiple Sequence Alignment and Phylogenetic Analysis. All protein sequences identified by BLASTP were aligned using ClustalW (http://www.ebi.ac.uk/Tools/msa/clustalw2/)

[38]. The final alignment was further refined after excluding the relatively poorly conserved regions at the protein ends and consisted of sequences spanning the conserved domains. 16S *rDNA* and CCD phylogenetic trees were constructed by the NJ (Neighbor-Joining) method, using the program MEGA 4.0 (http://www.megasoftware.net/) [39] and ML (Maximum-Likelihood) method, using the program PHYML (http://www.atgc-montpellier.fr/phyml/binaries.php/) [40], with bootstrap support values deriving from 1000 randomized and replicated datasets. The Le and Gascuel evolutionary model [41] was selected for the protein phylogenies assuming an estimated proportion of invariant sites and a gamma correction. Graphical representation and edition of the phylogenetic tree were performed with TreeDyn (v198.3) [42]. In the phylogenetic analysis of CCD, CCD enzymes of fungi were used as outgroup to root the tree.

2.3. Motif Scanning and Structure Domain Analysis. To identify conserved motifs, Multiple Expectation Maximization for Motif Elicitation (MEME) version 2.2 [43] was employed with a set of parameters as follows: number of repetitions-any, maximum number of motifs-100, number of sites ($\geq$5 and $\leq$50), and optimum motif width set to $\geq$6 and $\leq$200 [44, 45]. The Simple Modular Architecture Research Tool (SMART) (http://smart.embl-heidelberg.de/) and Conserved Domains Database (CDD) (http://www.ncbi.nlm.nih.gov/cdd/) were applied to predict the structure domains of these CCD proteins sequences, relying on hidden Markov models and Reverse Position-Specific BLAST separately [46, 47].

2.4. Tertiary Structure Prediction. To well understand the evolution of certain enzyme, protein structure was analyzed using homology modeling. The protein sequences of CCD from *Prochlorococcus marinus* MIT9312 (PMT9312_0282), *Nostoc punctiforme* ATCC 29133 (Npun_F0298), *Cyanothece* sp. PCC 7424 (PCC7424_5321), and *Anabaena* sp. PCC 7120 (all1106) were submitted to the protein model server: SWISS-MODEL Web server [48, 49] (http://swissmodel.expasy.org/) though Automated Model, respectively. All the manipulations were performed using Pdb-Viewer [50, 51].

3. Results

3.1. Identification of Putative CCD Proteins. The 37 completed cyanobacteria were used in this research, and the detail information about key features of these cyanobacterial species was summarized in Supplementary material Table 1. The phylogenetic tree based on 16S *rDNA* was showed in Figure 1. The candidate genes identified and the distribution across cyanobacterial strains in this study were listed in Table 2 and Figure 1. Totally 61 putative *ccd* genes were predicted and annotated from 37 completed cyanobacterial genomes using BLASTP programs with the query sequences. 50 of these genes were originally annotated as retinal pigment epithelial membrane protein (RPE), lignostilbene-α, β-dioxygenase (LSD), or carotenoid oxygenase (CO). The remaining 11 proteins were accepted as CCD enzymes in this research, including 5 hypothetical proteins, 4 undefined proteins, 1 apocarotenoid 15,15′-dioxygenase (APCO), and 1 similar to neoxanthin cleavage enzyme. Interestingly, RPE65 (retinal pigment epithelium 65-kDa protein) shows a significant degree of sequence homology to the CCD family. Thus, many CCD enzymes were originally annotated RPE in cyanobacteria.

Amid diverse cyanobacterial genomes, the number of *ccd* gene varies from 1 to 5. Within unicellular cyanobacteria, N_2-fixing *Cyanothece* sp. ATCC 51142 has 5 *ccd* genes, much more than other species. An animal-cyanobacterial symbiont (*Acaryochloris marina* MBIC11017) has 4 *ccd* genes. All of the *Synechococcus* and *Prochlorococcus marinus* strains have only one *ccd* gene except that *Synechococcus sp.* PCC 700 has two *ccd* genes. Two unicellular strains inhabit in freshwater (*Synechocystis* sp. PCC 6803 and *Microcystis aeruginosa* NIES-843) and one marine cyanobacteria (*Gloeobacter violaceus* PCC 7421) each contains two *ccd* genes. The *Cyanothece* group has two or more *ccd* genes, including *Cyanothece* sp. PCC 7425 (2), *Cyanothece* sp. PCC 8801 (3), and *Cyanothece* sp. PCC 7424 (4). Compared to unicellular cyanobacteria, filamentous N_2-fixing cyanobacteria have more *ccd* gene (3 for *Anabaena* sp. PCC 7120, *Anabaena variabilis* ATCC 29413, and *Nostoc punctiforme* ATCC 29133, 2 for *Trichodesmium erythraeum* IMS101). However, *Arthrospira platensis* NIES-39 contains only one *ccd* gene.

3.2. Phylogenetic Analysis of ccd Genes in Cyanobacteria. Considering the confusion created by unspecific annotation with possibly separate evolutionary histories, the translated CCD proteins from cyanobacteria and the characterized CCD enzymes from bacteria, eukaryotic algae, higher plants, and fungi were used to perform a phylogenetic analysis. The tree topology matches our Neighbor-Joining tree (data not shown), and the high branch support values coincide with high neighbor joining bootstrap values, suggesting that the CCD phylogeny is robust in different tree reconstruction methods. Therefore, only the ML trees were displayed in this paper. The CCD phylogenetic tree was rooted in the CCD of fungi. Observation of the tree revealed that all CCD enzymes fell into four clades (Figure 2): clade 1: CCD7, clade 2: APCO, clade 3: CCD1/NCED/CCD4, and clade 4: CCD8.

As shown in Figures 1 and 2, the first clade, *ccd7*, was composed of almost all the cyanobacterial strains (in this paper), except for filamentous nitrogen fixation *Trichodesmium erythraeum* IMS101. The *ccd7*-homologous genes from two unicellular marine cyanobacteria *Synechococcus* (except for *Synechococcus* sp. PCC 7002) and *Prochlorococcus* constituted the first subfamily. Amino acid sequence identity of genes from this subfamily ranged from 59% to 98% (Supplementary Material Table 2). In the second subfamily, *ccd7*-homologous genes from two hot-spring habitat cyanobacteria (*Synechococcus* sp. JA-3-3Ab and *Synechococcus* sp. JA-2-3B′a (2-13)) and a thylakoid-lacking cyanobacterium (*Gloeobacter* sp. PCC 7421) clustered into a group. Other cyanobacterial *ccd7* genes clustered into the last group, including filamentous cyanobacteria and unicellular cyanobacterial *Cyanothece.* Moreover, *ccd7*-homologous

TABLE 2: List of putative *ccd* genes identified across 37 cyanobacterial in this paper.

Species	Gene	Length	Annotation	Proposed function
P. marinus MED4	0280	507		CCD7
P. marinus MIT9312	0282	494	RPE	CCD7
P. marinus MIT9313	1879	507		CCD7
P. marinus SS120	0312	496		CCD7
P. marinus str. AS9601	03031	494	RPE	CCD7
P. marinus str. MIT 9211	03071	495	RPE	CCD7
P. marinus str. MIT 9215	03051	494	RPE	CCD7
P. marinus str. MIT 9301	03041	494	RPE	CCD7
P. marinus str. MIT 9303	25111	507	RPE	CCD7
P. marinus str. MIT 9515	03131	495	RPE	CCD7
P. marinus str. NATL1A	03601	497	RPE	CCD7
P. marinus str. NATL2A	1646	497	LSD	CCD7
A. platensis NIES-39	06860	492	LSD	CCD7
S. elongates PCC 6301	1315d	493	LSD	CCD7
S. elongates PCC 7942	0196	493	BCD	CCD7
S. sp. CC9311	0261	504	RPE	CCD7
S. sp. CC9605	0221	488	LSD	CCD7
S. sp. CC9902	0248	489	LSD	CCD7
S. sp. JA-2-3B′a (2-13)	0263	482	LSD	CCD7
S. sp. JA-3-3Ab	1068	482	LSD	CCD7
S. sp. PCC 7002	A1425	492	LSD	CCD7
S. sp. PCC 7002	A2200	484	RPE	APCO
S. sp. RCC307	2292	486	LSD	CCD7
S. sp. WH 7803	0270	495	LSD	CCD7
S. sp. WH8102	0227	489		CCD7
S. sp. PCC 6803	1541	490	HP	CCD7
S. sp. PCC 6803	1648	480	HP	APCO
T. elongatus BP-1	0015	487	LSD	CCD7
T. IMS101	3794	464	CO	CCD1
T. IMS101	3212	489	CO	NCED
G. violaceus PCC 7421	3689	482	LSD	CCD7
G. violaceus PCC 7421	2774	475	HP	APCO
M. aeruginosa NIES-843	09040	515	CO	CCD7
M. aeruginosa NIES-843	60600	490	HP	APCO
N. punctiforme ATCC 29133	2869	460	CO	CCD1
N. punctiforme ATCC 29133	0298	498	CO	CCD7
N. punctiforme ATCC 29133	0776	491	CO	APCO
C. sp. PCC 8801	0352	487	CO	CCD8
C. sp. PCC 8801	4340	495	CO	CCD7
C. sp. PCC 8801	0015	468	CO	APCO
C. sp. PCC 7424	4841	464	CO	CCD1
C. sp. PCC 7424	0935	488	CO	CCD8
C. sp. PCC 7424	0449	491	CO	CCD7
C. sp. PCC 7424	5321	477	CO	APCO
C. sp. PCC 7425	4222	501	CO	CCD7
C. sp. PCC 7425	3964	486	CO	APCO
C. sp. ATCC 51142	2215	459	RPE	CCD1
C. sp. ATCC 51142	3977	486	RPE	CCD8

Table 2: Continued.

Species	Gene	Length	Annotation	Proposed function
C. sp. ATCC 51142	2665	494	LSD	CCD7
C. sp. ATCC 51142	4927	481	CO	CCD7
C. sp. ATCC 51142	2515	468	CO	APCO
A. marina MBIC11017	0661	488	LSD	CCD7
A. marina MBIC11017	3102	483	RPE	CCD7
A. marina MBIC11017	1272	479	RPE	APCO
A. marina MBIC11017	0925	485	RPE	APCO
A. sp. PCC 7120	1106	475	NCE	CCD1
A. sp. PCC 7120	4284	497	LSD	CCD7
A. sp. PCC 7120	4895	472	HP	APCO
A. variabilis ATCC 29413	2047	765	RPE	CCD1
A. variabilis ATCC 29413	3710	463	RPE	CCD1
A. variabilis ATCC 29413	1236	510	RPE	CCD7

genes were discovered among eukaryotic microalgae (*Ostreococcus tauri*), bacteria (*Mycobacterium vanbaalenii* PYR-1), and higher plants (*Solanum lycopersicum* and *Arabidopsis thaliana*). All of these CCD7s are orthologues because of obvious evolutionary relationships with high bootstrap value.

All the *apco*-homologous genes clustered into clade 2, including filamentous cyanobacteria (*Anabaena variabilis* ATCC 29413, *Anabaena* sp. PCC 7120, and *Acaryochloris marina* MBIC11017) and unicellular cyanobacteria *Cyanothece*. In addition, *apco*-homologous genes from *Gloeobacter violaceus* PCC 7421, *Microcystis aeruginosa* NIES-843, *Synechocystis* sp. PCC 6803, and *Synechococcus* sp. PCC 7002 were also assembled into this group. Amino acid sequence identity of genes from this clade ranged from 43% to 76% (Supplementary Material Table 3). It is interesting that two copies of *apco*-homologous genes existed in animal-cyanobacterial symbiont (*Acaryochloris marina* MBIC11017). Moreover, *ccd7*-homologous gene was discovered in eukaryotic algae (*Ostreococcus tauri*) while was absent in higher plants. Results from BLASTP and phylogenetic tree suggest that a close evolutionary relationship exists between CCD7 enzyme and APCO enzyme, between CCD7 enzyme and APCO enzyme.

CCD1 enzyme from filamentous cyanobacteria, algae, as well as higher plant (*Zea mays*) stays in a sister group relation with NCED and CCD4 enzyme from higher plant. They build the third monophyletic group with NCED enzymes from *Phaeodactylum tricornutum* as the basal branch of this clade. Amino acid sequence identity of genes from CCD1, NCED, and CCD8 was summarized in Supplementary Material Table 4. This group included all filamentous cyanobacteria, except for *A. platensis* NIES-39. Surprisingly, the *ccd1*-homologous gene was absent in all unicellular cyanobacteria except for *Cyanothece* ATCC51142, suggesting this organism might acquire this gene by horizontal gene transfer. It is interesting that the protein sequence of Tery_3212 from *Trichodesmium erythraeum* IMS101 is highly similar with NCED from *Arabidopsis thaliana* (99%). It is worth considering that CCD4 enzyme from *Arabidopsis thaliana* form a monophyletic with NCED enzyme, indicating that they may origin from common ancestor and acquire different function under natural selection during the evolution. CCD8 enzymes are widely distributed in eukaryote (fungi, eukaryotic algae, and higher plants), whereas they are missing in all cyanobacteria except for *Cyanothece* (PCC 7424, 8801 and ATCC 51142), suggesting these cyanobacteria obtain this gene by horizontal gene transfer or produce it later under the natural selection during the evolution.

3.3. Conserved Motifs. According to CDD and SMART domain analyses, 8 protein sequences which were originally annotated as hypothesis protein or unidentified belonged to cyanobacterial CCD enzymes categories as well. All the protein sequences belonged to Pfam: RPE65 (IPR004294), which included β-carotene-15,15′-monooxygenase (BCDO1; EC 1.14.99.36), β-carotene-9′,10′-dioxygenase (BCDO2/CCD7), 9-*cis*-epoxycarotenoid dioxygenase (NCED), β-apocarotenoid-15,15′-oxygenase (APCO), and retinal pigment epithelial membrane protein (RPE). However, domain analyses failed to further distinguish them. To facilitate the classification of different types of CCD enzymes, conserved motifs were identified by MEME tool and multiple sequence alignments with ClustalW. There were two typical glycine-rich ($G_{X7}G_{X5}GP$ and HHPFDGDGMI) motifs and one leucine-rich ($LALWEA_{X2}P_{X4}P_{X2}P_{X4}P_{X2}L$) motif existed in all cyanobacterial CCD proteins (Figure 3). Moreover, CCD7 from *Synechococcus* and *Prochlorococcus* organisms shared 59–98% similarities (Supplementary material Table 2) and possessed conserved patterns ($EAFSAHP_{X2}D$, $NPLPF_{X2}G_{X2}GAAQCL_{X}S$). CCD7 from other cyanobacteria organisms shared 54–95% similarities (Supplementary Material Table 5) and possessed conserved patterns ($AP_{X}G_{X2}GEP_{X}F_{X}P_{X}P$ and $L_{X2}H_{X}PY_{X}LHG$). In addition, $RG_{X}FG_{X}4GG$, $F_{X}FIHDF_{X2}T_{X5}F$, and $FVFHH_{X}NA$ motifs also existed in CCD7 enzymes from distinct organisms (Figures 3(a) and 3(b)). It was interesting that protein sequences of CCD7 and CCD1 have the same $P_{X3}P_{X2}FH$ box in the N-terminal (Figures 3(b) and 3(d)). All these results suggested that all *ccd* genes have originated from

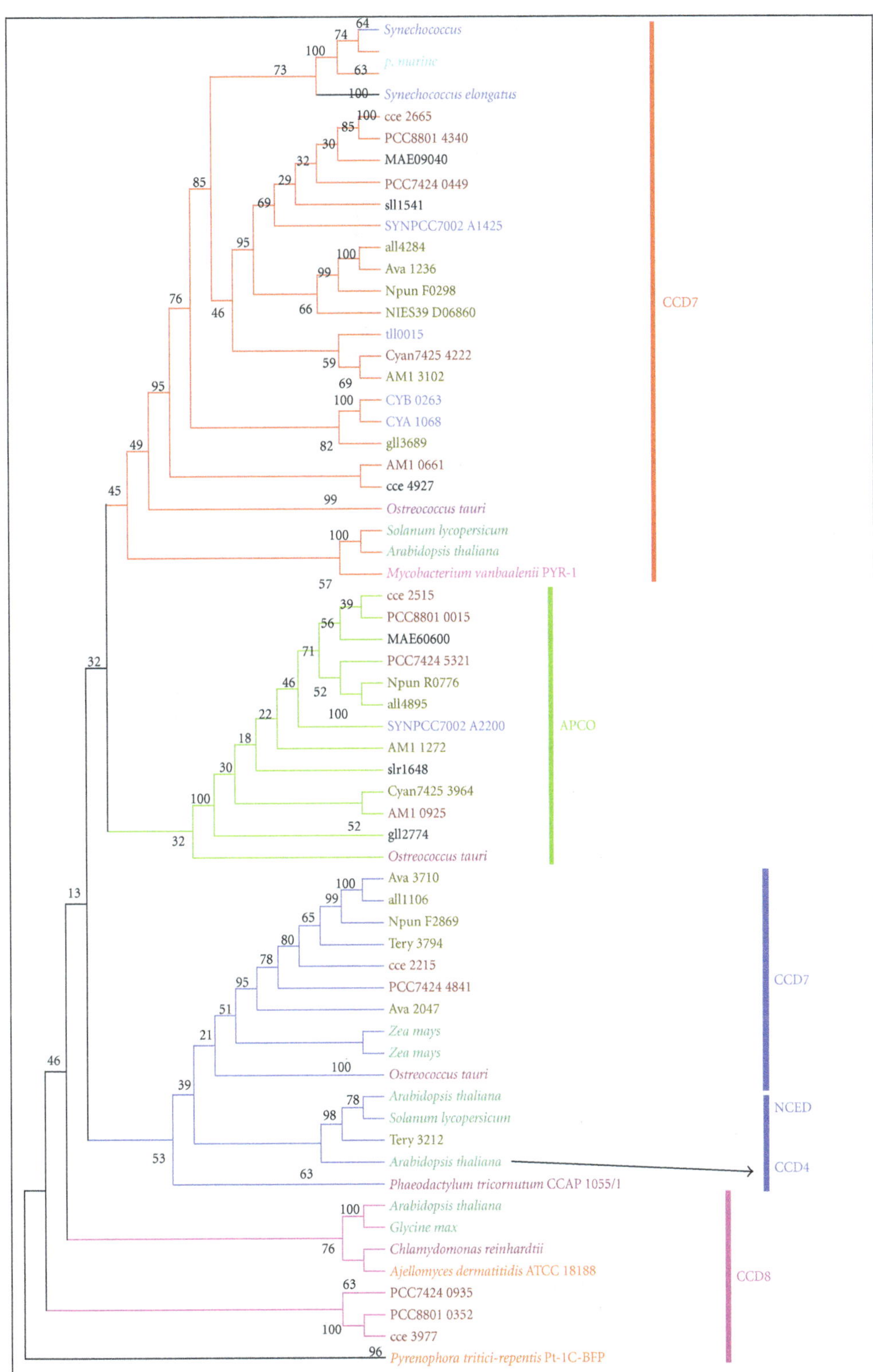

FIGURE 2: Maximum-likelihood tree of CCD enzymes of cyanobacteria, bacteria, eukaryotic algae, fungi, and higher plants. The model of LG+I+G was applied to construct ML tree using PHYML as described in the methods section. *Numbers* at the node indicated bootstrap values (%) for 1,000 replicates. Cyanobacterial CCD enzymes IDs and the strain names are as in Table 2 and Figure 1. Red line: CCD7, green: APCO, blue: CCD1, CCD4, and NCED, pink: CCD8. Species belonged to similar clade were represented by colored boxes. Orange: fungi, wheat: bacteria, purple: eukaryotic algae, barium: higher plants, cyan: cyanobacterial *Prochlorococcus marinus*, blue: *Thermosynechococcus elongates, Synechococcus elongates, and Synechococcus,* ruby: cyanobacterial *Cyanothece*, and smudge: filamentous cyanobacteria. Unlabeled species included *Synechocystis* sp. PCC 6803 (sll1541 and slr1648), *Gloeobacter violaceus* PCC 7421 (gll3689 and gll2774), *Microcystis aeruginosa* NIES-843 (MAE09040 and MAE60600).

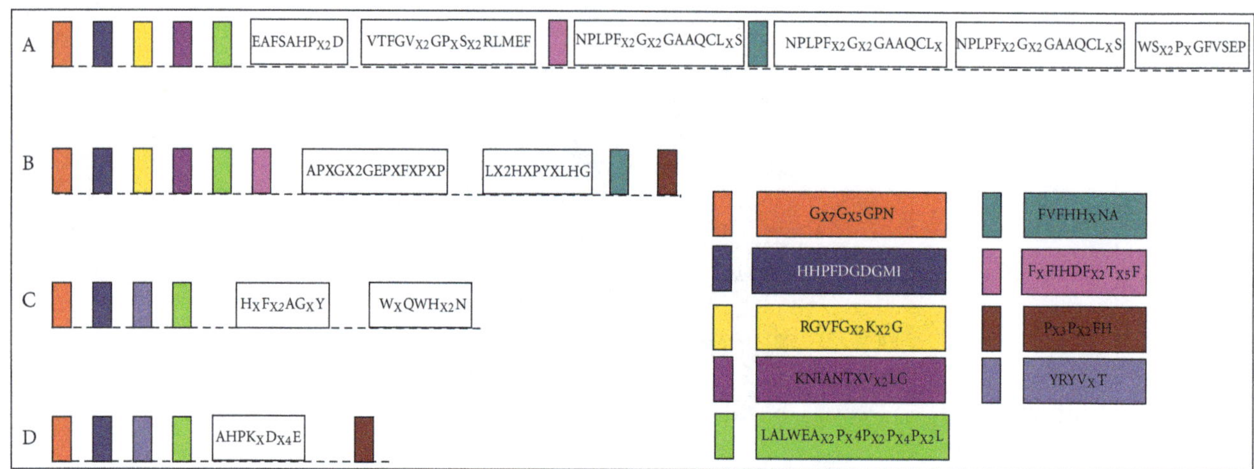

Figure 3: The conserved motifs of the cyanobacterial CCD enzymes. Schematic representation of motifs identified in cyanobacterial CCD enzymes using MEME motif search tool and ClustalW. The results were edited by hand. Length of box did not correspond to length of motif. Boxes represented the same or similar motifs that were labeled in the same color. (a) CCD7 from *Synechococcus* and *Prochlorococcus marinus*, (b) CCD7 from other cyanobacterial strains, (c) APCO, and (d) CCD1 and NCED.

a common ancestor and exhibited different functions under distinct natural selection during the evolution.

3.4. Structure of CCD Enzymes. To understand the evolution of cyanobacterial CCD enzymes, protein structures of CCD7 from *Prochlorococcus* MIT9312 (PMT9312_0282) and *Nostoc punctiforme* ATCC 29133 (Npun_F0298), APCO from *Cyanothece* sp. PCC 7424 (PCC7424_5321), and CCD1 from *Anabaena* sp. PCC 7120 (all1106) were analyzed using homology modeling as described material and methods. A comparable analysis for the tertiary structure of different type of CCD-encoding ORFs from cyanobacteria revealed that a single loop formed with seven β-strands and one helix has conserved in four models, which may be related to binding domain (Figure 4).

CCD7-encoding ORFs from different cyanobacterial organisms had conserved structure, which included α-helix (1), four-antiparallel β strand (4), two-, three-, and five-antiparallel β strand (1, 1, and 1), respectively. The additional sheets in the N- and C- terminal are the only difference (Figures 5(a) and 5(b)). As showed in Figures 5(c) and 5(d), five-antiparallel β strand existing in CCD7 was lacking in that of the APCO and CCD1. However, two α-helixes were present in CCD1-encoding ORFs. We supposed all the CCD-encoding ORFs in a given lineage may evolve through gene duplication that happened under natural selection during the evolution.

4. Discussion

The distribution of putative CCD encoding open reading frames (ORFs) in cyanobacteria is an integrated function of the genome sizes and the ecophysiological properties. In large extent, most cyanobacterial strains possess proportionate numbers of putative *ccd* genes with genome sizes, except for a few particular cases, for example, *Arthrospira platensis* NIES-39 and *Gloeobacter violaceus* PCC 7421. All of *Prochlorococcus* and most *Synechococcus* (except for *Synechococcus* sp. PCC 7002) strains, which live in the oligotrophic open ocean and have smaller genome sizes, maintain one CCD encoding ORF. Gene duplication event or larger genome size than other *Synechococcus* species may be responsible for two *ccd* genes existed in *Synechococcus* sp. PCC 7002. Metacaspases were found to be absent in all *Prochlorococcus* and marine *Synechococcus* strains, except *Synechococcus* sp. PCC 7002 [31]. Compared to unicellular cyanobacteria with smaller genome sizes, CCD-encoding ORFs are abundant in filamentous with larger genome sizes. On the other hand, filamentous heterocystous cyanobacteria in response to the absence of combined nitrogen and exhibiting ecological properties including broad symbiotic competence with plants and fungi are responsible for containing more putative CCD-encoding ORFs even after allowing for their larger genome sizes [32, 33]. Moreover, according to our results, we speculated that the diverse distributions of *ccd* genes may reflect various environmental selective pressures. Unicellular *Cyanothece* that inhabits in freshwater has two or more CCD encoding ORFs, which is beyond the scaling effect of genome size. Similar phenomenon occurred in *Synechocystis* sp. PCC 6803 and *Microcystis aeruginosa* NIES-843. Considering the similar genome size, environmental selective pressure may take responsibility for this difference. This parallel pattern of distribution was provided by other cyanobacterial systemic analysis. For example, metacaspases systerm, Serine/threonine kinases, and restriction modification system in cyanobacteria indicate remarkable reduction of these proteins and environmental stress response systems in the ocean [31–33]. Gene lost is revealed to facilitate these cyanobacteria (*Prochlorococcus*) to acclimatize to the oligotrophic environment. The major driving force was supposed to be "a selective process sustaining the adaptation of these cyanobacteria," which was discussed by Dufresne et al. (2005) in detail [52].

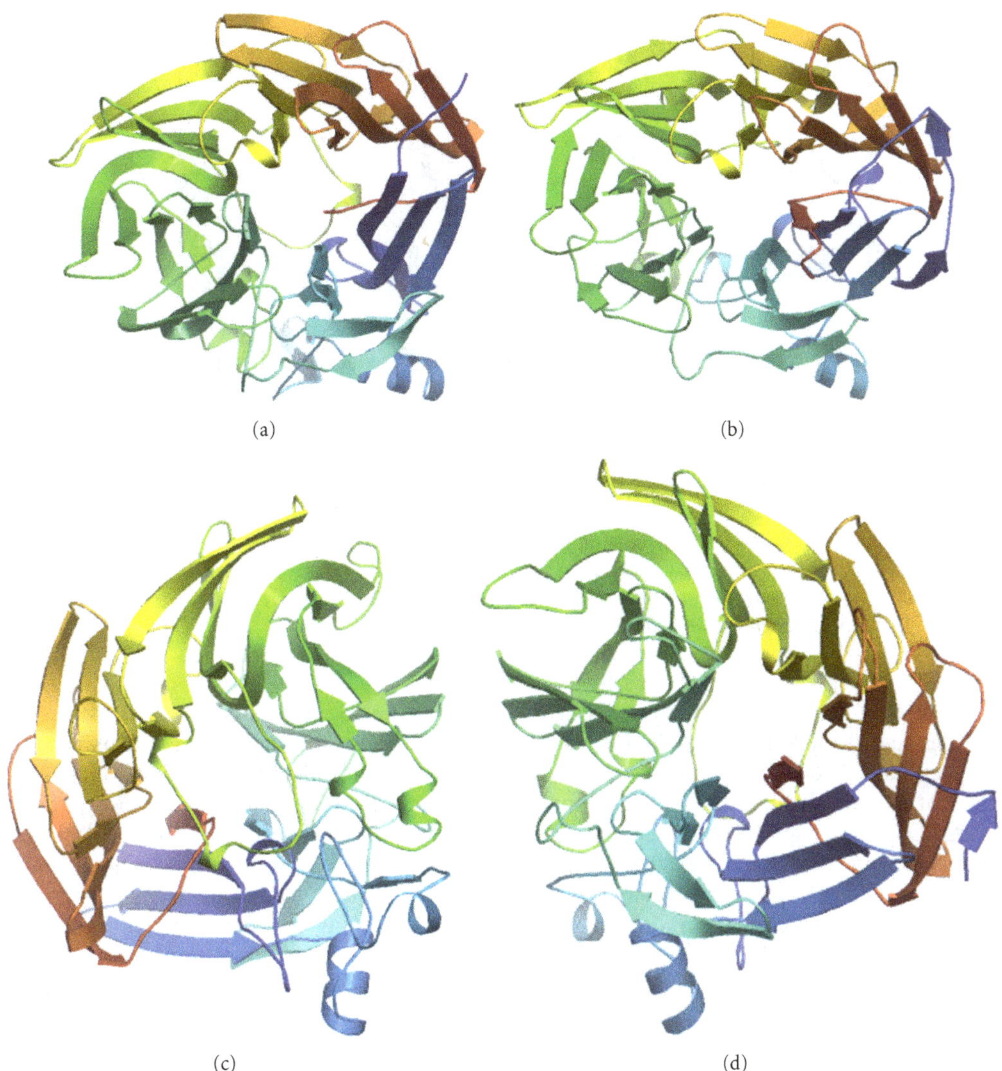

FIGURE 4: Model of structure of CCD enzymes from cyanobacteria strains. (a) The structure model of CCD7 from *Prochlorococcus* MIT9312 (PMT9312_0282), (b) the structure model of CCD7 from *Nostoc punctiforme* ATCC 29133 (Npun_F0298), (c) the structure model of APCO from *Cyanothece* sp. PCC 7424 (PCC7424_5321), (d) the structure model of CCD1 from *Anabaena* sp. PCC 7120 (all1106).

It is well accepted that there are NCED, CCD1, CCD4, CCD7, and CCD8 subfamily in higher plants [21]. Among them, NCED plays key role in cleavage of the 11,12 double bond of *9-cis* violaxanthin or *9′-cis* neoxanthin, which results in the formation of abscisic acid (C_{15}) [10]. Our results showed that *nced*-homologous genes were absent in all cyanobacteria except for *Trichodesmium erythraeum* ISM 101, suggesting they did not contain ABA. These results were consistence with Pryce (1972), who supposed that, as distinct from higher plants, algae and liverworts did not contain ABA and its functions were fulfilled by lunularic acid [53]. Surprisingly, this hormone has been found in green microalgae (*Chlorella* sp., *Dunaliella salina*, and *Haematococcus pluvialis*) [54, 55] and also in the thalli of brown macrophytes from the genus *Ascophyllum* and some species of *Laminaria* [56].

Converting C_{40} trans-carotenoids to C_{27} apocarotenoids and the following step C_{27} to C_{18} is catalyzed by the CCD7 and CCD8 enzymes, respectively. They involve in the biosynthesis of strigolactones (C_{18}) [57–59]. The recombinant CCD7 protein of *Arabidopsis* exhibited a specific 9′-10′ (but not 9-10/9′-10′) cleavage activity *in vitro* converting β-carotene to the C_{27} compound β-apo-10′-carotenal and the C_{13} compound β-ionone. The C_{18} compound was formed by a secondary cleavage of the C_{27} apocarotenoid generated by AtCCD7 [59]. These two types of CCDs have transit peptides indicative of their action in plastids [60]. In this paper, *ccd7*-homologous genes were present in all cyanobacteria except for *Trichodesmium erythraeum* ISM 101 and *ccd8*-homologous genes were discovered in *Cyanothece*. Our results strongly support a cyanobacterial origin of CCD7 proteins in eukaryotic algae and higher plants (Figure 2). However, further investigation is necessary to explain whether a cyanobacterial origin of CCD8 proteins.

The formation of C_{13} and C_{14} from C_{27} is catalyzed by CCD1 or CCD4 enzymes in the higher plant [61–63]. The recombinant CCD1 or CCD4 enzymes from several

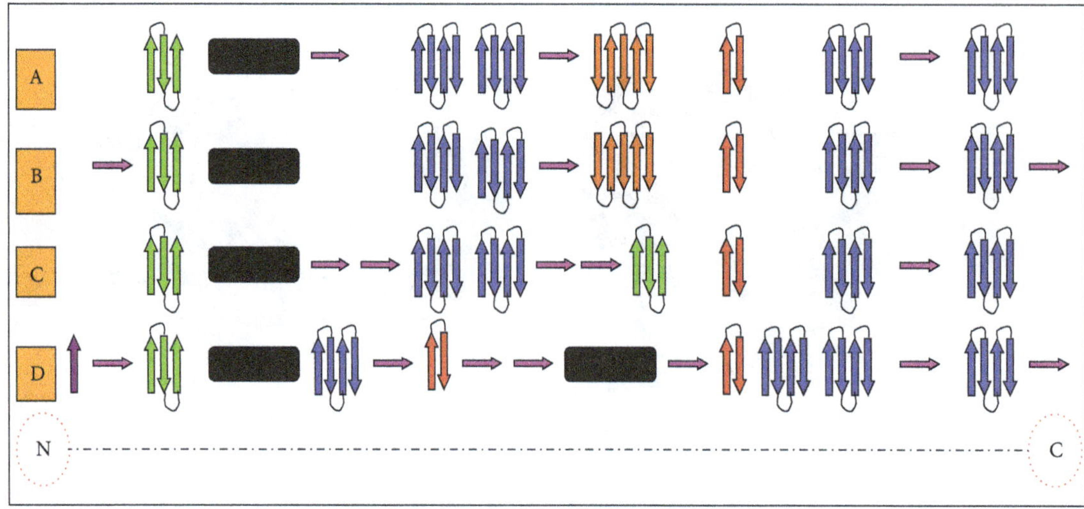

FIGURE 5: Sketch of the chain-fold of CCD enzyme showing the seven-bladed β-sheet as the basic motif. Distinct types of β-sheets were presented by colored arrowheads. Bottle green boxes stand for α-helix. (a) The second structure of CCD7 enzyme from *Prochlorococcus* MIT9312 (PMT9312_0282), (b) CCD7 from *Nostoc punctiforme* ATCC 29133 (Npun_F0298), (c) APCO from *Cyanothece* sp. PCC 7424 (PCC7424_5321), (d) CCD1 from *Anabaena* sp. PCC 7120 (all1106).

plants have been shown to preferentially catalyze a single-step symmetrical cleavage at the 9-10/9′-10′ double bonds of various C_{40} carotenoids [17, 60, 64, 65]. Interestingly, the CsCCD4 recombinant enzymes produced considerably more β-ionone than CsCCD1, implying that the CCD4 enzymes may have higher enzyme activities [66]. In *planta*, CCD1 and CCD4 differ in their subcellular location being cytosolic and plastidial, respectively. Surprisingly, *ccd1*-homologous genes were discovered in N_2-fixing cyanobacteria whereas *ccd4*-homologous gene was absent in all cyanobacteria (Figure 1). This relationship between the cyanobacterial, phototrophic eukaryotic algae, and higher plants strongly supports a cyanobacterial origin of CCD1 proteins (Figure 2). This conclusion has been demonstrated by Scherzinger and Al-Babili [29], who found that the NSC1 (NosCCD) enzyme is an ortholog of plant CCD1 enzymes [29]. Moreover, in *planta*, CCD1 and/or CCD4 might exhibit additional cleavage specificities including 7-8/7′-8′ cleavage in a certain subcellular environment or guided by a specific cofactor to provide the altered cleavage activity required for generation of crocetin and picrocrocin precursors. Recombinant CCD1 has been shown to exhibit 5-6/5′-6′ or 7-8/7′-8′ activities in addition to 9-10/9′-10′ under certain artificial conditions *in vitro* but only on acyclic or monocyclic substrates [67].

Apart from the numerous enzymes acting on C_{40} carotenoids, there are some special CCD enzymes converting only β-apocarotenals but not of bicyclic β-carotene. This has led to its classification as an apocarotenoid 15-15′-oxygenase (APCO). For example, APCO (apocarotenoid cleavage oxygenase) from the cyanobacterium *Synechocystis* sp. PCC 6803 cleaves the 15-15′double bond of various all-*trans*-apocarotenoids [20]. In this paper, *apco*-homologous genes were discovered by BLASTP analysis in some cyanobacteria and eukaryotic algae, which may be responsible for the production of apocarotenoids-derived volatile compounds in these organisms. For example, *Microcystis aeruginosa* blooms liberate β-ionone [20, 25]. Moreover, it was worthy consideration that *apco*-homologous genes were absent in higher plant and were evolutional closed with *ccd7* genes. Kiefer et al. [68] identified a new type of APCO, which catalyses the asymmetric cleavage of β-carotenoid at the C_9'-C_{10}' position and produces the volatile compound β-ionone (C_{13}) and apo-10′-carotenal (C_{27}) [68].

Considering the sites (chloroplast) of C_{40} carotenoids biosynthesis and the location of each CCD enzyme, some hypothesis were summarized: (1) the additional function of cleavage C_{40} carotenoids suggests that the *ccd1* and *ccd7* genes may origin in cyanobacteria via eukaryotic algae towards higher plants; (2) the *ccd1* gene transfers to cytosol by endosymbiotic gene transfer during the evolution in higher plants; (3) the CCD1, CCD7, and APCO enzymes occur earlier than other members of CCDs (CCD4, CCD8, and NCED) enzymes in cyanobacteria; (4) the *apco*-homologous gene was absent because of gene lost, whereas the *ccd4*-, *ccd8*-, and *nced*-homologous genes were present because of gene duplication in higher plant. These hypothesis were supported by one of our results that *ccd7*-, *apco*-, and *ccd1*-homologous genes were widespread in cyanobacterial strains, while *ccd8*- and *nced*-homologous genes only existed in some species, and *ccd4*-homologous genes were absent in all cyanobacteria (Figures 1 and 2).

Another interesting result was the highly similar tertiary structure for different types of CCD enzymes from distinct cyanobacterial strains (Figures 4 and 5). Indeed, sequence alignments within the CCD family clearly show that the most highly conserved regions are within the β-strands forming the propeller as demonstrated in Figures 3, 4, and 5. This strongly suggests that all members of the CCDs family have the same β propeller chain-fold and may therefore be modeled along the lines of the other CCD structures and share several same characteristics. Schwartz et al. [10], Redmond et al. [12], and Kiefer et al. [68] found that members of the CCD family share several characteristics: first, they require a Fe^{2+} for catalytic activity [10, 12, 68]; second, they contain four conserved histidines that are thought to coordinate iron binding; third, they contain a conserved peptide sequence at their carboxyl terminus that minimally constitutes a signature sequence for the family.

At present, the only established structure of a CCDs family member is apocarotenoids oxygenase (APCO) from *Synechocystis* PCC 6803 [69], and the mechanisms of choosing and modulating substrate specifically have been discussed [20, 70]. As observed in numerous β propeller structures, the active center of APCO is located near the propeller axis on the top side. The individual CCO family members choose and modulate substrate by changing length hand sequence of the connecting loops, while retaining the rigid propeller scaffold for stability [71–73]. More analysis is necessary to further delineate the mechanisms of choosing substrate exactly for each CCD family member. However, such an analysis is beyond the scope of this paper.

5. Conclusion

CCD enzymes play significant role in forming of carotenoids cleavage products, which are essential to several prokaryotes, eukaryotic algae, fungi, and higher plants. A total of 61 putative *ccd* genes have been identified across 37 species of cyanobacteria. The *ccd7*-, *ccd1*-, and *apco*-homologous genes are widespread in cyanobacteria and *ccd8*-homologous genes only exist in a few species. The distribution of *ccd* genes in unicellular and filamentous cyanobacteria relies on the genome size and ecological habitat. According to BLASTP results and phylogenetic tree of 16 s *rDNA* and CCD, it seems that CCD7, APCO, and CCD1 enzymes appeared earlier than other members of CCD enzymes. A slight difference exists between distinct types of CCD enzymes by motif scanning, while the secondary and tertiary structures are highly similar through homology modeling. All CCD enzymes share conserved basic configuration, which is constituted by a single loop formed with seven β-strands and one helix. This paper may provide new insight for the evolutional and functional investigation of CCD enzyme in cyanobacteria.

Abbreviations

CCD:	Carotenoid cleavage dioxygenase
CCD (1, 4, 7, and 8):	Carotenoid cleavage dioxygenase (1, 4, 7, and 8)
NCED:	Nine-*cis*-epoxycarotenoid dioxygenase
APCO:	Apocarotenoid 15,15′ oxygenase
RPE:	Retinal pigment epithelial membrane protein
CO:	Carotenoid oxygenase
LSD:	Lignostilbene-α, β-dioxygenase
HP:	Hypothetical proteins
UP:	Undefined proteins
NCE:	Neoxanthin cleavage enzyme
BS:	Bootstrap value.

Acknowledgments

The author thank Professor Fangqing Zhao (Beijing Institutes of Life Science, Chinese Academy of Sciences) for carefully modifying the draft. This work was supported by the National Natural Science Foundation of China (40876082), Chinese Academy of Sciences, international partnership program "Typical environment processes and their effects on resources," and Outstanding Young Scholars Fellowship of Shandong Province (Molecular Phycology JQ200914).

References

[1] R. Y. Stanier and G. Cohen-Bazire, "Phototrophic prokaryotes: the cyanobacteria," *Annual Review of Microbiology*, vol. 31, pp. 225–274, 1977.

[2] C. Xiaoyuan, Q. Yang, F. Zhao et al., "Comparative analysis of fatty acid desaturases in cyanobacterial genomes," *Comparative and Functional Genomics*, vol. 2008, Article ID 284508, 2008.

[3] D. Scanlan, "Cyanobacteria: ecology, niche adaptation and genomics," *Microbiology Today*, vol. 28, pp. 128–130, 2001.

[4] G. Rocap, F. W. Larimer, J. Lamerdin et al., "Genome divergence in two *Prochlorococcus* ecotypes reflects oceanic niche differentiation," *Nature*, vol. 424, no. 6952, pp. 1042–1047, 2003.

[5] A. Dufresne, M. Salanoubat, F. Partensky et al., "Genome sequence of the cyanobacterium *Prochlorococcus marinus* SS120, a nearly minimal oxyphototrophic genome," *Proceedings of the National Academy of Sciences of the United States of America*, vol. 100, no. 17, pp. 10020–10025, 2003.

[6] J. C. Meeks, J. Elhai, T. Thiel et al., "An overview of the Genome of *Nostoc punctiforme*, a multicellular, symbiotic Cyanobacterium," *Photosynthesis Research*, vol. 70, no. 1, pp. 85–106, 2001.

[7] F. Partensky, W. R. Hess, and D. Vaulot, "*Prochlorococcus*, a marine photosynthetic prokaryote of global significance," *Microbiology and Molecular Biology Reviews*, vol. 63, no. 1, pp. 106–127, 1999.

[8] Y. Nakamura, T. Kaneko, S. Sato et al., "Complete genome structure of *gloeobacter violaceus* PCC 7421, a cyanobacterium that lacks thylakoids (Supplement)," *DNA Research*, vol. 10, no. 4, pp. 181–201, 2003.

[9] B. C. Tan, S. H. Schwartz, J. A. D. Zeevaart, and D. R. Mccarty, "Genetic control of abscisic acid biosynthesis in maize," *Proceedings of the National Academy of Sciences of the United States of America*, vol. 94, no. 22, pp. 12235–12240, 1997.

[10] S. H. Schwartz, B. C. Tan, D. A. Gage, J. A. D. Zeevaart, and D. R. McCarty, "Specific oxidative cleavage of carotenoids by VP14 of maize," *Science*, vol. 276, no. 5320, pp. 1872–1874, 1997.

[11] J. Von Lintig and K. Vogt, "Filling the gap in vitamin A research. Molecular identification of an enzyme cleaving β-carotene to retinal," *Journal of Biological Chemistry*, vol. 275, no. 16, pp. 11915–11920, 2000.

[12] T. M. Redmond, S. Gentleman, T. Duncan et al., "Identification, expression, and substrate specificity of a mammalian β-carotene 15,15′-dioxygenase," *Journal of Biological Chemistry*, vol. 276, no. 9, pp. 6560–6565, 2001.

[13] A. Prado-Cabrero, D. Scherzinger, J. Avalos, and S. Al-Babili, "Retinal biosynthesis in fungi: characterization of the carotenoid oxygenase CarX from Fusarium fujikuroi," *Eukaryotic Cell*, vol. 6, no. 4, pp. 650–657, 2007.

[14] F. Bouvier, O. Dogbo, and B. Camara, "Biosynthesis of the food and cosmetic plant pigment bixin (annatto)," *Science*, vol. 300, no. 5628, pp. 2089–2091, 2003.

[15] F. Bouvier, C. Suire, J. Mutterer, and B. Camara, "Oxidative remodeling of chromoplast carotenoids: identification of the carotenoid dioxygenase *CsCCD* and *CsZCD* genes involved in *Crocus* secondary metabolite biogenesis," *Plant Cell*, vol. 15, no. 1, pp. 47–62, 2003.

[16] A. Prado-Cabrero, A. F. Estrada, S. Al-Babili, and J. Avalos, "Identification and biochemical characterization of a novel carotenoid oxygenase: elucidation of the cleavage step in the Fusarium carotenoid pathway," *Molecular Microbiology*, vol. 64, no. 2, pp. 448–460, 2007.

[17] M. E. Auldridge, D. R. McCarty, and H. J. Klee, "Plant carotenoid cleavage oxygenases and their apocarotenoid products," *Current Opinion in Plant Biology*, vol. 9, no. 3, pp. 315–321, 2006.

[18] B. Camara and F. Bouvier, "Oxidative remodeling of plastid carotenoids," *Archives of Biochemistry and Biophysics*, vol. 430, no. 1, pp. 16–21, 2004.

[19] H. Pfander and F. Wittwer, "Carotenoid composition in safran (author's transl)," *Helvetica Chimica Acta*, vol. 58, no. 7, pp. 2233–2236, 1975.

[20] D. P. Kloer and G. E. Schulz, "Structural and biological aspects of carotenoid cleavage," *Cellular and Molecular Life Sciences*, vol. 63, no. 19-20, pp. 2291–2303, 2006.

[21] D. S. Floss and M. H. Walter, "Role of carotenoid cleavage dioxygenase 1 (CCD1) in apocarotenoid biogenesis revisited," *Plant Signaling and Behavior*, vol. 4, no. 3, pp. 172–175, 2009.

[22] O. Ahrazem, A. Trapero, M. D. Gómez, A. Rubio-Moraga, and L. Gómez-Gómez, "Genomic analysis and gene structure of the plant carotenoid dioxygenase 4 family: a deeper study in *Crocus sativus* and its *allies*," *Genomics*, vol. 96, no. 4, pp. 239–250, 2010.

[23] B. C. Tan, L. M. Joseph, W. T. Deng et al., "Molecular characterization of the *Arabidopsis* 9-cis epoxycarotenoid dioxygenase gene family," *Plant Journal*, vol. 35, no. 1, pp. 44–56, 2003.

[24] A. Ohmiya, S. Kishimoto, R. Aida, S. Yoshioka, and K. Sumitomo, "Carotenoid cleavage dioxygenase (CmCCD4a) contributes to white color formation in Chrysanthemum petals," *Plant Physiology*, vol. 142, no. 3, pp. 1193–1201, 2006.

[25] G. Giuliano, S. Al-Babili, and J. Von Lintig, "Carotenoid oxygenases: cleave it or leave it," *Trends in Plant Science*, vol. 8, no. 4, pp. 145–149, 2003.

[26] A. Alder, I. Holdermann, P. Beyer, and S. Al-Babili, "Carotenoid oxygenases involved in plant branching catalyse a highly specific conserved apocarotenoid cleavage reaction," *Biochemical Journal*, vol. 416, no. 2, pp. 289–296, 2008.

[27] S. Ruch, P. Beyer, H. Ernst, and S. Al-Babili, "Retinal biosynthesis in Eubacteria: in vitro characterization of a novel carotenoid oxygenase from *Synechocystis* sp. PCC 6803," *Molecular Microbiology*, vol. 55, no. 4, pp. 1015–1024, 2005.

[28] E. K. Marasco, K. Vay, and C. Schmidt-Dannert, "Identification of carotenoid cleavage dioxygenases from *Nostoc* sp. PCC 7120 with different cleavage activities," *Journal of Biological Chemistry*, vol. 281, no. 42, pp. 31583–31593, 2006.

[29] D. Scherzinger and S. Al-Babili, "In vitro characterization of a carotenoid cleavage dioxygenase from *Nostoc* sp. PCC 7120 reveals a novel cleavage pattern, cytosolic localization and induction by highlight," *Molecular Microbiology*, vol. 69, no. 1, pp. 231–244, 2008.

[30] D. Scherzinger, S. Ruch, D. P. Kloer, A. Wilde, and S. Al-Babili, "Retinal is formed from apo-carotenoids in *Nostoc* sp. PCC7120: in vitro characterization of an apo-carotenoid oxygenase," *Biochemical Journal*, vol. 398, no. 3, pp. 361–369, 2006.

[31] Q. Jiang, S. Qin, and Q. Y. Wu, "Genome-wide comparative analysis of metacaspases in unicellular and filamentous cyanobacteria," *BMC Genomics*, vol. 11, no. 1, article 198, 2010.

[32] X. Zhang, F. Zhao, X. Guan, Y. Yang, C. Liang, and S. Qin, "Genome-wide survey of putative Serine/Threonine protein kinases in cyanobacteria," *BMC Genomics*, vol. 8, article 395, 2007.

[33] F. Zhao, X. Zhang, C. Liang, J. Wu, Q. Bao, and S. Qin, "Genome-wide analysis of restriction-modification system in unicellular and filamentous cyanobacteria," *Physiological Genomics*, vol. 24, no. 3, pp. 181–190, 2006.

[34] C. Liang, F. Zhao, W. Wei, Z. Wen, and S. Qin, "Carotenoid biosynthesis in cyanobacteria: structural and evolutionary scenarios based on comparative genomics," *International Journal of Biological Sciences*, vol. 2, no. 4, pp. 197–207, 2006.

[35] D. W. Mount, "Using the basic local alignment search tool (BLAST)," *CSH Protocols*, vol. 2007, 2007.

[36] S. F. Altschul, W. Gish, W. Miller, E. W. Myers, and D. J. Lipman, "Basic local alignment search tool," *Journal of Molecular Biology*, vol. 215, no. 3, pp. 403–410, 1990.

[37] S. F. Altschul, T. L. Madden, A. A. Schäffer et al., "Gapped BLAST and PSI-BLAST: q new generation of protein database search programs," *Nucleic Acids Research*, vol. 25, no. 17, pp. 3389–3402, 1997.

[38] J. D. Thompson, D. G. Higgins, and T. J. Gibson, "CLUSTAL W: improving the sensitivity of progressive multiple sequence alignment through sequence weighting, position-specific gap penalties and weight matrix choice," *Nucleic Acids Research*, vol. 22, no. 22, pp. 4673–4680, 1994.

[39] K. Tamura, J. Dudley, M. Nei, and S. Kumar, "MEGA4: Molecular Evolutionary Genetics Analysis (MEGA) software version 4.0," *Molecular Biology and Evolution*, vol. 24, no. 8, pp. 1596–1599, 2007.

[40] S. Guindon and O. Gascuel, "A Simple, Fast, and Accurate Algorithm to Estimate Large Phylogenies by Maximum Likelihood," *Systematic Biology*, vol. 52, no. 5, pp. 696–704, 2003.

[41] S. Q. Le and O. Gascuel, "An improved general amino acid replacement matrix," *Molecular Biology and Evolution*, vol. 25, no. 7, pp. 1307–1320, 2008.

[42] F. Chevenet, C. Brun, A. L. Bañuls, B. Jacq, and R. Christen, "TreeDyn: towards dynamic graphics and annotations for analyses of trees," *BMC Bioinformatics*, vol. 7, article 439, 2006.

[43] T. L. Bailey and C. Elkan, "The value of prior knowledge in discovering motifs with MEME," *Proceedings of International Conference on Intelligent Systems for Molecular Biology*, vol. 3, pp. 21–29, 1995.

[44] X. Chi, Q. Yang, Y. Lu et al., "Genome-wide analysis of fatty acid desaturases in Soybean (*Glycine max*)," *Plant Molecular Biology Reporter*, vol. 29, no. 4, pp. 769–783, 2011.

[45] T. L. Bailey and C. Elkan, "Fitting a mixture model by expectation maximization to discover motifs in biopolymers," *Proceedings of International Conference on Intelligent Systems for Molecular Biology*, vol. 2, pp. 28–36, 1994.

[46] A. Marchler-Bauer and S. H. Bryant, "CD-Search: protein domain annotations on the fly," *Nucleic Acids Research*, vol. 32, pp. W327–W331, 2004.

[47] I. Letunic, T. Doerks, and P. Bork, "SMART 6: recent updates and new developments," *Nucleic Acids Research*, vol. 37, no. 1, pp. D229–D232, 2009.

[48] K. Arnold, L. Bordoli, J. Kopp, and T. Schwede, "The SWISS-MODEL workspace: a web-based environment for protein

structure homology modelling," *Bioinformatics*, vol. 22, no. 2, pp. 195–201, 2006.

[49] T. Schwede, J. Kopp, N. Guex, and M. C. Peitsch, "SWISS-MODEL: an automated protein homology-modeling server," *Nucleic Acids Research*, vol. 31, no. 13, pp. 3381–3385, 2003.

[50] N. Guex and M. C. Peitsch, "SWISS-MODEL and the Swiss-PdbViewer: an environment for comparative protein modeling," *Electrophoresis*, vol. 18, no. 15, pp. 2714–2723, 1997.

[51] P. Benkert, M. Biasini, and T. Schwede, "Toward the estimation of the absolute quality of individual protein structure models," *Bioinformatics*, vol. 27, no. 3, pp. 343–350, 2011.

[52] A. Dufresne, L. Garczarek, and F. Partensky, "Accelerated evolution associated with genome reduction in a free-living prokaryote," *Genome Biology*, vol. 6, no. 2, p. R14, 2005.

[53] R. J. Pryce, "The occurrence of lunularic and abscisic acids in plants," *Phytochemistry*, vol. 11, no. 5, pp. 1759–1761, 1972.

[54] N. Tominaga, M. Takahata, and H. Tominaga, "Effects of NaCl and KNO_3 concentrations on the abscisic acid content of *Dunaliella* sp. (Chlorophyta)," *Hydrobiologia*, vol. 267, no. 1–3, pp. 163–168, 1993.

[55] M. Kobayashi, N. Hirai, Y. Kurimura, H. Ohigashi, and Y. Tsuji, "Abscisic acid-dependent algal morphogenesis in the unicellular green alga *Haematococcus pluvialis*," *Plant Growth Regulation*, vol. 22, no. 2, pp. 79–85, 1997.

[56] K. Nimura and H. Mizuta, "Inducible effects of abscisic acid on sporophyte discs from *Laminaria japonica* Areschoug (Laminariales, Phaeophyceae)," *Journal of Applied Phycology*, vol. 14, no. 3, pp. 159–163, 2002.

[57] C. F. Mouchel and O. Leyser, "Novel phytohormones involved in long-range signaling," *Current Opinion in Plant Biology*, vol. 10, no. 5, pp. 473–476, 2007.

[58] J. Booker, M. Auldridge, S. Wills, D. McCarty, H. Klee, and O. Leyser, "MAX3/CCD7 is a carotenoid cleavage dioxygenase required for the synthesis of a novel plant signaling molecule," *Current Biology*, vol. 14, no. 14, pp. 1232–1238, 2004.

[59] S. H. Schwartz, X. Qin, and M. C. Loewen, "The biochemical characterization of two carotenoid cleavage enzymes from *Arabidopsis* indicates that a carotenoid-derived compound inhibits lateral branching," *Journal of Biological Chemistry*, vol. 279, no. 45, pp. 46940–46945, 2004.

[60] M. E. Auldridge, A. Block, J. T. Vogel et al., "Characterization of three members of the Arabidopsis carotenoid cleavage dioxygenase family demonstrates the divergent roles of this multifunctional enzyme family," *Plant Journal*, vol. 45, no. 6, pp. 982–993, 2006.

[61] W. Maier, H. Peipp, J. Schmidt, V. Wray, and D. Strack, "Levels of a terpenoid glycoside (Blumenin) and cell wall-bound phenolics in some cereal mycorrhizas," *Plant Physiology*, vol. 109, no. 2, pp. 465–470, 1995.

[62] A. Klingner, H. Bothe, V. Wray, and F. J. Marner, "Identification of a yellow pigment formed in maize roots upon mycorrhizal colonization," *Phytochemistry*, vol. 38, no. 1, pp. 53–55, 1995.

[63] M. H. Walter, T. Fester, and D. Strack, "Arbuscular mycorrhizal fungi induce the non-mevalonate methylerythritol phosphate pathway of isoprenoid biosynthesis correlated with accumulation of the "yellow pigment" and other apocarotenoids," *Plant Journal*, vol. 21, no. 6, pp. 571–578, 2000.

[64] S. Mathieu, N. Terrier, J. Procureur, F. Bigey, and Z. Günata, "A Carotenoid Cleavage Dioxygenase from *Vitis vinifera* L.: functional characterization and expression during grape berry development in relation to C_{13}-norisoprenoid accumulation," *Journal of Experimental Botany*, vol. 56, no. 420, pp. 2721–2731, 2005.

[65] S. H. Schwartz, X. Qin, and J. A. D. Zeevaart, "Characterization of a novel carotenoid cleavage dioxygenase from plants," *Journal of Biological Chemistry*, vol. 276, no. 27, pp. 25208–25211, 2001.

[66] A. Rubio, J. L. Rambla, M. Santaella et al., "Cytosolic and plastoglobule-targeted carotenoid dioxygenases from *Crocus sativus* are both involved in β-ionone release," *Journal of Biological Chemistry*, vol. 283, no. 36, pp. 24816–24825, 2008.

[67] J. T. Vogel, B. C. Tan, D. R. McCarty, and H. J. Klee, "The carotenoid cleavage dioxygenase 1 enzyme has broad substrate specificity, cleaving multiple carotenoids at two different bond positions," *Journal of Biological Chemistry*, vol. 283, no. 17, pp. 11364–11373, 2008.

[68] C. Kiefer, S. Hessel, J. M. Lampert et al., "Identification and characterization of a mammalian enzyme catalyzing the asymmetric oxidative cleavage of orovitamin A," *Journal of Biological Chemistry*, vol. 276, no. 17, pp. 14110–14116, 2001.

[69] D. P. Kloer, S. Ruch, S. Al-Babili, P. Beyer, and G. E. Schulz, "The structure of a retinal-forming carotenoid oxygenase," *Science*, vol. 308, no. 5719, pp. 267–269, 2005.

[70] M. H. Walter and D. Strack, "Carotenoids and their cleavage products: biosynthesis and functions," *Natural Product Reports*, vol. 28, no. 4, pp. 663–692, 2011.

[71] Y. Takahashi, G. Moiseyev, Y. Chen, and J. X. Ma, "Identification of conserved histidines and glutamic acid as key residues for isomerohydrolase activity of RPE65, an enzyme of the visual cycle in the retinal pigment epithelium," *FEBS Letters*, vol. 579, no. 24, pp. 5414–5418, 2005.

[72] P. Ala-Laurila, K. Donner, and A. Koskelainen, "Thermal activation and photoactivation of visual pigments," *Biophysical Journal*, vol. 86, no. 6, pp. 3653–3662, 2004.

[73] G. Moiseyev, Y. Takahashi, Y. Chen et al., "RPE65 is an iron(II)-dependent isomerohydrolase in the retinoid visual cycle," *Journal of Biological Chemistry*, vol. 281, no. 5, pp. 2835–2840, 2006.

8

Intron Retention and TE Exonization Events in *ZRANB2*

Sang-Je Park,[1,2] Jae-Won Huh,[1] Young-Hyun Kim,[1,3] Heui-Soo Kim,[2] and Kyu-Tae Chang[1,3]

[1] *National Primate Research Center, Korea Research Institute of Bioscience and Biotechnology, Ochang, Chungbuk 363-883, Republic of Korea*
[2] *Department of Biological Sciences, College of Natural Sciences, Pusan National University, Busan 609-735, Republic of Korea*
[3] *National Primate Research Center, Korea Research Institute of Bioscience and Biotechnology, University of Science & Technology, Ochang, Chungbuk 363-883, Republic of Korea*

Correspondence should be addressed to Kyu-Tae Chang, changkt@kribb.re.kr

Academic Editor: Kyudong Han

The Zinc finger, RAN-binding domain-containing protein 2 (*ZRANB2*), contains arginine/serine-rich (RS) domains that mediate its function in the regulation of alternative splicing. The *ZRANB2* gene contains 2 LINE elements (L3b, Plat_L3) between the 9th and 10th exons. We identified the exonization event of a LINE element (Plat_L3). Using genomic PCR, RT-PCR amplification, and sequencing of primate DNA and RNA samples, we analyzed the evolutionary features of *ZRANB2* transcripts. The results indicated that 2 of the LINE elements were integrated in human and all of the tested primate samples (hominoids: 3 species; Old World monkey: 8 species; New World monkey: 6 species; prosimian: 1 species). Human, rhesus monkey, crab-eating monkey, African-green monkey, and marmoset harbor the exon derived from LINE element (Plat_L3). RT-PCR amplification revealed the long transcripts and their differential expression patterns. Intriguingly, these long transcripts were abundantly expressed in Old World monkey lineages (rhesus, crab-eating, and African-green monkeys) and were expressed via intron retention (IR). Thus, the *ZRANB2* gene produces 3 transcript variants in which the Cterminus varies by transposable elements (TEs) exonization and IR mechanisms. Therefore, *ZRANB2* is valuable for investigating the evolutionary mechanisms of TE exonization and IR during primate evolution.

1. Introduction

Zinc finger, RAN-binding domain-containing protein 2 (*ZRANB2*), also known as ZIS and ZNF265, lies on human chromosome 1p31 and was identified in rat renal juxtaglomerular (JG) cells [1], human [2], and mouse [3]. *ZRANB2* contains 2 zinc finger domains, a C-terminal RS (arginine/serine-rich) domain, a glutamic acid-rich domain, and a nuclear localization sequence [4]. It interacts with components of the splicing factors *U170K*, *U2AF35*, and *XE7* and regulates splicing of *GluR-B*, *SMN2*, and *Tra2β* [5–7]. The 2 zinc finger domains recognize single-stranded RNA (ssRNA) and bind to a consensus AGGUAA motif [8]. *ZRANB2*, thus, mediates alternative splicing of pre-mRNA, is ubiquitously expressed in various tissues, and is highly conserved from nematodes to humans [3, 5, 9].

Alternative splicing (AS) of premessenger RNAs (pre-mRNAs) is an important molecular mechanism that increases human transcriptome complexity and flexibility [10]. Genome-wide analyses of AS events suggest that 40–60% of human genes have alternatively spliced transcripts [11]. With the aid of accumulated transcriptome sequencing data, 5 distinct AS mechanisms have been identified, including exon skipping, alternative 5′ splice sites, alternative 3′ splice sites, intron retention, and mutual exclusion. Intron retention (IR) events are very rare, accounting for less than 3% of all AS events in the human and mouse genomes [10]. In humans, 14.8% of the 21,106 known genes showed at least 1 IR event, mostly involving untranslated regions (UTRs). Eighty-eight cases of IR events seem to be involved in several syndrome-associated genes and tumorigenic processes [12]. Although exonization is not categorized as 1 of the 5 distinct mechanisms, it represents an AS process by which new exons are acquired from intronic DNA sequences [13]. The generation of canonical splicing sites (splicing acceptor and donor sites) by genomic insertions/deletions or mutations could cause the exonization events [13]. Recent studies have indicated that exonization events are derived from transposable elements (TEs) such as LTR retrotransposons (e.g., human endogenous retroviruses (HERVs)) and non-LTR retrotransposons (e.g., short interspersed elements [SINEs] and long interspersed elements (LINEs)) in various species [14, 15]. Many TE sequences contain potential splice sites [16]. TEs are a major component, comprising more than 40% of the human genome [17]. LINEs and LTR retrotransposons insert themselves into new genomic positions through a copy and paste mechanism by encoding their own reverse transcriptase [18]. SINEs do not have their own reverse transcriptase domain; however, they are reverse transcribed and inserted into the genome by LINE element-enzymatic machinery [13]. TEs regulate gene activity by providing promoter, enhancer, exonization, and new polyadenylation signals [19]. Thus, various genes are regulated by TEs [20]. Nekrutenko and Li suggest that TEs are located in ~4% of protein-coding regions in the human genome [21]. According to the genome-wide survey by Sela et al., 1824 human genes have TE-derived exons; SINE and LINE elements comprise approximately 68% and 18% of exonized TEs, respectively, [22]. Although LINE elements are present in fewer copies and mediate exonizations at a lower frequency than SINE elements do, exonization events of LINEs frequently occur in the human genome. LINEs comprise up to 20% of the human genome [23]. LINE1, LINE2, and LINE3/CR1 are 3 distantly related LINE families that represent approximately 17%, 3%, and 0.3% of the human genome, respectively, [23]. Thus, transposable elements and intronic sequences may serve as transcript units to enrich the transcriptome with limited genomic resources [13].

We performed evolutionary and comparative analyses of rhesus, crab-eating, and African-green monkeys and marmosets to investigate exonization events derived from insertion of LINEs and IR in the protein-coding regions of human *ZRANB2*.

2. Materials and Methods

2.1. RNA and Genomic DNA Samples. Total RNA from human (*Homo sapiens* cerebrum, colon, liver, lung, kidney, and stomach) and rhesus monkey (*Macaca mulatta* cerebrum, colon, liver, lung, kidney, pancreas, and stomach) were purchased from Clontech. Total RNA from crab-eating monkey (*Macaca fascicularis* cerebrum, colon, liver, lung, kidney, pancreas, and stomach), African-green monkey (*Cercopithecus aethiops* cerebrum, colon, liver, lung, kidney, pancreas, and stomach), and marmoset (*Callithrix jacchus* cerebrum, colon, liver, lung, Kidney, and stomach) were extracted with the RNeasy mini-kit (Qiagen) and the RNase-Free DNase set (Qiagen). Total RNA samples were provided by the National Primate Research Center (NPRC) of Korea.

We used a standard protocol to isolate genomic DNA form heparinized blood samples from the following species: (1) HU: human (*Homo sapiens*); (2) hominoids, CH: chimpanzee (*Pan troglodytes*), BO: bonobo (*Pan paniscus*); (3) Old World monkeys, RH: rhesus monkey (*Macaca mulatta*), JA: Japanese monkey (*Macaca fuscata*), CR: crab-eating monkey (*Macaca fascicularis*), PI: pig-tail monkey (*Macaca nemestrina*), AF: African-green monkey (*Cercopithecus aethiops*), MA: mandrill (*Mandrillus sphinx*), CO: colobus (*Procolobus sp.*), LA: langur (*Trachypithecus sp.*); (4) New World monkeys, MAR: marmoset (*Callithrix jacchus*), TA: tamarin (*Saguinus midas*), CA: capuchin (*Cebus apella*), SQ: squirrel monkey (*Saimiri sciureus*), NI: night monkey (*Aotus nigriceps*), SP: spider monkey (*Ateles geoffroyi*); (5) prosimian, RL: ring-tailed lemur (*Lemur catta*).

2.2. RT-PCR and PCR Amplification. *ZRANB2* transcripts were analyzed by RT-PCR amplification. M-MLV reverse transcriptase with an annealing temperature of 42°C was used with an RNase inhibitor (Promega). We performed PCR amplification of pure mRNA samples without reverse transcription to demonstrate that the mRNA samples prepared did not contain genomic DNA (data not shown). As a standard control, *RPL32* was amplified by the primers RPL32-S (5′-CAA CAT TGG TTA TGG AAG CAA CA-3′) and RPL32-AS (5′-TGA CGT TGT GGA CCA GGA ACT-3′) from human and rhesus monkey. TE fusion transcript was amplified by the primer pair S1 (5′-GAA ATA TCC CGA CAG GGT TC-3′) and AS1 (5′-GCT GCT TTC TTC AAT GGT CTG-3′) derived from the *ZRANB2* gene (Genbank accession no. NM_005455.4). RT-PCR experiments were carried out for 30 cycles of 94°C for 30 s, 58°C for 30 s, and 72°C for 1 m 30 s. Genomic DNAs from various primates were PCR amplified. The Plat_L3 and L3b elements were amplified by primer pairs S2 (5′-CCC TGT GAC ACG GTG TAG AA-3′) and AS2 (5′-CAG ATC ATT GGG AAT CTG TCC-3′). The 9th exon boundary of *ZRANB2* was amplified by primers S3 (5′-ATA AAA ATC TAA CCC TTG ACT AGG AA-3′) and AS3 (5′-AAA CGT AAA GTC CTG TTA ATG CAG-3′). Genomic PCR experiments were carried out for 30 cycles of 94°C for 30 s, 58°C for 30 s, and 72°C for 30 s.

2.3. Molecular Cloning of Genomic PCR and RT-PCR Products and Sequencing Procedure. RT-PCR products were separated on a 1.2% agarose gel, purified with the Gel SV extraction kit (GeneAll), and cloned into the pGEM-T-easy vector (Promega). The cloned DNA was isolated using the Plasmid DNA mini-prep kit (GeneAll). Sequencing of primate DNA samples and alternative transcripts was performed by a commercial sequencing company (Macrogen).

3. Results

3.1. Structural Analysis of the ZRANB2 Gene. The *ZRANB2* gene shows two different transcripts (NM_203350.2 and NM_005455.4) in humans according to the GenBank database (Figure 1). Isoform a (NM_203350.2) is composed of 10 exons and is transcribed into a 3070-bp mRNA sequence with a 294-bp 5′ UTR, 993-bp coding sequence, and 1783-bp 3′ UTR. Isoform b (NM_005455.4) is composed of 11 exons and is transcribed into a 3145-bp mRNA with a 294-bp 5′ UTR, 963-bp-coding sequence, and 1888-bp 3′ UTR. The isoform b transcript includes a 196-bp Plat_L3 element

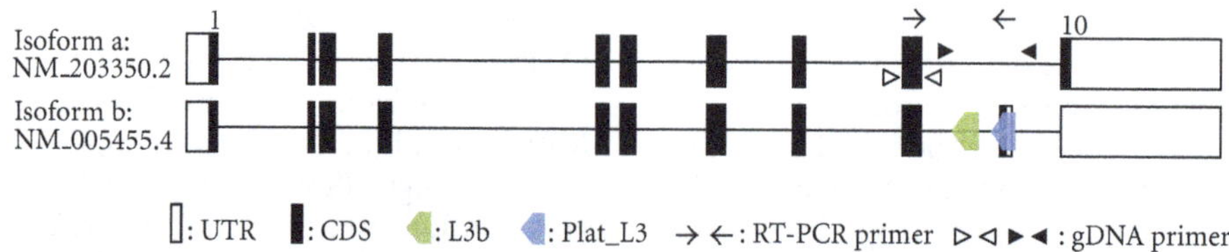

Figure 1: Structural analysis of human *ZRANB2* gene transcripts. The Plat_L3 and L3b elements are located in intron 9. Open and closed boxes represent the untranslated region of the exons and protein-coding region, respectively. Arrows indicate the RT-PCR primer position, and open and closed arrowheads indicate genomic PCR primer positions. These figures are structural illustrations, not to scale.

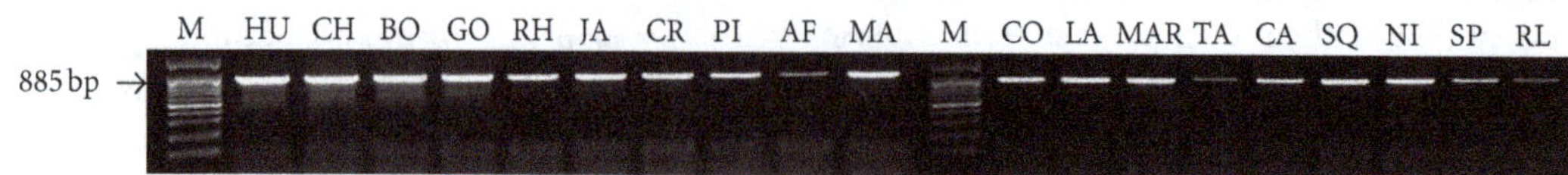

Figure 2: PCR amplification of Plat_L3 and L3b in various primates. Primate DNA samples were utilized for integration analyses of the Plat_L3 and L3b elements in the *ZRANB2* gene. M indicates the molecular size marker. Primate DNA samples are abbreviated as follows: (1) HU: human (*Homo sapiens*); (2) Hominoids: CH: chimpanzee (*Pan troglodytes*), BO: bonobo (*Pan paniscus*), GO: gorilla (*Gorilla gorilla*); (3) Old World monkey: RH: rhesus monkey (*Macaca mulatta*), JA: Japanese monkey (*Macaca fuscata*), CR: crab-eating monkey (*Macaca fascicularis*), PI: pig-tail monkey (*Macaca nemestrina*), AF: African green monkey (*Cercopithecus aethiops*), MA: mandrill (*Mandrillus sphinx*), CO: colobus (*Procolobus sp.*), LA: langur (*Trachypithecus sp.*); (4) New World monkey: MAR: Marmoset (*Callithrix jacchus*), TA: tamarin (*Saguinus midas*), CA: capuchin (*Cebus apella*), SQ: squirrel monkey (*Saimiri sciureus*), NI: night monkey (*Aotus nigriceps*), SP: spider monkey (*Ateles geoffroyi*); (5) prosimian: RL: ring-tailed lemur (*Lemur catta*).

belonging to the CR1/LINE3 family from *Ornithorhynchus*. This antisense-oriented Plat_L3 element is inserted between the 9th and 10th exons of *ZRANB2* and provides canonical splicing sites (splicing acceptor and donor site) that produce a new Plat_L3-derived exon via exonization. A premature termination codon (PTC) is generated in the Plat_L3-derived exon of isoform b. The isoform a transcript encodes 330 amino acids, but the isoform b transcript encodes 320 amino acids (Supplemantry Figure 1 in Supplementary Materials available online at doi: 10.1155/2012/170208). These transcript variants possess Ctermini with different amino acid compositions. Another L3b element of the CR1/LINE3 family is integrated in the 5′ upstream intronic region of the Plat_L3 element (Figure 3(f)).

3.2. Integration Time of Plat_L3 and L3b Elements. To determine when the Plat_L3 and L3b elements were integrated during primate radiation, we performed genomic PCR amplification using primer pairs specific to highly conserved regions in various primate samples (Figure 2). We validated the randomly selected amplified products by sequencing (see Supplementary Figure 2). The Plat_L3 and L3b elements in the *ZRANB2* gene were integrated in all tested primate lineages, including hominoids (human, chimpanzee, bonobo, and gorilla), Old World monkeys (rhesus monkey, Japanese monkey, crab-eating monkey, pig-tail monkey, African-green monkey, mandrill, colobus, and langur), New World monkeys (marmoset, tamarin, capuchin, squirrel monkey, night monkey, and spider monkey), and prosimian (ring-tailed lemur).

3.3. RT-PCR Amplification and Sequencing Analysis of ZRANB2 in Human and Monkeys. To investigate the expression pattern of isoform b transcript (containing the Plat_L3 element-derived exon), we performed comparative RT-PCR analysis in 6 human and marmoset tissues (cerebrum, colon, liver, lung, kidney, and stomach) and 7 rhesus monkey, crab-eating monkey, and African-green monkey tissues (cerebrum, colon, liver, lung, kidney, pancreas, and stomach). To amplify the isoform b-specific transcript in primate tissues, we designed primers specific to the 9th exon (sense primer) and Plat_L3-derived exon (antisense primer). We confirmed that these primer pair sequences were highly conserved in human and marmoset. The expected RT-PCR products (215 bp) were ubiquitously transcribed in all tested samples (Figures 3(a)–3(e)). Remarkably, an unexpected product (upper bands) was also ubiquitously expressed in rhesus monkey, crab-eating monkey, and African-green monkey. Although the unexpected bands were not detectable by gel electrophoresis of human and marmoset samples, we confirmed that the very weak bands were present in these species. Sequencing of the amplified product revealed that it was an intron-retained transcript variant (V1) (Figure 3(f)).

4. Discussion

4.1. TE-Exonization of LINE Elements during Primate Evolution. A structural analysis revealed that Plat_L3 is integrated between the 9th and 10th exons in an intronic region of *ZRANB2*, where it provides an alternative splicing site that generates a new Plat_L3-derived exon via exonization (Figures 1 and 4). To determine when the Plat_L3 element was integrated, PCR amplification was performed in various primate genomes (Figure 2). Plat_L3 was integrated in all primate lineages, suggesting that Plat_L3 was integrated in a common ancestor prior to the divergence of simian and prosimian, possibly more than 63 million years ago. RT-PCR showed that the Plat_L3-encoded exon is transcribed

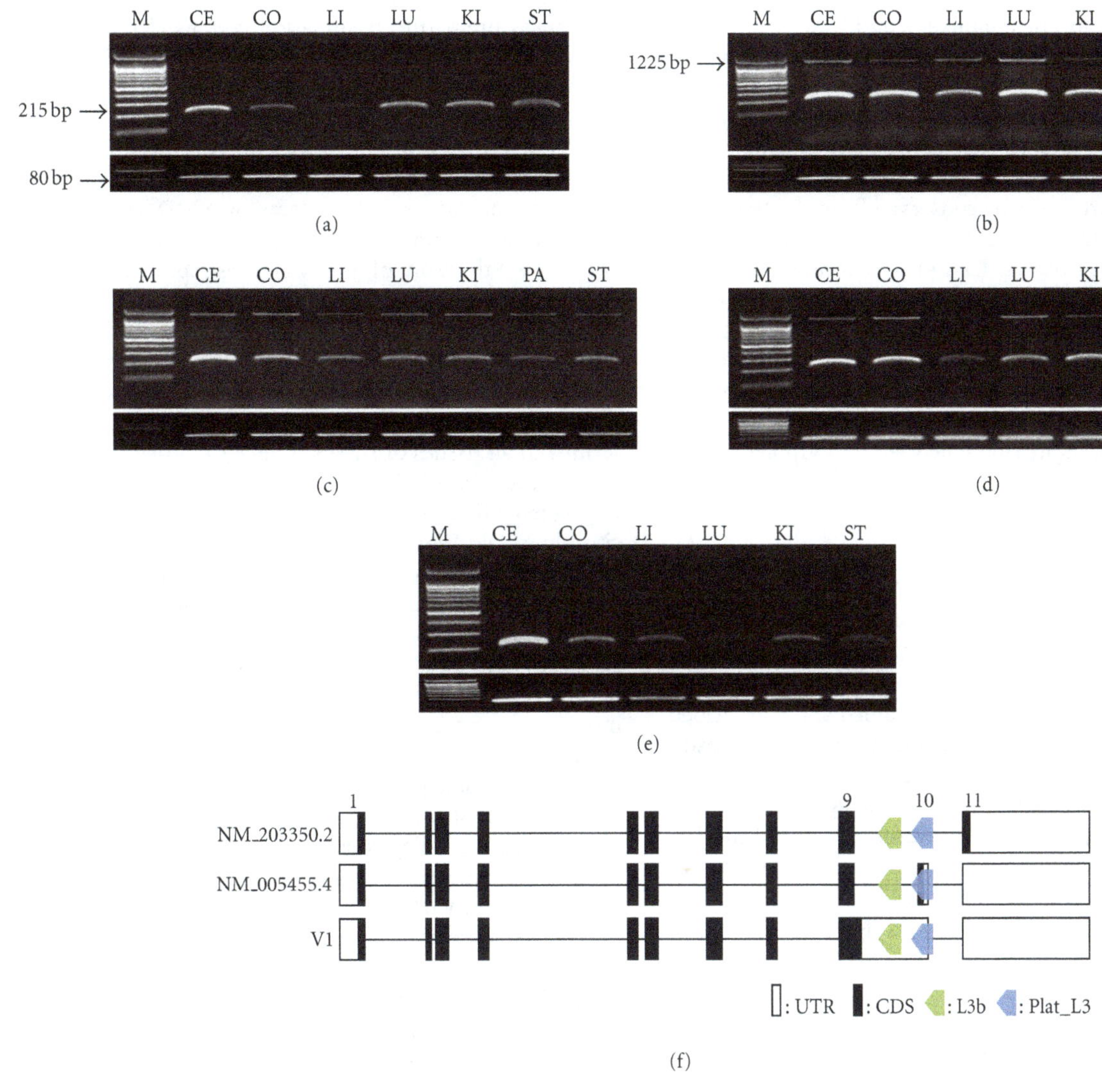

FIGURE 3: RT-PCR amplification and transcript variants of the *ZRANB2* gene. NM_005455.4 transcripts (215 bp) and V1 transcript (1225 bp) in various tissues of human (a), rhesus monkey (b), crab-eating monkey (c), African-green monkey (d), and marmoset (e). *RPL32* (80 bp) indicates the positive control. M indicates the size marker. Primate cDNA samples are abbreviated as follows: CE: cerebellum; CO: colon; LI: liver; LU: lung; KI: kidney; PA: pancreas; ST: stomach. Structural analysis of transcript variants (f). Open and closed boxes represent the untranslated region of the exons and protein-coding region, respectively.

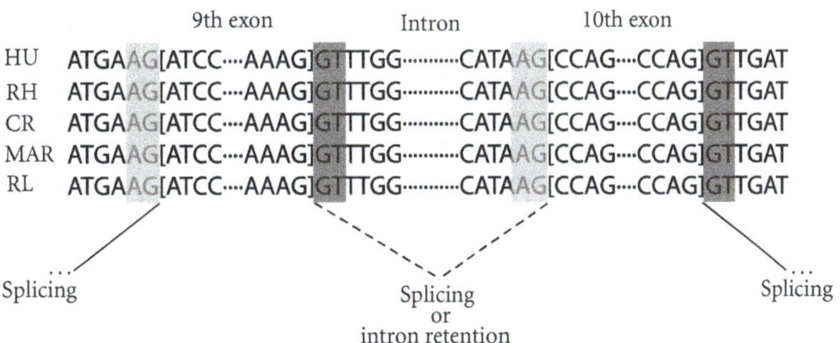

FIGURE 4: Boundary sequences of potential 3′ and 5′ alternative splice sites in various primates. Canonical splicing acceptor site "AG" and donor site "GT" are indicated in gray and dark gray, respectively. The sliced line indicates splicing, and the dotted line indicates splicing or retained intron.

in human, Old World monkey (rhesus monkey, crab-eating monkey, and African-green monkey), and New World monkey (marmoset) (Figures 3(a)–3(e)). Sequencing analysis showed that integrated Plat_L3 sequences are highly conserved in all primate lineages, in comparison to the adjacent intronic sequences and L3b element (Supplementary Figure 2). Moreover, canonical splicing sites (splicing acceptor and donor sites) are perfectly conserved from the hominoid to prosimian lineages (Figure 4). Therefore, perfectly conserved splicing sites of Plat_L3 and well-conserved Plat_L3 sequences could be exonized in *ZRANB2* gene transcripts in different primate lineages, including hominoids, Old World monkeys, and New World monkeys. We were unable to validate the Plat_L3-derived transcripts in prosimian samples, but, based on sequence analysis, we assume the existence of exonization events in ring-tailed lemur (Supplementary Figure 2).

4.2. Abundant IR Event in Old World Monkeys. An unexpected large band (1225 bp) detected by RT-PCR amplification was found in all tissues of Old World monkeys (rhesus monkey, crab-eating monkey, and African-green monkey) (Figure 3). Sequencing revealed this to be an alternatively spliced variant with the 9th intron retained and transcribed via an IR event [10]. Intriguingly, the intron-retained exon was detected at very low levels in human and marmoset monkey, unlike Old World monkeys. We suggest 2 alternative hypotheses of lineage-specific IR events and mixed mechanisms, including IR events and lineage-specific protection of nonsense-mediated mRNA decay (NMD).

In the previous model of IR events, high GC content in an intron sequence reduced the excision rate [24], and the retained introns are significantly shorter than nonretained introns [25]. However, we found that the 9th intron sequence of *ZRANB2* has a low GC content (approximately 30%) in human, rhesus monkey, and marmoset and was 1023 bp long, which is greater than the length of nonretained introns. Therefore, these older models could not explain the results of the IR event in ZRNAB2. The splicing acceptor and donor sites of the 9th and 10th exons were highly conserved in primates. Therefore, the IR event was not induced by weak signals of alternative splicing sites (Figure 4). Recent studies have shown that splicing is repressed by the binding of polypyrimidine tract-binding proteins (*PTB*) to specific sequence motifs (CUCUCU, UUCUCU, UUCCUU, and CUUCUUC), induced by IR events in *FOSB* [26]. We found that the *PTB*-binding consensus sequence UUCUCU 16 bp upstream from the 3′ end of the 9th intron was perfectly conserved from hominoid to prosimian (Supplementary Figure 2). Moreover, recent studies suggested regulation of *PTB* movement from the cytoplasm into the nucleus by phosphorylation [27]. These concepts are merged in the *ZRANB2* gene, where the IR event may occur via the *PTB*-binding sequence in human, rhesus monkey, crab-eating monkey, African-green monkey, and marmoset; however, the phosphorylation status of PTB may regulate the IR event in the primate lineage.

To explain the mixed mechanisms including IR events and lineage-specific protection of NMD, we first analyzed the relationship between IR events and NMD. In most cases, IR introduces premature termination codons (PTCs) into the mRNA, typically resulting in degradation by NMD [28]. The exon junction complex (EJC) NMD model is a well-known regulatory mechanism in mammals. The distance of PTC from the exon-exon junction is important; PTC located more than 50–55 nucleotides (nt) upstream of the last exon-exon junction causes mRNA decay by NMD, whereas PTC-located downstream of this boundary does not induce NMD [29]. The 9th exon of the V1 transcript containing the retained intron induced a PTC in the adjacent 5′ exonic region (Figure 3(f)). This PTC is located more than 50–55 nucleotides (nt) upstream of the last exon-exon junction. Theoretically, the V1 transcript should be degraded by the NMD mechanism in all tested primates; however, it was transcribed in all tissues of rhesus monkey, crab-eating monkey, and African-green monkey (Figures 3(a)–3(e)). Therefore, we suggest a lineage-specific protection event for NMD, specifically in Old World monkeys. Although a lineage-specific NMD protection mechanism has not been clearly established, a few studies have shown that the cytidine (C) to uridine (U) RNA-edited *APOB* mRNA was protected from NMD by the *APOBEC1-ACF*-editing complex [30]. The uORF-containing thrombopoietin (*TOP*) gene and nonsense mutation in the first exon of the β-globin (*HBB*) gene also escape NMD [31, 32]. Therefore, these elements specific to Old World monkeys may yield similar results. Supplementary Figure 2 illustrates the 7 Old World monkey-specific nucleotides. We believe that this sequence affects protection from NMD. However, further studies are needed to demonstrate the validity of this mechanism, including the relationship between specific nucleotides and NMD escape.

The species- and lineage-specific IR, NMD, and NMD escape mechanisms have not been demonstrated. A large number of IR studies have been performed in human and mouse. Therefore, experimental validation of our results in Old World monkeys (rhesus monkey, crab-eating monkey, and African-green monkey) suggesting that abundant IR events could be an attractive topic for IR and NMD research.

4.3. Three Isoforms of ZRANB2. Different transcripts are generated by the Plat_L3-derived exonization and intron retention events in *ZRANB2*. These transcripts encode Ctermini of varying size and amino-acid composition (Supplementary Figure 1). Isoform a (NM_203350.2), isoform b (NM_005455.4), and V1 encoded 330, 320, and 314 amino acids, respectively. The amino-acid sequences encoded by isoform b (SQVIGENTKQP) and V1 (FGFL) differ from the sequence in the C-terminus of isoform a. This region belongs to the SR domain, which is essential for nuclear localization of *ZRANB2* and is regulated by phosphorylation [33]. These functional features were demonstrated by an SR deletion study. However, more detailed deletion studies have not been performed. Although we did not perform a functional analysis, our results indicate that Plat_L3-exonization and intron retention events could yield various transcripts in the primate lineage. A previous study suggested that integrated full-length L1 elements in the intronic region of host genes could cause transcriptional interference (TI) by IR and exonization [34]. However, isoforms a and b and variant transcripts of the *ZRANB2* gene were not prematurely terminated by IR and exonization derived from the Plat_L3 element (Figure 3(f)). Our results also show that Plat_L3-exonized isoform b was transcribed in all tested human and monkey samples. The TI effect did not occur in *ZRANB2* by IR and exonization events derived from the Plat_L3 element. Therefore, we focused on the mechanism of gene diversity by IR and exonization.

5. Conclusion

We investigated and compared exonization derived from Plat_L3 element and IR events in the *ZRANB2* gene in human and monkeys (rhesus monkey, crab-eating monkey, African-green monkey, and marmoset). First, we confirmed that the Plat_L3 and L3b LINE elements in the intronic region of *ZRANB2* were integrated in human and all primate lineages (18 species, including hominoids, Old World monkeys, New World monkeys, and prosimian). RT-PCR experiments indicated that the Plat_L3-encoded exon was conserved in all tested tissues of human and the 4 monkeys; the IR event occurred only in Old World monkeys (rhesus monkey, crab-eating monkey, and African-green monkey). Transcript variants of *ZRANB2* genes derived from these events encoded different-sized products via 6-frame translation sequence analysis. Based on our results, we assume that IR and TE-derived exonization events are intriguing evolutionary factors that could enhance the transcriptome and protein diversity under limited genomic sources in the primate lineage.

Author's Contribution

Sang-Je Park, Jae-Won Huh, and Young-Hyun Kim are contributed equally to this work.

Acknowledgment

This research was supported by a Grant from the KRIBB Research Initiative Program (KGM4241231).

References

[1] E. A. Karginova, E. S. Pentz, I. G. Kazakova, V. F. Norwood, R. M. Carey, and R. A. Gomez, "Zis: a developmentally regulated gene expressed in juxtaglomerular cells," *American Journal of Physiology*, vol. 273, no. 5, pp. F731–F738, 1997.

[2] M. Nakano, K. I. Yoshiura, M. Oikawa et al., "Identification, characterization and mapping of the human ZIS (zinc- finger, splicing) gene," *Gene*, vol. 225, no. 1-2, pp. 59–65, 1998.

[3] D. J. Adams, L. van der Weyden, A. Kovacic et al., "Chromosome localization and characterization of the mouse and human zinc finger protein 265 gene," *Cytogenetics and Cell Genetics*, vol. 88, no. 1-2, pp. 68–73, 2000.

[4] C. A. Plambeck, A. H. Y. Kwan, D. J. Adams et al., "The structure of the zinc finger domain from human splicing factor ZNF265 fold," *The Journal of Biological Chemistry*, vol. 278, no. 25, pp. 22805–22811, 2003.

[5] D. J. Adams, L. van der Weyden, A. Mayeda, S. Stamm, B. J. Morris, and J. E. J. Rasko, "ZNF265—a novel spliceosomal protein able to induce alternative splicing," *Journal of Cell Biology*, vol. 154, no. 1, pp. 25–32, 2001.

[6] A. H. Mangs, H. J. L. Speirs, C. Goy, D. J. Adams, M. A. Markus, and B. J. Morris, "XE7: a novel splicing factor that interacts with ASF/SF2 and ZNF265," *Nucleic Acids Research*, vol. 34, no. 17, pp. 4976–4986, 2006.

[7] J. Li, X. H. Chen, P. J. Xiao et al., "Expression pattern and splicing function of mouse ZNF265," *Neurochemical Research*, vol. 33, no. 3, pp. 483–489, 2008.

[8] F. E. Loughlin, R. E. Mansfield, P. M. Vaz et al., "The zinc fingers of the SR-like protein ZRANB2 are single-stranded RNA-binding domains that recognize 5′ splice site-like sequences," *Proceedings of the National Academy of Sciences of the United States of America*, vol. 106, no. 14, pp. 5581–5586, 2009.

[9] M. Ladomery, R. Marshall, L. Arif, and J. Sommerville, "4SR, a novel zinc-finger protein with SR-repeats, is expressed during early development of Xenopus," *Gene*, vol. 256, no. 1-2, pp. 293–302, 2000.

[10] G. Ast, "How did alternative splicing evolve?" *Nature Reviews Genetics*, vol. 5, no. 10, pp. 773–782, 2004.

[11] B. Modrek and C. Lee, "A genomic view of alternative splicing," *Nature Genetics*, vol. 30, no. 1, pp. 13–19, 2002.

[12] P. A. F. Galante, N. J. Sakabe, N. Kirschbaum-Slager, and S. J. de Souza, "Detection and evaluation of intron retention events in the human transcriptome," *RNA*, vol. 10, no. 5, pp. 757–765, 2004.

[13] J. Schmitz and J. Brosius, "Exonization of transposed elements: a challenge and opportunity for evolution," *Biochimie*, vol. 93, no. 11, pp. 1928–1934, 2011.

[14] N. Sela, E. Kim, and G. Ast, "The role of transposable elements in the evolution of non-mammalian vertebrates and invertebrates," *Genome Biology*, vol. 11, no. 6, article R59, 2010.

[15] N. Sela, B. Mersch, A. Hotz-Wagenblatt, and G. Ast, "Characteristics of transposable element exonization within human and mouse," *PLoS ONE*, vol. 5, no. 6, Article ID e10907, 2010.

[16] W. Makalowski, G. A. Mitchell, and D. Labuda, "Alu sequences in the coding regions of mRNA: a source of protein variability," *Trends in Genetics*, vol. 10, no. 6, pp. 188–193, 1994.

[17] E. S. Lander, L. M. Linton, B. Birren et al., "Initial sequencing and analysis of the human genome," *Nature*, vol. 409, no. 6822, pp. 860–921, 2001.

[18] T. Wicker, F. Sabot, A. Hua-Van et al., "A unified classification system for eukaryotic transposable elements," *Nature Reviews Genetics*, vol. 8, no. 12, pp. 973–982, 2007.

[19] E. Gogvadze and A. Buzdin, "Retroelements and their impact on genome evolution and functioning," *Cellular and Molecular Life Sciences*, vol. 66, no. 23, pp. 3727–3742, 2009.

[20] A. Böhne, F. Brunet, D. Galiana-Arnoux, C. Schultheis, and J. N. Volff, "Transposable elements as drivers of genomic and biological diversity in vertebrates," *Chromosome Research*, vol. 16, no. 1, pp. 203–215, 2008.

[21] A. Nekrutenko and W. H. Li, "Transposable elements are found in a large number of human protein-coding genes," *Trends in Genetics*, vol. 17, no. 11, pp. 619–621, 2001.

[22] N. Sela, B. Mersch, N. Gal-Mark, G. Lev-Maor, A. Hotz-Wagenblatt, and G. Ast, "Comparative analysis of transposed element insertion within human and mouse genomes reveals Alu's unique role in shaping the human transcriptome," *Genome Biology*, vol. 8, no. 6, article R127, 2007.

[23] A. J. Gentles, M. J. Wakefield, O. Kohany et al., "Evolutionary dynamics of transposable elements in the short-tailed opossum Monodelphis domestica," *Genome Research*, vol. 17, no. 7, pp. 992–1004, 2007.

[24] G. J. Goodall and W. Filipowicz, "Different effects of intron nucleotide composition and secondary structure on pre-mRNA splicing in monocot and dicot plants," *The EMBO Journal*, vol. 10, no. 9, pp. 2635–2644, 1991.

[25] S. Stamm, J. Zhu, K. Nakai, P. Stoilov, O. Stoss, and M. Q. Zhang, "An alternative-exon database and its statistical analysis," *DNA and Cell Biology*, vol. 19, no. 12, pp. 739–756, 2000.

[26] V. Marinescu, P. A. Loomis, S. Ehmann, M. Beales, and J. A. Potashkin, "Regulation of retention of FosB intron 4 by PTB," *PLoS ONE*, vol. 2, no. 9, article e828, 2007.

[27] J. Xie, J. A. Lee, T. L. Kress, K. L. Mowry, and D. L. Black, "Protein kinase A phosphorylation modulates transport of the polypyrimidine tract-binding protein," *Proceedings of the National Academy of Sciences of the United States of America*, vol. 100, no. 15, pp. 8776–8781, 2003.

[28] N. A. Faustino and T. A. Cooper, "Pre-mRNA splicing and human disease," *Genes and Development*, vol. 17, no. 4, pp. 419–437, 2003.

[29] E. Nagy and L. E. Maquat, "A rule for termination-codon position within intron-containing genes: when nonsense affects RNA abundance," *Trends in Biochemical Sciences*, vol. 23, no. 6, pp. 198–199, 1998.

[30] A. Chester, A. Somasekaram, M. Tzimina et al., "The apolipoprotein B mRNA editing complex performs a multifunctional cycle and suppresses nonsense-mediated decay," *The EMBO Journal*, vol. 22, no. 15, pp. 3971–3982, 2003.

[31] C. Stockklausner, S. Breit, G. Neu-Yilik et al., "The uORF-containing thrombopoietin mRNA escapes nonsense-mediated decay (NMD)," *Nucleic Acids Research*, vol. 34, no. 8, pp. 2355–2363, 2006.

[32] G. Neu-Yilik, B. Amthor, N. H. Gehring et al., "Mechanism of escape from nonsense-mediated mRNA decay of human β-globin transcripts with nonsense mutations in the first exon," *RNA*, vol. 17, no. 5, pp. 843–854, 2011.

[33] S. Ohte, S. Kokabu, S. I. Iemura et al., "Identification and functional analysis of Zranb2 as a novel Smad-binding protein that suppresses BMP signaling," *Journal of Cellular Biochemistry*, vol. 113, no. 3, pp. 808–814, 2012.

[34] K. Kaer, J. Branovets, A. Hallikma, P. Nigumann, and M. Speek, "Intronic l1 retrotransposons and nested genes cause transcriptional interference by inducing intron retention, exonization and cryptic polyadenylation," *PLoS ONE*, vol. 6, no. 10, Article ID e26099, 2011.

9

Versatility of RNA-Binding Proteins in Cancer

Laurence Wurth

Gene Regulation Programme, Center for Genomic Regulation (CRG) and UPF, 08003 Barcelona, Spain

Correspondence should be addressed to Laurence Wurth, laurence.wurth@crg.es

Academic Editor: Armen Parsyan

Posttranscriptional gene regulation is a rapid and efficient process to adjust the proteome of a cell to a changing environment. RNA-binding proteins (RBPs) are the master regulators of mRNA processing and translation and are often aberrantly expressed in cancer. In addition to well-studied transcription factors, RBPs are emerging as fundamental players in tumor development. RBPs and their mRNA targets form a complex network that plays a crucial role in tumorigenesis. This paper describes mechanisms by which RBPs influence the expression of well-known oncogenes, focusing on precise examples that illustrate the versatility of RBPs in posttranscriptional control of cancer development. RBPs appeared very early in evolution, and new RNA-binding domains and combinations of them were generated in more complex organisms. The identification of RBPs, their mRNA targets, and their mechanism of action have provided novel potential targets for cancer therapy.

1. Introduction

Traditionally, it has been well accepted that cancer development is dictated in part by aberrant transcriptional events and signaling pathways. More recently, it has become clear that posttranscriptional regulation of gene expression also controls cell proliferation, differentiation, invasion, metastasis, apoptosis, and angiogenesis which influence initiation and progression of cancer [1–4]. Regulation of already transcribed messenger RNAs (mRNAs) is an efficient and rapid way to alter gene expression and plays a crucial role in tumorigenesis.

After transcription, nascent mRNAs undergo several processing steps including splicing, capping, 3′ end formation, surveillance, nucleocytoplasmic transport, and, for many transcripts, localization before being translated and finally degraded [5, 6]. The mRNA does not exist alone in the cell, and its metabolism is largely defined by bound RNA-binding proteins (RBPs). RBPs, which regulate all steps of RNA biogenesis, form dynamic units with the RNA, called ribonucleoprotein complexes (RNPs) [7]. Different sets of RBPs are associated to the mRNA at different time points and in different compartments, thereby regulating the fate of their target in a time- and space-dependent way. RBPs often provide a landing platform for the recruitment of additional factors and enzymes to the mRNA. RBPs are the master regulators of post-transcriptional gene expression and, thus, are expected to play important roles in cancer development [1]. Besides RBPs, the discovery of microRNAs (miRNA) was of great inspiration for the RNA field and provided a new powerful tool to regulate gene expression. miRNAs associate with RBPs to form microRNPs (miRNP) which regulate translation and RNA stability by binding to complementary sequences in target mRNAs. miRNPs have been found to regulate expression of factors implicated in tumorigenesis, but we will not discuss this mechanism here (for recent reviews see [8, 9]).

RBPs bind to specific sequences or secondary structures typically found in the untranslated regions (UTRs) but also in the open reading frame (ORF) of target mRNAs [10, 11]. UTRs in particular have offered more flexibility to evolution, as the constraints of encoding a protein product have not been imposed upon them. As a consequence, diverse and often conserved regulatory elements are present in the UTRs [12]. In the 5′UTR, ribose methylation of the cap structure as well as 5′ terminal polypyrimidine sequences or secondary structures such as internal ribosome entry sites (IRESs) control protein expression. Sequence elements in the 3′UTR regulate the stability of the mRNA, its translational efficiency and localization. Specific binding of regulatory proteins to these elements is achieved through RNA-binding domains (RBDs). More than 40 RBDs have been identified. Among them, the most prominent are the RNA recognition motif (RRM), K-homology domain (KH), double stranded RNA-binding domain (dsRBD), zinc finger, Arginine-rich domain, cold-shock domain (CSD), and the PAZ and PIWI domains [13]. An RNA-binding protein can contain combinations of different RBDs, which allow a high flexibility for interaction with different targets. RBP purification techniques followed by high throughput proteomics will hopefully allow us in the near future to identify new RNA-binding proteins as well as new RNA-binding domains. Powerful techniques like CLIP-seq (UV cross-linking and immunoprecipitation followed by high throughput sequencing) are helping to identify new RBP targets in a genome wide scale, as well as new RBP binding sites [14–16]. The list of RBPs, RBDs and their targets is far from being complete. New technology is proving helpful to unravel the complexity of post-transcriptional gene regulation.

In cancer cells, expression of numerous oncoproteins or tumor suppressors is under the control of specific RBPs. Splicing, stability, localization as well as translation of these mRNAs are highly regulated, often in a tissue-specific manner [6]. Many RBPs are aberrantly expressed in cancer cells and have thus a cancer-specific regulatory activity [1, 17, 18]. Deregulation of RBP expression in cancer may have its origin on epigenetic events or on miRNA-dependent controls, although the detailed molecular mechanisms are often obscure [19–21]. An additional layer of regulation is provided by signaling: the phosphorylation status of some RBPs is defined by signaling pathways that are deregulated in cancer, and this phosphorylation controls RBP activity and subsequently the expression of its target mRNAs [22, 23]. Signaling pathway alterations occur in different stages of tumor formation and are often correlated with tumor grade.

In this paper, we will summarize the different functions of RBPs in post-transcriptional gene regulation and the impact of aberrant regulation on tumorigenesis. In addition, we will discuss the conservation of specific RBPs across eukaryotes, which may yield hints on how diversity has been generated.

2. RNA-binding Proteins Implicated in Cancer Development

Post-transcriptional gene regulation implies factors which act at different levels of mRNA metabolism, including alternative splicing, localization, stability of the mRNA or cap-dependent and -independent translation. In this section I will introduce a subset of RBPs involved in cancer development which play key roles in each of the steps of RNA regulation, namely, Sam68, eIF4E, La, and HuR to illustrate the powerful RBP regulatory capacity in cancer.

2.1. Sam68 Regulates Alternative Splicing of Cancer-Related mRNAs. Sam68 belongs to the evolutionarily conserved signal transduction and activation of RNA (STAR) family of RBPs [4, 24, 25]. Sam68 is predominately nuclear but has also been detected in the cytoplasm and exerts multiple activities

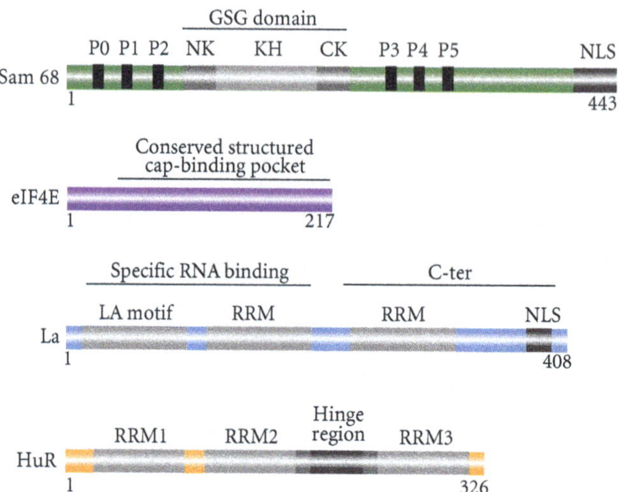

FIGURE 1: Schematic representation of the 4 RBPs discussed in this paper: Sam68, eIF4E, La, and HuR; RBDs are depicted in light gray. RRM: RNA recognition motif; GSG: GRP33/SAM68/GLD-1 domain, composed of a KH domain (KH) flanked by N-terminal (NK) and C-terminal (CK) extensions; LA: La motif. Phosphorylation sites of La are indicated in black (P0–P5). Nuclear localization signals (NLSs) are represented in dark gray. The number of amino acids of each protein is indicated.

in gene expression, from transcription and signaling to splicing regulation [4, 26]. RNA binding is achieved by a KH domain embedded in a highly conserved region called GSG (GRP/Sam68/GLD1) domain [27] (Figure 1). RNA binding is used for splicing regulation and is modulated by posttranslational modifications, such as phosphorylation or acetylation [22, 25, 28] (Figure 2).

The role of Sam68 in alternative splicing seems directly related to its oncogenic properties. Alternative splicing (AS) allows the majority of human genes to encode for multiple protein isoforms, which often play different or even opposite roles [29]. In addition to the spliceosome, a set of RBPs are necessary to control alternative splicing [7]. Aberrant expression of RBPs in cancer can lead to deregulation of splicing, and subsequent changes in the proteome [30]. The splicing targets of Sam68 support its involvement in tumor progression [4, 31]. Furthermore, the function of Sam68 in AS is regulated by signaling pathways which are often deregulated in cancer cells, establishing a link between signal transduction, alternative splicing, and gene expression during tumorigenesis [22, 32, 33] (Figure 2).

Sam68 is overexpressed in breast, prostate, renal, and cervical cancer cells [26, 34–36] and is also frequently upregulated in tumors [34, 37].

The first hard evidence that Sam68 is involved in regulation of alternative splicing with an impact on tumorigenesis was provided by the demonstration that it promotes inclusion of exon v5 in the CD44 pre-mRNA [33]. CD44 encodes a cell surface molecule involved in cancer cell proliferation. CD44 transcript isoforms are alternatively generated by the inclusion of 10 variant exons, which are decisive in tumor

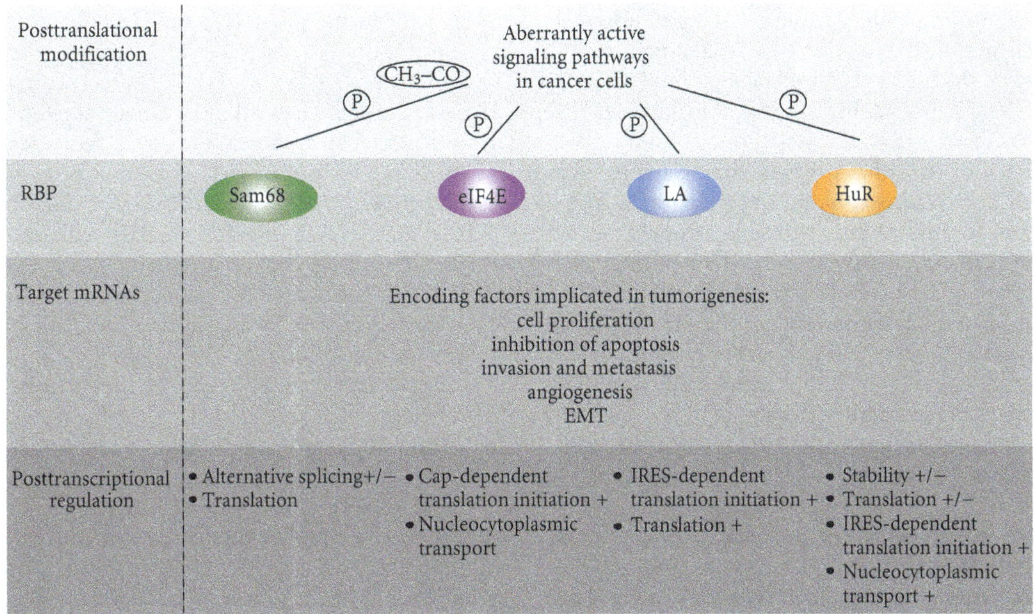

FIGURE 2: Overview of posttranscriptional gene regulation by Sam68, eIF4E, La, and HuR in tumorigenesis. In cancer cells, RBPs are posttranslationally modified by aberrantly active signaling pathways that activate their binding to targets encoding proteins implicated in tumorigenesis. The steps of mRNA metabolism regulated by RBPs are indicated. (+) and (−) specify up- or downregulation. P and CH_3–CO indicate phosphorylation and acetylation of RBPs. EMT: epithelial to mesenchymal transition.

progression [38]. Depletion of Sam68 strongly reduces the inclusion of several variable exons. Interestingly, Sam68 activity is controlled by the Ras signaling pathway, and Sam68 phosphorylation by ERK is needed to promote v5 inclusion [33].

Sam68 also regulates AS of cyclin D1, a protooncogene frequently deregulated in cancer cells [39, 40]. In addition, Sam68 promotes the generation of a stable SF2/ASF, isoform through regulation of splicing. The protooncogene SF2/ASF, also a splicing factor, is in turn responsible for processing of ΔRon pre-mRNA, which encodes a factor involved in EMT in colon cancer cells [41].

Another connection of Sam68 with cancer could be provided by the control of AS of the Bcl-x transcript. The Bcl-x gene can yield the antiapoptotic Bcl-x(L) factor or the proapoptotic Bcl-x(S) [22, 42]. Some studies have reported that Sam68 overexpression causes the accumulation of proapoptotic Bcl-x(s) in a manner that depends on the RNA-binding activity of Sam68 [22, 42]. However, the observation that Sam68 and the antiapoptotic Bcl-x(L) are upregulated in prostate cancer cells is at odds with a proposed activity of Sam68 in Bclx(S) upregulation [34, 43]. This apparent contradiction was resolved by the finding that the activity of Sam68 on Bcl-x AS depends on its phosphorylation status, which can switch Sam68 function from proapototic to anti-apoptotic in cancer cells. Indeed, Src-like kinase, which is often activated in cancer, phosphorylates Sam68 and thereby promotes splicing of the antiapoptotic Bcl-x(l) variant which inhibits cell death [22].

Intriguingly, in advanced breast and renal tumors, Sam68 was found to localize in the cytoplasm [26, 35]. These observations suggest a potential function of Sam68 in translational control in advanced stages of tumorigenesis. In accordance with a potential role of Sam68 in translation, it was previously proposed to regulate the translation of selected mRNAs in male germ cells and neurons [44, 45].

Other RBPs regulating splicing in cancer cells are hnRNPs (A/B) H, SR proteins (ASF/SF2), RBM5, HuR, and PTB. The interested reader can refer to the following reviews and articles [30, 46].

2.2. eIF4E Overexpression in Cancer Enhances Translation Initiation of Specific mRNAs. Translation initiation is a critical step of protein synthesis and is highly regulated [47]. One of the most crucial regulators is the cap-binding protein eIF4E (eukaryotic initiation factor 4E) [48]. In the cytoplasm, eIF4E binds directly to the m^7GTP-cap structure present at the 5′end of all mRNAs and interacts with eIF4G, which in turn recruits the 43S ribosomal complex during initiation of translation. eIF4E and eIF4G together with the RNA helicase eIF4A form the eIF4F complex, which is often targeted for translational regulation [47].

Early findings indicated that eIF4E overexpression leads to malignant transformation of fibroblasts [49, 50]. Since then, numerous studies have reported overexpression of eIF4E in different tumor types (e.g., breast, prostate, gastric colon, lung, skin, and lymphomas) [51]. Elevated expression of eIF4E often correlates with malignancy and poor prognosis [52, 53]. Surprisingly, overexpression of eIF4E does not induce a global increase in protein synthesis but augments translation of a subset of mRNAs encoding mostly prooncogenic proteins [2, 54] (Figure 2). mRNAs regulated by eIF4E overexpression include those encoding components of the cell cycle machinery (cyclin D1, CDK2, c-myc, RNR2, ODC, surviving, Mcl-1, Bcl-2) or factors implicated in angiogenesis (VEGF, FGF-2, PDGF) and invasion (MMP9) [2, 51, 55, 56].

It has been proposed that mRNAs coding for proteins upregulated in oncogenesis contain long and highly structured 5′UTRs [10]. mRNAs bearing stable secondary structures in the 5′UTR are poorly translated in normal conditions and may be particularly dependent on the eIF4F complex and the unwinding capacity of the eIF4A helicase to initiate translation. Thus eIF4E overexpression may lead to enhanced translation of otherwise inefficiently translated transcripts involved in tumorigenesis [2]. Interestingly, eIF4E seems to be implicated in nucleocytoplasmic transport of mRNAs (e.g., cyclin D) and thus may regulate expression of some genes in an initiation-independent way [57].

eIF4E activity is regulated by signaling pathways amplified in human cancers (Figure 2). The protein kinase mTor phosphorylates eIF4E-binding proteins (4E-BP). In their unphosphorylated state, 4E-BPs bind to eIF4E on the same site recognized by eIF4G, blocking the formation of the cap-binding complex. Phosphorylation of 4E-BP leads to loss of affinity for eIF4E and increases translation [2]. In addition, eIF4E phosphorylation by MAPK-integrating kinases MNK1 and MNK2 enhances cap-dependent initiation [47, 54].

Given the important role of eIF4E in tumorigenesis, reducing either eIF4E activity or levels in cancer cells has become an attractive anticancer strategy [51, 58]. Many compounds inhibiting mTor kinase activity have proven to be efficient. For example, PP242, Tonin1, and INK128 are ATP active site inhibitors of mTOR and block the phosphorylation of all mTor targets including 4E-BP [51]. Unfortunately, cells of some cancer types are insensitive to treatment with mTor inhibitors [59]. As an alternative strategy, inhibiting eIF4E expression with antisense oligonucleotides (AON) has given promising results in suppressing tumor growth in vivo [60].

2.3. La Is an ITAF Implicated in Cancer. The multifunctional RNA-binding protein La is primarily nuclear but can shuttle between the nucleus and the cytoplasm [61, 62]. According to its localization, La functions in small RNA processing [63] and in translation of mRNAs [64–66]. La can be divided into three regions: the N-terminus, which contains the conserved La motif; a less conserved RNA recognition motif (RRM); and a weakly conserved C terminus, which contains an RRM, and a nuclear localization signal (NLS) [67] (Figure 1). The La motif folds into an RRM and its high conservation suggests that it carries out a specific function [68, 69]. La interacts with cellular and viral mRNAs and regulates IRES and cap-dependent translation initiation [64, 66, 70–73]. An IRES is a nucleotide sequence folding in a specific secondary structure that recruits ribosomes independently of the cap structure [74]. During cellular stress, cap-dependent translation is downregulated, and IRES-dependent translation of many mRNAs is favored [75]. For example, under the hypoxic conditions usually found in the interior of a tumor, IRES-mediated translation of the angiogenic factor VEGF is favored leading to vascularization of the tumor [76]. Specific RNA-binding proteins termed IRES transacting factors (ITAFS) are required to regulate IRES-dependent translation in cancer development [74]. La is an ITAF that regulates the IRES-dependent translation of mRNAs involved in cell proliferation, angiogenesis and apoptosis [64, 77, 78] (Figure 2).

As an ITAF, La interacts directly with the IRES of the mRNA encoding the proapoptotic factor XIAP [64]. In addition, La regulates IRES-dependent translation of LamB1, a factor that drives invasion, angiogenesis and metastasis [79, 80]. La also binds to the IRES of cyclin D1 (CCND1) in cervical cancer tissues, and its overexpression correlates with upregulation of cyclin D1 while its depletion leads to a reduction of cyclin D1 levels and a defect in cell proliferation [77].

La is overexpressed in chronic myeloid leukemia, cervical cancer tissues, oral squamous cell carcinoma (SCC), and in a number of cancer cell lines compared to nontumorigenic cells [66, 77, 78, 81]. In SCC, La is required for expression of β-catenin and MMP-2, proteins implicated in cell-cell adhesion and cell motility, respectively [78]. In leukemia, increased levels of La correlate with upregulation of MDM2 (an oncogenic tyrosine kinase). La interacts directly with the 5′UTR of mdm2 mRNA and enhances its translation [66].

Using mouse glial progenitor cells, Brennet proposed that La functions as a translational regulator during KRas/Akt oncogenic signaling [62]. Ras and Akt pathways are aberrantly active in cancer cells and play a pivotal role in the formation and regulation of glioblastoma [82]. In this tumor type La is phosphorylated by Akt, and this changes its distribution from the nucleus to the cytoplasm leading to association of a subset of La-bound mRNAs to polysomes. Many of these mRNAs encode factors implicated in oncogenesis such as Cyclin G2, Bcl2, and PDGFA [62].

The number of known La mRNA targets is still limited and further studies are necessary to understand its function in tumorigenesis. However, La already represents a promising target for cancer therapy. As an example, La activity has been efficiently blocked by a synthetic peptide corresponding to amino acids 11 to 28 of La. By competition, the peptide inhibits IRES-driven translation of Hepatitis C without affecting cap-dependent translation of cellular mRNAs [83]. This peptide could also be used to block expression of cancer related mRNA targets of La.

Other ITAFs implicated in cancer are PTB, hnRNP A1, hnRNP E1, hnRNP E2, and YB1. The interested reader can refer to the following reviews and articles [84, 85].

2.4. HuR Regulates the Stability and Translation of Cancer-Related Transcripts. The human antigen R (HuR) is the most prominent RBP known to be implicated in tumorigenesis [3]. Overexpression of HuR has been observed in lymphomas, gastric, breast, pancreatic, prostate, oral, colon, skin, lung, ovarian, and brain cancers [86–91]. Elevated cytoplasmic accumulation of HuR correlates with high-grade malignancy and serves as a prognostic factor of poor clinical outcome in some cancer types [92–95]. Localized in the nucleus of normal cells, HuR often translocates to the cytoplasm in transformed cells [96, 97]. HuR's subcellular localization is regulated by posttranslational modifications, and the enzymes modifying HuR are all implicated in cancer [97] (Figure 2). In the cytoplasm, HuR binds to adenine- and uridine-rich elements (AU-rich elements or AREs) located in 3′UTR of target mRNAs [98]. AU-rich elements serve as binding sites for a variety of RBPs that modulate mRNA half-life [11]. An estimated 10% of all mRNAs bear AU-rich sequences [99]. The minimal functional ARE sequence is a nonamer UUAUUUAWW [100]. Most RBPs binding to AREs promote rapid deadenylation and degradation of substrate mRNAs by targeting them to the exosome (e.g., TTP, AUF1, CUGBP2) [101]. On the contrary, HuR most often enhances the stability of its target mRNAs [3]. In addition, HuR can also regulate the splicing of a certain number of targets [102].

HuR is a member of the embryonic lethal abnormal vision (ELAV) family of proteins and contains three RRMs that provide high-affinity RNA binding [103] (Figure 1). HuR target mRNAs encode products that promote proliferation, inhibit apoptosis, increase angiogenesis, and facilitate invasion and metastasis. For an extensive list of HuR targets, see [3]. Below I will give an overview of HuR targets and will summarize the different mechanisms by which HuR regulates their expression.

Upon binding to the 3′UTR, HuR stabilizes the mRNAs coding for cyclins (cyclin D1, E1, A2, B1), favoring cell cycle progression and promoting proliferation of cancer cells [104–106]. HuR also promotes cancer cell survival by stabilizing transcripts encoding antiapoptotic factors like Bcl-2, Mcl-1, SIRT1, and p21 [90, 107–110]. mRNAs coding for proteins implicated in invasion and metastasis (MMP-9) [111, 112], cell migration and adhesion (Urokinase A and uPA receptor) [113] or EMT (snail) are also stabilized by HuR [114]. Expression of the proangiogenic factors VEGF and HIF-1α is controlled by HuR. Regulation of HIF-1α mRNA is interesting, as HuR binds to both the 5′ and 3′UTRs and promotes translation and stability [115, 116]. The mechanism by which HuR stabilizes its targets is still unclear, but recent studies have proposed an interplay between HuR and miRNAs [117]. HuR is able to suppress activity of miRNAs, by inhibiting their recruitment to the mRNA or even by promoting their downregulation. Some examples of cross-talk between HuR and miRNAs will be given in the next paragraph.

ERBB-2 overexpression is associated with development and progression of prostate cancer. HuR enhances ERBB-2 expression using a miRNA-dependent mechanism. HuR binds to a uridinerich element (URE) in the 3′UTR of ERBB-2 and inhibits action of miR-331-3p to a nearby site [118]. The presence of HuR on the mRNA does not alter miR-331-3p binding, which leads to the hypothesis that HuR may rather reduce association between ERBB-2 mRNA and the RNA silencing complex [118]. In colorectal cancer, HuR overexpression and localization in the cytoplasm correlate with decreased levels of miR-16, a miRNA that binds to the 3′UTR of COX-2 mRNA and inhibits its expression by mRNA decay [119]. Intriguingly, HuR interacts with miR-16 and promotes its downregulation in an mRNA ligand-dependent manner. Thus, HuR stabilizes COX-2 mRNA by binding to the ARE and by downregulating miR-16 [119].

Interestingly, HuR is able to repress the translation of the proapoptotic factor c-Myc by recruiting the let-7 miRNP to the 3′UTR [120]. HuR is not the only RBP which assists in targeting miRNPs to the 3′UTR of mRNAs, as was shown with the example of TTP [121].

HuR also represses the translation of some of its targets by binding to the 5′UTR. This is the case for p27, which prevents cell proliferation [122].

It has been recently shown that HuR can act as an ITAF binding to the IRES of XIAP mRNA, which encodes an anti-apoptotic factor [123]. HuR stimulates the translation of XIAP mRNA by binding to XIAP IRES and enhancing its recruitment into polysomes.

Interestingly in the case of the antiapoptotic factor prothymosin alpha (ProTα), HuR binding to its 3′UTR enhances nuclear export of the mRNA followed by induced translation upon UV irradiation [124].

In summary, the majority of HuR mRNA targets are stabilized upon binding, and translation is enhanced. As an ITAF, HuR binds to IRES structures and enhances translation. HuR is also able to inhibit translation by binding to 5′UTR or by recruiting miRNPs to the 3′UTR. On the other hand, HuR also inhibits miRNA binding to the 3′UTR of its target mRNAs. Finally, HuR is increasing cytoplasmic abundance of target mRNAs probably via enhanced mRNA nulear export. These examples illustrate the complexity of HuR regulatory activity.

The large spectrum of mRNA targets regulated by HuR confirms its potential to coordinate nearly all steps of tumorigenesis. Overexpressed in a high number of cancer types, HuR provides a good candidate for therapy design. Surprisingly, however, a recent study showed that elevated levels of HuR may be advantageous for cancer therapy. In pancreatic ductal adenocarcinoma, HuR levels modulate the therapeutic activity of gemcitabine (GEM), a common chemotherapeutic agent [125]. GEM exposure to cancer cells increases the amount of cytoplasmic HuR and promotes its association with dCK mRNA, which encodes the enzyme that activates GEM, establishing a positive feedback loop that improves its therapeutic efficacy. This example shows that therapies that reduce the level of HuR have to be designed carefully, and perhaps in a tumor type-dependent manner [126].

Besides HuR, a number of other factors can regulate the stability and expression of mRNAs bearing AREs [127]. The TIS11 family of RBPs composed of Tristetraprolin (TTP) and butyrate response factors 1 and 2 (BRF-1 and-2) bind and target ARE-containing mRNAs for rapid degradation [101]. AUF1 is able to stabilize or destabilize ARE-containing mRNAs [128]. The CELF family of RNA-binding proteins is composed of 6 members, which promote either mRNA decay or translation of its target mRNAs [129, 130]. For example CUGBP2 binds COX-2 mRNA which is then stabilized but translationally repressed [131]. T-cell intracellular antigen-1 (TIA-1) and TIA-1-related (TIAR) proteins are translational silencers [132]. Some of these factors have common targets and compete for binding depending on cellular conditions.

3. Conservation of RBPs across Eukaryotes

Post-transcriptional gene regulation is a coordinated, efficient, rapid and flexible mechanism to control the proteome of the cell in response to different physiological conditions. It is thus not surprising that some organisms have become highly dependent on post-transcriptional mechanisms to regulate gene expression, like, for example, the protozoan parasite, trypanosome [133–135]. The trypanosome genome encodes very few potential regulatory transcription factors, and gene regulation relies mostly on RNA-binding proteins [136]. It has been proposed that 3–11% of the proteome in bacteria, archea and eukaryotes are putative RNA-binding proteins [137]. The large number of RBPs suggests that RNA metabolism may be a central and evolutionarily conserved contributor to cell physiology. Most of the RNA-binding domains known today are present in early stages of evolution. Interestingly, several new eukaryotic-specific RNA-binding domains have emerged, like the RRM, which suggests that post-transcriptional gene regulation became more complex with evolution [137].

The RNA-binding proteins described in this paper are widely conserved across eukaryotes (Figure 3). We could detect homologues of HuR only in metazoa and not in fungi and plants. Human HuR is the most divergent family member of the ELAV proteins. While the other members, HuD, HuC, and Hel-N1, present a neuron- and brain-specific expression, where they are mostly implicated in alternative splicing, HuR is ubiquitously expressed and fulfills numerous functions [138, 139].

Sam68 homologues exist in all eukaryotes except fungi (Figure 3). In the STAR protein family, the Sam68 subfamily is composed of Sam68 (SRC-associated in mitosis, 68 kd) and the Sam68-like mammalian proteins 1 and 2 (SLM-1 and SLM-2, also named T-STAR in humans) [140–143]. As in the case of HuR, Sam68 is ubiquitously expressed, whereas SLM-1 and SLM-2 expression is restricted to few cell types or tissues [144]. In humans, Sam68 has acquired a larger spectrum of functions and plays a major role in signaling and splicing in different tissues.

Contrary to HuR and Sam68, La homologues can be identified in all three phyla: metazoa, fungi, and plants (Figure 3). La was first characterized as a human protein, and homologues have been identified in a wide variety of other eukaryotes [63]. The N-terminal part containing the La motif is highly conserved, in contrast to the C-terminal domain which varies both in size and sequence between species, ranging from 70 amino acids in the yeasts *S. cerevisiae* and *S. pombe* to more than 220 amino acids in vertebrates. Human La is phosphorylated at different sites, all located in the C terminus [63] (Figure 2). Interestingly these sites are only conserved in vertebrate La proteins. The presence of an additional C-terminal region including different functional domains and phosphorylation sites shows that La has evolved to a highly regulated and multifunctional factor in vertebrates.

The translation initiation factor eIF4E is highly conserved across eukaryotes. Sequence comparisons revealed a phylogenetically conserved 182 amino acid C-terminal

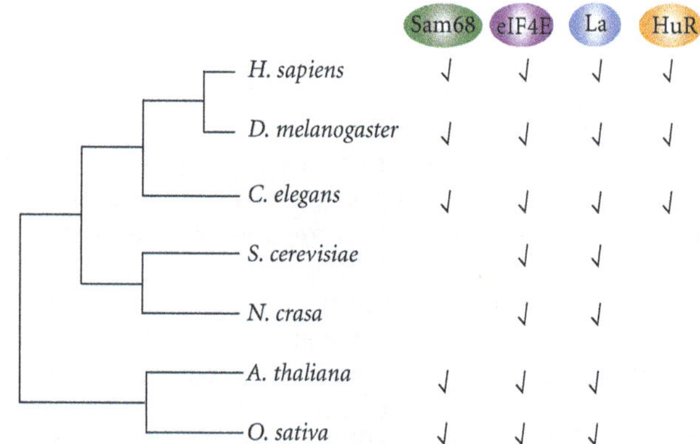

Figure 3: Conservation of Sam68, eIF4E, La, and HuR in different phyla. Phylogenetic tree of the RBPs described in this paper. The presence of homologues is indicated.

region [145, 146]. In contrast, the N-terminal region is poorly conserved and is not required for cap-dependent translation [145]. Functional conservation has also been demonstrated, as mammalian eIF4E can rescue the lethality caused by disruption of the yeast eIF4E gene [147]. The crystallographic structure of eIF4E in mouse, yeast, human, and wheat has been solved [145, 148–150]. The three-dimensional structure of the C-terminal part of murine eIF4E demonstrates that the surface of the molecule resembles a cupped hand that contains a narrow cap-binding slot. The remarkable level of sequence identity across phylogeny suggests that all known eIF4Es share the same structure in their conserved C-terminal region [145]. eIF4E thus does not contain a canonical RBD but adopts a conserved three-dimensional structure which interacts with the cap.

Interestingly most eukaryotic organisms express multiple eIF4E family members, and it has been proposed that a ubiquitously expressed member of the family may be implicated in general translation initiation while others could be involved in specialized functions [151, 152]. eIF4E family members may provide an additional layer of control in translation and may regulate specific subsets of mRNAs, which could be linked to cancer development.

4. Concluding Remarks

In cancer research, the impact of post-transcriptional gene regulation has been considered only since a few years. Today, it is well established that a subset of RBPs are key regulators of processes involved in tumorigenesis. The genome wide analysis of RBPs and their RNA targets has allowed a better understanding of the complex world of mRNA metabolism and the connections existing between different RBPs. According to the "RNA operon" concept, mRNAs encoding functionally related proteins are coregulated by specific RBPs, ensuring an efficient, flexible, and coordinated response to cellular need [144, 153, 154]. RNA operons can be interconnected. HuR and eIF4E for example, share common mRNA targets like c-myc, cyclin D1 and VEGF, suggesting an orchestrated regulation of the expression of genes implicated in tumorigenesis [155, 156]. In addition, HuR regulates expression of eIF4E in cancer cells [156]. These observations show that post-transcriptional regulation events are highly linked and provide a powerful mechanism to control the fate of a cell.

RBPs are highly versatile factors that can bind to multiple RNA targets and regulate their fate by a variety of mechanisms. The fact that every step of the mRNA life cycle is narrowly controlled allows RBPs to fine tune expression in a very precise manner. The conservation of RBPs across eukaryotes and the emergence of more complexity along evolution also point to an essential role of RBPs. Post-transcriptional gene regulation is a central mechanism of emerging importance in cancer research which is expected to provide novel targets for therapy design.

Acknowledgments

The author would like to thank Fátima Gebauer for suggestions and critical reading of the paper. He also thanks Margarita Meer and Fyodor Kondrashov for performing phylogenetic analysis. L. Wurth is supported by the National Research Fund, Luxembourg, and cofounded under the Marie Curie Actions of European Commission (FP7-COFUND) and work in Fátima Gebauer's lab is supported by grants BFU2009-08243 and Consolider CSD2009-00080 from MICINN.

References

[1] M. Y. Kim, J. Hur, and S. Jeong, "Emerging roles of RNA and RNA-binding protein network in cancer cells," *BMB Reports*, vol. 42, no. 3, pp. 125–130, 2009.

[2] D. Silvera, S. C. Formenti, and R. J. Schneider, "Translational control in cancer," *Nature Reviews Cancer*, vol. 10, no. 4, pp. 254–266, 2010.

[3] K. Abdelmohsen and M. Gorospe, "Posttranscriptional regulation of cancer traits by HuR," *Wiley Interdisciplinary Reviews*, vol. 1, no. 2, pp. 214–229, 2011.

[4] P. Bielli, R. Busa, M. P. Paronetto, and C. Sette, "The RNA-binding protein Sam68 is a multifunctional player in human cancer," *Endocrine-Related Cancers*, vol. 18, no. 4, pp. R91–R102, 2011.

[5] J. D. Keene, "Ribonucleoprotein infrastructure regulating the flow of genetic information between the genome and the proteome," *Proceedings of the National Academy of Sciences of the United States of America*, vol. 98, no. 13, pp. 7018–7024, 2001.

[6] M. J. Moore, "From birth to death: the complex lives of eukaryotic mRNAs," *Science*, vol. 309, no. 5740, pp. 1514–1518, 2005.

[7] M. C. Wahl, C. L. Will, and R. Lührmann, "The Spliceosome: design Principles of a Dynamic RNP Machine," *Cell*, vol. 136, no. 4, pp. 701–718, 2009.

[8] M. van Kouwenhove, M. Kedde, and R. Agami, "MicroRNA regulation by RNA-binding proteins and its implications for cancer," *Nature Reviews Cancer*, vol. 11, pp. 644–656, 2011.

[9] A. L. Kasinski and F. J. Slack, "MicroRNAs en route to the clinic: progress in validating and targeting microRNAs for cancer therapy," *Nature Reviews Cancer*, vol. 11, no. 12, pp. 849–864, 2011.

[10] B. M. Pickering and A. E. Willis, "The implications of structured 5′ untranslated regions on translation and disease," *Seminars in Cell and Developmental Biology*, vol. 16, no. 1, pp. 39–47, 2005.

[11] Y. Audic and R. S. Hartley, "Post-transcriptional regulation in cancer," *Biology of the Cell*, vol. 96, no. 7, pp. 479–498, 2004.

[12] L. Duret, F. Dorkeld, and C. Gautier, "Strong conservation of non-coding sequences during vertebrates evolution: potential involvement in post-transcriptional regulation of gene expression," *Nucleic Acids Research*, vol. 21, no. 10, pp. 2315–2322, 1993.

[13] B. M. Lunde, C. Moore, and G. Varani, "RNA-binding proteins: modular design for efficient function," *Nature Reviews Molecular Cell Biology*, vol. 8, no. 6, pp. 479–490, 2007.

[14] J. Ule, K. Jensen, A. Mele, and R. B. Darnell, "CLIP: a method for identifying protein-RNA interaction sites in living cells," *Methods*, vol. 37, no. 4, pp. 376–386, 2005.

[15] D. D. Licatalosi, A. Mele, J. J. Fak et al., "HITS-CLIP yields genome-wide insights into brain alternative RNA processing," *Nature*, vol. 456, no. 7221, pp. 464–469, 2008.

[16] S. Lebedeva, M. Jens, K. Theil et al., "Transcriptome-wide analysis of regulatory interactions of the RNA-binding protein HuR," *Molecular Cell*, vol. 43, no. 3, pp. 340–352, 2011.

[17] K. E. Lukong, K. W. Chang, E. W. Khandjian, and S. Richard, "RNA-binding proteins in human genetic disease," *Trends in Genetics*, vol. 24, no. 8, pp. 416–425, 2008.

[18] E. Ortiz-Zapater, D. Pineda, N. Martinez-Bosch et al., "Key contribution of CPEB4-mediated translational control to cancer progression," *Nature Medicine*, vol. 18, pp. 83–90, 2012.

[19] F. Xu, X. Zhang, Y. Lei et al., "Loss of repression of HuR translation by miR-16 may be responsible for the elevation of HuR in human breast carcinoma," *Journal of Cellular Biochemistry*, vol. 111, no. 3, pp. 727–734, 2010.

[20] B. H. Sohn, I. Y. Park, J. J. Lee et al., "Functional switching of TGF-beta1 signaling in liver cancer via epigenetic modulation of a single CpG site in TTP promoter," *Gastroenterology*, vol. 138, no. 5, pp. 1898–e12, 2010.

[21] C. A. Gebeshuber, K. Zatloukal, and J. Martinez, "miR-29a suppresses tristetraprolin, which is a regulator of epithelial polarity and metastasis," *EMBO Reports*, vol. 10, no. 4, pp. 400–405, 2009.

[22] M. P. Paronetto, T. Achsel, A. Massiello, C. E. Chalfant, and C. Sette, "The RNA-binding protein Sam68 modulates the alternative splicing of Bcl-x," *Journal of Cell Biology*, vol. 176, no. 7, pp. 929–939, 2007.

[23] H. H. Kim, K. Abdelmohsen, A. Lal et al., "Nuclear HuR accumulation through phosphorylation by Cdk1," *Genes and Development*, vol. 22, no. 13, pp. 1804–1815, 2008.

[24] C. Vernet and K. Artzt, "STAR, a gene family involved in signal transduction and activation of RNA," *Trends in Genetics*, vol. 13, no. 12, pp. 479–484, 1997.

[25] K. E. Lukong and S. Richard, "Sam68, the KH domain-containing superSTAR," *Biochimica et Biophysica Acta*, vol. 1653, no. 2, pp. 73–86, 2003.

[26] Z. Zhang, J. Li, H. Zheng et al., "Expression and cytoplasmic localization of SAM68 is a significant and independent prognostic marker for renal cell carcinoma," *Cancer Epidemiology Biomarkers and Prevention*, vol. 18, no. 10, pp. 2685–2693, 2009.

[27] Q. Lin, S. J. Taylor, and D. Shalloway, "Specificity and determinants of Sam68 RNA binding. Implications for the biological function of K homology domains," *The Journal of Biological Chemistry*, vol. 272, no. 43, pp. 27274–27280, 1997.

[28] I. Babic, A. Jakymiw, and D. J. Fujita, "The RNA binding protein Sam68 is acetylated in tumor cell lines, and its acetylation correlates with enhanced RNA binding activity," *Oncogene*, vol. 23, no. 21, pp. 3781–3789, 2004.

[29] B. Hartmann and J. Valcárcel, "Decrypting the genome's alternative messages," *Current Opinion in Cell Biology*, vol. 21, no. 3, pp. 377–386, 2009.

[30] C. J. David and J. L. Manley, "Alternative pre-mRNA splicing regulation in cancer: pathways and programs unhinged," *Genes and Development*, vol. 24, no. 21, pp. 2343–2364, 2010.

[31] K. E. Lukong and S. Richard, "Targeting the RNA-binding protein Sam68 as a treatment for cancer?" *Future Oncology*, vol. 3, no. 5, pp. 539–544, 2007.

[32] R. B. Irby and T. J. Yeatman, "Role of Src expression and activation in human cancer," *Oncogene*, vol. 19, no. 49, pp. 5636–5642, 2000.

[33] N. Matter, P. Herrlich, and H. König, "Signal-dependent regulation of splicing via phosphorylation of Sam68," *Nature*, vol. 420, no. 6916, pp. 691–695, 2002.

[34] R. Busà, M. P. Paronetto, D. Farini et al., "The RNA-binding protein Sam68 contributes to proliferation and survival of human prostate cancer cells," *Oncogene*, vol. 26, no. 30, pp. 4372–4382, 2007.

[35] L. Song, L. Wang, Y. Li et al., "Sam68 up-regulation correlates with, and its down-regulation inhibits, proliferation and tumourigenicity of breast cancer cells," *Journal of Pathology*, vol. 222, no. 3, pp. 227–237, 2010.

[36] Z. Li, C. P. Yu, Y. Zhong et al., "Sam68 expression and cytoplasmic localization is correlated with lymph node metastasis as well as prognosis in patients with early-stage cervical cancer," *Annals of Oncology*, vol. 23, no. 3, pp. 638–646, 2012.

[37] P. Rajan, L. Gaughan, C. Dalgliesh et al., "Regulation of gene expression by the RNA-binding protein Sam68 in cancer," *Biochemical Society Transactions*, vol. 36, no. 3, pp. 505–507, 2008.

[38] D. L. Cooper, "Retention of CD44 introns in bladder cancer: understanding the alternative splicing of pre-mRNA opens new insights into the pathogenesis of human cancers," *Journal of Pathology*, vol. 177, no. 1, pp. 1–3, 1995.

[39] M. P. Paronetto, M. Cappellari, R. Busà et al., "Alternative splicing of the cyclin D1 proto-oncogene is regulated by the RNA-binding protein Sam68," *Cancer Research*, vol. 70, no. 1, pp. 229–239, 2010.

[40] E. S. Knudsen and K. E. Knudsen, "Retinoblastoma tumor suppressor: where cancer meets the cell cycle," *Experimental Biology and Medicine*, vol. 231, no. 7, pp. 1271–1281, 2006.

[41] C. Valacca, S. Bonomi, E. Buratti et al., "Sam68 regulates EMT through alternative splicing-activated nonsense-mediated mRNA decay of the SF2/ASF proto-oncogene," *Journal of Cell Biology*, vol. 191, no. 1, pp. 87–99, 2010.

[42] L. H. Boise, M. Gonzalez-Garcia, C. E. Postema et al., "bcl-x, A bcl-2-related gene that functions as a dominant regulator of apoptotic cell death," *Cell*, vol. 74, no. 4, pp. 597–608, 1993.

[43] D. R. Mercatante, J. L. Mohler, and R. Kole, "Cellular response to an antisense-mediated shift of Bcl-x pre-mRNA splicing and antineoplastic agents," *The Journal of Biological Chemistry*, vol. 277, no. 51, pp. 49374–49382, 2002.

[44] M. P. Paronetto, F. Zalfa, F. Botti, R. Geremia, C. Bagni, and C. Sette, "The nuclear RNA-binding protein Sam68 translocates to the cytoplasm and associates with the polysemes in mouse spermatocytes," *Molecular Biology of the Cell*, vol. 17, no. 1, pp. 14–24, 2006.

[45] J. Grange, V. Boyer, R. Fabian-Fine, N. B. Fredj, R. Sadoul, and Y. Goldberg, "Somatodendritic Localization and mRNA Association of the Splicing Regulatory Protein Sam68 in the Hippocampus and Cortex," *Journal of Neuroscience Research*, vol. 75, no. 5, pp. 654–666, 2004.

[46] J. M. Izquierdo, N. Majos, and S. Bonnal, "Regulation of Fas alternative splicing by antagonistic effects of TIA-1 and PTB on exon definition," *Molecular Cell*, vol. 19, no. 4, pp. 475–484, 2005.

[47] N. Sonenberg and A. G. Hinnebusch, "Regulation of translation initiation in eukaryotes: mechanisms and biological targets," *Cell*, vol. 136, no. 4, pp. 731–745, 2009.

[48] I. Topisirovic, Y. V. Svitkin, N. Sonenberg, and A. J. Shatkin, "Cap and cap-binding proteins in the control of gene expression," *Wiley Interdisciplinary Reviews*, vol. 2, no. 2, pp. 277–298, 2011.

[49] A. Lazaris-Karatzas, K. S. Montine, and N. Sonenberg, "Malignant transformation by a eukaryotic initiation factor subunit that binds to mRNA 5′ cap," *Nature*, vol. 345, no. 6275, pp. 544–547, 1990.

[50] S. G. Zimmer, A. Debenedetti, and J. R. Graff, "Translational control of malignancy: the mRNA cap-binding protein, eIF-4E, as a central regulator of tumor formation, growth, invasion and metastasis," *Anticancer Research*, vol. 20, no. 3, pp. 1343–1351, 2000.

[51] A. C. Hsieh and D. Ruggero, "Targeting eukaryotic translation initiation factor 4E (eIF4E) in cancer," *Clinical Cancer Research*, vol. 16, no. 20, pp. 4914–4920, 2010.

[52] J. R. Graff, B. W. Konicek, R. L. Lynch et al., "eIF4E activation is commonly Elevated in advanced human prostate cancers and significantly related to reduced patient survival," *Cancer Research*, vol. 69, no. 9, pp. 3866–3873, 2009.

[53] L. J. Coleman, M. B. Peter, T. J. Teall et al., "Combined analysis of eIF4E and 4E-binding protein expression predicts breast cancer survival and estimates eIF4E activity," *British Journal of Cancer*, vol. 100, no. 9, pp. 1393–1399, 2009.

[54] H. G. Wendel, R. L. A. Silva, A. Malina et al., "Dissecting eIF4E action in tumorigenesis," *Genes and Development*, vol. 21, no. 24, pp. 3232–3237, 2007.

[55] Y. Mamane, E. Petroulakis, L. Rong, K. Yoshida, L. W. Ler, and N. Sonenberg, "eIF4E—from translation to transformation," *Oncogene*, vol. 23, no. 18, pp. 3172–3179, 2004.

[56] A. De Benedetti and J. R. Graff, "eIF-4E expression and its role in malignancies and metastases," *Oncogene*, vol. 23, no. 18, pp. 3189–3199, 2004.

[57] B. Culjkovic, I. Topisirovic, L. Skrabanek, M. Ruiz-Gutierrez, and K. L. B. Borden, "eIF4E promotes nuclear export of cyclin D1 mRNAs via an element in the 3′UTR," *Journal of Cell Biology*, vol. 169, no. 2, pp. 245–256, 2005.

[58] A. Malina, R. Cencic, and J. Pelletier, "Targeting translation dependence in cancer.," *Oncotarget*, vol. 2, no. 1-2, pp. 76–88, 2011.

[59] M. A. Bjornsti and P. J. Houghton, "The TOR pathway: a target for cancer therapy," *Nature Reviews Cancer*, vol. 4, no. 5, pp. 335–348, 2004.

[60] J. R. Graff, B. W. Konicek, T. M. Vincent et al., "Therapeutic suppression of translation initiation factor eIF4E expression reduces tumor growth without toxicity," *Journal of Clinical Investigation*, vol. 117, no. 9, pp. 2638–2648, 2007.

[61] S. A. Rutjes, P. J. Utz, A. Van Der Heijden, C. Broekhuis, W. J. Van Venrooij, and G. J. M. Pruijn, "The La (SS-B) autoantigen, a key protein in RNA biogenesis, is dephosphorylated and cleaved early during apoptosis," *Cell Death and Differentiation*, vol. 6, no. 10, pp. 976–986, 1999.

[62] F. Brenet, N. D. Socci, N. Sonenberg, and E. C. Holland, "Akt phosphorylation of La regulates specific mRNA translation in glial progenitors," *Oncogene*, vol. 28, no. 1, pp. 128–139, 2009.

[63] S. L. Wolin and T. Cedervall, "The La protein," *Annual Review of Biochemistry*, vol. 71, pp. 375–403, 2002.

[64] M. Holcik and R. G. Korneluk, "Functional characterization of the X-linked inhibitor of apoptosis (XIAP) internal ribosome entry site element: role of la autoantigen in XIAP translation," *Molecular and Cellular Biology*, vol. 20, no. 13, pp. 4648–4657, 2000.

[65] C. Crosio, P. P. Boyl, F. Loreni, P. Pierandrei-Amaldi, and F. Amaldi, "La protein has a positive effect on the translation of TOP mRNAs in vivo," *Nucleic Acids Research*, vol. 28, no. 15, pp. 2927–2934, 2000.

[66] R. Trotta, T. Vignudelli, O. Candini et al., "BCR/ABL activates mdm2 mRNA translation via the La antigen," *Cancer Cell*, vol. 3, no. 2, pp. 145–160, 2003.

[67] D. J. Van Horn, C. J. Yoo, D. Xue, H. Shi, and S. L. Wolin, "The La protein in Schizosaccharomyces pombe: a conserved yet dispensable phosphoprotein that functions in tRNA maturation," *RNA*, vol. 3, no. 12, pp. 1434–1443, 1997.

[68] R. J. Maraia and R. V. A. Intine, "Recognition of nascent RNA by the human La antigen: conserved and divergent features of structure and function," *Molecular and Cellular Biology*, vol. 21, no. 2, pp. 367–379, 2001.

[69] D. J. Kenan and J. D. Keene, "La gets its wings," *Nature Structural and Molecular Biology*, vol. 11, no. 4, pp. 303–305, 2004.

[70] R. S. McLaren, N. Caruccio, and J. Ross, "Human la protein: a stabilizer of histone mRNA," *Molecular and Cellular Biology*, vol. 17, no. 6, pp. 3028–3036, 1997.

[71] F. Brenet, N. Dussault, J. Borch et al., "Mammalian peptidylglycine α-amidating monooxygenase mRNA expression can be modulated by the La autoantigen," *Molecular and Cellular Biology*, vol. 25, no. 17, pp. 7505–7521, 2005.

[72] K. Spångberg, L. Wiklund, and S. Schwartz, "Binding of the La autoantigen to the hepatitis C virus 3′ untranslated region protects the RNA from rapid degradation in vitro," *Journal of General Virology*, vol. 82, no. 1, pp. 113–120, 2001.

[73] Y. K. Kim, S. H. Back, J. Rho, S. H. Lee, and S. K. Jang, "La autoantigen enhances translation of Bip mRNA," *Nucleic Acids Research*, vol. 29, no. 24, pp. 5009–5016, 2001.

[74] A. A. Komar and M. Hatzoglou, "Cellular IRES-mediated translation: the war of ITAFs in pathophysiological states," *Cell Cycle*, vol. 10, no. 2, pp. 229–240, 2011.

[75] K. A. Spriggs, M. Bushell, and A. E. Willis, "Translational Regulation of Gene Expression during Conditions of Cell Stress," *Molecular Cell*, vol. 40, no. 2, pp. 228–237, 2010.

[76] I. Stein, A. Itin, P. Einat, R. Skaliter, Z. Grossman, and E. Keshet, "Translation of vascular endothelial growth factor mRNA by internal ribosome entry: implications for translation under hypoxia," *Molecular and Cellular Biology*, vol. 18, no. 6, pp. 3112–3119, 1998.

[77] G. Sommer, C. Rossa, A. C. Chi, B. W. Neville, and T. Heise, "Implication of RNA-binding protein La in proliferation, migration and invasion of lymph node-metastasized hypopharyngeal SCC cells," *PLoS ONE*, vol. 6, no. 10, Article ID e25402, 2011.

[78] G. Sommer, J. Dittmann, J. Kuehnert et al., "The RNA-binding protein la contributes to cell proliferation and CCND1 expression," *Oncogene*, vol. 30, no. 4, pp. 434–444, 2011.

[79] M. Petz, N. Them, H. Huber, H. Beug, and W. Mikulits, "La enhances IRES-mediated translation of laminin B1 during malignant epithelial to mesenchymal transition," *Nucleic Acids Research*, vol. 40, no. 1, pp. 290–302, 2012.

[80] X. Sanjuan, P. L. Fernandez, R. Miquel et al., "Overexpression of the 67-kD laminin receptor correlates with tumour progression in human colorectal carcinoma," *The Journal of Pathology*, vol. 179, no. 4, pp. 376–380, 1996.

[81] F. Al-Ejeh, J. M. Darby, and M. P. Brown, "The La autoantigen is a malignancy-associated cell death target that is induced by DNA-damaging drugs," *Clinical Cancer Research*, vol. 13, no. 18, pp. 5509s–5518s, 2007.

[82] E. C. Holland, J. Celestino, C. Dai, L. Schaefer, R. E. Sawaya, and G. N. Fuller, "Combined activation of Ras and Akt in neural progenitors induces glioblastoma formation in mice," *Nature Genetics*, vol. 25, no. 1, pp. 55–57, 2000.

[83] R. E. Izumi, S. Das, B. Barat, S. Raychaudhuri, and A. Dasgupta, "A peptide from autoantigen La blocks poliovirus and hepatitis C virus cap-independent translation and reveals a single tyrosine critical for La RNA binding and translation stimulation," *Journal of Virology*, vol. 78, no. 7, pp. 3763–3776, 2004.

[84] L. C. Cobbold, L. A. Wilson, K. Sawicka et al., "Upregulated c-myc expression in multiple myeloma by internal ribosome entry results from increased interactions with and expression of PTB-1 and YB-1," *Oncogene*, vol. 29, no. 19, pp. 2884–2891, 2010.

[85] Y. Shi, P. J. Frost, B. Q. Hoang et al., "IL-6-induced stimulation of c-Myc translation in multiple myeloma cells is mediated by myc internal ribosome entry site function and the RNA-binding protein, hnRNP A1," *Cancer Research*, vol. 68, no. 24, pp. 10215–10222, 2008.

[86] J. Bergalet, M. Fawal, C. Lopez et al., "HuR-Mediated control of C/EBPβ mRNA stability and translation in ALK-Positive anaplastic large cell lymphomas," *Molecular Cancer Research*, vol. 9, no. 4, pp. 485–496, 2011.

[87] W. Kakuguchi, T. Kitamura, T. Kuroshima et al., "HuR knockdown changes the oncogenic potential of oral cancer cells," *Molecular Cancer Research*, vol. 8, no. 4, pp. 520–528, 2010.

[88] S. L. Nowotarski and L. M. Shantz, "Cytoplasmic accumulation of the RNA-binding protein HuR stabilizes the ornithine decarboxylase transcript in a murine nonmelanoma skin cancer model," *The Journal of Biological Chemistry*, vol. 285, no. 41, pp. 31885–31894, 2010.

[89] J. Wang, B. Wang, J. Bi, and C. Zhang, "Cytoplasmic HuR expression correlates with angiogenesis, lymphangiogenesis, and poor outcome in lung cancer," *Medical Oncology*, vol. 28, pp. 577–585, 2010.

[90] N. Filippova, X. Yang, Y. Wang et al., "The RNA-binding protein HuR promotes glioma growth and treatment resistance," *Molecular Cancer Research*, vol. 9, no. 5, pp. 648–659, 2011.

[91] F. Bolognani, A. I. Gallani, and L. Sokol, "mRNA stability alterations mediated by HuR are necessary to sustain the fast growth of glioma cells," *Journal of Neuro-Oncology*, vol. 106, no. 3, pp. 531–542, 2012.

[92] P. S. Yoo, C. A. W. Sullivan, S. Kiang et al., "Tissue microarray analysis of 560 patients with colorectal adenocarcinoma: high expression of HuR predicts poor survival," *Annals of Surgical Oncology*, vol. 16, no. 1, pp. 200–207, 2009.

[93] M. Heinonen, R. Fagerholm, K. Aaltonen et al., "Prognostic role of HuR in hereditary breast cancer," *Clinical Cancer Research*, vol. 13, no. 23, pp. 6959–6963, 2007.

[94] C. Denkert, W. Weichert, K. J. Winzer et al., "Expression of the ELAV-like protein HuR is associated with higher tumor grade and increased cyclooxygenase-2 expression in human

breast carcinoma," *Clinical Cancer Research*, vol. 10, no. 16, pp. 5580–5586, 2004.

[95] C. Denkert, W. Weichert, S. Pest et al., "Overexpression of the embryonic-lethal abnormal vision-like protein HuR in ovarian carcinoma is a prognostic factor and is associated with increased cyclooxygenase 2 expression," *Cancer Research*, vol. 64, no. 1, pp. 189–195, 2004.

[96] H. Hasegawa, W. Kakuguchi, T. Kuroshima et al., "HuR is exported to the cytoplasm in oral cancer cells in a different manner from that of normal cells," *British Journal of Cancer*, vol. 100, no. 12, pp. 1943–1948, 2009.

[97] A. Doller, J. Pfeilschifter, and W. Eberhardt, "Signalling pathways regulating nucleo-cytoplasmic shuttling of the mRNA-binding protein HuR," *Cellular Signalling*, vol. 20, no. 12, pp. 2165–2173, 2008.

[98] C. M. Brennan and J. A. Steitz, "HuR and mRNA stability," *Cellular and Molecular Life Sciences*, vol. 58, no. 2, pp. 266–277, 2001.

[99] A. S. Halees, R. El-badrawi, and K. S. A. Khabar, "ARED Organism: expansion of ARED reveals AU-rich element cluster variations between human and mouse," *Nucleic Acids Research*, vol. 36, no. 1, pp. D137–D140, 2008.

[100] C. A. Lagnado, C. Y. Brown, and G. J. Goodali, "AUUUA is not sufficient to promote poly(A) shortening and degradation of an mRNA: the functional sequence within AU-rich elements may be UUAUUUA(U/A)(U/A)," *Molecular and Cellular Biology*, vol. 14, no. 12, pp. 7984–7995, 1994.

[101] S. Sanduja and D. A. Dixon, "Tristetraprolin and E6-AP: killing the messenger in cervical cancer," *Cell Cycle*, vol. 9, no. 16, pp. 3135–3136, 2010.

[102] J. M. Izquierdo, "Hu antigen R (HuR) functions as an alternative pre-mRNA splicing regulator of Fas apoptosis-promoting receptor on exon definition," *The Journal of Biological Chemistry*, vol. 283, no. 27, pp. 19077–19084, 2008.

[103] T. D. Levine, F. Gao, P. H. King, L. G. Andrews, and J. D. Keene, "Hel-N1: an autoimmune RNA-binding protein with specificity for 3' uridylate-rich untranslated regions of growth factor mRNAs," *Molecular and Cellular Biology*, vol. 13, no. 6, pp. 3494–3504, 1993.

[104] A. Lal, K. Mazan-Mamczarz, T. Kawai, X. Yang, J. L. Martindale, and M. Gorospe, "Concurrent versus individual binding of HuR and AUF1 to common labile target mRNAs," *EMBO Journal*, vol. 23, no. 15, pp. 3092–3102, 2004.

[105] X. Guo and R. S. Hartley, "HuR contributes to cyclin E1 deregulation in MCF-7 breast cancer cells," *Cancer Research*, vol. 66, no. 16, pp. 7948–7956, 2006.

[106] W. Wang, M. C. Caldwell, S. Lin, H. Furneaux, and M. Gorospe, "HuR regulates cyclin A and cyclin B1 mRNA stability during cell proliferation," *EMBO Journal*, vol. 19, no. 10, pp. 2340–2350, 2000.

[107] D. Ishimaru, S. Ramalingam, T. K. Sengupta et al., "Regulation of Bcl-2 expression by HuR in HL60 leukemia cells and A431 carcinoma cells," *Molecular Cancer Research*, vol. 7, no. 8, pp. 1354–1366, 2009.

[108] K. Abdelmohsen, A. Lal, H. K. Hyeon, and M. Gorospe, "Posttranscriptional orchestration of an anti-apoptotic program by HuR," *Cell Cycle*, vol. 6, no. 11, pp. 1288–1292, 2007.

[109] K. Abdelmohsen, R. Pullmann Jr., A. Lal et al., "Phosphorylation of HuR by Chk2 Regulates SIRT1 Expression," *Molecular Cell*, vol. 25, no. 4, pp. 543–557, 2007.

[110] S. J. Cho, J. Zhang, and X. Chen, "RNPC1 modulates the RNA-binding activity of, and cooperates with, HuR to regulate p21 mRNA stability," *Nucleic Acids Research*, vol. 38, no. 7, Article ID gkp1229, pp. 2256–2267, 2010.

[111] E. S. Akool, H. Kleinert, F. M. A. Hamada et al., "Nitric oxide increases the decay of matrix metalloproteinase 9 mRNA by inhibiting the expression of mRNA-stabilizing factor HuR," *Molecular and Cellular Biology*, vol. 23, no. 14, pp. 4901–4916, 2003.

[112] A. Huwiler, E. S. Akool, A. Aschrafi, F. M. A. Hamada, J. Pfeilschifter, and W. Eberhardt, "ATP Potentiates Interleukin-1β-induced MMP-9 Expression in Mesangial Cells via Recruitment of the ELAV Protein HuR," *The Journal of Biological Chemistry*, vol. 278, no. 51, pp. 51758–51769, 2003.

[113] H. Tran, F. Maurer, and Y. Nagamine, "Stabilization of urokinase and urokinase receptor mRNAs by HuR is linked to its cytoplasmic accumulation induced by activated mitogen-activated protein kinase-activated protein kinase 2," *Molecular and Cellular Biology*, vol. 23, no. 20, pp. 7177–7188, 2003.

[114] R. Dong, J. G. Lu, Q. Wang et al., "Stabilization of Snail by HuR in the process of hydrogen peroxide induced cell migration," *Biochemical and Biophysical Research Communications*, vol. 356, no. 1, pp. 318–321, 2007.

[115] S. Galban, Y. Kuwano, and R. Pullmann Jr., "RNA-binding proteins HuR and PTB promote the translation of hypoxia-inducible factor 1alpha," *Molecular and Cellular Biology*, vol. 28, no. 1, pp. 93–107, 2008.

[116] L. G. Sheflin, A. P. Zou, and S. W. Spaulding, "Androgens regulate the binding of endogenous HuR to the AU-rich 3′UTRs of HIF-1alpha and EGF mRNA," *Biochemical and Biophysical Research Communications* , vol. 322, no. 2, pp. 644–651, 2004.

[117] N. C. Meisner and W. Filipowicz, "Properties of the regulatory RNA-binding protein HuR and its role in controlling miRNA repression," *Advances in experimental medicine and biology*, vol. 700, pp. 106–123, 2010.

[118] M. R. Epis, A. Barker, and K. M. Giles, "The RNA-binding protein HuR opposes the repression of ERBB-2 gene expression by microRNA miR-331-3p in prostate cancer cells," *The Journal of Biological Chemistry*, vol. 286, no. 48, pp. 41442–41454, 2011.

[119] L. E. Young, A. E. Moore, L. Sokol, N. Meisner-Kober, and D. A. Dixon, "The mRNA stability factor HuR inhibits micro-RNA-16 targeting of cyclooxygenase-2," *Molecular Cancer Research*, vol. 10, no. 1, pp. 167–180, 2012.

[120] H. K. Hyeon, Y. Kuwano, S. Srikantan, K. L. Eun, J. L. Martindale, and M. Gorospe, "HuR recruits let-7/RISC to repress c-Myc expression," *Genes and Development*, vol. 23, no. 15, pp. 1743–1748, 2009.

[121] Q. Jing, S. Huang, S. Guth et al., "Involvement of microRNA in AU-Rich element-mediated mRNA instability," *Cell*, vol. 120, no. 5, pp. 623–634, 2005.

[122] M. Kullmann, U. Göpfert, B. Siewe, and L. Hengst, "ELAV/Hu proteins inhibit p27 translation via an IRES element in the p27 5′UTR," *Genes and Development*, vol. 16, no. 23, pp. 3087–3099, 2002.

[123] D. Durie, S. M. Lewis, U. Liwak, M. Kisilewicz, M. Gorospe, and M. Holcik, "RNA-binding protein HuR mediates cytoprotection through stimulation of XIAP translation," *Oncogene*, vol. 30, no. 12, pp. 1460–1469, 2011.

[124] A. Lal, T. Kawai, X. Yang, K. Mazan-Mamczarz, and M. Gorospe, "Antiapoptotic function of RNA-binding protein HuR effected through prothymosin α," *EMBO Journal*, vol. 24, no. 10, pp. 1852–1862, 2005.

[125] C. L. Costantino, A. K. Witkiewicz, Y. Kuwano et al., "The role of HuR in gemcitabine efficacy in pancreatic cancer: HuR up-regulates the expression of the gemcitabine metabolizing

enzyme deoxycytidine kinase," *Cancer Research*, vol. 69, no. 11, pp. 4567–4572, 2009.

[126] J. R. Brody and G. E. Gonye, "HuR's role in gemcitabine efficacy: an exception or opportunity?" *Wiley Interdisciplinary Reviews*, vol. 2, no. 3, pp. 435–444, 2011.

[127] I. Lopez de Silanes, M. P. Quesada, and M. Esteller, "Aberrant regulation of messenger RNA 3′-untranslated region in human cancer," *Cellular Oncology*, vol. 29, no. 1, pp. 1–17, 2007.

[128] F. M. Gratacos and G. Brewer, "The role of AUF1 in regulated mRNA decay," *Wiley Interdisciplinary Reviews*, vol. 1, no. 3, pp. 457–473, 2011.

[129] C. Barreau, L. Paillard, A. Méreau, and H. B. Osborne, "Mammalian CELF/Bruno-like RNA-binding proteins: molecular characteristics and biological functions," *Biochimie*, vol. 88, no. 5, pp. 515–525, 2006.

[130] I. A. Vlasova and P. R. Bohjanen, "Posttranscriptional regulation of gene networks by GU-rich elements and CELF proteins," *RNA Biology*, vol. 5, no. 4, pp. 201–207, 2008.

[131] D. Mukhopadhyay, C. W. Houchen, S. Kennedy, B. K. Dieckgraefe, and S. Anant, "Coupled mRNA stabilization and translational silencing of cyclooxygenase-2 by a novel RNA binding protein, CUGBP2," *Molecular Cell*, vol. 11, no. 1, pp. 113–126, 2003.

[132] D. A. Dixon, G. C. Balch, N. Kedersha et al., "Regulation of cyclooxygenase-2 expression by the translational silencer TIA-1," *Journal of Experimental Medicine*, vol. 198, no. 3, pp. 475–481, 2003.

[133] M. Ouellette and B. Papadopoulou, "Coordinated gene expression by post-transcriptional regulons in African trypanosomes," *Journal of Biology*, vol. 8, no. 11, article 100, 2009.

[134] C. Clayton and M. Shapira, "Post-transcriptional regulation of gene expression in trypanosomes and leishmanias," *Molecular and Biochemical Parasitology*, vol. 156, no. 2, pp. 93–101, 2007.

[135] S. Haile and B. Papadopoulou, "Developmental regulation of gene expression in trypanosomatid parasitic protozoa," *Current Opinion in Microbiology*, vol. 10, no. 6, pp. 569–577, 2007.

[136] J. B. Palenchar and V. Bellofatto, "Gene transcription in trypanosomes," *Molecular and Biochemical Parasitology*, vol. 146, no. 2, pp. 135–141, 2006.

[137] V. Anantharaman, E. V. Koonin, and L. Aravind, "Comparative genomics and evolution of proteins involved in RNA metabolism," *Nucleic Acids Research*, vol. 30, no. 7, pp. 1427–1464, 2002.

[138] W. J. Ma, S. Cheng, C. Campbell, A. Wright, and H. Furneaux, "Cloning and characterization of HuR, a ubiquitously expressed Elav-like protein," *The Journal of Biological Chemistry*, vol. 271, no. 14, pp. 8144–8151, 1996.

[139] D. Antic and J. D. Keene, "Embryonic lethal abnormal visual RNA-binding proteins involved in growth, differentiation, and posttranscriptional gene expression," *American Journal of Human Genetics*, vol. 61, no. 2, pp. 273–278, 1997.

[140] S. Fumagalli, N. F. Totty, J. J. Hsuan, and S. A. Courtneidge, "A target for Src in mitosis," *Nature*, vol. 368, no. 6474, pp. 871–874, 1994.

[141] S. J. Taylor and D. Shalloway, "An RNA-binding protein associated with Src through its SH2 and SH3 domains in mitosis," *Nature*, vol. 368, no. 6474, pp. 867–871, 1994.

[142] M. Di Fruscio, T. Chen, and S. Richard, "Characterization of Sam68-like mammalian proteins SLM-1 and SLM-2: SLM-1 is a Src substrate during mitosis," *Proceedings of the National Academy of Sciences of the United States of America*, vol. 96, no. 6, pp. 2710–2715, 1999.

[143] J. P. Venables, C. Vernet, S. L. Chew et al., "T-STAR/ETOILE: a novel relative of SAM68 that interacts with an RNA-binding protein implicated in spermatogenesis," *Human Molecular Genetics*, vol. 8, no. 6, pp. 959–969, 1999.

[144] M. P. Paronetto and C. Sette, "Role of RNA-binding proteins in mammalian spermatogenesis," *International Journal of Andrology*, vol. 33, no. 1, pp. 2–12, 2009.

[145] J. Marcotrigiano, A. C. Gingras, N. Sonenberg, and S. K. Burley, "Cocrystal structure of the messenger RNA 5′ cap-binding protein (eIF4E) bound to 7-methyl-GDP," *Cell*, vol. 89, no. 6, pp. 951–961, 1997.

[146] A. C. Gingras, B. Raught, and N. Sonenberg, "eIF4 initiation factors: effectors of mRNA recruitment to ribosomes and regulators of translation," *Annual Review of Biochemistry*, vol. 68, pp. 913–963, 1999.

[147] M. Altmann, P. P. Muller, J. Pelletier, N. Sonenberg, and H. Trachsel, "A mammalian translation initiation factor can substitute for its yeast homologue in vivo," *The Journal of Biological Chemistry*, vol. 264, no. 21, pp. 12145–12147, 1989.

[148] H. Matsuo, H. Li, A. M. McGuire et al., "Structure of translation factor elF4E bound to m7GDP and interaction with 4E-binding protein," *Nature Structural Biology*, vol. 4, no. 9, pp. 717–724, 1997.

[149] K. Tomoo, X. Shen, K. Okabe et al., "Crystal structures of 7-methylguanosine 5′-triphosphate (m7GTP)- and P1-7-methylguanosine-P3-adenosine-5′, 5′-triphosphate (m7GpppA)-bound human full-length eukaryotic initiation factor 4E: biological importance of the C-terminal flexible region," *Biochemical Journal*, vol. 362, no. 3, pp. 539–544, 2002.

[150] A. F. Monzingo, S. Dhaliwal, A. Dutt-Chaudhuri et al., "The structure of eukaryotic translation initiation factor-4E from wheat reveals a novel disulfide bond," *Plant Physiology*, vol. 143, no. 4, pp. 1504–1518, 2007.

[151] R. E. Rhoads, "EIF4E: new family members, new binding partners, new roles," *The Journal of Biological Chemistry*, vol. 284, no. 25, pp. 16711–16715, 2009.

[152] G. Hernández and P. Vazquez-Pianzola, "Functional diversity of the eukaryotic translation initiation factors belonging to eIF4 families," *Mechanisms of Development*, vol. 122, no. 7-8, pp. 865–876, 2005.

[153] A. R. Morris, N. Mukherjee, and J. D. Keene, "Systematic analysis of posttranscriptional gene expression," *Wiley Interdisciplinary Reviews*, vol. 2, no. 2, pp. 162–180, 2010.

[154] S. C. Janga and N. Mittal, "Construction, structure and dynamics of post-transcriptional regulatory network directed by RNA-binding proteins," *Advances in Experimental Medicine and Biology*, vol. 722, pp. 103–117, 2011.

[155] I. Topisirovic, N. Siddiqui, and K. L. B. Borden, "The eukaryotic translation initiation factor 4E (eIF4E) and HuR RNA operons collaboratively regulate the expression of survival and proliferative genes," *Cell Cycle*, vol. 8, no. 7, pp. 960–961, 2009.

[156] I. Topisirovic, N. Siddiqui, S. Orolicki et al., "Stability of eukaryotic translation initiation factor 4E mRNA is regulated by HuR, and this activity is dysregulated in cancer," *Molecular and Cellular Biology*, vol. 29, no. 5, pp. 1152–1162, 2009.

10

Differential Expression ESTs Associated with Fluorosis in Rats Liver

Y. Q. He,[1,2] Y. Pan,[1] L. J. Ying,[1] and R. Zhao[3]

[1] *The Laboratory Animal Research Center, Jiangsu University, Zhenjiang 212013, China*
[2] *College of Food Science and Biological Engineering, Jiangsu University, Zhenjiang 212013, China*
[3] *School of Clinical Medicine, Jiangsu University, Zhenjiang 212013, China*

Correspondence should be addressed to Y. Q. He, yqhe@ujs.edu.cn

Academic Editor: Ferenc Olasz

The fluoride has volcanic activity and abundantly exists in environment combining with other elements as fluoride compounds. Recent researches indicated that the molecular mechanisms of intracellular fluoride toxicity were very complex. However, the molecular mechanisms underlying the effects on gene expression of chronic fluoride-induced damage is unknown, especially the detailed regulatory process of mitochondria. In the present study, we screened the differential expression ESTs associated with fluorosis by DDRT-PCR in rat liver. We gained 8 genes, 3 new ESTs, and 1 unknown function sequence and firstly demonstrated that microsomal glutathione S-transferase 1 (MGST1), ATP synthase H^+ transporting mitochondrial F_0 complex subunit C1, selenoprotein S, mitochondrial IF1 protein, and mitochondrial succinyl-CoA synthetase alpha subunit were participated in mitochondria metabolism, functional and structural damage process caused by chronic fluorosis. This information will be very helpful for understanding the molecular mechanisms of fluorosis.

1. Introduction

The fluoride has volcanic activity and abundantly exists in environment combining with other elements as fluoride compounds. With the development of the world economy, more and more organofluorine compounds are increasingly used. These compounds have a wide range of functions and can serve as agrochemicals, pharmaceuticals, refrigerants, pesticides, surfactants, fire-extinguishing agents, fibers, membranes, ozone depletors, and insulating materials [1, 2]. At the same time, in last decade, the fluoride effects have resurfaced due to the awareness that this element interacts with cellular systems even at low doses. Excessive fluoride intake over a long period of time may result in a serious public health problem called fluorosis, which is characterized by dental mottling and skeletal manifestations such as crippling deformities, osteoporosis, and osteosclerosis [3].

In recent years, metabolic, functional, and structural damage caused by chronic fluorosis have been reported in many tissues. Fluoride can induce oxidative stress and modulate intracellular redox homeostasis, lipid peroxidation, and protein carbonyl content, as well as alter gene expression and cause apoptosis. Genes modulated by fluoride include those related to the stress response metabolic enzymes, the cell-cell communications, and signal transduction [1]. In most cases, fluoride acts as an enzyme inhibitor, but fluoride ions can occasionally stimulate enzyme activity. Research data strongly suggest that fluoride inhibits protein secretion and/or synthesis and that it influences distinct signaling pathways involved in proliferation and apoptosis including the mitogen-activated protein kinase (MAPK), p53, activator protein-1 (AP-1), and nuclear factor kappa B (NF-κB) pathways [4–6]. However, information about the molecular mechanism of fluoride-induced tissue damage is almost unknown. Some studies point out that the mitochondria are the key intracellular targets for different stressors including fluoride [7]. Fluoride alters the activity of many mitochondria-rich cells such as those of the human kidney and the rat liver [8, 9], but the molecular mechanisms of chronic fluoride-induced mitochondrial damage are scarce. As an important detoxification organ, liver plays a pivotal role on resolving fluorosis effects. In the present study, in order to get a deeper understanding of the molecular mechanisms underlying the effects of fluoride on mitochondrial gene expression and metabolism, we screened differential display genes or expressed sequence tags (ESTs) involved in Sprague-Dawley rats with fluorosis.

2. Materials and Methods

2.1. Animals and Fluoride Exposures. The protocol for the study was reviewed and approved by the Institutional Animal Use and Care Committees of Jiangsu University. The male Sprague-Dawley rats ($n = 30$) were received three week after weaning, an age at which all regions of the central nervous system are rapidly developing. All animals were housed in stainless-steel cages suspended in stainless-steel racks with the humidity ranged from 30 to 55%, and the room temperature remained between 22 and 25°C, fed with a standard pellet diet, and given distilled water ad libitum. The animals were allowed to acclimatize to the laboratory conditions for four days before experiments began.

The rats were weighed and randomly divided into 2 groups (15/group). The first group animals were injected i.p. with aqueous NaF (20 mg/kg/body weight/day) selected on the basis of the LD50 value of fluoride, which is 51.6 mg/kg body weight/day in mice and maintained for 30 days. The second group served as control and was injected mammalian physiological saline. The NaF solution (30 mL) was prepared fresh each day in double distilled water. At the end of the 30-day treatment, the animals were killed by decapitation, the brains were rapidly dissected, and the hippocampus removed. The tissue was immediately frozen in liquid nitrogen and stored at −80°C until use. The experiments were performed in accordance with the regional legal regulations. The body weight of each animal was weighed before treatment, and weighed every 10 days. The F-levels in serum were determined [10] by using a CSB-F-I fluoride ion electrode (Changsa Analysis Instrumentation, China) to confirm chronic fluorosis.

2.2. Total RNA Extraction and RNA Pools. Total RNA of liver samples was extracted using Trizol kit (Invitrogen, Carlsbad, USA) according to the manufacturer's instructions. To remove any contaminated genomic DNA from the RNA samples, 25 μL of total RNA was treated with an equal volume of RNase-free DNase-I mixture (Promega, Madison, WI, USA), and the whole mixture was incubated at 37°C for 30 min. Subsequently, the mixture was subjected to phenol-chloroform extraction followed by ethanol precipitation. The concentration of purified RNA was determined by absorbance at 260 nm. An RNA pool containing equal amount RNA of three individuals from the same breed was established. The RNA samples were stored at −70°C.

2.3. Primers. The cDNA primers of three anchored primers of H-T_{11}A, H-T_{11}G and H-T_{11}C and eight arbitrary primers of P1–P8 were designed according to the third generation primers of GenHunter Company. The 3′ end anchored primers were

H-T_{11}G: 5′-AAGCTTTTTTTTTTTG-3′;
H-T_{11}C: 5′-AAGCTTTTTTTTTTTC-3;
H-T_{11}A: 5′-AAGCTTTTTTTTTTTA-3.
The 5′ end arbitrary oligonucleotide primers were:
P1: 5′-TGCCGAAGCTTTGGTGTC-3′;
P2: 5′-TGCCGAAGCTTTGGTACC-3′;
P3: 5′-TGCCGAAGCTTTGGTAGC-3′;
P4: 5′-TGCCGAAGCTTTGGTATG-3′;
P5: 5′-TGCCGAAGCTTTGGTCAC-3′;
P6: 5′-TGCCGAAGCTTTGGTCAG-3′;
P7: 5′-TGCCGAAGCTTTGGTCTG-3′;
P8: 5′-TGCCGAAGCTTTGGTCTC-3′.

2.4. Reverse Transcription. A 1 μL of 3′ anchored primer (1 μmol/L) was combined with 4 μL (1 μg/μL) RNase-free DNase I-treated RNA samples and 1 μL of 10 mmol/L dNTP mix, the mixture was adjusted to a final volume of 11 μL by adding DEPC-treated water. Then it was incubated for 5 min at 65°C followed by cooling on ice for 2 min. First-strand synthesis was initiated by mixing the mixture with 1 μL (50 units/μL) M-MLV reverse transcriptase (Invitrogen) in a final volume of 8 μL mix containing 2 μL of 10 × RT buffer (200 mmol/L Tris-HCl (pH 8.4), 500 mmol/L KCl), 4 μL of 25 mmol/L $MgCl_2$, and 2 μL of 0.1 mol/L DTT. The reverse transcription reaction was carried out at 42°C for 1 h, then terminated by heating at 70°C in a water bath for 15 min. The reverse transcription product of cDNA was stored at −20°C.

2.5. DD-PCR. A 1 μL of the reverse transcription product was added to 24 μL of PCR solution containing 20 μmol/L dNTPs, 1.5 mmol/L $MgCl_2$, 1 U of Taq DNA polymerase (Promega), 20 μmol/L H-T_{11}N (H-T_{11}G/H-T_{11}C/H-T_{11}A) primer used for RT, and 10 μmol/L of one of eight arbitrary primers. The PCR reaction was carried out on Mastercycler 5333 (Eppendorf AG, Hamburg, Germany). The PCR parameters were as follows: initial denaturation at 94°C for 5 min; followed by one cycle of denaturation at 94°C for 30 s, annealing at 40°C for 2 min, extension at 72°C for 2 min; followed by 35 cycles of denaturation at 94°C for 30 s, annealing at 60°C for 2 min, extension at 72°C for 2 min; with a final extension step at 72°C for 5 min. The amplified fragments of DDRT-PCR were separated by electrophoresis using 8% polyacrylamide gels (acrylamide : bisacrylamide = 39 : 1). Then, the PCR products were visualized by silver staining and photographed and analyzed using an AlphaImager 2200 and 1220 Documentation and Analysis Systems (Alpha Innotech Corporation, San Leandro, CA, USA).

2.6. Reverse Northern Dot Blot Analysis. Reverse Northern dot blotting was carried out according to the instruction manual of DIG-HIGH Prime DNA labeling and detection starter kit I (Roche, Penzberg, Germany). In brief, 1 μg of template DNA was denatured by heating in a boiling water bath for 10 min and quickly chilling in an ice/water bath, then DIG-HIGH Prime was mixed and incubated at 37°C for 1 h, and the reaction was stopped by adding 2 μL of 0.2 mol/L EDTA (pH 8.0). Subsequently, the RT products were fixed on the nylon membranes positively charged by UV crosslinking and were then hybridized in DIG Easy Hyb (10 mL/100 cm^2) containing DIG-labeled DNA probe at appropriate hybridization temperature for 20 h. After hybridization and stringency washes, the nylon membranes were incubated in 100 mL blocking solution for 30 min and in 20 mL antibody solution for 30 min. Then it was detected in freshly color substrate solution in the dark after washing two times in 100 mL washing buffer for 15 min, and the reaction will be completed after 16 h and terminated by 50 mL TE buffer for 5 min.

2.7. Purification, Sequencing, and Analysis of Differential Bands. Differential fragments from 100 bp to 1200 bp were excised from the gels, extracted by boiling in 100 μL of distilled water, purified with a DNA purification system (Promega), reamplified using the same primer corresponding to cDNA display. The PCR amplification conditions were as follows: initial denaturation at 94°C for 3 min; followed by 33 cycles of denaturation at 94°C for 2 min, annealing at 55°C for 2 min, extension at 72°C for 2 min; with a final extension at 72°C for 5 min. The 25 μL of the reamplified PCR sample along with 5 μL of gel loading buffer was analyzed on 1.5% agarose/EtBr gel in 1 × TAE (40 mmol/L Tris base, 20 mmol/L acetic acid, 2 mmol/L EDTA pH 8.0). The PCR bands of interest were cut from the gels under UV illumination. The DNA was purified using DNA extraction kit (Qiagen, Hilden, Germany). The purified differential bands were ligated into the pGEM-T Easy vector (Promega) according to manufacturer's instructions and transformed into *Escherichia coli* DH5α competence cells. The positive clones were sequenced using an automatic ABI 3730 sequencer (Perkin Elmer Applied Biosystems, Foster City, CA, USA) by Shanghai Invitrogen Biotechnology Ltd. Co. (Shanghai, China) and analyzed in GenBank by BLAST.

3. Results

3.1. Extraction and Reverse Transcription of Total RNA. We collected the livers at the same conditions except fluoride treatment for extracting the RNA. The gel profile of RNA showed that two bands of 18S and 28S were clear and could be used for reverse transcription.

3.2. DD-PCR. We displayed the different bands of two kinds of cDNA using 24 primer pairs consisting of 3 anchored primers and 8 arbitrary primers. The partial results were shown in Figure 1. At last 1532 bands were detected on statistics, and 36 bands were recovered and sequenced based on their differences (lack, density, and size). The recovered differential bands were reamplified with the same anchor primer and arbitrary primer using the recovered products as templates. However, some of the recovered bands could not be perfectly amplified because of the false positive.

3.3. Reverse Northern Dot Blot. In order to eliminate the disruption of false positive bands, reverse northern was carried out to validate differential bands. 36 differential bands of PCR amplifications were blotted on two nylon membranes and hybridized with probes of liver cDNA. The results showed that 13 positive blots were detected (Figure 2).

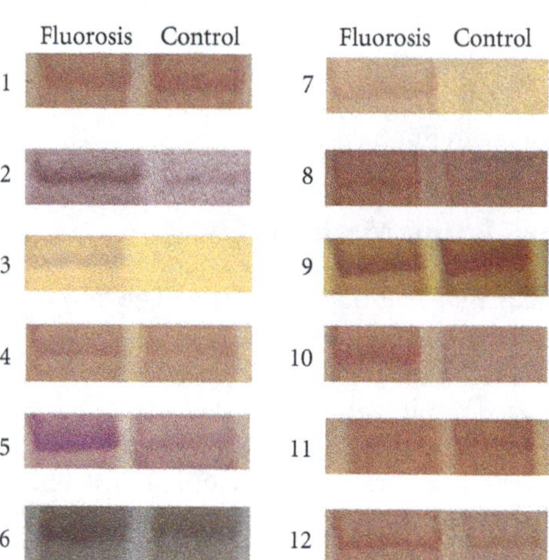

FIGURE 1: Differential display profiles from different primer combinations of rat liver. 1–12: correspond to the no. of Table 1.

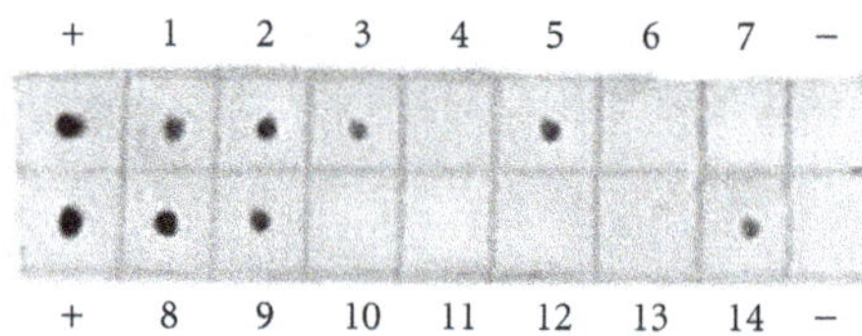

FIGURE 2: Partial results of reverse northern dot blots for differential display bands. +: positive control; –: false control. 1–14: the results of dot blot of bands, visible dot represents positive result, and no dot represents false result.

3.4. Sequence Analysis of Differential Bands. All the13 positive bands detected by reverse northern dot blot were sequenced and analyzed in GenBank by BLAST. The alignment standards include that the ESTs length should be more than 100 bp, the *E* value should be less than 0, and the identity should exceed 90%. The results were listed in Table 1. The BLAST results demonstrated that 8 genes including microsomal glutathione S-transferase 1, acyl-CoA synthetase long-chain family member, ATP synthase H^+ transporting mitochondrial F_0 complex subunit C1, haptoglobin, selenoprotein S, mitochondrial IF1 protein, acyl-CoA synthetase long-chain family member, mitochondrial succinyl-CoA synthetase alpha subunit, 3 new ESTs, and 1 unknown function sequence were differential expressed.

4. Discussion

Several molecular strategies including microarray analysis, in vivo expression technology (IVET), DDRT-PCR, and signature-tagged mutagenesis have been used to identify differential expression genes in all kinds of organs, tissues, and individuals. Although DDRT-PCR was originally known

Table 1: Differential expression genes of DDRT-PCR for fluorosis in rat liver.

No.	GenBank accession number	bp	Copies	BLAST results	Score	*E* value	Identities
(1)	NM_134349.3	183	1	Microsomal glutathione S-transferase 1	307 bits (166)	1*e*−80	170/172 (99%)
(2)	NM_012820.1	324	1	Acyl-CoA synthetase long-chain family member	540 bits (292)	2*e*−150	297/299 (99%)
(3)	New EST	176	1	No significant similarity found	/	/	/
(4)	AI029143.1	317	1	UI-R-C0-ip-g-08-0-UI	549 bits (297)	4*e*−153	304/307 (99%)
(5)	NM_017311.1	266	1	ATP synthase, H^+ transporting, mitochondrial F_0 complex, subunit C1	444 bits (240)	2*e*−121	240/240 (100%)
(6)	NM_012582.2	332	1	Haptoglobin	586 bits (317)	3*e*−164	319/320 (99%)
(7)	NM_173120.2	318	2	Selenoprotein S	558 bits (302)	6*e*−156	309/312 (99%)
(8)	L07806.1	483	1	Mitochondrial IF1 protein	761 bits (412)	0.0	412/412 (100%)
(9)	New EST	283	1	No significant similarity found	/	/	/
(10)	NM_012820.1	196	1	Acyl-CoA synthetase long-chain family member	540 bits (292)	2*e*−150	297/299 (99%)
(11)	New EST	311	1	No significant similarity found	/	/	/
(12)	J03621.1	455	1	Mitochondrial succinyl-CoA synthetase alpha subunit	797 bits (431)	0.0	433/434 (99%)

that it had highly false positive bands, recent modifications have allowed this technique to be used again perfectly. In this study, by using a novel and modified DDRT-PCR technique, the primers were the third generation DDRT-PCR primers consisting of both anchored and arbitrary primers, which have Hind enzyme site. The number of primers also reduces to three anchored primers and eight arbitrary primers. This primer combination had proved by computer analysis to possess the ability to include all mRNAs in certain stage of some cell development, which makes it easier for next operation and treatment. As in most described DDRT-PCR protocols, total RNA was used as the starting material. In the present study, to exclude partial degradation, we tested total RNA preparation with DNase I to reduce artifacts caused by possible contamination with genomic DNA. To reduce the rate of mismatches that occur with low stringency annealing during subsequent PCR cycles, we followed the strategy published by Zhao et al. [11], the cycle procedure was only one initial cycle with low stringency annealing temperature (40°C), and followed by 35 high stringency PCR cycles with annealing temperature (60°C). This method significantly improved the reproducibility of band patterns.

According to the results of the sequence BLAST, we gained 8 genes, 3 new ESTs, and 1 unknown function sequence and firstly demonstrated that microsomal glutathione S-transferase 1 (MGST1), ATP synthase H^+ transporting mitochondrial F_0 complex subunit C1, selenoprotein S, mitochondrial IF1 protein, and mitochondrial succinyl-CoA synthetase alpha subunit were participated in mitochondria metabolism, functional and structural damage process caused by chronic fluorosis.

Membrane-bound microsomal GST (MGST1), genetically different from the cytosolic GSTs, is a homotrimer with one thiol group (Cys49) per subunit and is activated by thiol alkylation, by disulfide bond formation or sulfenic acid formation, and by other mechanisms [10, 12–15]. mtMGST1, a similar type of microsomal MGST1, is also activated by modification of thiol to a mixed disulfide bond (S-glutathionylation) and by a disulfide-linked mtMGST1 dimer in oxidative stress in vivo and in vitro. It was also clarified that the activation of mtMGST1 contributes to cytochrome c release from mitochondria [16]. MGST1 is activated by oxidative stress, and mtMGST1 is activated through mixed disulfide bond formation that contributes to cytochrome c release from mitochondria through the mitochondrial permeability transition (MPT) pore [16]. Fluoride can induce an increase in the release of cytochrome c from the mitochondria to the cytosol and straightly enter in mitochondria metabolism. In addition, NADPH, and ATP-depended activation of MGST1, which showed the relationship between MGST1 and cellular respiration [17]. According to these references, we could demonstrate that MGST1 participated in the cellular effects by fluorosis on cellular respiration and inner membrane permeability and membrane potential.

ATP synthase H^+ transporting mitochondrial F_0 complex subunit C1 is an isoform of mtATP synthase subunit 9. F_0F_1-ATP synthase, also known as ATP synthase, is composed of three parts: the catalytic sector F_1, consisting of five subunits (3α; 3β; γ; δ; ε), the H^+ translocating sector F_0, and a stalk connecting F_1 with F_0, both consisting of a variable number of subunits [18–20]. In the F_0F_1-ATP synthase complex, in addition to the F_1-γ and the F_0I-PVP (b) subunits [19], also the oligomycin-sensitivity-conferring protein (OSCP) is involved in the coupling of the F_1 catalytic activity to transmembrane proton translocation by F_0 [21]. It has been shown that the F_0F_1-ATP synthase represents one of the sites of T3 action [22, 23]. It is no doubt that ATP synthase expression will be in differential level between fluorosis rat and control rat. Each process including cellular respiration, ROS generation, inner membrane permeability, calcium and phosphate transportation, glucose transport, and cell apoptosis need ATP synthase and energy. Obviously we should know its importance in resistant to fluorosis.

Another gene we screened, mitochondrial IF1 protein, is associated to ATPase activity. IF1 is present in animals [24, 25], yeasts [26, 27], and plants [28, 29]. Its structure is essentially α-helical [30]. Pioneering investigations have shown that bovine IF1 (bIF1) lacking the first 16 residues after partial enzymatic digestion retains the capacity to inhibit ATPase activity [31]. It has been later reported that the minimal inhibitory sequence consists of residues 22–46 [32] or 14–47 [23] in bovine and of residues 17–41 in yeasts [32]. Deletion of the first 17 residues in mammal IF1 resulted in a still inhibitory peptide but with a very low affinity for ATP synthase, whereas conflicting results were reported concerning the removal of the first 21 residues [32–34]. Importantly, the visible part in N-terminus of IF1 (residues 8–21) appears wrapped around the γ subunit and also interacting with the α-subunit. This finding raises the problem of the role of the Nter part of IF1 in the different steps of the inhibitory process. IF1 peptides partially truncated in Nter have been shown to inhibit ATPase activity after several minutes of incubation [33, 34], but in vivo the inhibitory peptide very rapidly responds to pmf variations, as shown by the absence of ATP hydrolysis activity immediately after adding an uncoupler to mitochondria [35]. IF1 appears to limit mitochondrial ROS generation, limiting autophagy which is increased by IF1 knockdown [36].

We screened mitochondrial succinyl-CoA synthetase alpha subunit (mtSCS), and its function is also correlated to ATPase activity. SCS is a Krebs cycle enzyme that catalyzes substrate-level phosphorylation in the forward direction and replenishes succinyl-CoA for ketone body catabolism and porphyrin biosynthesis in the reverse direction [37–39]. In mammals two isoforms of SCS have been identified: one specific for ADP/ATP and another specific for GDP/GTP. In some species (i.e., *Escherichia coli*), SCS is not specific, using either the ATP or GTP [39]. SCSα has a well-documented histidine phosphorylation site but has also been shown to bind phosphate in its dephosphorylated form as a means of stabilizing the complex [39]. Furthermore, SCS can generate ATP in the absence of a proton motive force in the inner membrane, potentially playing a role in maintaining matrix ATP levels under energy-limited conditions, such as transient hypoxia. Inorganic phosphate (Pi), a putative cytosolic signaling molecule, plays a multifaceted role in the regulation of mitochondrial metabolism. Pi can alter the free concentration of Mg^{2+} and Ca^{2+} ions, increase mitochondrial volume, influence the mitochondrial transition pore, and directly modify the activity of several Krebs cycle dehydrogenases [40–43].

Selenoprotein S has recently been described as an endoplasmic reticulum (ER) and plasma membrane-located selenoprotein involved in the physiologic adaptation to ER stress [44, 45]. The SELS gene is known to be expressed in a wide variety of tissues and cell types, including tissues important for glycemic control such as adipose tissue, muscle, and liver [44, 45]. Some studies have shown that SELS genotype is associated with circulating levels of proinflammatory cytokines, such as tumor necrosis factor-α (TNF-α) and interleukin-1β (IL-1β), suggesting that SELS plays a role in inflammation [46].

5. Conclusion

Microsomal glutathione S-transferase 1 (MGST1), ATP synthase H^+ transporting mitochondrial F_0 complex subunit C1, selenoprotein S, mitochondrial IF1 protein, and mitochondrial succinyl-CoA synthetase alpha subunit were participated in mitochondria metabolism, functional and structural damage process caused by chronic fluorosis. This information will be very helpful for understanding the molecular mechanisms of fluorosis.

Acknowledgments

This work was supported by Jiangsu Natural Science Foundation of China (no. BK2007561) and by Jiangsu University Fundation of China (no. 07JGD16).

References

[1] O. Barbier, L. Arreola-Mendoza, and L. M. Del Razo, "Molecular mechanisms of fluoride toxicity," *Chemico-Biological Interactions*, vol. 188, no. 2, pp. 319–333, 2010.

[2] L. H. Weinstein and A. Davidson, *Fluorides in the Environment. Effects on Plants and Animals*, CABI Publishing, Oxford, UK, 2004.

[3] E. T. Urbansky, "Fate of fluorosilicate drinking water additives," *Chemical Reviews*, vol. 102, no. 8, pp. 2837–2854, 2002.

[4] H. Karube, G. Nishitai, K. Inageda, H. Kurosu, and M. Matsuoka, "NaF activates MAPKs and induces apoptosis in odontoblast-like cells," *Journal of Dental Research*, vol. 88, no. 5, pp. 461–465, 2009.

[5] Y. Zhang, W. Li, H. S. Chi, J. Chen, and P. K. DenBesten, "JNK/c-Jun signaling pathway mediates the fluoride-induced downregulation of MMP-20 in vitro," *Matrix Biology*, vol. 26, no. 8, pp. 633–641, 2007.

[6] M. Zhang, A. Wang, T. Xia, and P. He, "Effects of fluoride on DNA damage, S-phase cell-cycle arrest and the expression of NF-κB in primary cultured rat hippocampal neurons," *Toxicology Letters*, vol. 179, no. 1, pp. 1–5, 2008.

[7] C. D. Anuradha, S. Kanno, and S. Hirano, "Oxidative damage to mitochondria is a preliminary step to caspase-3 activation in fluoride-induced apoptosis in HL-60 cells," *Free Radical Biology and Medicine*, vol. 31, no. 3, pp. 367–373, 2001.

[8] M. L. Cittanova, B. Lelongt, M. C. Verpont et al., "Fluoride ion toxicity in human kidney collecting duct cells," *Anesthesiology*, vol. 84, no. 2, pp. 428–435, 1996.

[9] E. Dabrowska, M. Balunowska, R. Letko, and B. Szynaka, "Ultrastructural study of the mitochondria in the submandibular gland, the pancreas and the liver of young rats, exposed to NaF in drinking water," *Roczniki Akademii Medycznej w Bialymstoku*, vol. 49, pp. 180–181, 2004.

[10] N. Imaizumi, S. Miyagi, and Y. Aniya, "Reactive nitrogen species derived activation of rat liver microsomal glutathione S-transferase," *Life Sciences*, vol. 78, no. 26, pp. 2998–3006, 2006.

[11] S. Zhao, S. L. Ooi, and A. B. Pardee, "New primer strategy improves precision of differential display," *BioTechniques*, vol. 18, no. 5, pp. 842–850, 1995.

[12] Q. S. Hossain, E. Ulziikhishig, K. K. Lee, H. Yamamoto, and Y. Aniya, "Contribution of liver mitochondrial membrane-bound glutathione transferase to mitochondrial permeability transition pores," *Toxicology and Applied Pharmacology*, vol. 235, no. 1, pp. 77–85, 2009.

[13] C. Andersson, E. Mosialou, R. Weinander, and R. Morgenstern, "Enzymology of microsomal glutathione S-transferase," *Advances in Pharmacology*, vol. 27, no. C, pp. 19–35, 1994.

[14] Y. Aniya and M. W. Anders, "Regulation of rat liver microsomal glutathione S-tranferase activity by thiol/disulfide exchange," *Archives of Biochemistry and Biophysics*, vol. 270, no. 1, pp. 330–334, 1989.

[15] M. Shimoji, Y. Aniya, and R. Morgenstern, "Activation of microsomal glutathione transferase 1," in *Toxicology of Glutathione Transferases*, Y. C. Awasthi, Ed., pp. 293–319, Taylor & Francis, New York, NY, USA, 2007.

[16] K. K. Lee, M. Shimoji, Q. S. Hossain, H. Sunakawa, and Y. Aniya, "Novel function of glutathione transferase in rat liver mitochondrial membrane: role for cytochrome c release from mitochondria," *Toxicology and Applied Pharmacology*, vol. 232, no. 1, pp. 109–118, 2008.

[17] R. Rinaldi, Y. Aniya, R. Svensson et al., "NADPH dependent activation of microsomal glutathione transferase," *Chemico-Biological Interactions*, vol. 147, no. 2, pp. 163–172, 2004.

[18] R. Mangiullo, A. Gnoni, F. Damiano et al., "3,5-diiodo-L-thyronine upregulates rat-liver mitochondrial F_0F_1-ATP synthase by GA-binding protein/nuclear respiratory factor-2," *Biochimica et Biophysica Acta*, vol. 1797, no. 2, pp. 233–240, 2010.

[19] S. Papa, F. Guerrieri, F. Zanotti, M. Fiermonte, G. Capozza, and E. Jirillo, "The γ subunit of F_1 and the PVP protein of F_0 (F_0I) are components of the gate of mitochondrial F_0F_1H+-ATP synthase," *FEBS Letters*, vol. 272, no. 1-2, pp. 117–120, 1990.

[20] J. A. Berden and A. F. Hartog, "Analysis of the nucleotide binding sites of mitochondrial ATP synthase provides evidence for a two-site catalytic mechanism," *Biochimica et Biophysica Acta*, vol. 1458, no. 2-3, pp. 234–251, 2000.

[21] T. Xu, F. Zanotti, A. Gaballo, G. Raho, and S. Papa, "F_1 and F_0 connections in the bovine mitochondrial ATP synthase. The role of the of α subunit N-terminus, oligomycin-sensitivity conferring protein (OCSP) and subunit d," *European Journal of Biochemistry*, vol. 267, no. 14, pp. 4445–4455, 2000.

[22] R. P. Hafner, G. C. Brown, and M. D. Brand, "Thyroid-hormone control of state-3 respiration in isolated rat liver mitochondria," *Biochemical Journal*, vol. 265, no. 3, pp. 731–734, 1990.

[23] F. Guerrieri, M. Kalous, E. Adorisio et al., "Hypothyroidism leads to a decreased expression of mitochondrial F_0F_1- ATP synthase in rat liver," *Journal of Bioenergetics and Biomembranes*, vol. 30, no. 3, pp. 269–276, 1998.

[24] T. Andrianaivomananjaona, M. Moune-Dimala, S. Herga, V. David, and F. Haraux, "How the N-terminal extremity of Saccharomyces cerevisiae IF1 interacts with ATP synthase: a kinetic approach," *Biochimica et Biophysica Acta*, vol. 1807, no. 2, pp. 197–204, 2011.

[25] M. E. Pullman and G. C. Monroy, "A naturally occurring inhibitor of mitochondrial adenosine," *The Journal of Biological Chemistry*, vol. 238, pp. 3762–3769, 1963.

[26] M. Satre, M. B. de Jerphanion, J. Huet, and P. V. Vignais, "ATPase inhibitor from yeast mitochondria. Purification and properties," *Biochimica et Biophysica Acta*, vol. 387, no. 2, pp. 241–255, 1975.

[27] H. Matsubara, T. Hase, T. Hashimoto, and K. Tagawa, "Amino acid sequence of an intrinsic inhibitor of mitochondrial ATPase from yeast," *Journal of Biochemistry*, vol. 90, no. 4, pp. 1159–1165, 1981.

[28] B. Norling, C. Tourikas, B. Hamasur, and E. Glaser, "Evidence for an endogenous ATPase inhibitor protein in plant mitochondria. Purification and characterization," *European Journal of Biochemistry*, vol. 188, no. 2, pp. 247–252, 1990.

[29] K. E. Polgreen, J. Featherstone, A. C. Willis, and D. A. Harris, "Primary structure and properties of the inhibitory protein of the mitochondrial ATPase (H+-ATP synthase) from potato," *Biochimica et Biophysica Acta*, vol. 1229, no. 2, pp. 175–180, 1995.

[30] E. Cabezón, M. J. Runswick, A. G. W. Leslie, and J. E. Walker, "The structure of bovine IF1, the regulatory subunit of mitochondrial F-ATPase," *EMBO Journal*, vol. 20, no. 24, pp. 6990–6996, 2002.

[31] A. C. Dianoux, A. Tsugita, G. Klein, and P. V. Vignais, "Effects of proteolytic fragmentations on the activity of the mitochondrial natural ATPase inhibitor," *FEBS Letters*, vol. 140, no. 2, pp. 223–228, 1982.

[32] J. S. Stout, B. E. Partridge, D. A. Dibbern, and S. M. Schuster, "Peptide analogs of the beef heart mitochondrial F1-ATPase inhibitor protein," *Biochemistry*, vol. 32, no. 29, pp. 7496–7502, 1993.

[33] M. J. van Raaij, G. L. Orriss, M. G. Montgomery et al., "The ATPase inhibitor protein from bovine heart mitochondria: the minimal inhibitory sequence," *Biochemistry*, vol. 35, no. 49, pp. 15618–15625, 1996.

[34] M. S. Lebowitz and P. L. Pedersen, "Protein inhibitor of mitochondrial ATP synthase: relationship of inhibitor structure to pH-dependent regulation," *Archives of Biochemistry and Biophysics*, vol. 330, no. 2, pp. 342–354, 1996.

[35] R. Venard, D. Brèthes, M. F. Giraud, J. Vaillier, J. Velours, and F. Haraux, "Investigation of the role and mechanism of IF1 and STF1 proteins, twin inhibitory peptides which interact with the yeast mitochondrial ATP synthase," *Biochemistry*, vol. 42, no. 24, pp. 7626–7636, 2003.

[36] M. Campanella, A. Seraphim, R. Abeti, E. Casswell, P. Echave, and M. R. Duchen, "IF1, the endogenous regulator of the F_1F_o-ATPsynthase, defines mitochondrial volume fraction in HeLa cells by regulating autophagy," *Biochimica et Biophysica Acta*, vol. 1787, no. 5, pp. 393–401, 2009.

[37] D. Phillips, A. M. Aponte, S. A. French, D. J. Chess, and R. S. Balaban, "Succinyl-CoA synthetase is a phosphate target for the activation of mitochondrial metabolism," *Biochemistry*, vol. 48, no. 30, pp. 7140–7149, 2009.

[38] S. Kaufman, C. Gilvarg, O. Cori, and S. Ochoa, "Enzymatic oxidation of alpha-ketoglutarate and coupled phosphorylation," *Journal of Biological Chemistry*, vol. 203, pp. 869–888, 1953.

[39] M. E. Fraser, M. N. G. James, W. A. Bridger, and W. T. Wolodko, "Phosphorylated and dephosphorylated structures of pig heart, GTP-specific succinyl-CoA synthetase," *Journal of Molecular Biology*, vol. 299, no. 5, pp. 1325–1339, 2000.

[40] S. J. Holt and D. L. Riddle, "SAGE surveys C. elegans carbohydrate metabolism: evidence for an anaerobic shift in the long-lived dauer larva," *Mechanisms of Ageing and Development*, vol. 124, no. 7, pp. 779–800, 2003.

[41] J. M. Weinberg, M. A. Venkatachalam, N. F. Roeser, and I. Nissim, "Mitochondrial dysfunction during hypoxia/reoxygenation and its correction by anaerobic metabolism of citric acid cycle intermediates," *Proceedings of the National Academy of Sciences of the United States of America*, vol. 97, no. 6, pp. 2826–2831, 2000.

[42] E. Basso, V. Petronilli, M. A. Forte, and P. Bernardi, "Phosphate is essential for inhibition of the mitochondrial permeability transition pore by cyclosporin A and by cyclophilin D

ablation," *Journal of Biological Chemistry*, vol. 283, no. 39, pp. 26307–26311, 2008.

[43] J. S. R. Zavala, J. P. Pardo, and R. M. Sanchez, "Modulation of 2-oxoglutarate dehydrogenase complex by inorganic phosphate, Mg2+, and other effectors," *Archives of Biochemistry and Biophysics*, vol. 379, no. 1, pp. 78–84, 2000.

[44] M. Olsson, B. Olsson, P. Jacobson et al., "Expression of the selenoprotein S (SELS) gene in subcutaneous adipose tissue and SELS genotype are associated with metabolic risk factors," *Metabolism: Clinical and Experimental*, vol. 60, no. 1, pp. 114–120, 2011.

[45] K. H. Kim, Y. Gao, K. Walder, G. R. Collier, J. Skelton, and A. H. Kissebah, "SEPS1 protects RAW264.7 cells from pharmacological ER stress agent-induced apoptosis," *Biochemical and Biophysical Research Communications*, vol. 354, no. 1, pp. 127–132, 2007.

[46] J. E. Curran, J. B. M. Jowett, K. S. Elliott et al., "Genetic variation in selenoprotein S influences inflammatory response," *Nature Genetics*, vol. 37, no. 11, pp. 1234–1241, 2005.

11

Genotype-dependent Burst of Transposable Element Expression in Crowns of Hexaploid Wheat (*Triticum aestivum L.*) during Cold Acclimation

Debbie Laudencia-Chingcuanco[1] and D. Brian Fowler[2]

[1] *Genomics and Gene Discovery Unit, USDA-ARS WRRC, 800 Buchanan Street, Albany, CA 94710, USA*
[2] *Department of Plant Sciences, University of Saskatchewan, 51 Campus Drive, Saskatoon, SK, Canada S7N 5A8*

Correspondence should be addressed to Debbie Laudencia-Chingcuanco, debbie.laudencia@ars.usda.gov

Academic Editor: Soraya E. Gutierrez

The expression of 1,613 transposable elements (TEs) represented in the Affymetrix Wheat Genome Chip was examined during cold treatment in crowns of four hexaploid wheat genotypes that vary in tolerance to cold and in flowering time. The TE expression profiles showed a constant level of expression throughout the experiment in three of the genotypes. In winter Norstar, the most cold-hardy of the four genotypes, a subset of the TEs showed a burst of expression after vernalization saturation was achieved. About 47% of the TEs were expressed, and both Class I (retrotransposons) and Class II (DNA transposons) types were well represented. *Gypsy* and *Copia* were the most represented among the retrotransposons while *CACTA* and *Mariner* were the most represented DNA transposons. The data suggests that the *Vrn-A1* region plays a role in the stage-specific induction of TE expression in this genotype.

1. Introduction

Transposable elements (TEs) are DNA sequences that can move or transpose to new locations within the genome. Transposable elements are generally classified based on their mechanism of transposition. Class I TE, or retrotransposons, move by a "copy and paste" mechanism whereby the TE is transcribed into an RNA intermediate that is converted to DNA by RNA-dependent DNA polymerases before it reinserts itself in the genome. Class II TE, or DNA transposons, utilize the "cut and paste" mechanism wherein the TE element DNA itself is excised from the genome and reinserted in a new position. TE transposition has been shown to generate mutations that can alter gene expression, and to create gene deletions and duplications, chromosome breaks, and genome rearrangements [1]. Thus, TEs can be a powerful force for species adaptation to adverse biotic and abiotic challenges and a facilitator of speciation by creating potentially advantageous genetic variations upon which natural selection can act [2, 3].

Transposable elements are ubiquitous and can be found in both prokaryotic and eukaryotic genomes. In plants (especially cereals), TEs make up a large portion of the genome. In wheat, repetitive and transposable elements comprise more than 80% of the genomic sequence [4]. To prevent the potentially harmful effects of TE transposition, the host plant has evolved several mechanisms to repress TE expression [5, 6]. TE can be silenced before or after transcription. Transcriptional silencing of TE includes DNA methylation and chromatin remodeling, which renders the elements unavailable for transcription. On the other hand, posttranscription silencing involves small noncoding RNA (sRNA) which directs sequence-specific degradation of transcripts.

Several events, which cause what Barbara McClintock called "genome shock," have been shown to release TEs from repression [1, 7]. In plants, these activating events include genome hybridization [8], nutrient deprivation [9], infection [10, 11], and abiotic stresses like drought [12] and high temperature [10, 13]. One report showed a cold-induced activation of a family of retrotransposons in alfalfa [14].

Table 1: Vernalization genotype and minimum cold tolerance of near isogenic lines used in these studies.

Line	Genotype	LT50
Winter Norstar	vrn-A1, vrn-B1, vrn-D1	−23a
Spring Manitou	*Vrn-A1*, vrn-B1, vrn-D1	−8.3c
Winter Manitou	vrn-A1, vrn-B1, vrn-D1	−13.3b
Spring Norstar	*Vrn-A1*, vrn-B1, vrn-D1	−13b

LT50 indicates the lowest temperature by which 50% of the treated samples survived after acclimation; a to c: within columns, mean numbers followed by the same letter are not different as determined by Duncan's new multiple range test ($P < 0.05$).

In this paper we explored the response of TEs in wheat in response to cold.

We previously carried out a genome-wide expression analysis of the response to cold treatment of the four hexaploid wheat genotypes considered in this study [15]. In these genotypes, *Vrn-A1*, a major regulator of the transition of the shoot apex from vegetative to reproductive meristem in response to cold treatment in wheat, was swapped between a highly cold tolerant line, winter Norstar, and a cold-sensitive cultivar spring Manitou (see Table 1). Allelic variations in the *Vrn-A1* locus are the main determinants of the winter and spring habit in wheat [16, 17]. The swapped region is less than 37 cM and does not include *Frost resistance-2* (*Fr-2*), a major locus that controls tolerance to low temperature located about 40 cM proximal to *Vrn-A1* in chromosome 5AL [18]. We reported the identification of 2771 differentially expressed genes that could be involved in the development of low-temperature tolerance [15]. We have now mined this same dataset to investigate the expression of transposable elements during cold treatment in hexaploid wheat. We show that members of both Class I and Class II types of transposable elements in wheat are expressed during cold treatment. Furthermore, we showed that a genotype-dependent burst of TE expression in the crown occurs after vernalization saturation has been achieved and implicates the winter *vrn-A1* region to play a role in this event.

2. Materials and Methods

2.1. Plant Materials and Growth Conditions. Wheat (*Triticum aestivum* L.) cultivars and near isogenic lines (NILs) used in this study were developed and characterized as described in a previous microarray experiment [15, 18]. The lines differ in tolerance to low temperature and in flowering time. The major vernalization locus, *Vrn-A1*, which is a major determinant of flowering time in wheat, was swapped in the winter cultivar "Norstar" (*vrn-A1*) and the spring cultivar "Manitou" (*Vrn-A1*). The initial hybrid was backcrossed to each parent ten times prior to selfing to produce homozygous NILs (spring Norstar and winter Manitou) that are theoretically ~99.9% genetically similar to the parental lines. For each backcross, a phenotype-based selection of progeny with *Vrn-A1* or *vrn-A1* locus was used to ensure that the donor parent allele was incorporated into the genetic background of the recurrent parent.

For these studies, seeds were imbibed in the dark for 2 days at 4°C and then transferred to an incubator and allowed to germinate for 3 days at 22°C. Actively germinating seeds were transferred, embryo down, to plexiglass trays with holes backed by a 1.6 mm mesh screen and grown for 10 days in hydroponic tanks filled with continuously aerated one-half strength modified Hoagland's solution [19] at 20°C in 16-hour days at 320 μmol m^{-2} s^{-1} PPFD, by which time they had 3-4 fully developed leaves. The seedlings were then transferred to 6°C chambers (measured at crown level) under 16-hour photoperiod and 220 μmol m^{-2} s^{-1} PPFD and sampled at regular intervals. Details on phenological development and cold acclimation of these lines when grown under these conditions have been given in an earlier publication [15].

2.2. Global Gene Expression Profiling. The microarray study was performed as previously reported [15]. The experimental design included 4 genotypes (winter and spring Norstar NIL and spring and winter Manitou NIL) acclimated at 0, 2, 14, 21, 35, 42, 56, and 70 days. Crown tissue was harvested after each acclimation period at the same time each day (4 hours after dawn) to neutralize circadian rhythm effect. RNA was isolated from three biological replicate samples (pool of 25 crowns/sample) for each acclimation period. RNA labelling and Affymetrix Wheat Genome GeneChip array hybridization were performed according to the manufacturer's instructions (Affymetrix Inc, Santa Clara, CA, USA). Microarray data were extracted from scanned GeneChip images and analyzed using the Affymetrix Microarray Suite (MAS) version 5.0. Probe set signal normalization and summarization was carried out using the Robust Multi-array normalization algorithm (RMA), as implemented in GeneSpring software (Agilent, Santa Clara, CA, USA). RMA-normalized values were filtered for those present (as determined by the Affymetrix MAS probe summarization protocol) in 2 out of 3 of the biological replicates for each time point. In this paper, a probe set will be deemed to represent a potentially unique wheat gene or gene encoding a transposable element. The accumulation of its transcripts as measured by the signal intensities in each probe set represents the "expression" of the gene. Clustering of gene expression profiles using the *k*-means algorithm was implemented in the software Genesis [20]. The most recent annotations for the probesets were verified using wheat data in Plant Expression Database (http://www.plexdb.org/). Microarray data have been deposited to the NCBI Gene Expression Omnibus (GEO) database with accession number GSE23889.

2.3. Probesets for Wheat Transposable Elements. The sequences of transposable elements identified from wheat were downloaded from the Triticeae Repeat Element (TREP; release #10, July 10, 2008) database in GrainGenes (http://wheat.pw.usda.gov/ITMI/Repeats/) and used to query the target sequences used to generate the Affymetrix Wheat Genome Chip using BLASTN [21].

2.4. TE Expression during Drought Stress. The RMA-normalized data (TA23_RMA_tmt_mean.txt) for the genome-wide

Table 2: Transposable elements represented on the array.

DNA transposons	On array	Cluster 1	Cluster 2	Cluster 3	Cluster 4
Helitron	3	0	0	0	0
HAT	2	0	0	1	0
CACTA	236	33	9	59	14
Harbinger	12	2	0	6	1
Mutator-like	11	0	0	5	1
Mariner	111	1	0	55	24
Unknown	4	0	0	3	0
Subtotal	379 (23%)				
Retrotransposons					
Copia	243	31	9	56	4
Gypsy	896	102	46	216	29
LINE	27	1	0	0	0
Unknown	43	2	0	31	2
Subtotal	1209 (75%)				
Unclassified	25	2	0	12	3
Total	1613	174	64	444	78

expression of wheat genes during drought stress using the Affymetrix Wheat Genome GeneChip [12] was downloaded from the Plant Expression Database (http://www.plexdb.org/).

3. Results

3.1. Identification of Transposable Elements Represented on the Array. The nonredundant sequences for the identified transposable elements (TEs) in wheat were downloaded from GrainGenes nrTREP database and used to query the Affymetrix wheat target sequences for homology with a cut-off of *e*-10. The two classes of transposable elements were well represented in the 476 unique TE sequences in the database that included 218 retrotransposons 182 DNA transposons, and 76 unclassified TEs.

The query identified 1,613 probesets on the wheat array representing 251 unique TEs with 75% retrotransposons and 23% DNA transposons (Table 2); we will refer to each probeset from now on to encode a TE. Among the retrotransposons, *Gypsy* and *Copia* were the most represented while the *CACTA* and *Mariner* were the most represented DNA transposons. Both *Gypsy* and *Copia* belong to the long-terminal repeat (LTR) containing family of retrotransposons wherein the autonomous element LTRs flank several genes needed for transposition. *CACTA* and Mariner are TEs flanked by short-terminal inverted repeats (TIR).

3.2. Expression of Wheat Transposable Elements in Response to Cold Treatment. Of the 1,613 TEs represented, 760 were expressed (Table 2), about 47% of all the TE probesets on the array. The majority of the expressed TEs were retrotransposons (529), 74% of which were of the *Gypsy* type. Of the 214 expressed DNA transposons 54% were *CACTA* elements.

Clustering of the profiles of the expressed TEs into 4 groups using the *k*-means algorithm showed that the majority of the TEs were expressed at a constant level throughout the duration of the experiment in three of the genotypes (Figure 1). In winter Norstar about 31% of the expressed TEs showed a burst of activity at 56 days of cold treatment, just after vernalization saturation was reached. This burst of TE expression, however, was rapidly repressed to basal level. TEs in cluster 1 showed a burst in relative expression between 2x and 4x above the basal level while Cluster 2 showed a burst of expression between 4x and 70x above the basal level. Expression of TEs in clusters 3 and 4 was similar to the other genotypes.

The TEs in clusters 1 and 2 (listed in Supplemental data 1 available online at doi:10.1155/2012/232530) included representatives from both Class I (45 members) and II (191 members). Of the Class I TEs, the superfamily *CACTA* had 14 families represented with family *Jorge* having the highest number of members (17). Among the Class II TEs, the superfamily *Gypsy* had 22 families represented. Four of the *Gypsy* families had 12 or more members (*Sabrina*-27, *Sumana*-21, *Wilma*-17, and *Fatima*-12). A member of the family *Sakura* had the highest change in expression (70x).

3.3. TE Expression and the Vrn-A1 Locus. The spring Norstar genome is ~99.9% identical to winter Norstar, yet its TEs in cluster 1 and 2 did not exhibit the same burst in activity. The main difference between these two genotypes is the swapped *Vrn-A1* locus. Winter Norstar has the winter allele *vrn-A1* gene that requires cold treatment to be expressed, whereas spring Norstar has the spring allele *Vrn-A1* that is constitutively expressed and does not require vernalization to flower. This suggests that the *Vrn-A1* locus could be playing a role in the burst in TE activity in cluster 1 and 2 in winter Norstar.

Winter Manitou and winter Norstar share the swapped region that includes the winter allele *vrn-A1*. Similar to winter Norstar, winter Manitou requires cold treatment to flower and reaches vernalization saturation after 42 days.

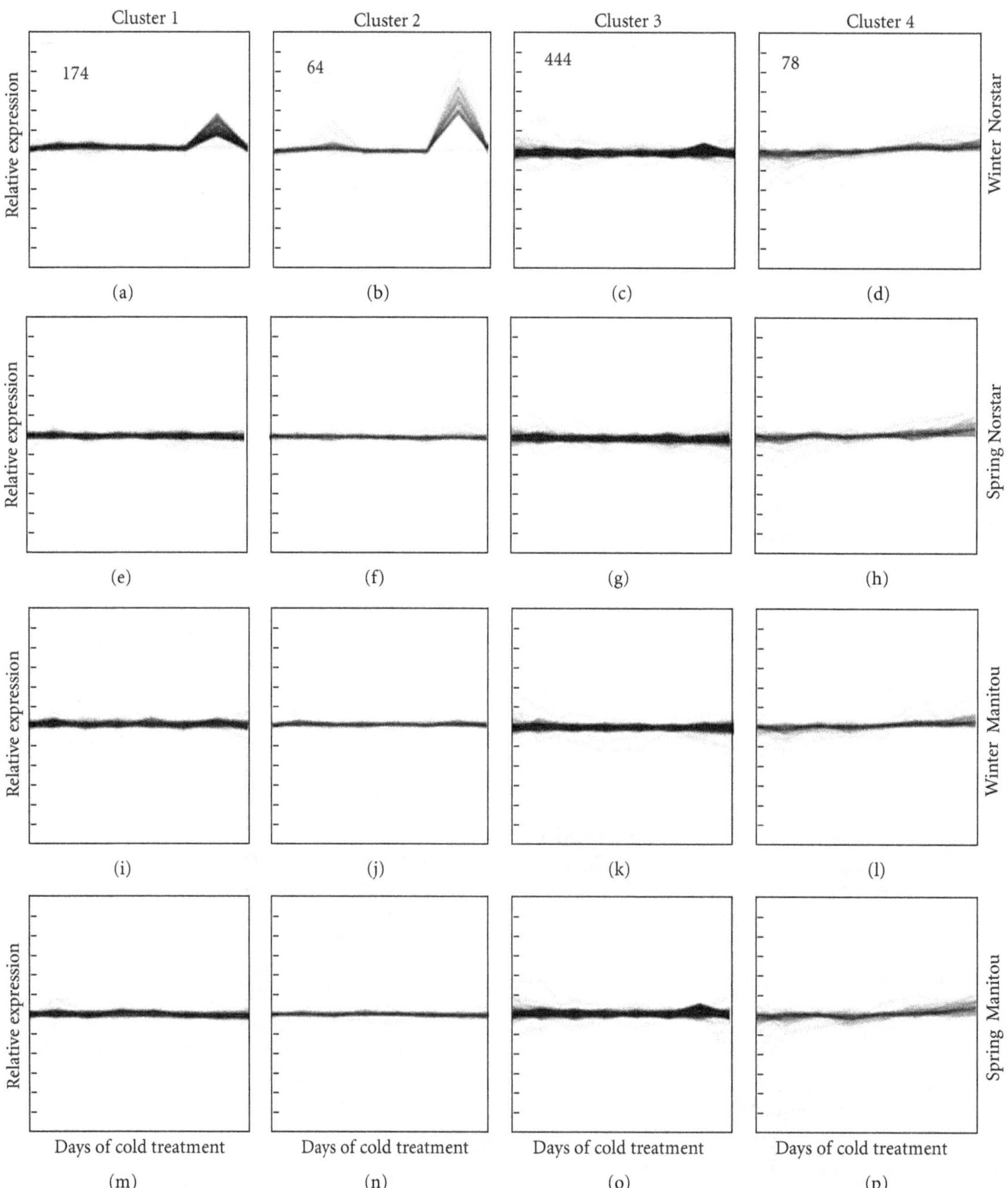

FIGURE 1: Pattern of TE gene expression during cold acclimation. The expression profiles of 760 TE genes across all 8 time points were grouped into 4 clusters using the k-means algorithm. Different numbers of groups were tested, but we found the grouping into 4 gave the best representation of the different profiles. The mean-centered relative gene expression value (in $\log_2$ scale) for each gene was plotted on the y-axis, and the time of cold treatment was plotted on the x-axis. Tick marks on the x-axis represent 0, 2, 14, 21, 28, 35, 42, 56, and 70 days of cold treatment. Tick marks for 0 and 70 days overlap with the sides of the cluster box. The value at the upper left corner on each of the 4 top panels indicates the number of TEs in the cluster.

However, the winter Manitou TEs in cluster 1 and 2 did not show the burst of activity observed in winter Norstar indicating that other factor(s) outside of the swapped region are required for the burst in TE expression.

3.4. Cluster 1 and 2 TE Expression Activated during Drought Stress. In hexaploid wheat, a genomewide expression analysis during drought stress was previously carried out by Aprile et al. (2009 [12]) using the same Affymetrix genome chips used in our study. Their data showed high expression of a cluster of 91 genes only in the Chinese Spring 5AL-10 deletion line but not in the two other lines examined: wild type bread wheat Chinese Spring and durum wheat Cresco [12]. Of the 91 genes in the cluster, 13 were annotated as TE while the others were mainly of unknown function.

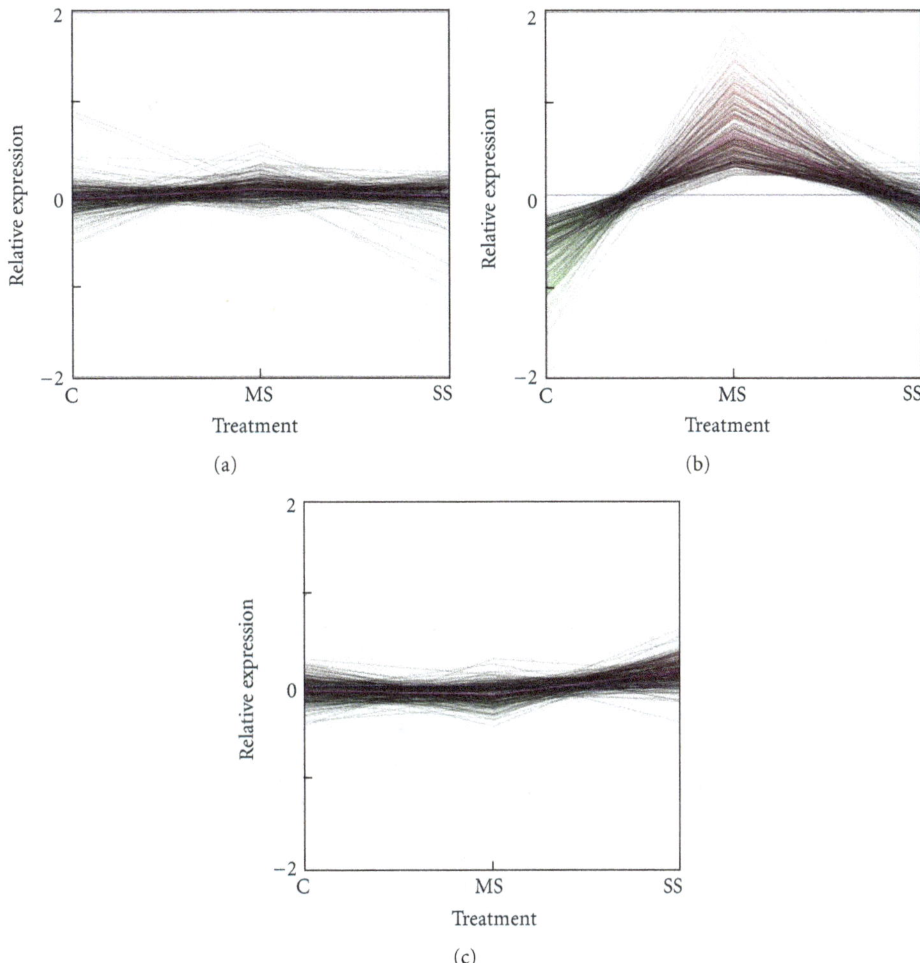

Figure 2: TE expression profiles during drought treatment. The expression profiles of 218 TE genes during drought stress in (a) bread wheat Chinese Spring, (b) Chinese Spring 5AL deletion line, and (c) durum wheat Cresco were compared. The mean-centered relative gene expression value (in $\log_2$ scale) for each gene was plotted on the y-axis and the level of drought treatment was indicated on the x-axis. Tick marks on the x-axis C, MS, and SS represent control, moderate stress, and severe stress, respectively.

Comparison of the probesets of the 13 transposable elements observed to be upregulated in the Chinese Spring 5AL-10 deletion line showed that these were a subset of the same transposable elements that were upregulated in winter Norstar in our study. We therefore examined the expression of the same 238 TEs that were upregulated in winter Norstar in the genotypes examined during drought stress using their dataset. The microarray data from the work of Aprile et al. was downloaded from the Plant Expression Database and reanalyzed to determine the expression profiles of the TEs in cluster 1 and 2 in our study. As shown in Figure 2, more than 90% of the TEs that were upregulated during cold treatment in our study were also upregulated in the Chinese Spring 5AL-10 deletion line during moderate drought treatment in a genotype-dependent manner. Chinese Spring 5AL-10 is derived from Chinese Spring wherein a segment of the long arm of chromosome 5 from 0.56 region to the end of the telomere was deleted. The deleted region contains several genes including those that affect low-temperature tolerance and flowering time, including the *Vrn-A1* locus. Thus, this data not only provides support to the genotype-dependent expression of TEs we observed in winter Norstar but further implicates the region containing the *Vrn-A1* locus in this process.

4. Discussion

4.1. Validation of the Burst of TE Expression in Winter Norstar. Several lines of evidence support or validate the stage-specific burst of expression in winter Norstar. First, the signal and background levels in the biological samples at 56 days after treatment were similar to the other slides in the time-series experiment (see Supplemental data 2). The mean signal for each of the biological replicates at 56 days after cold treatment was between 6.13 and 6.14 ($\log_2$), similar to the rest of the samples in the whole experiment indicating that the increase in TE expression at this time point was not due to technical experimental errors. Furthermore, majority of the transposable elements in winter Norstar showed constant expression that was similar to the other genotypes. Only

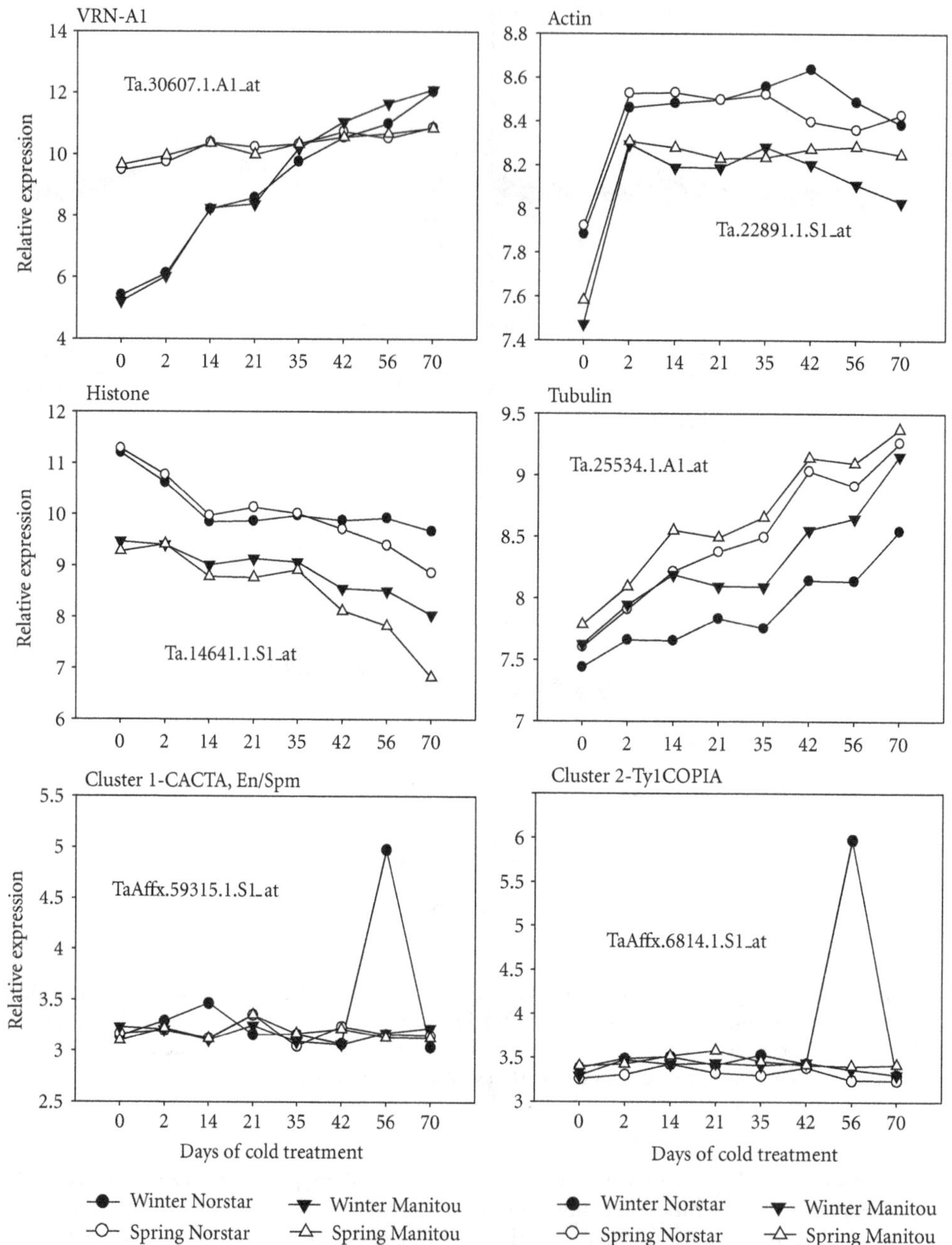

FIGURE 3: Expression profiles of selected TE and non-TE genes. The y-axis shows the values for the relative expression of the vernalization gene *VRN-A1*, selected housekeeping genes for actin, histone, and tubulin and representative TE from cluster 1 and 2 given as mean of $\log_2$-transformed signal intensities. The x-axis shows the days of cold treatment. The identification of each probeset is displayed on each chart.

a subset of the transposons appears to show a burst of expression at 56 days after cold treatment. If the high signal observed in winter Norstar was an artifact, it is expected that most, if not all, of the TEs would show a higher signal at 56 days of cold treatment. Second, as we previously reported, analysis of the expression of known cold responsive genes using the same dataset gave expression profiles consistent with other independently published reports [15]. For example, the expression of the *Vrn-A1* gene in the swapped *Vrn-A1* locus showed the predicted swapped expression profiles in the spring and winter genotypes (Figure 3). As expected the spring *Vrn-A1* allele was constitutively expressed in the spring habit genotypes, whereas the *vrn-A1* winter allele showed a cold-inducible expression profile in the winter genotypes. Third, at 56 days after cold treatment in winter Norstar, the expression profiles of selected housekeeping genes commonly used as controls in other transcript level measurement methods were consistent with the trends in other genotypes (Figure 3). Lastly, genotype- and treatment-stage-specific induction of TEs have been previously reported in wheat (see above) and in other plants [14, 22, 23].

4.2. Genotype, and Developmental-Stage-Specific TE Expression in Plants. Recent reports have shown instances of genotype- and stage-specific expression of TEs in other plants. In alfalfa, a genotype-specific induction of a family of TEs during cold treatment was exhibited only in the most cold-hardy line investigated [14]. TEs in Arabidopsis pollen were also shown to be reactivated and transposed only in the vegetative nucleus but not in the sperm nuclei [22]. In rice, spaceflight-induced stress due to cosmic irradiation, microgravity, and space magnetic field [24] resulted in new *mPing* TE transpositions in two of the three cultivars grown in space [25]. The rice transposable element *mPing* has also been shown to be preferentially amplified in the genome of the rice strain EG4 compared to that of Nipponbare [26]. Furthermore, a survey of TE expression in rice during development revealed that the heading stage panicle had the highest detected expression, whereas the somatic shoot tissue had the lowest [23]. Thus, the observed genotype-dependent burst of TE expression after vernalization saturation in the crowns of cold-treated winter Norstar wheat in our report is not an isolated case.

4.3. Epigenetic Regulation of Vrn1 and TEs. Flowering time is an epigenetically controlled process in higher plants [27–29]. In Arabidopsis, the model dicot plant system where the process has been investigated more intensively, epigenetic control includes both DNA methylation and chromatin remodeling. Recent advances indicate that vernalization, the process wherein plant exposure to non-freezing temperature accelerates the competence to flower, involves the Polycomb group (PcG) of genes (reviewed in [28]) in the silencing of FLOWERING LOCUS C (*FLC*). The expression of *FLC*, the critical gene in the vernalization response in Arabidopsis, is reduced as the level of methylation in lysine 27 of histone3 (H3K27me3) is progressively increased. In cereals, the *Vrn-A1* gene which controls flowering time is also epigenetically regulated [28, 30]. In barley, it has been shown that the repressed HvVRN1 has high level of H3K27me3 before vernalization. Vernalization-induced activation of the gene correlated with an increased level of methylation in lysine 4 of histone 3 (H3K4me3) and a decrease in H3K27me3 in the HvVRN1 chromatin. It has been postulated that the epigenetic memory of vernalization in cereals is based on the maintenance of an active chromatin state of the *Vrn1* locus [28].

Flowering time is also accelerated by environmental stress [29, 31]. Like TE activation, the epigenetic mechanisms of flowering time acceleration include DNA methylation, histone modifications, and microRNAs (reviewed in [30]). In this study, the burst in TE expression during cold treatment in the crowns of winter Norstar coincided with the stage after vernalization saturation has been achieved. Vernalization saturation is reached when the shoot apex has transitioned from vegetative to reproductive state (VTR); thus, additional cold treatment no longer accelerates flowering. Do the mechanisms that control flowering time during stress overlap with the activation of TEs? Is the observed burst of TE reactivation simply a response to the same epigenetic mechanisms that control flowering time?

Does the *Vrn-A1* locus play a role in the observed burst in TE expression? If so, then why was the same phenomenon not observed in winter Manitou which shares the same winter allele *vrn-A1* locus as winter Norstar? It is possible that both the region encoding the winter allele *vrn-A1* and genetic factors specific to Norstar background but outside the swapped *vrn-A1* locus are required for this burst of TE expression. Thus, the burst of TE expression was not observed even in spring Norstar which shares ~99.9% of winter Norstar genome but carries the spring *Vrn-A1* allele. The genotype-dependent reactivation of the same TEs during drought stress in Chinese Spring 5AL-10 line where the *Vrn-A1* was deleted supports the potential involvement of *Vrn-A1* locus in this phenomenon. Other factors that maybe involved in the epigenetic regulation of TE expression aside from the *Vrn-A1* locus may include those that encode for DNA-methylases, DNA-demethylases, small RNAs, and the proteins involved in small RNA processing, all of which have been implicated in chromatin remodeling [1]. Genetic variations in these loci between Norstar and winter Manitou background could be responsible for the differences in the expression of TEs during cold acclimation. Clearly, more tests are needed to determine whether these factors and the *Vrn1* locus are linked to the observed genotype-dependent TE burst of activation in wheat.

4.4. TE and Evolution. The stress-associated reactivation of TEs is postulated to play a role in host genome plasticity to survive adverse environments [32, 33]. Stress-induced TE mobility increases the generation of genetic variability that can be useful for adaption. In support of this, the rice *mPing* MITEs have been shown to be preferentially amplified in domesticated cultivars adapted to environmental extremes [34]. Furthermore, promoters of some transposable elements show similarity to those found in regulatory regions of host defense genes [5].

The molecular bases of the mechanisms involved in TE repression and reactivation in plants are beginning to unfold but are still not clearly understood. Transposable elements make up a substantial fraction of plant genomes and have been shown to serve as a major contributor of genetic variations. How the host suppresses and reactivates the expression of TEs in its genome is a key issue in genome biology. It has been proposed that the theory of punctuated equilibria [35, 36] in evolutionary biology results from the epigenetic control of TE expression [33]. Our data provides insights into how TEs are regulated *in planta*. Our result indicates that TE expression in wheat could be induced in a genotype and developmental stage specific manner during cold treatment and suggests a potential role of the *Vrn-A1* locus in this process.

Disclosure

USDA is an equal opportunity provider and employer. Specific product names mentioned in this paper do not constitute an endorsement and do not imply a recommendation over other suitable products.

Acknowledgments

This work was supported by the Genome Canada/Genome Prairie (DBF), Ducks Unlimited (DBF) and USDA-ARS CRIS Project 5325-21000-015D (DLC). The authors thank Drs. Roger Thilmony and Grace Chen for their critical reading of the manuscript.

References

[1] J. A. Shapiro, "Mobile DNA and evolution in the 21st century," *Mobile DNA*, vol. 1, no. 1, article no. 4, 2010.

[2] C. Feschotte, "Transposable elements and the evolution of regulatory networks," *Nature Reviews Genetics*, vol. 9, no. 5, pp. 397–405, 2008.

[3] R. Rebollo, B. Horard, B. Hubert, and C. Vieira, "Jumping genes and epigenetics: towards new species.," *Gene*, vol. 454, no. 1-2, pp. 1–7, 2010.

[4] R. B. Flavell, M. D. Bennett, J. B. Smith, and D. B. Smith, "Genome size and the proportion of repeated nucleotide sequence DNA in plants," *Biochemical Genetics*, vol. 12, no. 4, pp. 257–269, 1974.

[5] R. K. Slotkin and R. Martienssen, "Transposable elements and the epigenetic regulation of the genome," *Nature Reviews Genetics*, vol. 8, no. 4, pp. 272–285, 2007.

[6] H. Okamoto and H. Hirochika, "Silencing of transposable elements in plants," *Trends in Plant Science*, vol. 6, no. 11, pp. 527–534, 2001.

[7] P. Capy, G. Gasperi, C. Biémont, and C. Bazin, "Stress and transposable elements: co-evolution or useful parasites?" *Heredity*, vol. 85, no. 2, pp. 101–106, 2000.

[8] K. Kashkush and B. Yaakov, "Methylation, transcription, and rearrangements of transposable elements in synthetic allopolyploids," *International Journal of Plant Genomics*, vol. 2011, Article ID 569826, 7 pages, 2011.

[9] L. Li, X. Wang, M. Xia et al., "Tiling microarray analysis of rice chromosome 10 to identify the transcriptome and relate its expression to chromosomal architecture.," *Genome biology*, vol. 6, no. 6, article R52, 2005.

[10] M. A. Grandbastien, C. Audeon, E. Bonnivard et al., "Stress activation and genomic impact of Tnt1 retrotransposons in Solanaceae," *Cytogenetic and Genome Research*, vol. 110, no. 1–4, pp. 229–241, 2005.

[11] D. Melayah, E. Bonnivard, B. Chalhoub, C. Audeon, and M. A. Grandbastien, "The mobility of the tobacco Tnt1 retrotransposon correlates with its transcriptional activation by fungal factors," *Plant Journal*, vol. 28, no. 2, pp. 159–168, 2001.

[12] A. Aprile, A. M. Mastrangelo, A. M. De Leonardis et al., "Transcriptional profiling in response to terminal drought stress reveals differential responses along the wheat genome," *BMC Genomics*, vol. 10, article no. 279, 2009.

[13] L. W. Young, R. H. Cross, S. A. Byun-McKay, R. W. Wilen, and P. C. Bonham-Smith, "A high- and low-temperature inducible Arabidopsis thaliana HSP101 promoter located in a nonautonomous Mutator-like element," *Genome*, vol. 48, no. 3, pp. 547–555, 2005.

[14] S. Ivashuta, M. Naumkina, M. Gau et al., "Genotype-dependent transcriptional activation of novel repetitive elements during cold acclimation of alfalfa (Medicago sativa)," *The Plant Journal*, vol. 31, no. 5, pp. 615–627, 2002.

[15] D. Laudencia-Chingcuanco, S. Ganeshan, F. You, B. Fowler, R. Chibbar, and O. Anderson, "Genome-wide gene expression analysis supports a developmental model of low temperature tolerance gene regulation in wheat (*Triticum aestivum* L.)," *BMC Genomics*, vol. 12, article 299, 2011.

[16] L. Yan, M. Helguera, K. Kato, S. Fukuyama, J. Sherman, and J. Dubcovsky, "Allelic variation at the VRN-1 promoter region in polyploid wheat," *Theoretical and Applied Genetics*, vol. 109, no. 8, pp. 1677–1686, 2004.

[17] D. Fu, P. Szucs, L. Yan et al., "Large deletions within the first intron in VRN-1 are associated with spring growth habit in barley and wheat," *Molecular Genetics and Genomics*, vol. 273, no. 1, pp. 54–65, 2005.

[18] M. Båga, S. V. Chodaparambil, A. E. Limin, M. Pecar, D. B. Fowler, and R. N. Chibbar, "Identification of quantitative trait loci and associated candidate genes for low-temperature tolerance in cold-hardy winter wheat," *Functional and Integrative Genomics*, vol. 7, no. 1, pp. 53–68, 2007.

[19] D. B. Fowler, "Cold acclimation threshold induction temperatures in cereals," *Crop Science*, vol. 48, no. 3, pp. 1147–1154, 2008.

[20] A. Sturn, J. Quackenbush, and Z. Trajanoski, "Genesis: cluster analysis of microarray data," *Bioinformatics*, vol. 18, no. 1, pp. 207–208, 2002.

[21] S. F. Altschul, T. L. Madden, A. A. Schäffer et al., "Gapped BLAST and PSI-BLAST: a new generation of protein database search programs," *Nucleic Acids Research*, vol. 25, no. 17, pp. 3389–3402, 1997.

[22] R. K. Slotkin, M. Vaughn, F. Borges et al., "Epigenetic reprogramming and small RNA silencing of transposable elements in pollen," *Cell*, vol. 136, no. 3, pp. 461–472, 2009.

[23] Y. Jiao and X. Deng, "A genome-wide transcriptional activity survey of rice transposable element-related genes," *Genome Biology*, vol. 8, no. 2, article no. R28, 2007.

[24] A. Mashinsky, I. Ivanova, T. Derendyaeva, G. Nechitailo, and F. Salisbury, ""From seed-to-seed" experiment with wheat plants under space-flight conditions," *Advances in Space Research*, vol. 14, no. 11, pp. 13–19, 1994.

[25] L. Long, X. Ou, J. Liu, X. Lin, L. Sheng, and B. Liu, "The spaceflight environment can induce transpositional activation of multiple endogenous transposable elements in a genotype-dependent manner in rice," *Journal of Plant Physiology*, vol. 166, no. 18, pp. 2035–2045, 2009.

[26] K. Naito, E. Cho, G. Yang et al., "Dramatic amplification of a rice transposable element during recent domestication," *Proceedings of the National Academy of Sciences of the United States of America*, vol. 103, no. 47, pp. 17620–17625, 2006.

[27] R. Muller and J. Goodrich, "Sweet memories: epigenetic control in flowering," *F1000 Biol Rep*, vol. 3, p. 13, 2011.

[28] S. N. Oliver, E. J. Finnegan, E. S. Dennis, W. J. Peacock, and B. Trevaskis, "Vernalization-induced flowering in cereals is associated with changes in histone methylation at the VERNALIZATION 1 gene," *Proceedings of the National Academy of Sciences of the United States of America*, vol. 106, no. 20, pp. 8386–8391, 2009.

[29] M. W. Yaish, J. Colasanti, and S. J. Rothstein, "The role of epigenetic processes in controlling flowering time in plants exposed to stress," *Journal of Experimental Botany*, vol. 62, no. 11, pp. 3727–3735, 2011.

[30] B. Trevaskis, "Goldacre Paper: The central role of the VERNALIZATION1 gene in the vernalization response of cereals," *Functional Plant Biology*, vol. 37, no. 6, pp. 479–487, 2010.

[31] K. C. Wada and K. Takeno, "Stress-induced flowering," *Plant Signaling and Behavior*, vol. 5, no. 8, pp. 944–947, 2010.

[32] S. R. Wessler, "Plant retrotransposons: turned on by stress," *Current Biology*, vol. 6, no. 8, pp. 959–961, 1996.

[33] B. McClintock, "The significance of responses of the genome to challenge," *Science*, vol. 226, no. 4676, pp. 792–801, 1984.
[34] N. Jiang, Z. Bao, X. Zhang et al., "An active DNA transposon family in rice," *Nature*, vol. 421, no. 6919, pp. 163–167, 2003.
[35] N. Eldredge, S. J. Gould, J. A. Coyne, and B. Charlesworth, "On punctuated equilibria," *Science*, vol. 276, no. 5311, pp. 338–341, 1997.
[36] D. W. Zeh, J. A. Zeh, and Y. Ishida, "Transposable elements and an epigenetic basis for punctuated equilibria," *BioEssays*, vol. 31, no. 7, pp. 715–726, 2009.

12

On the Diversification of the Translation Apparatus across Eukaryotes

Greco Hernández,[1] Christopher G. Proud,[2] Thomas Preiss,[3] and Armen Parsyan[4,5]

[1] *Division of Basic Research, National Institute for Cancer (INCan), Avenida San Fernando No. 22, Col. Sección XVI, Tlalpan, 14080 Mexico City, Mexico*
[2] *Centre for Biological Sciences, University of Southampton, Life Sciences Building (B85), Southampton SO17 1BJ, UK*
[3] *Genome Biology Department, The John Curtin School of Medical Research, The Australian National University, Building 131, Garran Road, Acton, Canberra, ACT 0200, Australia*
[4] *Goodman Cancer Centre and Department of Biochemistry, Faculty of Medicine, McGill University, 1160 Pine Avenue West, Montreal, QC, Canada H3A 1A3*
[5] *Division of General Surgery, Department of Surgery, Faculty of Medicine, McGill University Health Centre, Royal Victoria Hospital, 687 Pine Avenue West, Montreal, QC, Canada H3A 1A1*

Correspondence should be addressed to Greco Hernández, ghernandezr@incan.edu.mx

Academic Editor: Brian Wigdahl

Diversity is one of the most remarkable features of living organisms. Current assessments of eukaryote biodiversity reaches 1.5 million species, but the true figure could be several times that number. Diversity is ingrained in all stages and echelons of life, namely, the occupancy of ecological niches, behavioral patterns, body plans and organismal complexity, as well as metabolic needs and genetics. In this review, we will discuss that diversity also exists in a key biochemical process, translation, across eukaryotes. Translation is a fundamental process for all forms of life, and the basic components and mechanisms of translation in eukaryotes have been largely established upon the study of traditional, so-called model organisms. By using modern genome-wide, high-throughput technologies, recent studies of many nonmodel eukaryotes have unveiled a surprising diversity in the configuration of the translation apparatus across eukaryotes, showing that this apparatus is far from being evolutionarily static. For some of the components of this machinery, functional differences between different species have also been found. The recent research reviewed in this article highlights the molecular and functional diversification the translational machinery has undergone during eukaryotic evolution. A better understanding of all aspects of organismal diversity is key to a more profound knowledge of life.

1. Protein Synthesis Is a Fundamental Process of Life

Proteins are one of the elementary components of life and account for a large fraction of mass in the biosphere. They catalyze most reactions that sustain life and play structural, transport, and regulatory roles in all living organisms. Hence, "translation," that is, the synthesis of proteins by the ribosome using messenger (m)RNA as the template, is a fundamental process for all forms of life, and a large proportion of an organism's energy is committed to translation [1, 2]. Accordingly, regulating protein synthesis is crucial for all organisms. Indeed, many mechanisms to control gene expression at the translational level have evolved in eukaryotes [3]. These mechanisms have endowed eukaryotes with the potential to rapidly and reversibly respond to stress or sudden environmental changes [1, 2, 4]. Translational control also plays a crucial role in tissues and developmental processes where transcription is quiescent, or where asymmetric spatial localization of proteins is required, such as early embryogenesis, learning and memory, neurogenesis, and gametogenesis [5–10]. Moreover, recent global gene expression measurements have shown that the cellular abundance of proteins in mammalian cells is predominantly controlled at the level of translation [11, 12].

Eukaryotic translation is a sophisticated, tightly regulated, multistep process, the basic steps of which are conserved in all eukaryotes. It is performed by the ribosome together with multiple auxiliary "translation" factors (proteins) and is divided into four steps: initiation, elongation, termination, and recycling. These basic processes of translation were established experimentally in eukaryotes some decades ago, and many regulatory mechanisms have been subsequently elucidated [13, 14]. However, it was only recently that, with the use of powerful genome-wide sequencing, proteomics and bioinformatics-based technologies, a surprising diversity in components of the translation apparatus across eukaryotes was unveiled. In some cases, even functional differences between same molecules from different species have also been identified. Additionally, there is evidence that even the genetic code itself has continued to evolve in some phyla. These findings indicate that after eukaryotes emerged, the translational apparatus further evolved during eukaryotic diversification. In this article, we will review recent research revealing the diversification that the genetic code and many components of the translational machinery have undergone across eukaryotes.

2. Overview of the Translation Process in Eukaryotes

2.1. Initiation. The aim of the initiation step is both to ensure the recruitment of the mRNA to the ribosome and the positioning the ribosome in the proper frame at the start codon, which is achieved in a set of steps mediated by eukaryotic initiation factors (eIF). For most eukaryotic mRNAs, this happens by the so-called cap-dependent mechanism (Figure 1) [15–18]. It begins with the dissociation of the ribosome into its 60S and 40S subunits by eIF6. Free 40S subunit, which is stabilized by eIF3, eIF1, and eIF1A, binds to a ternary complex (consisting of eIF2 bound to an initiator Met-tRNA$_i^{Met}$ and GTP) to form a 43S preinitiation complex. On the other hand, the cap structure (m^{7}GpppN, where N is any nucleotide) of the mRNA is recognized by eIF4E in complex with the scaffold protein eIF4G. Then, the 43S preinitiation complex is recruited to the 5′ end of the mRNA, a process that is coordinated by eIF4E through its interactions with eIF4G and the 40S ribosomal subunit-associated eIF3. The ribosomal complex then scans in a 5′ $\longrightarrow$ 3′ direction along the 5′-untranslated region (UTR) through interactions with the eIF4G-bound RNA helicase eIF4A and eIF4B to reach the start codon, usually an AUG. During scanning, eIF4B stimulates the activity of eIF4A which unwinds secondary RNA structures in the mRNA. eIF1, eIF1A, and eIF5 assist in the positioning of the 40S ribosomal subunit at the correct start codon so that eIF2 can deliver the anti-codon of the initiator Met-tRNA$_i^{Met}$ as the cognate partner for the start codon, directly to the peptidyl (P)-site of the 40S ribosomal subunit. Once the ribosomal subunit is placed on the start codon, a 48S pre-initiation complex is formed. Then, eIF5 promotes GTP hydrolysis by eIF2 to release the eIF proteins. Finally, the GTPase eIF5B is required for the joining of the 60S ribosomal subunit to the 40S subunit to form an 80S initiation complex. The poly A-binding protein (PABP) is able to interact with the 3′-poly(A) tail and eIF4G promoting circularization of the mRNA and increasing the efficiency of subsequent rounds of initiation (Figure 1) [15–20].

In the case of some viral and cellular mRNAs, 5′-UTR recognition by the 40S ribosomal subunit happens without involvement of eIF4E and is, instead, driven by RNA structures located in *cis* within the mRNA itself. Such structures are operationally defined as internal ribosome entry site (IRES) and are located in the proximity of the start codon ([21–23]; Martinez-Salas et al. this issue).

2.2. Elongation. After initiation, the 80S ribosome is assembled at the start codon of the mRNA containing a Met-tRNA$_i^{Met}$ in the P-site. Then, elongation takes place (Figure 1); this is the process of decoding codons and formation of peptide bonds sequentially to add amino acid residues to the carboxy-terminal end of the nascent peptide [16, 24–26]. This process is assisted by elongation factors (eEF) and involves four major steps. (1) Formation of the ternary complex eEF1A·GTP·aminoacyl-tRNA and delivery of the first elongator aminoacyl-tRNAs to an empty ribosomal tRNA-binding site called the A-(acceptor) site. It is in the A-site where codon/anticodon decoding takes place. (2) Interaction of the ribosome with the mRNA-tRNA. This duplex activates eEF1A·GTP hydrolysis and guanine nucleotide exchange on eEF1A. (3) Peptide bond formation then occurs between the P-site peptidyl-tRNA and the incoming aminoacyl moiety of an A-site aminoacyl-tRNA. This reaction is catalyzed by the peptidyl transferase center of the 60S ribosomal subunit, and the products comprise of a new peptidyl-tRNA that is one amino acid residue longer and a deacylated (discharged) tRNA. (4) Binding of eEF2·GTP and GTP hydrolysis promotes the translocation of the mRNA such that the deacylated tRNA moves to the E-(exit) site, the peptidyl-tRNA is in the P-site, and the mRNA moves by three nucleotides to place the next mRNA codon into the A-site. The deacylated tRNA in E-site is then ejected from the ribosome. The whole process is repeated along the mRNA sequence until a stop codon is reached and the process of termination is initiated [16, 24–26].

2.3. Termination. Translation termination is mediated by two polypeptide chain-release factors, eRF1 and eRF3 (Figure 1). When any of the termination codons (UAA, UAG, and UGA) is exposed in the A-site, eRF1 recognizes the codon, binds the A-site, and triggers the release of the nascent polypeptide from the ribosome by hydrolysing the ester bond linking the polypeptide chain to the P-site tRNA. This reaction leaves the P-site tRNA in a deacylated state, leaving it to be catalyzed by the peptidyl transferase center of the ribosome. eRF1 recognizes stop signals and functionally acts as a tRNA-mimic, whereas eRF3 is a ribosome- and eRF1-dependent GTPase that, by forming a stable complex with eRF1, stimulates the termination process [16, 27, 28].

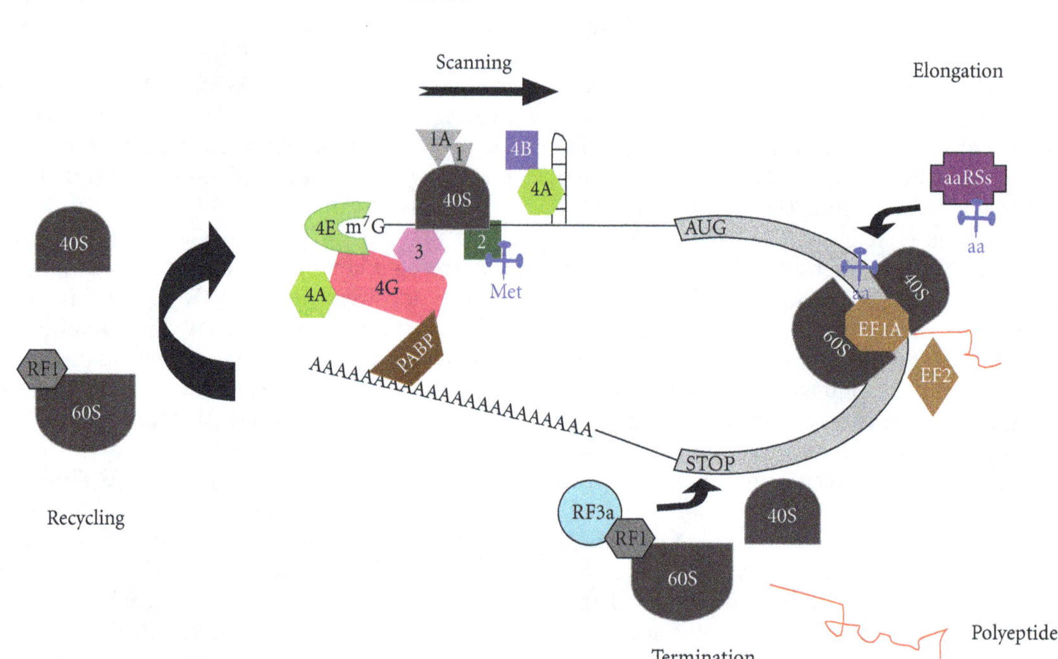

Figure 1: The general process of translation in eukaryotes. A typical eukaryotic mRNA is represented. The cap structure (m^7G), the open reading frame (light gray box) and the poly(A) tail are depicted. During *Initiation*, most eukaryotic mRNAs are translated by the cap-dependent mechanism, which requires recognition by eIF4E (green crescent) complexed with eIF4G (red) and eIF4A (light green)—the so-called eIF4F complex—of the cap structure at the 5′ end. A 43S preinitiation complex (consisting in a 40S ribosomal subunit (dark gray) loaded with eIF3 (pink), eIF1 and eIF1A (light grey), initiator Met-tRNA$_i^{Met}$ (blue clover), eIF2 (dark green), and GTP binds the eIF4F-mRNA complex and scans along the 5′-UTR of the mRNA to reach the start codon (usually an *AUG* triplet). During the scanning eIF4A, stimulated by eIF4B (dark blue), unwinds secondary RNA structure in an ATP-dependent manner. The poly A-binding protein (PABP, dark brown) binds both the poly(A) tail and eIF4G promoting mRNA circularization. *Elongation* is assisted by elongation factors eEF1A and eEF2 (light brown). During this step, aminoacyl-tRNA synthetases (aaRSs, purple) catalyze the binding of amino acids (aa) to cognate tRNAs. *Termination* is mediated by the release factors eRF1 (gray) and eRF3 (light blue) and happens when a termination codon (*STOP*) of the mRNA is exposed in the A-site of the ribosome. In this step, the completed polypeptide (red) is released. During *Recycling*, which is required to allow further rounds of translation, both ribosomal subunits dissociate from the mRNA. eRF1 remains associated with the posttermination complexes after polypeptide release.

2.4. Recycling. In the recycling step, both ribosomal subunits are dissociated, releasing the mRNA and deacetylated tRNA, so that both ribosomal subunits can be used for another round of initiation [16, 27, 28] (Figure 1). The closed-loop model proposes that, during translation, cross-talk occurs between both ends due to the circular conformation of the mRNA. According to this model, termination and recycling may not release the 40S ribosomal subunit back into the cytoplasm. Instead, this subunit may be passed from the poly(A) tail back to the 5′-end of the mRNA, so that a new round of initiation can be started [16, 27].

3. Divergence in the Genetic Code

The deciphering of the genetic code in the early 1960's established one of the basic foundations of modern biology. Soon after, the essential universality of the genetic code was recognized, that is, the assignment of 20 amino acids to 64 codons and two punctuation marks (start and stop signals) is substantially the same for all extant forms of life on earth [29]. Nevertheless, variations to the "universal" genetic code, wherein the meaning of a "universal" codon is changed to a different one, have recently been uncovered in a wide range of bacteria, organelles, and the nuclear genome of eukaryotes, revealing that the genetic code is still evolving in some lineages [30–33]. In eukaryotes, deviations from the standard nuclear genetic code have arisen independently multiple times in unicellular organisms of five lineages, namely, ciliates, Diplomonads, fungi (in the genus *Candida* and some ascomycetes), polymastigid oxymonads, and green algae (in Dasycladales and Cladophorales) [30, 31, 33–39]. Most codon variations in eukaryotes are found to be the reassignment of the stop codons UAG and UAA to glutamine, and the stop codon UGA to tryptophan or cysteine (Figure 2). All reported code variations in ciliates, Diplomonads, and green algae belong to this kind. In contrast, *Candida* ambiguously utilizes the codon CUG (universally used for leucine) for both serine and leucine. The underlying mechanisms of codon reassignment are mutations in tRNA genes that affect decoding, RNA editing, or mutations in eRF1 [30, 31, 34–39].

The observation that the same codon reassignments have occurred independently in closely related species (within the yeasts, green algae, and ciliate taxa) supports the notion that these changes provide a selective advantage in similar ecological niches [30]. Whether there is a restriction for the genetic code to change in multicellular organisms is not known.

4. Diversity in the Initiation Step

4.1. Functional Divergence of eIF Proteins. While the fundamental principles of translation are well conserved across all forms of life, in eukaryotes the initiation step has undergone substantial increase in complexity as compared to prokaryotes [3, 22, 40–44]. Most evidence for molecular and functional diversification among the translation components has been found in the eIF4 proteins (Figure 2). Most eukaryotic phyla possess several paralog genes for members of the eIF4 families, with well-documented differential expression patterns and variable biochemical properties among paralogs of the same organism [45–72]. For eIF4E and eIF4G cognates, even evidence of physiological specialization has been found among both unicellular and multicellular organisms (Table 1). These findings support the hypothesis that in organisms with several paralogs, an ubiquitous set of eIF4 factors supports global translation initiation whereas other paralogs perform their activity in specific cellular processes [45]. In some cases, eIF4E cognates have evolved towards translational repressors. Class 2 eIF4Es are exemplified by eIF4E-homolog protein (4E-HP) in human, eIF4E-2 in mouse [63], eIF4E-8 in *Drosophila* [52, 58, 73], IF4 in *C. elegans* [74, 75], and nCBP in *A. thaliana* [76], and they can bind the 5′ cap structure of mRNA but do not bind eIF4G [58, 77], thereby acting as a translational repressors of mRNAs associated with it [73, 78]. Class 2 eIF4Es are widespread across metazoa, plants, and some fungi although absent in the model ascomycetes *S. cerevisiae* and *S. pombe* [46]. Since the *Arabidopsis* [76] and *Caenorhabditis* [74] orthologs promote translation of some mRNAs, it seems most likely 4E-HP diverged from a widespread ancestral eIF4E to form a translational repressor in metazoa [3]. A similar example is eIF4E-1B, which emerged only in vertebrates as a translational repressor of a subset of oocyte mRNAs [57, 59, 79], and *Leishmania* eIF4E-1, which under heat shock conditions binds to a *Leishmania*-specific 4E-BP and becomes translationally inactive [71]. In other cases, eIF4E cognates have evolved towards a new molecular function not related to translation. This is the case with *Trypanosoma* eIF4E-1 and eIF4E-2, which are essential nuclear and cytoplasmic proteins, respectively [49], and *Giardia* (eIF4E-2), which binds only to nuclear noncoding small RNAs [64]. However, it is also possible that this was an ancestral function of eIF4E [22, 40].

Whereas the need for distinct eIF4 proteins in different tissues may have been the driving force behind the evolution of various paralogs in multicellular organisms, in unicellular eukaryotes different paralogs may be differentially needed during distinct life stages [49]. Specific features of mRNA metabolism in some phyla also might have driven the evolution of eIF4Es in specific organisms, such as the use of different cap structures (usually mono- and trimethylated) in mRNAs from worms of the phylum Nematoda [50, 51, 54, 80], and flagellate protists of the order Kinetoplastida [49, 65, 66]. These mRNAs result from the *trans*-splicing process to produce mature mRNAs.

Other eIFs have also undergone molecular diversification across eukaryotes, including the multisubunit eIF3 whose subunit composition ranges from 5 to 13 nonidentical polypeptides in different phyla [99], and eIF6 that is duplicated into two or three paralogs in plants [100]. However, the functional relevance of these phenomena (if any) is not known.

4.2. Multiple RNA Helicases for Translation Initiation. The evolution of cap-dependent translation has led to a dependency on RNA helicase activity to unwind the 5′-UTR secondary structure during the scanning [22, 40]. The DEAD-box RNA helicase/ATPase eIF4A is the main helicase thought to perform this activity. Recently, other RNA helicases from diverse organisms have also been found to facilitate translation of specific mRNAs with structured 5′-UTRs (Figure 2). Such is the case of the mammalian, *Drosophila* and yeast DEAD-box helicases DDX3 and Ded1, as well as the human DExH-box helicases RHA and DHX29 [101–103]. In *Drosophila*, the DEAD-box helicase Vasa interacts with eIF5B and regulates the translation of *gurken* and *mei-P26* mRNAs. Evidence supports the idea that Vasa is a translational activator of specific mRNAs involved in germline development [6, 7]. In contrast, orthologs of the *Xenopus* helicase Xp54 (DEAD-box, DDX6-like helicases) in a spectrum of organisms, including *Drosophila* Me31B, *Saccharomyces* Dhh1, human rck/p54, and *Caenorhabditis* CGH-1 have been found to repress translation of stored mRNAs and promote aggregation into germplasm-containing structures [104].

Most RNA helicases involved in translation also play a variety of roles in other processes of RNA metabolism, including mRNA RNP assembly, RNA degradation, RNA export, and splicing [103]. This functional versatility of RNA helicases leads us to speculate that a wider diversity of other, yet unidentified, helicases might be involved in translation in all eukaryotes. This could be the case of the *Arabidopsis* eIF4F complex, which contains eIF4A in proliferating cells but different RNA helicases in quiescent cells [105]. Whether these helicases play a role in translation is not known.

4.3. Divergence in the Regulation of Initiation: Diversity of eIF4E-Binding Proteins. Almost twenty years ago, it was discovered that eIF4E is negatively regulated in mammalian cells by three related proteins, the eIF4E-binding proteins (4E-BPs) 1, 2, and 3. These proteins share with eIF4G the motif YXXXXLϕ (where X is any amino acid and ϕ is a hydrophobic residue) that interacts with the convex dorsal surface of eIF4E, so binding of 4E-BPs to eIF4E precludes its association with eIF4G and represses cap-dependent

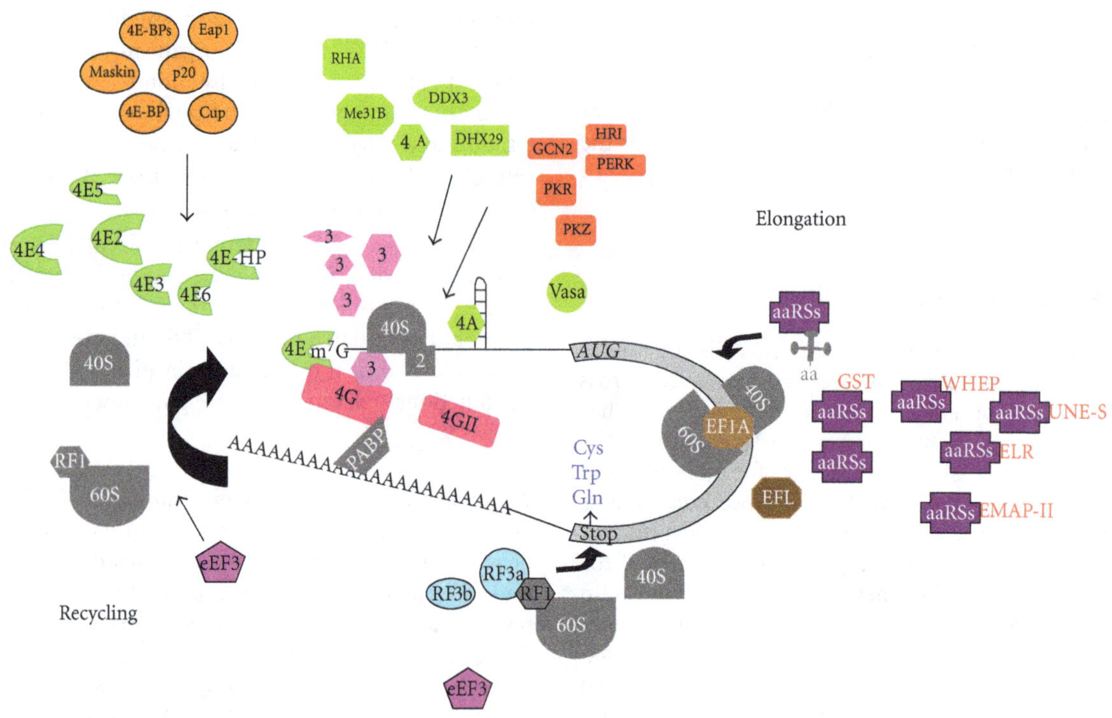

Figure 2: Diversity in the configuration of the translation apparatus across eukaryotes. The different components of the translation machinery that show diversity in different phyla are shown in colors. Components with some diversity that is not discussed here are depicted in gray. Several copies of eIF4E (green crescent) and eIF4G (red) have been found in plants, metazoan, and protists. In some cases, eIF4E cognates have evolved towards translational repressors (4E-HP is an example). Many 4E-binding proteins (orange) have been discovered in species from metazoan, fungi and protists. The subunit composition of eIF3 (pink) ranges from 5 to 13 nonidentical polypeptides in different phyla. There is, however, a core of five homolog subunits shared by most eukaryotes. Several RNA helicases (light green) from diverse organisms have been found to be involved in *Initiation*. A family of five kinases (*HRI*, *PERK*, *GCN2*, *PKR*, and *PKZ*, red) phosphorylate the alpha subunit of eIF2 to inhibit global translation under stress conditions. The presence of eIF2alpha kinases varies in different lineages. Different domains (red), such as *WHEP*, *EMAPII*, *ELR*, *GST*, and *UNE-S*, have been added to different aminoacyl-tRNA synthetases (*aaRSs*, purple) in distinct phyla of multicellular species. For *Elongation* to happen, a number of protist, algae and fungi (most of them unicellular organisms) lack eEF1A (light brown) and instead possess the related factor elongation factor-like (EFL, dark brown). For *Termination*, most organisms only contain a single eRF3 (light blue). In contrast, mammalian species express two eRF3s (viz. eRF3a and eRF3b). Ribosomes from all eukaryotes perform *Elongation* with eEF1A and eEF2. However, the yeast *S. cerevisiae* requires an additional essential factor, eEF3 (light purple), for *Elongation* to proceed. Genes encoding eEF3 have been found exclusively in many species of fungi. Evidence supports the notion that eEF3 activity promotes ribosome recycling. Variations to the "universal" genetic code, wherein the meaning of a "universal" codon is changed to a different one, exist in several species of in unicellular eukaryotes. Most codon variations are the reassignment of the stop codons UAG and UAA to glutamine, and the stop codon UGA to tryptophan or cysteine.

translation [8, 106]. In the last years, a myriad of 4E-binding proteins has been discovered in species from distantly related taxa, including mammals, plants, *Drosophila*, *Caenorhabditis*, yeast [3, 8, 106], and *Leishmania* [71] (Figure 2). Interestingly, most 4E-BPs are phylogenetically unrelated to each other and control translation in disparate, species-specific processes, such as embryogenesis in *Drosophila*, neurogenesis in mammals, or pseudohyphal growth in yeast. Moreover, some 4E-BPs utilize non-canonical motifs to bind eIF4E. These observations support the idea that binding to eIF4E evolved independently in multiple taxonomic groups [3].

4.4. Divergence in the Regulation of Initiation: The Case of eIF4E Phosphorylation. In mammalian cells, the kinases ERK or p38MAPK phosphorylate and activate the MAPK-interacting kinases (Mnkl/2). Mnk interacts with the carboxy-terminal part of eIF4G to directly phosphorylate eIF4E on Ser-209. This phosphorylation appears to regulate the function of eIF4E although the precise consequences are unclear [107–110]. Mammals possess two Mnk genes (*MKNK1/2*) which in humans, but not mice, give rise to four Mnk isoforms by alternative splicing; these isoforms have distinct properties in terms of activity, regulation, and subcellular localization [111]. In *Drosophila*, the single Mnk orthologue, LK6, also phosphorylates eIF4E-1 at a serine residue corresponding to mammalian Ser-209, a phosphorylation that is critical for development and cell growth [112–115]. However, the effects of phosphorylation on eIF4E activity and its physiological relevance are different across eukaryotes. Indeed, a residue equivalent to Ser-209 is present in metazoan eIF4Es but is absent in different fungi, protists and plants ([67];R. Jagus et al., this issue). Accordingly, Mnk is conserved among metazoans, but no Mnk ortholog exists in *S. cerevisiae* or plants, whose eIF4Gs

Table 1: Specialized activities of eIF4 proteins.

Protein[a]	Activity	Reference
eIF4E cognates		
Dm eIF4E-1, M eIF4E-1, Ce IFE-3, Sp eIF4E-1, Sc eIF4E, Plant eIF4E and eIF(iso)4E, Z eIF4E-1A, Gl eIF4E-2; Tb eIF4E-3 and eIF4E-4; Lm eIF4E-1 and eIF4E-4	Supports general cap-dependent initiation of translation. Essential gene.	[49, 54, 55, 57, 58, 62, 64, 65, 67, 72, 81–84]
M eIF4E-1	mRNA nucleocytoplasm transport.	[85]
Dm eIF4E-1	Involved in *sex-lethal* (*Sxl*)-dependent female-specific alternative splicing of male specific lethal-2 (msl-2) mRNA and *Sxl* pre-mRNAs.	[86]
Sp eIF4E-2	Supports cap-dependent translation initiation during stress response.	[62]
Ce IFE-1	Required for gametogenesis.	[87–89]
Ce IFE-2	Involved in chromosome segregation at meiosis at elevated temperatures.	[90]
Ce IFE-4	Promotes expression of specific mRNAs involved in egg lying. Nonessential gene.	[74]
Dm eIF4E-3	Testis-specific protein, essential for spermatogenesis.	[91]
La eF4E-4	Supports translation in promastigotes stage.	[71]
Dm 4E-HP, M 4E-HP	Negative regulator of translation.	[58, 73, 77, 78]
Xl eIF4E-1B	Negative regulator of translation.	[57, 79]
La eIF4E-1	Represses translation under heat shock conditions.	[71]
Gl eIF4E-1	Involved in nuclear snRNAs metabolism and play no role in translation.	[64]
Tb eIF4E-1 and eIF4E-2	Essential genes. Play no role in translation.	[49]
eIF4G cognates		
M eIF4G-I and eIF4G-II, Dm eIF4G, Sc eIF4G-I and eIF4G-II, plant eIF4G and Plant eIF(iso)4G, Ce p170 of IFG-1,	Scaffold protein. Supports general cap- and IRES-dependent initiation of translation.	[55, 60, 67, 71, 92–96]
Dm eIF4G-2	Support translation initiation in testis.	[47, 48]
M eIF4G-2	Involved in hematopoietic cell differentiation.	[97]
M eIF4G-3	Essential for spermatogenesis.	[98]
Ce IFG-1	p130 of *ifg-1* gene is involved in mitotic and early meiotic germ cell development.	[93]
La eIF4G-3	Supports translation in promastigotes stage.	[71]

[a]At, *Arabidopsis thaliana*; Ce, *Caenorhabditis elagans*; Dm, *Drosophila melanogaster*; Lm, *Leishmania major*; La, *Leishmania amazonensis*; M, *mammalian*; Nt, *N. tabacum*; Sc, *Saccharomyces cerevisiae*; Sp, *Schizosaccharomyces pombe*; W, *wheat germ*; Xl, *Xenopus laevis*; Z, *zebra fish*; Gl, *Giardia lamblia*; Tb, *Trypanosoma brusei*.

lack a Mnk-binding domain ([67]; R. M. Patrick and K. S. Browning, this issue). Moreover, *Trypanosoma* eIF4E-3 [116] and *S. cerevisiae* eIF4E [117] are phosphorylated on residues which are not equivalent to mammalian Ser-209, and *S. cerevisiae* cells expressing a nonphosphorylatable version as sole source of eIF4E do not display any evident defect on global protein synthesis or cell growth [117]. These observations support the idea that eIF4E phosphorylation at Ser-209 by the MAPK-Mnk signaling pathway evolved only in metazoans and that, perhaps, alternative mechanisms regulate eIF4E in nonmetazoan eukaryotes [3].

4.5. Diversity in the Regulation of Initiation: The Case of eIF2alpha Phosphorylation. Under different stress conditions, general protein synthesis is inhibited through phosphorylation of the alpha subunit of eIF2 at Ser-51 by a family of kinases that are present in widely scattered lineages. They include the double-stranded RNA protein kinase (PKR) that is activated during viral infection, the heme-regulated inhibitor kinase (HRI) that is activated under heme deprivation or arsenite exposure, the PKR-like endoplasmatic reticulum kinase (PERK) that is activated by unfolded proteins in the lumen of the endoplasmic reticulum, and the general control nonderepressible 2 (GCN2) that is activated by uncharged tRNA and thus senses amino acid starvation [107, 118] (Figure 2). The presence of eIF2alpha kinases varies in different lineages; while GCN2 is present in all eukaryotes; PERK is found in only metazoans; HRI is found in vertebrates, the dipteran *Anopheles*, the fungi *Schizosaccharomyces*, and the echinoderm *Strongylocentrotus*; PKR is only found in vertebrates [53, 118]. Interestingly, in some teleost fishes, PKR has undergone further duplication into PKR and PKZ, which perhaps led teleost fishes to respond to an extended range of viral infections [119].

5. Diversity in the Elongation Step

5.1. Divergence in the Aminoacyl-tRNA Synthetases. The process of elongation is highly conserved among all forms of life [16, 24, 25]. Key molecules for elongation are aminoacyl-tRNA synthetases (aaRSs), which catalyze the aminoacylation reaction whereby an amino acid is attached to the cognate tRNA. aaRSs are the only components of the gene expression machinery that function at the interface between nucleic acids and proteins. Thus, by performing their activity, aaRSs establishes the fundamental rules of the universal genetic code and, thus, of translation. aaRSs constitute a family of 20 essential cellular enzymes that are grouped into two classes: class I, in which the aminoacylation domain has a Rossmann nucleotide-binding fold, and class II, in which this domain is a seven-stranded beta-sheet with flanking alpha-helices. The conservation of the genetic code suggests that aaRSs evolved very early before the emergence of the last universal common ancestor [120, 121].

Throughout evolution of multicellularity, different domains, such as the WHEP domain, the oligonucleotide binding fold-containing EMAPII domain, the tripeptide ELR (Glu-Leu-Arg), the glutathione S-transferase (GST) domain and a specialized amino-terminal helix (N-helix), have been progressively added to different aaRSs in distinct phyla (Figure 2). The tripeptide ELR and the EMAPII domain were incorporated simultaneously to TyrRSs in metazoans starting from insects; the WHEP domain is present in TrpRS only in chordates; a unique sequence motif, UNE-S, became fused to the C-terminal of SerRS of all vertebrates [120, 121]. In bilaterian animals, the glutamylRS and prolylRS were linked via WHEP domains giving rise to a bifunctional glutamyl-prolylRS [120, 121]. It was recently found that this fused enzyme is also present in the cnidarian *Nematostella*, which pushes the origin of glutamyl-prolylRS back to the cnidaria-bilaterian ancestor [122], and suggests that this enzyme further underwent fission in the nematode *C. elegans* where glutamylRS and prolylRS enzymes are separated. GlutamylRS and prolylRS are also separate in plants and fungi [120–122].

It has been found that the function of the aaRSs was either increased or impaired by the addition of the new domains. Whereas the WHEP domain regulates interaction of TrpRS with its cognate receptor, with MetRS this domain plays a tRNA-sequestering function. The Leu-zipper motif in ArgRS is important for the formation of multi aaRSs complex (MSC), which enhances channeling of tRNA to the ribosome. Moreover, different aaRSs play diverse roles in cellular activities beyond translation, such as stress response, plant and animal embryogenesis, cell death, immune responses, transcriptional regulation, and RNA splicing [120, 121, 123]. It was found that the incorporation of domains to aaRSs correlates positively with the increase in organism's complexity. For example, the number of aaRSs carrying the GST domain increases from two in fungi to four in insects, to five in fish, and six in humans [121]. Thus, it has been proposed that the newly fused aaRSs domains triggered the appearance of new biological functions for these proteins in different lineages and that the fusion of domains to aaRSs could have played an important part in expanding the complexity of newly emerging metazoan phyla [121].

5.2. Divergence in Elongation Factors. eEF1A plays a critical role in translation. It binds and delivers aa-tRNAs to the A-site of ribosomes during the elongation step. Because homologs of this essential protein occur in all domains of life, it was thought to exist in all eukaryotes. Strikingly, a recent genome-wide survey revealed that a number of lineages lack eEF1A and instead possess a related factor called elongation factor-like (EFL) protein that retains the residues critical for eEF1A function [124] (Figure 2). It was later found that EFL-encoding species are scattered widely across eukaryotes and that *eEF1A* and *EFL* genes display mutually exclusive phylogenetic distributions. Thus, it is assumed that eEF1A and EFL are functionally equivalent [124–132]. Since EFL is present only in eukaryotes, it is thought that eEF1A is ancestral to all extant eukaryotes and that a single duplication event in a specific lineage gave rise to EFL. EFL genes were then spread to other lineages via multiple independent lateral gene transfer events, where EFL took over the original eEF1A function resulting in secondary loss of the endogenous eEF1A. It is thought that both genes coexisted for some time before one or the other was lost. Indeed, the diatom *Thalassiosira* bears both *EFL* and *eEF1A* genes [129] and might be an example of this situation. It is also possible that there was a single gain of EFL early in evolution followed by differential loss of it [124, 128, 129, 131, 132]. So far, EFL genes have been identified in widespread taxa, including diatoms, green and red algae, fungi, euglenozoans, foraminiferans, cryptophytes, goniomonads, katablepharid, chlorarachniophytes, oomycetes, dinoflagellates, choanozoans, centrohelids, and haptophytes [124–132]. Most of them are unicellular organisms. In contrast, eEF1A is found in most eukaryotes, and multiple copies of this gene have been found in some insect orders, including Coleoptera, Hymenoptera, Diptera, Thysanoptera, and Hemiptera [133].

The eEF1A activity is modulated by diverse posttranslational modifications, including phosphorylation, lysine methylation, and methyl-esterification. eEF1A also undergoes modification by covalent binding of ethanolamine phosphoglycerol (EPG), whose function is not known and for whom the number of moieties attached varies in different eukaryotes [134]. Moreover, in addition to its role in translation, eEF1A has been reported to play several "moonlighting" functions, including binding to cytoskeletal proteins, signal transduction, protein nuclear export and import of tRNAs into mitochondria [134]. It is not known whether EFL undergoes the same posttranslational modifications as eEF1A does and whether it also displays non-translational activities.

6. Divergence in the Termination Step

The termination of protein synthesis is governed by eRF1, which is a monophyletic and highly conserved protein that is universally present in eukaryotes. Comprehensive analyses of genomic datasets show that eRF1 was inherited by eukaryotes from archaeal ancestors and that most eukaryotes encode only one eRF1. Known exceptions are *Arabidopsis thaliana*, which possesses three *eRF1* genes, and the ciliates *Tetrahymena, Oxytricha, Nyctotherus, Oxytricha, Euplotes,* and *Paramecium* which have two *eRF1* genes [135–138]. Interestingly, unusually high rates of eRF1 evolution have been found in organisms with variant genetic codes, especially in the N-terminal domain, which is responsible for stop-codon recognition [30, 34, 135, 136, 138, 139]. eRF1 displays structural similarity to tRNA molecules and mimics its activity during binding of ribosomal A-site during recognition of a stop codon [34, 139–141]. Since mutations in eRF1 N-terminal domain switch from omnipotent to bipotent mode for stop-codon specificity [35–38, 141], most likely the accelerated evolution of eRF1 in organisms with variations to the nuclear genetic code has been driven mainly to accommodate these variations [30, 31, 34–38, 135, 138–141].

eRF3 is a GTPase that stimulates the activity of eRF1 during the translation termination process. eRF3 arose in early eukaryotes by the duplication of the GTPase eEF1A. Consistent with this, eRF3 binds and transports eRF1, a structural mimic of tRNA, to the ribosomal A-site, similar to the role of eEF1A in binding and delivering aminoacyl-tRNAs to the same site during translation elongation [142, 143]. eRF3 is much more divergent than eRF1, especially in its N-terminal domain. In addition, *eRF3* is universal among eukaryotes, and most organisms only contain single-copies of this gene [137, 143]. In contrast, mammalian species express two eRF3s (viz. eRF3a and eRF3b; Figure 2). They possess different N regions and display drastically different tissue distribution and expression profiles during the cell cycle [143, 144]. Moreover, eRF3b but not eRF3a can substitute for yeast eRF3 in translation termination [145]. These observations indicate duplication and further functional divergence of eRF3 proteins in this lineage.

7. Divergence in the Recycling Step

Ribosomes from all eukaryotes perform elongation with eEF1A and eEF2. Interestingly, it has been known for some time that the yeast *Saccharomyces cerevisiae* requires an additional essential factor, eEF3, for the elongation cycle to proceed [146]. Genes encoding eEF3 were subsequently identified exclusively in other fungi (both yeasts and filamentous), including *Candida, Pneumocystis, Neurospora, Aspergillus,* and *Mucor* [147–150] (Figure 2). eEF3 is an ATPase that interacts with both ribosomal subunits and stimulates binding of aminoacyl-tRNA to the ribosomal A-site by enhancing the rate of deacylated tRNA dissociation from the E-site. Because E-site release is needed for efficient A-site binding of aminoacyl-tRNA, it was thought that eEF3 functions as a so-called "E-site" factor [16, 151]. Most recently, it was shown that post-termination complex, consisting of a ribosome, mRNA, and tRNA, is disassembled into single components by ATP and eEF3. Because the release of mRNA and deacylated tRNA and ribosome dissociation takes place simultaneously and no 40S—mRNA complexes remain, it is proposed that eEF3 activity promotes ribosome recycling [152]. "What were the evolutionary forces that led to the emergence of eEF3 exclusively in fungi?" is a very interesting, still open question.

8. Concluding Remarks

One of the most conspicuous features of life is its prominent ability to diversify. Current assessments of the biodiversity on Earth reaches 2 million species, although the true number of living organisms could easily be four times that number and likely much higher [153, 154]. The diversification of life has occurred at different levels, including the occupancy of ecological niches, behavioral patterns, body plans, and organismal complexity, and metabolic needs and capabilities. More recently, intensive whole-genome shotgun sequencing of microbial communities from different environments has unveiled a vast profusion of diversification also at the genetic level [155–157]. We have discussed that diversity also exists in the machinery that performs a fundamental process, translation, across eukaryotes. We speculate that the molecular diversification of the translation apparatus is among the basis that provided to early eukaryotes the scope to invade new ecological niches and overcome the different environmental and biological challenges this represented. Different evolutionary mechanisms might have been the driving forces leading to this molecular diversification in different lineages, including natural selection, sexual selection, genetic drift and neutral evolution. However, at this point, we can be nothing but speculative on the biological meaning of the molecular diversification reviewed here.

Traditional studies on so-called model organisms have taught us the global processes of eukaryotic translation. In the last years, the use of modern genome-wide, high-throughput technologies to study many non-model eukaryotes from different taxa has unveiled that diversification of the translation machinery configuration is far more expansive than previously thought. Collectively, these studies show that the translation apparatus in eukaryotes is far from being evolutionarily static. Therefore, we anticipate that, as more organisms are studied, additional diversification of components of the translation apparatus will be revealed. We believe that a better understanding of the diversity of all levels of organism will provide us a more profound understanding of Life.

Acknowledgments

The authors are very thankful to Rosemary Jagus for critical review of the manuscript, and to Michelle Kowanda for proofreading the manuscript. G. Hernández is supported by the National Institute for Cancer (Instituto Nacional de Cancerología, México) and Consejo Nacional de Ciencia y Tecnología (CONACyT). C. G. Proud is supported by AstraZeneca, The Biotechnology and Biological Sciences Research Council, the British Heart Foundation, Cancer Research UK, Janssen, the Kerkut Trust, the Medical Research Council, The Royal Society, and The Wellcome Trust. The work in the laboratory of T. Preiss is supported by grants from the Australian Research Council and the National Health and Medical Research Council of Australia. In many cases, the authors cite reviews, not original research articles. They apologise to many authors for not citing their original literature.

References

[1] M. B. Mathews, N. Sonenberg, and J. W. B. Hershey, "Origins and principles of translational control," in *Translational Control of Gene Expression*, N. Sonenberg, J. W. B. Hershey, and M. B. Mathews, Eds., Cold Spring Harbor Laboratory Press, Cold Spring Harbor, NY, USA, 2000.

[2] M. B. Mathews, N. Sonenberg, and J. W. B. Hershey, "Origins and principles of translational control," in *Translational Control in Biology and Medicine*, M. B. Mathews, N. Sonenberg, and J. W. B. Hershey, Eds., Cold Spring Harbor Laboratory Press, Cold Spring Harbor, NY, USA, 2007.

[3] G. Hernández, M. Altmann, and P. Lasko, "Origins and evolution of the mechanisms regulating translation initiation in eukaryotes," *Trends in Biochemical Sciences*, vol. 35, no. 2, pp. 63–73, 2010.

[4] B. Mazumder, V. Seshadri, and P. L. Fox, "Translational control by the 3′-UTR: the ends specify the means," *Trends in Biochemical Sciences*, vol. 28, no. 2, pp. 91–98, 2003.

[5] R. Renkawitz-Pohl, L. Hempel, M. Hollman, and M. A. Schafer, "Spermatogenesis," in *Comprehensive Molecular Insect Science*, L. I. Gilbert, K. Iatrou, and S. S. Gill, Eds., Elsevier Pergamon, 2005.

[6] J. D. Richter and P. Lasko, "Translational control in oocyte development," *Cold Spring Harbor Perspectives in Biology*, vol. 3, no. 9, Article ID a002758, 2011.

[7] P. Lasko, "Translational control during early development," in *Progress in Molecular Biology and Translational Science*, J. W. B. Hershey, Ed., Academic Press, Burlington, Vt, USA, 2009.

[8] N. Sonenberg and A. G. Hinnebusch, "New modes of translation control in development, behavior, and disease," *Molecular Cell*, vol. 28, no. 5, pp. 721–729, 2007.

[9] B. Thompson, M. Wickens, and J. Kimble, "Translational control in development," in *Translational Control in Biology and Medicine*, M. B. Mathews, N. Sonenberg, and J. W. B. Hershey, Eds., Cold Spring Harbor Laboratory Press, Cold Spring Harbor, NY, USA, 2007.

[10] M. Costa-Mattioli, W. S. Sossin, E. Klann, and N. Sonenberg, "Translational control of long-lasting synaptic plasticity and memory," *Neuron*, vol. 61, no. 1, pp. 10–26, 2009.

[11] B. Schwanhüusser, D. Busse, N. Li et al., "Global quantification of mammalian gene expression control," *Nature*, vol. 473, no. 7347, pp. 337–342, 2011.

[12] C. Vogel, R. De Sousa Abreu, D. Ko et al., "Sequence signatures and mRNA concentration can explain two-thirds of protein abundance variation in a human cell line," *Molecular Systems Biology*, vol. 6, article no. 400, 2010.

[13] M. B. Mathews, N. Sonenberg, and J. W. B. Hershey, Eds., *Translational Bontrol in Biology and Medicine*, Cold Spring Harbor Laboratory Press, Cold Spring Harbor, NY, USA, 2007.

[14] N. Sonenberg, J. W. B. Hershey, and M. B. Mathews, Eds., *Translational Control of Gene Expression*, Cold Spring Harbor Laboratory Press, Cold Spring Harbor, NY, USA.

[15] J. W. B. Hershey and W. C. Merrick, "Pathway and mechanism of initiation of protein synthesis," in *Translational Control of Gene Expression*, N. Sonenberg, J. W. B. Hershey, and M.B. Mathews, Eds., Cold Spring Harbor Laboratory Press, Cold Spring Harbor, NY, USA, 2000.

[16] L. D. Kapp and J. R. Lorsch, "The molecular mechanics of eukaryotic translation," *Annual Review of Biochemistry*, vol. 73, pp. 657–704, 2004.

[17] N. Sonenberg and A. G. Hinnebusch, "Regulation of translation initiation in eukaryotes: mechanisms and biological targets," *Cell*, vol. 136, no. 4, pp. 731–745, 2009.

[18] R. J. Jackson, C. U. T. Hellen, and T. V. Pestova, "The mechanism of eukaryotic translation initiation and principles of its regulation," *Nature Reviews Molecular Cell Biology*, vol. 11, no. 2, pp. 113–127, 2010.

[19] T. Preiss and M. W. Hentze, "Starting the protein synthesis machine: eukaryotic translation initiation," *BioEssays*, vol. 25, no. 12, pp. 1201–1211, 2003.

[20] F. Gebauer and M. W. Hentze, "Molecular mechanisms of translational control," *Nature Reviews Molecular Cell Biology*, vol. 5, no. 10, pp. 827–835, 2004.

[21] J. A. Dounda and P. Sarnow, "Translation initiation by viral internal ribosome entry sites," in *Translational Control in Biology and Medicine*, M. B. Mathews, N. Sonenberg, and J. W. B. Hershey, Eds., Cold Spring Harbor Laboratory Press, Cold Spring Harbor, NY, USA, 2007.

[22] G. Hernández, "Was the initiation of translation in early eukaryotes IRES-driven?" *Trends in Biochemical Sciences*, vol. 33, no. 2, pp. 58–64, 2008.

[23] A. Pacheco and E. Martinez-Salas, "Insights into the biology of IRES elements through riboproteomic approaches.," *Journal of Biomedicine and Biotechnology*, vol. 2010, Article ID 458927, 12 pages, 2010.

[24] D. J. Taylor, J. Frank, and T. G. Kinzy, "Structure and function of the eukaryotic ribosome and elongation factors," in *Translational Control in Biology and Medicine*, M. B. Mathews, N. Sonenberg, and J. W. B. Hershey, Eds., Cold Spring Harbor Laboratory Press, Cold Spring Harbor, NY, USA, 2007.

[25] G. R. Andersen, P. Nissen, and J. Nyborg, "Elongation factors in protein biosynthesis," *Trends in Biochemical Sciences*, vol. 28, no. 8, pp. 434–441, 2003.

[26] T. P. Herbert and C. G. Proud, "Regulation of translation elongation and the cotranslational protein targeting pathway," in *Translational Control in Biology and Medicine*, M. B. Mathews, N. Sonenberg, and J. W. B. Hershey, Eds., Cold Spring Harbor Laboratory Press, Cold Spring Harbor, NY, USA, 2007.

[27] M. Ehrenberg, V. Hauryliuk, C. G. Crist, and Y. Nakamura, "Translation termination, the prion [PSI+], and ribosomal recycling," in *Translational Control in Biology and Medicine*, M. B. Mathews, N. Sonenberg, and J. W. B. Hershey, Eds., Cold Spring Harbor Laboratory Press, Cold Spring Harbor, NY, USA, 2007.

[28] R. J. Jackson, C. U. T. Hellen, and T. V. Pestova, "Termination and post-termination events in eukaryotic translation," *Advances in Protein Chemistry and Structural Biology*, vol. 86, pp. 45–93, 2012.

[29] M. Szymański and J. Barciszewski, "The genetic code—40 years on," *Acta Biochimica Polonica*, vol. 54, no. 1, pp. 51–54, 2007.

[30] R. D. Knight, S. J. Freeland, and L. F. Landweber, "Rewiring the keyboard: evolvability of the genetic code," *Nature Reviews Genetics*, vol. 2, no. 1, pp. 49–58, 2001.

[31] A. V. Lobanov, A. A. Turanov, D. L. Hatfield, and V. N. Gladyshev, "Dual functions of codons in the genetic code," *Critical Reviews in Biochemistry and Molecular Biology*, vol. 45, no. 4, pp. 257–265, 2010.

[32] E. V. Koonin and A. S. Novozhilov, "Origin and evolution of the genetic code: the universal enigma," *IUBMB Life*, vol. 61, no. 2, pp. 99–111, 2009.

[33] M. A. S. Santos, G. Moura, S. E. Massey, and M. F. Tuite, "Driving change: the evolution of alternative genetic codes," *Trends in Genetics*, vol. 20, no. 2, pp. 95–102, 2004.

[34] C. A. Lozupone, R. D. Knight, and L. F. Landweber, "The molecular basis of nuclear genetic code change in ciliates," *Current Biology*, vol. 11, no. 2, pp. 65–74, 2001.

[35] B. Eliseev, P. Kryuchkova, E. Alkalaeva, and L. Frolova, "A single amino acid change of translation termination factor eRF1 switches between bipotent and omnipotent stop-codon specificity," *Nucleic Acids Research*, vol. 39, no. 2, pp. 599–608, 2011.

[36] S. Lekomtsev, P. Kolosov, L. Bidou, L. Frolova, J. P. Rousset, and L. Kisselev, "Different modes of stop codon restriction by the Stylonychia and Paramecium eRF1 translation termination factors," *Proceedings of the National Academy of Sciences of the United States of America*, vol. 104, no. 26, pp. 10824–10829, 2007.

[37] Y. Inagaki, C. Blouin, W. F. Doolittle, and A. J. Roger, "Convergence and constraint in eukaryotic release factor 1 (eRF1) domain 1: the evolution of stop codon specificity," *Nucleic Acids Research*, vol. 30, no. 2, pp. 532–544, 2002.

[38] A. Seit-Nebi, L. Frolova, and L. Kisselev, "Conversion of omnipotent translation termination factor eRF1 into ciliate-like UGA-only unipotent eRF1," *EMBO Reports*, vol. 3, no. 9, pp. 881–886, 2002.

[39] E. Cocquyt, G. H. Gile, F. Leliaert, H. Verbruggen, P. J. Keeling, and O. De Clerck, "Complex phylogenetic distribution of a non-canonical genetic code in green algae," *BMC Evolutionary Biology*, vol. 10, no. 1, article no. 327, 2010.

[40] G. Hernández, "On the origin of the cap-dependent initiation of translation in eukaryotes," *Trends in Biochemical Sciences*, vol. 34, no. 4, pp. 166–175, 2009.

[41] N. C. Kyrpides and C. R. Woese, "Universally conserved translation initiation factors," *Proceedings of the National Academy of Sciences of the United States of America*, vol. 95, no. 1, pp. 224–228, 1998.

[42] P. Londei, "Evolution of translational initiation: new insights from the archaea," *FEMS Microbiology Reviews*, vol. 29, no. 2, pp. 185–200, 2005.

[43] L. Aravind and E. V. Koonin, "Eukaryote-specific domains in translation initiation factors: implications for translation regulation and evolution of the translation system," *Genome Research*, vol. 10, no. 8, pp. 1172–1184, 2000.

[44] D. Benelli and P. Londei, "Begin at the beginning: evolution of translational initiation," *Research in Microbiology*, vol. 160, no. 7, pp. 493–501, 2009.

[45] G. Hernández and P. Vazquez-Pianzola, "Functional diversity of the eukaryotic translation initiation factors belonging to eIF4 families," *Mechanisms of Development*, vol. 122, no. 7-8, pp. 865–876, 2005.

[46] B. Joshi, K. Lee, D. L. Maeder, and R. Jagus, "Phylogenetic analysis of eIF4E-family members," *BMC Evolutionary Biology*, vol. 5, article no. 48, 2005.

[47] C. C. Baker and M. T. Fuller, "Translational control of meiotic cell cycle progression and spermatid differentiation in male germ cells by a novel eIF4G homolog," *Development*, vol. 134, no. 15, pp. 2863–2869, 2007.

[48] T. M. Franklin-Dumont, C. Chatterjee, S. A. Wasserman, and S. DiNardo, "A novel eIF4G homolog, off-schedule, couples translational control to meiosis and differentiation in Drosophila spermatocytes," *Development*, vol. 134, no. 15, pp. 2851–2861, 2007.

[49] E. R. Freire, R. Dhalia, D. M. N. Moura et al., "The four trypanosomatid eIF4E homologues fall into two separate groups, with distinct features in primary sequence and biological properties," *Molecular and Biochemical Parasitology*, vol. 176, no. 1, pp. 25–36, 2011.

[50] B. D. Keiper, B. J. Lamphear, A. M. Deshpande et al., "Functional characterization of five eIF4E isoforms in Caenorhabditis elegans," *Journal of Biological Chemistry*, vol. 275, no. 14, pp. 10590–10596, 2000.

[51] H. Miyoshi, D. S. Dwyer, B. D. Keiper, M. Jankowska-Anyszka, E. Darzynkiewicz, and R. E. Rhoads, "Discrimination between mono- and trimethylated cap structures by two isoforms of Caenorhabditis elegans eIF4E," *EMBO Journal*, vol. 21, no. 17, pp. 4680–4690, 2002.

[52] P. Lasko, "The Drosophila melanogaster genome: translation factors and RNA binding proteins," *Journal of Cell Biology*, vol. 150, no. 2, pp. F51–F56, 2000.

[53] J. Morales, O. Mulner-Lorillon, B. Cosson et al., "Translational control genes in the sea urchin genome," *Developmental Biology*, vol. 300, no. 1, pp. 293–307, 2006.

[54] M. Jankowska-Anyszka, B. J. Lamphear, E. J. Aamodt et al., "Multiple isoforms of eukaryotic protein synthesis initiation factor 4E in Caenorhabditis elegans can distinguish between mono- and trimethylated mRNA cap structures," *Journal of Biological Chemistry*, vol. 273, no. 17, pp. 10538–10541, 1998.

[55] K. S. Browning, "Plant translation initiation factors: it is not easy to be green," *Biochemical Society Transactions*, vol. 32, no. 4, pp. 589–591, 2004.

[56] L. K. Mayberry, M. Leah Allen, M. D. Dennis, and K. S. Browning, "Evidence for variation in the optimal translation initiation complex: plant eIF4B, eIF4F, and eIF(iso)4F differentially promote translation of mRNAs," *Plant Physiology*, vol. 150, no. 4, pp. 1844–1854, 2009.

[57] J. Robalino, B. Joshi, S. C. Fahrenkrug, and R. Jagus, "Two zebrafish eIF4E family members are differentially expressed

and functionally divergent," *Journal of Biological Chemistry*, vol. 279, no. 11, pp. 10532–10541, 2004.

[58] G. Hernández, M. Altmann, J. M. Sierra et al., "Functional analysis of seven genes encoding eight translation initiation factor 4E (eIF4E) isoforms in Drosophila," *Mechanisms of Development*, vol. 122, no. 4, pp. 529–543, 2005.

[59] A. V. Evsikov and C. Marín de Evsikova, "Evolutionary origin and phylogenetic analysis of the novel oocyte-specific eukaryotic translation initiation factor 4E in Tetrapoda," *Development Genes and Evolution*, vol. 219, no. 2, pp. 111–118, 2009.

[60] C. Goyer, M. Altmann, H. S. Lee et al., "TIF4631 and TIF4632: two yeast genes encoding the high-molecular-weight subunits of the cap-binding protein complex (eukaryotic initiation factor 4F) contain an RNA recognition motif-like sequence and carry out an essential function," *Molecular and Cellular Biology*, vol. 13, no. 8, pp. 4860–4874, 1993.

[61] M. Wakiyama, A. Suzuki, M. Saigoh et al., "Analysis of the isoform of Xenopus euakryotic translation initiation factor 4E," *Bioscience, Biotechnology and Biochemistry*, vol. 65, no. 1, pp. 232–235, 2001.

[62] M. Ptushkina, K. Berthelot, T. Von der Haar, L. Geffers, J. Warwicker, and J. E. G. McCarthy, "A second eIF4E protein in Schizosaccharomyces pombe has distinct eIF4G-binding properties," *Nucleic Acids Research*, vol. 29, no. 22, pp. 4561–4569, 2001.

[63] B. Joshi, A. Cameron, and R. Jagus, "Characterization of mammalian eIF4E-family members," *European Journal of Biochemistry*, vol. 271, no. 11, pp. 2189–2203, 2004.

[64] L. Li and C. C. Wang, "Identification in the ancient protist Giardia lamblia of two eukaryotic translation initiation factor 4E homologues with distinctive functions," *Eukaryotic Cell*, vol. 4, no. 5, pp. 948–959, 2005.

[65] Y. Yoffe, J. Zuberek, A. Lerer et al., "Binding specificities and potential roles of isoforms of eukaryotic initiation factor 4E in Leishmania," *Eukaryotic Cell*, vol. 5, no. 12, pp. 1969–1979, 2006.

[66] R. Dhalia, C. R. S. Reis, E. R. Freire et al., "Translation initiation in Leishmania major: characterisation of multiple eIF4F subunit homologues," *Molecular and Biochemical Parasitology*, vol. 140, no. 1, pp. 23–41, 2005.

[67] D. R. Gallie, "Translational control in plants and chloroplasts," in *Translational Control in Biology and Medicine*, M. B. Mathews, N. Sonenberg, and J. W. B. Hershey, Eds., Cold Spring Harbor Laboratory Press, Cold Spring Harbor, NY, USA, 2007.

[68] G. Hernández, P. Vázques-Pianzola, A. Zurbriggen, M. Altmann, J. M. Sierra, and R. Rivera-Pomar, "Two functionally redundant isoforms of Drosophila melanogaster eukaryotic initiation factor 4B are involved in cap-dependent translation, cell survival, and proliferation," *European Journal of Biochemistry*, vol. 271, no. 14, pp. 2923–2936, 2004.

[69] G. Hernández, M. Del Mar Castellano, M. Agudo, and J. M. Sierra, "Isolation and characterization of the cDNA and the gene for eukaryotic translation initiation factor 4G from Drosophila melanogaster," *European Journal of Biochemistry*, vol. 253, no. 1, pp. 27–35, 1998.

[70] G. W. Owttrim, T. Mandel, H. Trachsel, A. A. Thomas, and C. Kuhlemeier, "Characterization of the tobacco eIF-4A gene family," *Plant Molecular Biology*, vol. 26, no. 6, pp. 1747–1757, 1994.

[71] A. Zinoviev, M. Leger, G. Wagner, and M. Shapira, "A novel 4E-interacting protein in Leishmania is involved in stage-specific translation pathways," *Nucleic Acids Research*, vol. 39, no. 19, pp. 8404–8415, 2011.

[72] Y. Yoffe, M. Léger, A. Zinoviev et al., "Evolutionary changes in the Leishmania eIF4F complex involve variations in the eIF4E-eIF4G interactions," *Nucleic Acids Research*, vol. 37, no. 10, pp. 3243–3253, 2009.

[73] P. F. Cho, F. Poulin, Y. A. Cho-Park et al., "A new paradigm for translational control: inhibition via 5′-3′ mRNA tethering by Bicoid and the eIF4E cognate 4EHP," *Cell*, vol. 121, no. 3, pp. 411–423, 2005.

[74] T. D. Dinkova, B. D. Keiper, N. L. Korneeva, E. J. Aamodt, and R. E. Rhoads, "Translation of a small subset of Caenorhabditis elegans mRNAs is dependent on a specific eukaryotic translation initiation factor 4E isoform," *Molecular and Cellular Biology*, vol. 25, no. 1, pp. 100–113, 2005.

[75] R. E. Rhoads, "EIF4E: new family members, new binding partners, new roles," *Journal of Biological Chemistry*, vol. 284, no. 25, pp. 16711–16715, 2009.

[76] K. A. Ruud, C. Kuhlow, D. J. Goss, and K. S. Browning, "Identification and characterization of a novel cap-binding protein from Arabidopsis thaliana," *Journal of Biological Chemistry*, vol. 273, no. 17, pp. 10325–10330, 1998.

[77] E. Rom, H. C. Kim, A. C. Gingras et al., "Cloning and characterization of 4EHP, a novel mammalian eIF4E-related cap-binding protein," *Journal of Biological Chemistry*, vol. 273, no. 21, pp. 13104–13109, 1998.

[78] J. C. Villaescusa, C. Buratti, D. Penkov et al., "Cytoplasmic Prep1 interacts with 4EHP inhibiting Hoxb4 translation," *PLoS One*, vol. 4, no. 4, Article ID e5213, 2009.

[79] N. Minshall, M. H. Reiter, D. Weil, and N. Standart, "CPEB interacts with an ovary-specific eIF4E and 4E-T in early Xenopus oocytes," *Journal of Biological Chemistry*, vol. 282, no. 52, pp. 37389–37401, 2007.

[80] S. Lall, C. C. Friedman, M. Jankowska-Anyszka, J. Stepinski, E. Darzynkiewicz, and R. E. Davis, "Contribution of trans-splicing, 5′-leader length, cap-poly(A) synergism, and initiation factors to nematode translation in an Ascaris suum embryo cell-free system," *Journal of Biological Chemistry*, vol. 279, no. 44, pp. 45573–45585, 2004.

[81] M. Altmann, C. Handschin, and H. Trachsel, "mRNA cap-binding protein: cloning of the gene encoding protein synthesis initiation factor eIF-4E from Saccharomyces cerevisiae," *Molecular and Cellular Biology*, vol. 7, no. 3, pp. 998–1003, 1987.

[82] F. G. Maroto and J. M. Sierra, "Purification and characterization of mRNA cap-binding protein from Drosophila melanogaster embryos," *Molecular and Cellular Biology*, vol. 9, no. 5, pp. 2181–2190, 1989.

[83] C. M. Rodriguez, M. A. Freire, C. Camilleri, and C. Robaglia, "The Arabidosis thaliana cDNAs encoding for eIF4E and eIF(iso)4E are not functionally equivalent for yeast complementation and are differentially expressed during plant development," *Plant Journal*, vol. 13, no. 4, pp. 465–473, 1998.

[84] N. Sonenberg, "eIF4E, the mRNA cap-binding protein: from basic discovery to translational research," *Biochemistry and Cell Biology*, vol. 86, no. 2, pp. 178–183, 2008.

[85] L. Rong, M. Livingstone, R. Sukarieh et al., "Control of eIF4E cellular localization by eIF4E-binding proteins, 4E-BPs," *RNA*, vol. 14, no. 7, pp. 1318–1327, 2008.

[86] P. L. Graham, J. L. Yanowitz, J. K. M. Penn, G. Deshpande, and P. Schedl, "The translation initiation factor eif4e regulates the Sex-Specific expression of the master switch gene

Sxl in Drosophila melanogaster," *PLoS Genetics*, vol. 7, no. 7, Article ID e1002185, 2011.

[87] A. Amiri, B. D. Keiper, I. Kawasaki et al., "An isoform of eIF4E is a component of germ granules and is required for spermatogenesis in *C. elegans*," *Development*, vol. 128, no. 20, pp. 3899–3912, 2001.

[88] M. A. Henderson, E. Croniand, S. Dunkelbarger, V. Contreras, S. Strome, and B. D. Keiper, "A germline-specific isoform of eIF4E (IFE-1) is required for efficient translation of stored mRNAs and maturation of both oocytes and sperm," *Journal of Cell Science*, vol. 122, no. 10, pp. 1529–1539, 2009.

[89] I. Kawasaki, M. H. Jeong, and Y. H. Shim, "Regulation of sperm-specific proteins by IFE-1, a germline-specific homolog of eIF4E, in C. elegans," *Molecules and Cells*, vol. 31, no. 2, pp. 191–197, 2011.

[90] A. Song, S. Labella, N. L. Korneeva et al., "A *C. elegans* eIF4E-family member upregulates translation at elevated temperatures of mRNAs encoding MSH-5 and other meiotic crossover proteins," *Journal of Cell Science*, vol. 123, no. 13, pp. 2228–2237, 2010.

[91] G. Hernández, V. Gandin, H. Han, T. Ferreira, N. Sonenberg, and P. Lasko, "Translational control by Drosophila eIF4E-3 is essential for cell differentiation during spermiogenesis," *Development*. In press.

[92] J. M. Zapata, M. A. Martinez, and J. M. Sierra, "Purification and characterization of eukaryotic polypeptide chain initiation factor 4F from Drosophila melanogaster embryos," *Journal of Biological Chemistry*, vol. 269, no. 27, pp. 18047–18052, 1994.

[93] V. Contreras, M. A. Richardson, E. Hao, and B. D. Keiper, "Depletion of the cap-associated isoform of translation factor eIF4G induces germline apoptosis in *C. elegans*," *Cell Death and Differentiation*, vol. 15, no. 8, pp. 1232–1242, 2008.

[94] D. Prévôt, J. L. Darlix, and T. Ohlmann, "Conducting the initiation of protein synthesis: the role of eIF4G," *Biology of the Cell*, vol. 95, no. 3-4, pp. 141–156, 2003.

[95] T. V. Pestova, I. N. Shatsky, and C. U. T. Hellen, "Functional dissection of eukaryotic initiation factor 4F: the 4A subunit and the central domain of the 4G subunit are sufficient to mediate internal entry of 43S preinitiation complexes," *Molecular and Cellular Biology*, vol. 16, no. 12, pp. 6870–6878, 1996.

[96] A. Gradi, H. Imataka, Y. V. Svitkin et al., "A novel functional human eukaryotic translation initiation factor 4G," *Molecular and Cellular Biology*, vol. 18, no. 1, pp. 334–342, 1998.

[97] S. Caron, M. Charon, E. Cramer, N. Sonenberg, and I. Dusanter-Fourt, "Selective modification of eukaryotic initiation factor 4F (eIF4F) at the onset of cell differentiation: recruitment of eIF4GII and long-lasting phosphorylation of eIF4E," *Molecular and Cellular Biology*, vol. 24, no. 11, pp. 4920–4928, 2004.

[98] F. Sun, K. Palmer, and M. A. Handel, "Mutation of Eif4g3, encoding a eukaryotic translation initiation factor, causes male infertility and meiotic arrest of mouse spermatocytes," *Development*, vol. 137, no. 10, pp. 1699–1707, 2010.

[99] A. G. Hinnebusch, "eIF3: a versatile scaffold for translation initiation complexes," *Trends in Biochemical Sciences*, vol. 31, no. 10, pp. 553–562, 2006.

[100] J. Guo, Z. Jin, X. Yang, J. F. Li, and J. G. Chen, "Eukaryotic initiation factor 6, an evolutionarily conserved regulator of ribosome biogenesis and protein translation," *Plant Signaling and Behavior*, vol. 6, no. 5, pp. 766–771, 2011.

[101] A. L. Stevenson and J. E. G. McCarthy, "Found in translation: another RNA helicase function," *Molecular Cell*, vol. 32, no. 6, pp. 755–756, 2008.

[102] A. Parsyan, Y. Svitkin, D. Shahbazian et al., "MRNA helicases: the tacticians of translational control," *Nature Reviews Molecular Cell Biology*, vol. 12, no. 4, pp. 235–245, 2011.

[103] P. Linder and E. Jankowsky, "From unwinding to clamping—the DEAD box RNA helicase family," *Nature Reviews Molecular Cell Biology*, vol. 12, no. 8, pp. 505–516, 2011.

[104] A. Weston and J. Sommerville, "Xp54 and related (DDX6-like) RNA helicases: roles in messenger RNP assembly, translation regulation and RNA degradation," *Nucleic Acids Research*, vol. 34, no. 10, pp. 3082–3094, 2006.

[105] M. S. Bush, A. P. Hutchins, A. M. E. Jones et al., "Selective recruitment of proteins to 5′ cap complexes during the growth cycle in Arabidopsis," *Plant Journal*, vol. 59, no. 3, pp. 400–412, 2009.

[106] J. D. Richter and N. Sonenberg, "Regulation of cap-dependent translation by eIF4E inhibitory proteins," *Nature*, vol. 433, no. 7025, pp. 477–480, 2005.

[107] C. G. Proud, "Signalling to translation: how signal transduction pathways control the protein synthetic machinery," *Biochemical Journal*, vol. 403, no. 2, pp. 217–234, 2007.

[108] B. Raught and A. C. Gingras, "Signaling to translation initiation," in *Translational Control in Biology and Medicine*, M. B. Mathews, N. Sonenberg, and J. W. B. Hershey, Eds., Cold Spring Harbor Laboratory Press, Cold Spring Harbor, NY, USA, 2007.

[109] L. Furic, L. Rong, O. Larsson et al., "EIF4E phosphorylation promotes tumorigenesis and is associated with prostate cancer progression," *Proceedings of the National Academy of Sciences of the United States of America*, vol. 107, no. 32, pp. 14134–14139, 2010.

[110] G. C. Scheper and C. G. Proud, "Does phosphorylation of the cap-binding protein eIF4E play a role in translation initiation?" *European Journal of Biochemistry*, vol. 269, no. 22, pp. 5350–5359, 2002.

[111] M. Buxade, J. L. Parra-Palau, and C. G. Proud, "The Mnks: MAP kinase-interacting kinases (MAP kinase signal-integrating kinases).," *Frontiers in Bioscience*, vol. 13, pp. 5359–5373, 2008.

[112] J. H. Reiling, K. T. Doepfner, E. Hafen, and H. Stocker, "Diet-dependent effects of the Drosophila Mnk1/Mnk2 homolog Lk6 on growth via eIF4E," *Current Biology*, vol. 15, no. 1, pp. 24–30, 2005.

[113] P. E. D. Lachance, M. Miron, B. Raught, N. Sonenberg, and P. Lasko, "Phosphorylation of eukaryotic translation initiation factor 4E is critical for growth," *Molecular and Cellular Biology*, vol. 22, no. 6, pp. 1656–1663, 2002.

[114] N. Arquier, M. Bourouis, J. Colombani, and P. Léopold, "Drosophila Lk6 kinase controls phosphorylation of eukaryotic translation initiation factor 4E and promotes normal growth and development," *Current Biology*, vol. 15, no. 1, pp. 19–23, 2005.

[115] J. L. Parra-Palau, G. C. Scheper, D. E. Harper, and C. G. Proud, "The Drosophila protein kinase LK6 is regulated by ERK and phosphorylates the eukaryotic initiation factor eIF4E in vivo," *Biochemical Journal*, vol. 385, no. 3, pp. 695–702, 2005.

[116] I. R. E. Nett, D. M. A. Martin, D. Miranda-Saavedra et al., "The phosphoproteome of bloodstream form Trypanosoma brucei, causative agent of African sleeping sickness," *Molecular and Cellular Proteomics*, vol. 8, no. 7, pp. 1527–1538, 2009.

[117] N. I. T. Zanchin and J. E. G. McCarthy, "Characterization of the in vivo phosphorylation sites of the mRNA·Cap- binding complex proteins eukaryotic initiation factor-4E and p20 in Saccharomyces cerevisiae," *Journal of Biological Chemistry*, vol. 270, no. 44, pp. 26505–26510, 1995.

[118] T. E. Dever, A. C. Dar, and F. Sicheri, "The eIF2a kinases," in *Translational Control in Biology and Medicine*, M. B. Mathews, N. Sonenberg, and J. W. B. Hershey, Eds., Cold Spring Harbor Laboratory Press, Cold Spring Harbor, NY, USA, 2007.

[119] S. Rothenburg, N. Deigendesch, M. Dey, T. E. Dever, and L. Tazi, "Double-stranded RNA-activated protein kinase PKR of fishes and amphibians: varying the number of double-stranded RNA binding domains and lineage-specific duplications.," *BMC biology*, vol. 6, article 12, 2008.

[120] M. Szymański, M. Deniziak, and J. Barciszewski, "The new aspects of aminoacyl-tRNA synthetases," *Acta Biochimica Polonica*, vol. 47, no. 3, pp. 821–834, 2000.

[121] M. Guo, X. L. Yang, and P. Schimmel, "New functions of aminoacyl-tRNA synthetases beyond translation," *Nature Reviews Molecular Cell Biology*, vol. 11, no. 9, pp. 668–674, 2010.

[122] P. S. Ray, J. C. Sullivan, J. Jia, J. Francis, J. R. Finnerty, and P. L. Fox, "Evolution of function of a fused metazoan tRNA synthetase," *Molecular Biology and Evolution*, vol. 28, no. 1, pp. 437–447, 2011.

[123] P. L. Fox, P. S. Ray, A. Arif, and J. Jia, "Noncanonical functions of aminoacyl-tRNA synthetases in translational control," in *Translational Control in Biology and Medicine*, M. B. Mathews, N. Sonenberg, and J. W. B. Hershey, Eds., Cold Spring Harbor Laboratory Press, Cold Spring Harbor, NY, USA, 2007.

[124] P. J. Keeling and Y. Inagaki, "A class of eukaryotic GTPase with a punctate distribution suggesting multiple functional replacements of translation elongation factor 1α," *Proceedings of the National Academy of Sciences of the United States of America*, vol. 101, no. 43, pp. 15380–15385, 2004.

[125] M. Sakaguchi, K. Takishita, T. Matsumoto, T. Hashimoto, and Y. Inagaki, "Tracing back EFL gene evolution in the cryptomonads-haptophytes assemblage: separate origins of EFL genes in haptophytes, photosynthetic cryptomonads, and goniomonads," *Gene*, vol. 441, no. 1-2, pp. 126–131, 2009.

[126] G. H. Gile, P. M. Novis, D. S. Cragg, G. C. Zuccarello, and P. J. Keeling, "The distribution of elongation factor-1 alpha (EF-1α), elongation factor-like (EFL), and a non-canonical genetic code in the ulvophyceae: Discrete genetic characters support a consistent phylogenetic framework," *Journal of Eukaryotic Microbiology*, vol. 56, no. 4, pp. 367–372, 2009.

[127] E. Cocquyt, H. Verbruggen, F. Leliaert, F. W. Zechman, K. Sabbe, and O. De Clerck, "Gain and loss of elongation factor genes in green algae," *BMC Evolutionary Biology*, vol. 9, no. 1, article no. 39, 2009.

[128] G. P. Noble, M. B. Rogers, and P. J. Keeling, "Complex distribution of EFL and EF-1α proteins in the green algal lineage," *BMC Evolutionary Biology*, vol. 7, article no. 82, 2007.

[129] R. Kamikawa, Y. Inagaki, and Y. Sako, "Direct phylogenetic evidence for lateral transfer of elongation factor-like gene," *Proceedings of the National Academy of Sciences of the United States of America*, vol. 105, no. 19, pp. 6965–6969, 2008.

[130] R. Kamikawa, A. Yabuki, T. Nakayama, K. I. Ishida, T. Hashimoto, and Y. Inagaki, "Cercozoa comprises both EF-1α-containing and EFL-containing members," *European Journal of Protistology*, vol. 47, no. 1, pp. 24–28, 2011.

[131] R. Kamikawa, M. Sakaguchi, T. Matsumoto, T. Hashimoto, and Y. Inagaki, "Rooting for the root of elongation factor-like protein phylogeny," *Molecular Phylogenetics and Evolution*, vol. 56, no. 3, pp. 1082–1088, 2010.

[132] G. H. Gile, D. Faktorová, C. A. Castlejohn et al., "Distribution and phylogeny of EFL and EF1alpha in Euglenozoa suggest ancestral co-occurrence followed by differential loss," *PLoS One*, vol. 4, no. 4, Article ID e5162, 2009.

[133] M. Djernaes and J. Damgaard, "Exon-intron structure, paralogy and sequenced regions of elongation factor-1 alpha in Hexapoda," *Arthropod Systematics and Phylogeny*, vol. 64, no. 1, pp. 45–52, 2006.

[134] E. Greganova, M. Altmann, and P. Bütikofer, "Unique modifications of translation elongation factors," *FEBS Journal*, vol. 278, no. 15, pp. 2613–2624, 2011.

[135] O. T. P. Kim, K. Yura, N. Go, and T. Harumoto, "Newly sequenced eRF1s from ciliates: the diversity of stop codon usage and the molecular surfaces that are important for stop codon interactions," *Gene*, vol. 346, pp. 277–286, 2005.

[136] D. Moreira, S. Kervestin, O. Jean-Jean, and H. Philippe, "Evolution of eukaryotic translation elongation and termination factors: variations of evolutionary rate and genetic code deviations," *Molecular Biology and Evolution*, vol. 19, no. 2, pp. 189–200, 2002.

[137] G. C. Atkinson, S. L. Baldauf, and V. Hauryliuk, "Evolution of nonstop, no-go and nonsense-mediated mRNA decay and their termination factor-derived components," *BMC Evolutionary Biology*, vol. 8, no. 1, article no. 290, 2008.

[138] Y. Inagaki and W. F. Doolittle, "Class I release factors in ciliates with variant genetic codes," *Nucleic Acids Research*, vol. 29, no. 4, pp. 921–927, 2001.

[139] H. Song, P. Mugnier, A. K. Das et al., "The crystal structure of human eukaryotic release factor eRF1—mechanism of stop codon recognition and peptidyl-tRNA hydrolysis," *Cell*, vol. 100, no. 3, pp. 311–321, 2000.

[140] P. Kolosov, L. Frolova, A. Seit-Nebi et al., "Invariant amino acids essential for decoding function of polypeptide release factor eRF1," *Nucleic Acids Research*, vol. 33, no. 19, pp. 6418–6425, 2005.

[141] K. Ito, L. Frolova, A. Seit-Nebi, A. Karamyshev, L. Kisselev, and Y. Nakamura, "Omnipotent decoding potential resides in eukaryotic translation termination factor eRF1 of variant-code organisms and is modulated by the interactions of amino acid sequences within domain 1," *Proceedings of the National Academy of Sciences of the United States of America*, vol. 99, no. 13, pp. 8494–8499, 2002.

[142] Y. Inagaki and W. F. Doolittle, "Evolution of the eukaryotic translation termination system: origins of release factors," *Molecular Biology and Evolution*, vol. 17, no. 6, pp. 882–889, 2000.

[143] G. Zhouravleva, V. Schepachev, A. Petrova, O. Tarasov, and S. Inge-Vechtomov, "Evolution of translation termination factor eRF3: is GSPT2 generated by retrotransposition of GSPT1's mRNA?" *IUBMB Life*, vol. 58, no. 4, pp. 199–202, 2006.

[144] S. I. Hoshino, M. Imai, M. Mizutani et al., "Molecular cloning of a novel member of the eukaryotic polypeptide chain-releasing factors (eRF): its identification as eRF3 interacting with eRF1," *Journal of Biological Chemistry*, vol. 273, no. 35, pp. 22254–22259, 1998.

[145] C. L. Goff, O. Zemlyanko, S. Moskalenko et al., "Mouse GSPT2, but not GSPT1, can substitute for yeast eRF3 in vivo," *Genes to Cells*, vol. 7, no. 10, pp. 1043–1057, 2002.

[146] L. Skogerson and E. Wakatama, "A ribosome dependent GTPase from yeast distinct from elongation factor 2," *Proceedings of the National Academy of Sciences of the United States of America*, vol. 73, no. 1, pp. 73–76, 1976.

[147] M. F. Ypma-Wong, W. A. Fonzi, and P. S. Sypherd, "Fungus-specific translation elongation factor 3 gene present in *Pneumocystis carinii*," *Infection and Immunity*, vol. 60, no. 10, pp. 4140–4145, 1992.

[148] S. Qin, A. Xie, M. C. M. Bonato, and C. S. McLaughlin, "Sequence analysis of the translational elongation factor 3 from Saccharomyces cerevisiae," *Journal of Biological Chemistry*, vol. 265, no. 4, pp. 1903–1912, 1990.

[149] B. J. Di Domenico, J. Lupisella, M. Sandbaken, and K. Chakraburtty, "Isolation and sequence analysis of the gene encoding translation elongation factor 3 from Candida albicans," *Yeast*, vol. 8, no. 5, pp. 337–352, 1992.

[150] L. Skogerson, "Separation and characterization of yeast elongation factors," *Methods in Enzymology*, vol. 60, no. C, pp. 676–685, 1979.

[151] K. Chakraburtty and F. J. Triana-Alonso, "Yeast elongation factor 3: structure and function," *Biological Chemistry*, vol. 379, no. 7, pp. 831–840, 1998.

[152] S. Kurata, K. H. Nielsen, S. F. Mitchell, J. R. Lorsch, A. Kaji, and H. Kaji, "Ribosome recycling step in yeast cytoplasmic protein synthesis is catalyzed by eEF3 and ATP," *Proceedings of the National Academy of Sciences of the United States of America*, vol. 107, no. 24, pp. 10854–10859, 2010.

[153] F. A. Bisby, Y. R. Roskov, T. M. Orrell, D. Nicolson, L. E. Paglinawan et al., "Species 2000 & ITIS Catalogue of Life: 2010 Annual Checklist," 2010, http://www.catalogueoflife.org/annual-checklist/2010.

[154] C. Mora, D. P. Titterson, S. Adl, A. G. B. Simpson, and B. Worm, "How many species are there on Earth and in the ocean," *PLoS Biology*, vol. 9, no. 8, Article ID e1001127, 2011.

[155] S. G. Tringe, C. Von Mering, A. Kobayashi et al., "Comparative metagenomics of microbial communities," *Science*, vol. 308, no. 5721, pp. 554–557, 2005.

[156] A. C. McHardy and I. Rigoutsos, "What's in the mix: phylogenetic classification of metagenome sequence samples," *Current Opinion in Microbiology*, vol. 10, no. 5, pp. 499–503, 2007.

[157] J. C. Venter, K. Remington, J. F. Heidelberg et al., "Environmental genome shotgun sequencing of the Sargasso Sea," *Science*, vol. 304, no. 5667, pp. 66–74, 2004.

13

Epigenetic Alterations in Muscular Disorders

Chiara Lanzuolo

CNR Institute of Cellular Biology and Neurobiology, IRCCS Santa Lucia Foundation, Via Del Fosso di Fiorano 64, 00143 Rome, Italy

Correspondence should be addressed to Chiara Lanzuolo, chiara.lanzuolo@inmm.cnr.it

Academic Editor: Daniela Palacios

Epigenetic mechanisms, acting via chromatin organization, fix in time and space different transcriptional programs and contribute to the quality, stability, and heritability of cell-specific transcription programs. In the last years, great advances have been made in our understanding of mechanisms by which this occurs in normal subjects. However, only a small part of the complete picture has been revealed. Abnormal gene expression patterns are often implicated in the development of different diseases, and thus epigenetic studies from patients promise to fill an important lack of knowledge, deciphering aberrant molecular mechanisms at the basis of pathogenesis and diseases progression. The identification of epigenetic modifications that could be used as targets for therapeutic interventions could be particularly timely in the light of pharmacologically reversion of pathological perturbations, avoiding changes in DNA sequences. Here I discuss the available information on epigenetic mechanisms that, altered in neuromuscular disorders, could contribute to the progression of the disease.

1. Introduction

Although every cell within our body bears the same genetic information, only a small subset of genes is transcribed in a given cell at a given time. The distinct gene expression of genetically identical cells is responsible for cell phenotype and depends on the epigenome, which involve all structural levels of chromosome organization from DNA methylation and histone modifications up to nuclear compartmentalization of chromatin [1–5]. Enormous progress over the last few years in the field of epigenetic regulation indicated that the primary, monodimensional structure of genetic information is insufficient for a complete understanding of how the networking among regulatory regions actually works. The contribution of additional coding levels hidden in the three-dimensional structure of the chromosome and nuclear structures appears to be a fundamental aspect for the control of the quality and stability of genetic programs. Damage or perturbation of epigenetic components may lead to deviations from a determined cellular program, resulting in severe developmental disorders and tumour progression [6, 7]. Moreover, for human complex diseases, the phenotypic differences and the severity of the disease observed among patients could be attributable to inter-individual epigenomic variation. Unravelling the intricacies of the epigenome will be a complex process due to the enormity and dynamic nature of the epigenomic landscape but is essential to gain insights into the aetiology of complex diseases.

2. The Complexity of the Epigenome

The epigenome consists of multiple mechanisms of transcriptional regulation that establish distinct layers of genome organization and includes covalent modification of DNA and histones, packaging of DNA around nucleosomes, higher-order chromatin interactions, and nuclear positioning [4]. The first layer of epigenetic control is the DNA methylation, an heritable epigenetic mark typically associated with a repressed chromatin state [8], which seems to play a role, together with other histone modifications, in preventing gene reactivation [9]. Vertebrate genomes are predominantly methylated at cytosine of the dinucleotide sequence CpG (for a review see [3]). Despite the high level of CpG methylation, some regions of mammalian genomes are refractory to this modification [10]. These regions, called CpG islands, contain high levels of CpG dinucleotides [11] and localize at or near gene promoters [12], suggesting a strong correlation between differential methylation of CpG islands and flanking promoter activity. From the mechanistic point of view, DNA methylation can inhibit gene expression by blocking the access of transcriptional activators to their binding site on DNA or by recruiting chromatin modifying activities to DNA (for a review see [3]). For long time, DNA methylation was considered as a stable epigenetic mark. However, recently it has been shown that methylated cytosines could be converted to 5-hydroxymethylcytosines (5hmeC) by Tet (Ten eleven Translocation) family proteins [13–15] and the generation of 5hmeC is a necessary intermediate step preceding active demethylation of DNA [16]. The second level of epigenetic regulation occurs through posttranslational histone modification. Histone proteins assemble into a complex that associates with DNA forming the elementary unit of chromatin packaging: the nucleosome. The amino and carboxy termini of the histones (histone tails), protruding from the nucleosome, play an essential role in controlling gene expression, being the target for posttranscriptional modifications, including acetylation, methylation, phosphorylation, ubiquitylation, biotinylation, and several others (for a review see [2, 17]). Multiple histone modifications can also coexist on the same tail, dictating specific biological readouts [18–24]. In addition to histone modifications, a fraction of chromatin contains one or more variant isoforms of the canonical histones that can be incorporated into specific regions of the genome throughout the cell cycle and are essential for the epigenetic control of gene expression and other cellular responses (for a review see [25]). Combinatorial histone modifications and variants play an important role in folding nucleosomal arrays into higher-order chromatin structures, creating local structural and functional diversity and delimiting chromatin subdomains then subjected to a specific protein environment. Chromatin higher-order structures established at DNA level give signals that are recognized by specific binding proteins that in turn influence gene expression and other chromatin functions [1, 26]. This represents an additional layer of epigenetic gene regulation and includes factors, such as transcriptional repressors or activators, that recognizing specific chromatin patterns regulate the folding or modulate the activity of RNA Polymerase II (Pol II).

The topological organization of chromatin and the association of regulatory elements with specific components of the eukaryotic nucleus is another parameter to be considered in the complexity of the epigenetic information. It is now clear that specific chromosomal conformations, mediated by cis-trans interactions, are associated with distinct transcriptional states in many organisms, allowing the establishment of chromatin boundaries between promoters and regulatory element (for a review see [27, 28]). The nuclear localization also influences gene expression, regulating its access to specific machinery responsible for specific functions, such as transcription or replication [29, 30]. In addition, due to its highly dynamic nature, the genome moves in the nucleus driving specific genomic regions toward nuclear compartments defined by a high concentration of specific factors and substrates that facilitate more efficient biological reactions [31]. This constant motion plays also a role in coordinating the expression of coregulated genes, separated by longer chromosomal regions or located on different chromosomes [32].

The evolutionarily conserved Polycomb group of proteins (PcG) are multiprotein complexes that play a central role during development [1]. The most characterized PcG-encoded protein complexes are Polycomb Repressive Complex 1 (PRC1) and 2 (PRC2). Three other complexes were characterized in *Drosophila*, PHO-repressive complex (PhoRC), dRing-associated factors (dRAF) complex, and Polycomb repressive deubiquitinase (PR-DUB), and their components have orthologues in mammals [33, 34]. PcG complexes mediate gene silencing by regulating different levels of chromatin structures. Biochemical studies revealed that Enhancer of zeste 2 (EZH2), the Histone Methyl Transferase (HMTase) subunit of PRC2, marks lysine 27 of histone H3 [35–38] and PRC1 complex monoubiquitylates Lys 119 of histone H2A [39]. Moreover, the H3K27me3 mark constitutes a docking site for the chromodomain present in PRC1 components [35], determining a sequential recruitment of PRC complexes, although recent chromatin profiling studies evidenced that PRC1 and PRC2 also have targets independent of each other [40, 41]. Examination of the localization of PcG proteins in the nucleus has revealed that they are organized into distinct domains called Polycomb or PcG bodies, which are often localized, closed to pericentric heterochromatin [42]. PcG targets are frequently localized in PcG bodies in the tissue where they are repressed, suggesting that such nuclear localization may be required for efficient silencing [43, 44]. However, the number of PcG bodies is less than the number of PcG target genes, implying that several PcG targets share the same body. FISH studies together with Chromosome Conformation Capture (3C) analysis have confirmed this coassociation [43–46] and revealed that PcG-dependent higher-order structures organization is conserved in mammals [47–49]. The characteristic feature of the PcG memory system is inheritability of gene expression patterns throughout the cell cycle, ensured by the PcG capability to bind its own methylation mark [50, 51] and specific cell cycle-dependent dynamics [52–55]. Besides their extensively described role in development, in the last years emerging evidence has shown PcG involvement in several other biological processes, such as X chromosome inactivation, differentiation, and reprogramming (reviewed in [56–58]). The highly variability of PcG functions and the fine quantitative and qualitative tuning of their activities is generated by the association of different PcG proteins and their coregulators in a combinatorial fashion and/or by the regulation of their recruitment at specific chromatin sites (reviewed in [1, 59, 60]). One recent example is a genomewide study of TET complex localization, in murine Embryonic Stem (ES) cells. This complex, responsible for 5hmeC generation, colocalizes with a subpopulation of Polycomb-repressed genes, contributing to gene transcription control [61–63].

3. Muscle Diseases

Skeletal muscles are composed by multiple aligned multinucleated cells, the myofibers, wrapped in a plasma membrane called sarcolemma. Inside the sarcolemma and all around the myofibers, there is a specialized cytoplasm, the sarcoplasm, that contains the usual subcellular elements [64]. A plethora of structural molecules and cellular proteins connecting all the fibers components together with specialized signalling pathways and transcription factors are required for a correct muscle formation and function. Dysfunction or lack of any component of the skeletal muscle could lead to a muscular disorder, the muscular dystrophy (MD), clinically characterized by muscle weakness and skeletal muscle degeneration [64]. Some dystrophies arise from mutations of molecules that play a role outside the nucleus while other dystrophies derive from dysfunction of the nucleus or its membrane. The nonnuclear dystrophies include Duchenne MD (DMD), Becker MD (BMD), and all MD affecting proteins working in the sarcoplasm. DMD is the most severe form of muscular dystrophy and is caused by mutations that preclude the production of the essential cytoskeletal muscle protein dystrophin, which anchors proteins from the internal cytoskeleton to a complex of proteins (dystrophin-associated protein complex, DAPC) on the membrane of muscle fibers [65]. This interaction is important for the structural stabilization of the sarcolemma [66]. Interestingly, recent reports highlighted the influence of epigenetic mechanisms regulating histone deacetylation (HDAC) pathways in the development of this disease [67–69] and the reversion of some DMD-associated phenotypes in presence of inhibitors of HDACs [70, 71].

The nuclear dystrophies include all MD generated by a dysfunction of nuclear membrane (laminopathies) or by expansion or contraction of nucleotide repeats, not necessarily contained in a coding region, which affect nuclear function. Myotonic dystrophy is the most common MD in adult and is a complex multisystemic inherited muscle degenerative disorder caused by a pathogenic expansion of microsatellite repeats within noncoding elements of dystrophia myotonica protein kinase (*DMPK*) or zinc finger protein 9 (*ZNF9*) genes [72]. These expansions, although transcribed into RNA, do not affect the protein-coding region of any other gene. However, it has been shown that transcripts accumulate in the nucleus and interfere with protein families that regulate alternative splicing during development [64, 73]. In this paper, I will describe the contribution of epigenetic mechanisms mediated by Polycomb group of proteins to two human nuclear muscular dystrophies, facioscapulohumeral muscular dystrophy (FSHD), and laminopathies.

4. Polycomb Group of Protein as Epigenetic Regulators of Muscle Differentiation

PcG proteins regulate large numbers of target genes, primarily those involved in differentiation and development [74–80]. During cell differentiation the progressive restriction of the developmental potential and increased structural and functional specialization of cells ensure the formation of tissues and organs [57]. Myogenesis is a multistep process that starts with the commitment of multipotent mesodermal precursor cells. Upon appropriate stimuli these cells differentiate and fuse into multinucleated myotubes, giving rise to the myofibers. In mammals, PcG proteins are primarily involved in muscle differentiation by binding and repressing muscle-specific gene regulatory regions in undifferentiated myoblasts to prevent premature transcription. During myogenesis progression, PcG binding and H3K27me3 are lost at muscle-specific loci, resulting in appropriate muscle gene expression [81–84]. Interestingly, artificial modulation of EZH2 levels, either by depletion or overexpression, consistently affects normal muscle differentiation, accelerating or delaying, respectively, muscle cell fate determination [82, 83, 85]. Although emerging evidence suggested a key role for epigenetic mechanisms in muscular diseases [68, 71, 86–89], the precise contribution of Polycomb proteins to the pathology and progression remains largely unexplored.

5. Facioscapulohumeral Muscular Dystrophy (FSHD)

Facioscapulohumeral muscular dystrophy (FSHD) is a frequent (1 : 15.000) dominant autosomal miopathy that is characterized by progressive, often asymmetric weakness and wasting of facial (facio), shoulder, and upper arm (scapulohumeral) muscles [90]. Monozygotic twins with different penetrance of FSHD have been described, suggesting a strong epigenetic contribution to the pathology [91, 92]. Genetically, FSHD1, one of the two forms of FSHD, is caused by a contraction of the highly polymorphic D4Z4 macrosatellite repeat in chromosome 4q [93]. In the general population, this repeat array varies between 11 and 100 units of 3,3 kb each, ordered head to tail [94]. Most patients with FSHD1 present a partial deletion of the D4Z4 array, which leaves 1–10 units on the affected allele [93]. Although a linear negative correlation between repeat size and clinical severity has not been observed, some findings indicated that smaller D4Z4 arrays result in earlier disease onset and enhanced severity in patients [95–97]. Interestingly, at least one D4Z4 unit is necessary to develop FSHD, as monosomy of 4q does not cause the disease [98]. In addition to polymorphism associated with D4Z4 repeat number, two allelic variants of the 4q subtelomere, termed 4qA and 4qB, have been identified. These variants differ for the presence of a β satellite repeat immediately distal to the D4Z4 array on 4qA allele [99]. Whereas 4qA and 4qB chromosomes are almost equally common in the population, FSHD arises mainly from 4qA haplotype [99–102]. D4Z4 repeat arrays are not restricted to chromosome 4q, but homologous sequences have been identified on many chromosomes [103]. In particular, the subtelomere of chromosome 10q is almost identical to the region in 4q containing D4Z4 repeats, containing highly homologous and equally polymorphic repeat arrays [104, 105]. However, chromosome 10 with less than 11 repeat units does not cause FSHD1 [106], suggesting that the chromatin environment associated with chromosome 4q and/or 4q-specific DNA sequences could contribute to FSHD development. In agreement with this observation, the relatively gene-poor region flanking D4Z4 repeats on chromosome 4q contains two attractive candidates that have been characterized for their contribution to disease development: FRG1 (FSHD Region Gene 1) and the double-homeobox transcription factor DUX4. FRG1 is highly conserved in both vertebrates and invertebrates and it has been found overexpressed in some FSHD samples [107, 108]. Moreover, transgenic mice overexpressing FRG1 develop, selectively in the skeletal muscle, pathologies with physiological, histological, ultrastructural, and molecular features that mimic human FSHD [109]. However, FRG1 overexpression in FSHD samples is not a uniform finding [110, 111] and thus the contribution of the FRG1 gene to the FSHD phenotype needs further validation. Although some evidence suggests a role for FRG1 in pre-mRNA splicing [109, 112, 113], to date the mechanism of action and the role of FRG1 in FSHD onset and development is largely unknown.

Aberrant production of DUX4, the gene present in the D4Z4 array, was detected in both FSHD1 and FSHD2 muscle biopses [114], suggesting that D4Z4 could affect disease progression [115]. However, D4Z4 array has a complex transcriptional profile that includes sense and antisense transcripts and RNA processing [116]. The DUX4 mRNA is generated by transcription of the last, most distal, unit of the array, including a region named pLAM, which contains a polyadenylation signal, necessary for DUX4 transcript stabilization [115]. The absence of this polyadenylation signal on chromosome 10 suggests its involvement in FSHD development [100]. The DUX4 pre-mRNA can be alternatively spliced [116] and there has been found a DUX4 mRNA isoform encoding for the full-length protein, expressed in FSHD muscle, whereas healthy subjects present an alternative splicing mRNA encoding for a truncated protein [114].

DUX4 RNA and protein levels have been arguments of debate in the field for several years. Previous works demonstrated a proapoptotic function for DUX4 [117] and DUX4 overexpression was found to have dramatically toxic effect on cell growth [118]. On the other hand extremely low levels of DUX4 were found in FSHD muscles raising some doubts on the role of this gene in FSHD development [114, 119]. In a recent report, overexpression of DUX4 mRNA in human primary myoblasts followed by gene expression analysis showed deregulation of several genes involved in RNA splicing and processing, immune response pathways, and gametogenesis [119]. These genes were found aberrantly expressed in both FSHD1 and FSHD2 muscles while a partial recovery of the repressed state occurs upon depletion of endogenous DUX4 mRNA. Although no direct evidence was presented about the role of deregulated genes in the FSHD development, these findings suggest that critical DUX4 protein and RNA levels could be responsible for gene transcription deregulation in FSHD [119].

Aside putative genes involved in the FSHD development there is a general consensus in the field in supporting the view that epigenetic mechanisms are important players in FSHD, affecting the severity of the disease, its rate in progression, and the distribution of muscle weakness [120, 121]. Increasing evidence suggested that, in patients, chromatin conformation of FSHD locus is altered at multiple levels, from DNA methylation up to higher-order chromosome structures, resulting in perturbation of heterochromatic gene silencing in the subtelomeric domain of the long arm of chromosome 4. As stated previously, DNA methylation is associated with gene silencing and defects in methylation are generally associated with deregulation of transcriptional programs and disease [6]. D4Z4 is overall very GC-rich, having characteristics of CpG islands [122], and in healthy subjects is methylated, while contracted D4Z4 is always associated with an hypomethylation [123, 124]. Interestingly, FSHD2 patients, which phenotypically show FSHD though lacking D4Z4 contractions, display general D4Z4 hypomethylation [123], indicating an important epigenetic condition necessary to develop or generate the disease.

Combination of posttranslation histone modifications establishes a specific code that recruits nuclear factors responsible of several functions such as transcriptional or replication control. The D4Z4 repeat array is enriched of two repressive marks: trimethylation of lysine 9 or 27 of histone H3 (H3K9me3 and H3K27me3, resp.). The first, generally associated with constitutive heterochromatin, is deposited by the histone methyltransferase SUV39 and is responsible for HP1 repressor recruitment [125]. H3K27me3 is characteristic of facultative heterochromatin, is deposited by the PRC2 subunit EZH2, and in turn recruits PRC1 and PRC2 to establish transcriptionally repressed domains. It has been shown that H3K27me3 and the two Polycomb proteins YY1 and EZH2 are bound to D4Z4 and FRG1 promoter in myoblasts [107, 108] and are reduced during myogenic differentiation [108]. Interestingly, DNA association studies, by using 3C technologies [126], revealed that D4Z4 physically interacts with FRG1 promoter and this DNA loop is reduced upon differentiation. These epigenetic signatures dynamics during myogenesis are accompanied by a gradual upregulation of FRG1 [108]. Conversely, in FSHD1 myoblasts the D4Z4-FRG1 promoter interaction is reduced and FRG1 expression is anticipated during differentiation, suggesting an alteration of epigenetic signatures dynamics occurring when the differentiation starts. Notably, H3K27me3 can still be detected by ChIP at D4Z4 repeats in FSHD1 myoblasts, although by 3D immuno-FISH it was found specifically reduced on D4Z4 on 4q chromosome in FSHD1 myoblasts compared with controls [108]. This apparent inconsistency is justified by the extensive duplication of D4Z4 sequences in the human genome and the limitation of ChIP assay to distinguish specific 4q D4Z4 repeat. In addition to the complex heterochromatic features found at D4Z4 locus, there has been shown the presence of histone marks associated with transcriptional activation in the first proximal D4Z4 unit of the array, such as acetylation of histone H4 and dimethylation of Lys 4 of histone H3 [110, 125]. This could reflect the complexity of bidirectional transcriptional activity at the locus and could suggest the potential presence of noncoding RNA that further regulate the transcription.

As stated before, epigenetic chromatin regulation depends also on appropriate intranuclear positioning. Most nuclear events do not occur randomly in the nucleoplasm, rather regulatory proteins are spatially clustered in specific territories, and the position of chromosomal region in the nucleus influences its transcriptional activity. The 4q subtelomere is preferentially localized in the nuclear periphery in both controls and FSHD patients [127, 128], and this localization is evolutionary conserved [129]. In FSHD1 cells, this localization depends on a sequence within D4Z4 unit that tethers the subtelomere in the nuclear periphery in a CTCF and Lamin-A-dependent manner [130]. Although intranuclear positioning of 4q subtelomere does not change during muscle differentiation, when several epigenetic modifications take place [108], it has been shown that the nuclear periphery localization in controls and FSHD1 cells can be directed by different sequences, proximal or within D4Z4 repeat, respectively. This suggests that the nuclear environment of FSHD locus in normal or affected subjects could be different and could contribute to the disease development [130].

In summary, the epigenetic analysis suggests that probably the presence of more than ten D4Z4 repeats provides a physiological heterochromatization and repression of the subtelomeric region, due to the saturating levels of epigenetic repressors. In this view, less than ten D4Z4 repeats could be considered as *border line* genotype, because the correct heterochromatin formation is not ensured, determining a predisposition to the disease and also explaining the high variability in disease severity even in the same genetic background. This hypothesis is reinforced by the evidence that patients with less than 3 repeats have more chances to develop FSHD1 disease and that asymptomatic carriers of D4Z4 deletion are increasingly evident in FSHD [131]. Another complex issue about FSHD is the requirement for at least one D4Z4 repeat for the development of the disease, suggesting a *gain of function* effect, where the presence of an aberrant transcription of coding or noncoding RNA or dysregulated binding of epigenetic factors recruited by the D4Z4 array could be necessary for disease development. Systematic analysis of epigenetic modifications across the entire genome in FSHD1 and FSHD2 patients will be crucial to dissect epigenetic mechanisms acting specifically on D4Z4 locus and involved in FSHD pathogenesis and progression.

6. Laminopathies

The nuclear scaffold (or nuclear matrix) is the network of fibers found inside a cell nucleus. The lamina is the major component of nuclear matrix and is constituted by a complex meshwork of proteins closely associated with the inner nuclear membrane [132]. In vertebrates, lamins have been divided into A and B types, based on sequence homologies. All A-type lamins, A, C, C2, and Δ10, are encoded by alternative splicing of a single gene (LMNA) while two major mammalian B-type lamins, B1 and B2, are encoded by different genes (LMNB1 and LMNB2) [133]. All major lamins terminate with a CAAX-box that is involved in numerous posttranslational modifications including the farnesylation of the cysteine, removal of the-AAX, and carboxymethylation of the cysteine [134]. These modifications are thought to be important for the efficient targeting of the lamins to the inner nuclear membrane [135]. Moreover, Lamin A is further processed by the zinc metalloproteinase, Zmpste24/FACE1, which catalyzes the removal of additional 15 residues from Lamin A C-terminus including the farnesylated and carboxymethylated cysteine [136]. Expression of the A- and B-type lamins is developmentally regulated in mammals, resulting in cell type-specific complements of lamins [137]. In the last years, genome wide studies describing lamin bound chromosomal regions were focalized specifically on B-type [138, 139]. However, it is becoming increasingly evident that A-type lamins are scaffolds for proteins that regulate DNA synthesis, responses to DNA damage, chromatin organization, gene transcription, cell cycle progression, cell differentiation, cancer invasiveness, and epigenetic regulation of chromatin [140–143]. In line with this observation, lamin distribution in the nucleus is type specific, with Lamin B being predominantly present at inner nuclear membrane and Lamin A also present in lower concentrations, throughout the nucleoplasm [144], suggesting, for the latter, a role beyond the maintenance of mechanical stability of the nucleus. Genetic studies confirmed this hypothesis, showing that A- or B-type lamin mutations have different impacts on organisms. Mutations in genes encoding B-type lamins are not frequently connected to diseases in human and Lamin B1 null mice die during early postnatal life with severe defects in their lung and bones [145] while mice lacking Lamin B2 die shortly after birth with severe brain abnormalities. Taken together, these findings indicate that B-type lamins play a structural role in the nucleus essential for cell and tissue function. On the other hand, mice lacking A-type lamins have apparently normal embryonic development [146], but postnatal growth is delayed and they develop abnormalities of cardiac and skeletal muscle. This is in line with studies in human, where a large number of mutations of Lamin A/C (LMNA) were found, causing a wide range of human disorders, including lipodystrophy, neuropathies, autosomal dominant Emery-Dreifuss muscular dystrophy (EDMD), and progeria. The latter includes Hutchinson-Gilford progeria syndrome (HGPS), atypical Werner syndrome, restrictive dermopathy, and mandibuloacral dysplasia type A (MADA) [147]. Collectively, these degenerative disorders with a wide spectrum of clinical phenotypes are known as the laminopathies. To date, despite the identification of several mutations on Lamin A causing these disorders, it is difficult to correlate phenotype to genotype in laminopathies. It is still unclear how specific mutations result in a particular tissue-specific laminopathy phenotype [148] or why a single mutation in Lamin A gene can result in different phenotypes [149]. This suggests an involvement of the individual epigenetic background to the disease. Studies in HGPS cells confirmed this hypothesis finding several epigenetic alterations. In particular there has been shown a decrease of the heterochromatin mark H3K9me3 in pericentric regions and a downregulation of the PRC2 component EZH2, accompanied by a loss of H3K27me3 on the inactive X chromosome (Xi), which leads to some decondensation of the Xi [150]. Notably, it is not clear if observed epigenetic defects are cause or consequence of the irreversible cascade of cellular mechanisms dysfunction accompanying HGPS progression. Most inherited LMNA mutations in humans cause disorders that selectively affect striated muscle, determining decreased levels of A-type lamins. This was confirmed in the Lmna null mice, which develop abnormalities of cardiac and skeletal muscle reminiscent of those seen in human subjects [146]. Remarkably, in humans, decreased lamin A levels observed in some laminopathies could also be dependent on dominant negative effect caused by an aberrant form of Lamin A. Indeed, overexpression in transgenic mice of a human lamin A variant responsible for Emery-Dreifuss muscular dystrophy determines severe heart damage [151]. Several hypotheses have been proposed to explain molecular mechanisms underlying muscular dystrophies caused by lamin A mutations. The current model suggests that the prolonged exposure to mechanical stress of muscle cells determines the tissue-specific degeneration observed in laminopathies. This model takes into consideration only the structural role of lamin A, neglecting its functional role in chromatin organization and gene expression control. There is a growing body of evidence indicating that several signalling pathways, such as pRb, MyoD, Wnt-β catenin, and TGF-β, are altered in laminopathies [152, 153]. The Rb-MyoD-crosstalk is one of the most described pathways altered in laminopathies. MyoD is a master transcription factor of muscle differentiation that activates muscle-specific genes. Its levels are modulated by dephosphorilated pRb, which takes part in the acetylation and expression of MyoD [154]. Lamin A controls Rb levels favouring its dephosphorilation [155]. Thus in the absence of Lamin A the level of hypophosphorilated Rb and consequently the level of MyoD are reduced, determining a defect in muscle cells' differentiation [156]. This was confirmed by a decreased number of MyoD positive nuclei observed in skeletal muscle from laminopathy patients [157]. Given its role in muscle-specific genes regulation, PcG protein could be involved in aberrant gene expression observed in lamin A defective background. Several indirect evidences support this hypothesis indicating a potential crosstalk between PcG proteins and Lamin A. As mentioned previously the nuclear positioning of the PcG-regulated FSHD locus, responsible for the described neuromuscular disorder, is altered in human Lamin A/C null cells [127]. However, while the role of PcG proteins in governing local chromatin higher-order structures was extensively addressed [44, 47, 49], it is still unknown if they also control the chromosomal position in the nucleus and if the peripheral localization of FSHD is dependent on PcG proteins. Recently, it has been suggested that nuclear position of PcG proteins could be crucial for muscle differentiation [158]. In this work, Wang and colleagues have shown that the localization of PRC2 complex at the nuclear periphery is mediated by the myogenic regulator, Msx1, and is required for a correct repression of Msx1 target genes. This localization occurs in myoblasts and is necessary for a proper muscle differentiation [158]. The importance of chromatin architecture dynamics during muscle differentiation was further confirmed by studies performed by Mattout et al. in *C. elegans* [159]. Using ablation of the unique lamin gene in worm they found that lamin is necessary for perinuclear positioning of heterochromatin. Then, to test the physiological relevance of this association in developing animals, they monitored tissue-specific changes in nuclear position of specific genomic regions in worms that express a dominant mutant form of lamin, which mimics the human Emery-Dreifuss muscular dystrophy. They found that in lamin defective background, muscle-specific genes are not able to relocalize from the nuclear periphery to a more internal location and this determines loss of muscle integrity. Although there has been extensively shown the crucial role of Polycomb proteins in mediating nuclear chromatin architecture, to date no evidence supports a direct involvement of PcG in muscle genes relocalization during normal differentiation. Further studies are needed to determine if physiological epigenetic dynamics that ensure a correct myogenesis are altered in lamin defective background and the role of Polycomb proteins in this process.

7. Conclusions

In the last years the study of the epigenome and its role in human disease progression has attracted considerable interest. The insurgence of epigenetic deregulation in human pathologies suggests that specific diseases might benefit from epigenetic-targeted therapies and this type of drug therapy is becoming a reality in clinical settings [160, 161]. Notably, epigenetic variation could arise as a consequence of the disease. Distinguishing epigenetic variations causing or contributing to the disease process is not straightforward but is nevertheless crucial to elucidate the functional role of the disease-associated epigenetic variation and to optimize their utility in terms of diagnostics or therapeutics. Recent advances in genomic technologies, by the expanding use of next-generation DNA sequencing (ChIP-seq) to assess the genomic distribution of histone modifications, histone variants, DNA methylation, and epigenetic factors, will be helpful to study human disease-associated epigenetic variation at genomewide level. Combined with appropriate statistical and bioinformatic tools [162], these methods will give us a more complete picture of all the loci that are epigenetically altered, although they will not resolve the *cause or consequence* issue. Then, the functional characterization of the variety of epigenetic modifications at specific loci could provide insight into the function of these modifications in normal development and in subsequent transition to disease states. These studies could ultimately lead to the future development of more effective epigenetic-based therapies, although treatment with these classes of drugs should be carefully examined to determine whether the therapeutic benefits outweigh the potential adverse effects [163, 164].

Acknowledgments

The author would like to thank Valerio Orlando for his mentorship, encouragement, and for kindly providing lab space; and Federica Lo Sardo and Beatrice Bodega for constructive criticisms about this manuscript. C. Lanzuolo is supported by the Italian Ministry of Research and University (Futuro in Ricerca RBFR106S1Z_001).

References

[1] C. Beisel and R. Paro, "Silencing chromatin: comparing modes and mechanisms," *Nature Reviews Genetics*, vol. 12, no. 2, pp. 123–135, 2011.

[2] E. I. Campos and D. Reinberg, "Histones: annotating chromatin," *Annual Review of Genetics*, vol. 43, pp. 559–599, 2009.

[3] A. M. Deaton and A. Bird, "CpG islands and the regulation of transcription," *Genes and Development*, vol. 25, no. 10, pp. 1010–1022, 2011.

[4] C. Lanzuolo and V. Orlando, "The function of the epigenome in cell reprogramming," *Cellular and Molecular Life Sciences*, vol. 64, no. 9, pp. 1043–1062, 2007.

[5] Y. S. Mao, B. Zhang, and D. L. Spector, "Biogenesis and function of nuclear bodies," *Trends in Genetics*, vol. 27, no. 8, pp. 295–306, 2011.

[6] S. B. Baylin and P. A. Jones, "A decade of exploring the cancer epigenome—biological and translational implications," *Nature Reviews Cancer*, vol. 11, pp. 726–773, 2011.

[7] S. R. Bhaumik, E. Smith, and A. Shilatifard, "Covalent modifications of histones during development and disease pathogenesis," *Nature Structural and Molecular Biology*, vol. 14, no. 11, pp. 1008–1016, 2007.

[8] Z. Siegfried, S. Eden, M. Mendelsohn, X. Feng, B. Z. Tsuberi, and H. Cedar, "DNA methylation represses transcription in vivo," *Nature Genetics*, vol. 22, no. 2, pp. 203–206, 1999.

[9] Y. Q. Feng, R. Desprat, H. Fu et al., "DNA methylation supports intrinsic epigenetic memory in mammalian cells," *PLoS Genetics*, vol. 2, no. 4, article e65, 2006.

[10] A. Bird, M. Taggart, M. Frommer, O. J. Miller, and D. Macleod, "A fraction of the mouse genome that is derived from islands of nonmethylated, CpG-rich DNA," *Cell*, vol. 40, no. 1, pp. 91–99, 1985.

[11] R. S. Illingworth and A. P. Bird, "CpG islands—"a rough guide"," *FEBS Letters*, vol. 583, no. 11, pp. 1713–1720, 2009.

[12] R. S. Illingworth, U. Gruenewald-Schneider, S. Webb et al., "Orphan CpG Islands Identify numerous conserved promoters in the mammalian genome," *PLoS Genetics*, vol. 6, no. 9, article e1001134, 2010.

[13] S. Kriaucionis and N. Heintz, "The nuclear DNA base 5-hydroxymethylcytosine is present in purkinje neurons and the brain," *Science*, vol. 324, no. 5929, pp. 929–930, 2009.

[14] M. Tahiliani, K. P. Koh, Y. Shen et al., "Conversion of 5-methylcytosine to 5-hydroxymethylcytosine in mammalian DNA by MLL partner TET1," *Science*, vol. 324, no. 5929, pp. 930–935, 2009.

[15] S. Ito, A. C. D'Alessio, O. V. Taranova, K. Hong, L. C. Sowers, and Y. Zhang, "Role of tet proteins in 5mC to 5hmC conversion, ES-cell self-renewal and inner cell mass specification," *Nature*, vol. 466, no. 7310, pp. 1129–1133, 2010.

[16] P. Zhang, L. Su, Z. Wang, S. Zhang, J. Guan et al., "The involvement of 5-hydroxymethylcytosine in active DNA demethylation in mice," *Biology of Reproduction*, vol. 86, no. 4, p. 104, 2012.

[17] A. Munshi, G. Shafi, N. Aliya, and A. Jyothy, "Histone modifications dictate specific biological readouts," *Journal of Genetics and Genomics*, vol. 36, no. 2, pp. 75–88, 2009.

[18] M. Altaf, N. Saksouk, and J. Côté, "Histone modifications in response to DNA damage," *Mutation Research*, vol. 618, no. 1-2, pp. 81–90, 2007.

[19] O. Bell, M. Schwaiger, E. J. Oakeley et al., "Accessibility of the Drosophila genome discriminates PcG repression, H4K16 acetylation and replication timing," *Nature Structural and Molecular Biology*, vol. 17, no. 7, pp. 894–900, 2010.

[20] B. E. Bernstein, M. Kamal, K. Lindblad-Toh et al., "Genomic maps and comparative analysis of histone modifications in human and mouse," *Cell*, vol. 120, no. 2, pp. 169–181, 2005.

[21] M. L. Eaton, J. A. Prinz, H. K. MacAlpine, G. Tretyakov, P. V. Kharchenko, and D. M. MacAlpine, "Chromatin signatures of the Drosophila replication program," *Genome Research*, vol. 21, no. 2, pp. 164–174, 2011.

[22] G. J. Filion, J. G. van Bemmel, U. Braunschweig, W. Talhout, and J. Kind, "Systematic protein location mapping reveals five principal chromatin types in Drosophila cells," *Cell*, vol. 143, pp. 212–224, 2010.

[23] V. Krishnan, M. Z. Y. Chow, Z. Wang et al., "Histone H4 lysine 16 hypoacetylation is associated with defective DNA repair and premature senescence in Zmpste24-deficient mice," *Proceedings of the National Academy of Sciences of the United States of America*, vol. 108, no. 30, pp. 12325–12330, 2011.

[24] D. Schübeler, D. M. MacAlpine, D. Scalzo et al., "The histone modification pattern of active genes revealed through genome-wide chromatin analysis of a higher eukaryote," *Genes and Development*, vol. 18, no. 11, pp. 1263–1271, 2004.

[25] S. Henikoff and K. Ahmad, "Assembly of variant histones into chromatin," *Annual Review of Cell and Developmental Biology*, vol. 21, pp. 133–153, 2005.

[26] B. Schuettengruber, A. M. Martinez, N. Iovino, and G. Cavalli, "Trithorax group proteins: switching genes on and keeping them active," *Nature Reviews Molecular Cell Biology*, vol. 12, pp. 799–814, 2011.

[27] E. Splinter and W. de Laat, "The complex transcription regulatory landscape of our genome: control in three dimensions," *EMBO Journal*, vol. 30, pp. 4345–4355, 2011.

[28] G. Li and D. Reinberg, "Chromatin higher-order structures and gene regulation," *Current Opinion in Genetics and Development*, vol. 21, no. 2, pp. 175–186, 2011.

[29] P. Fraser and W. Bickmore, "Nuclear organization of the genome and the potential for gene regulation," *Nature*, vol. 447, no. 7143, pp. 413–417, 2007.

[30] C. Lanctôt, T. Cheutin, M. Cremer, G. Cavalli, and T. Cremer, "Dynamic genome architecture in the nuclear space: regulation of gene expression in three dimensions," *Nature Reviews Genetics*, vol. 8, no. 2, pp. 104–115, 2007.

[31] E. Soutoglou and T. Misteli, "Mobility and immobility of chromatin in transcription and genome stability," *Current Opinion in Genetics and Development*, vol. 17, no. 5, pp. 435–442, 2007.

[32] S. Schoenfelder, I. Clay, and P. Fraser, "The transcriptional interactome: gene expression in 3D," *Current Opinion in Genetics and Development*, vol. 20, no. 2, pp. 127–133, 2010.

[33] T. Klymenko, B. Papp, W. Fischle et al., "A polycomb group protein complex with sequence-specific DNA-binding and selective methyl-lysine-binding activities," *Genes and Development*, vol. 20, no. 9, pp. 1110–1122, 2006.

[34] J. C. Scheuermann, A. G. de Ayala Alonso, K. Oktaba et al., "Histone H2A deubiquitinase activity of the Polycomb repressive complex PR-DUB," *Nature*, vol. 465, no. 7295, pp. 243–247, 2010.

[35] R. Cao, L. Wang, H. Wang et al., "Role of histone H3 lysine 27 methylation in polycomb-group silencing," *Science*, vol. 298, no. 5595, pp. 1039–1043, 2002.

[36] B. Czermin, R. Melfi, D. McCabe, V. Seitz, A. Imhof, and V. Pirrotta, "Drosophila enhancer of Zeste/ESC complexes

have a histone H3 methyltransferase activity that marks chromosomal Polycomb sites," *Cell*, vol. 111, no. 2, pp. 185–196, 2002.

[37] A. Kuzmichev, K. Nishioka, H. Erdjument-Bromage, P. Tempst, and D. Reinberg, "Histone methyltransferase activity associated with a human multiprotein complex containing the enhancer of zeste protein," *Genes and Development*, vol. 16, no. 22, pp. 2893–2905, 2002.

[38] J. Müller, C. M. Hart, N. J. Francis et al., "Histone methyltransferase activity of a Drosophila Polycomb group repressor complex," *Cell*, vol. 111, no. 2, pp. 197–208, 2002.

[39] H. Wang, L. Wang, H. Erdjument-Bromage et al., "Role of histone H2A ubiquitination in Polycomb silencing," *Nature*, vol. 431, no. 7010, pp. 873–878, 2004.

[40] M. Ku, R. P. Koche, E. Rheinbay et al., "Genomewide analysis of PRC1 and PRC2 occupancy identifies two classes of bivalent domains," *PLoS Genetics*, vol. 4, no. 10, article e1000242, 2008.

[41] S. Schoeftner, A. K. Sengupta, S. Kubicek et al., "Recruitment of PRC1 function at the initiation of X inactivation independent of PRC2 and silencing," *EMBO Journal*, vol. 25, no. 13, pp. 3110–3122, 2006.

[42] D. Cmarko, P. J. Verschure, A. P. Otte, R. van Driel, and S. Fakan, "Polycomb group gene silencing proteins are concentrated in the perichromatin compartment of the mammalian nucleus," *Journal of Cell Science*, vol. 116, no. 2, pp. 335–343, 2003.

[43] F. Bantignies, V. Roure, I. Comet et al., "Polycomb-dependent regulatory contacts between distant hox loci in drosophila," *Cell*, vol. 144, no. 2, pp. 214–226, 2011.

[44] C. Lanzuolo, V. Roure, J. Dekker, F. Bantignies, and V. Orlando, "Polycomb response elements mediate the formation of chromosome higher-order structures in the bithorax complex," *Nature Cell Biology*, vol. 9, no. 10, pp. 1167–1174, 2007.

[45] F. Cléard, Y. Moshkin, F. Karch, and R. K. Maeda, "Probing long-distance regulatory interactions in the Drosophila melanogaster bithorax complex using Dam identification," *Nature Genetics*, vol. 38, no. 8, pp. 931–935, 2006.

[46] T. Sexton, E. Yaffe, E. Kenigsberg, F. Bantignies, and B. Leblanc, "Three-dimensional folding and functional organization principles of the Drosophila genome," *Cell*, vol. 148, pp. 458–472, 2012.

[47] S. Kheradmand Kia, P. Solaimani Kartalaei, E. Farahbakhshian, F. Pourfarzad, and M. von Lindern, "EZH2-dependent chromatin looping controls INK4a and INK4b, but not ARF, during human progenitor cell differentiation and cellular senescence," *Epigenetics & Chromatin*, vol. 2, no. 1, p. 16, 2009.

[48] D. Noordermeer, M. Leleu, E. Splinter, J. Rougemont, and W. De Laat, "The dynamic architecture of Hox gene clusters," *Science*, vol. 334, pp. 222–225, 2011.

[49] V. K. Tiwari, K. M. McGarvey, J. D. F. Licchesi et al., "PcG proteins, DNA methylation, and gene repression by chromatin looping," *PLoS Biology*, vol. 6, no. 12, pp. 2911–2927, 2008.

[50] K. H. Hansen, A. P. Bracken, D. Pasini et al., "A model for transmission of the H3K27me3 epigenetic mark," *Nature Cell Biology*, vol. 10, no. 11, pp. 1291–1300, 2008.

[51] R. Margueron, N. Justin, K. Ohno et al., "Role of the polycomb protein EED in the propagation of repressive histone marks," *Nature*, vol. 461, no. 7265, pp. 762–767, 2009.

[52] S. Chen, L. R. Bohrer, A. N. Rai et al., "Cyclin-dependent kinases regulate epigenetic gene silencing through phosphorylation of EZH2," *Nature Cell Biology*, vol. 12, no. 11, pp. 1108–1114, 2010.

[53] S. Kaneko, G. Li, J. Son et al., "Phosphorylation of the PRC2 component Ezh2 is cell cycle-regulated and up-regulates its binding to ncRNA," *Genes and Development*, vol. 24, no. 23, pp. 2615–2620, 2010.

[54] C. Lanzuolo, F. Lo Sardo, A. Diamantini, and V. Orlando, "PcG complexes set the stage for epigenetic inheritance of gene silencing in early S phase before replication," *PLoS Genetics*, vol. 7, article e1002370, 2011.

[55] C. Lanzuolo, F. Lo Sardo, and V. Orlando, "Concerted epigenetic signatures inheritance at PcG targets through replication," *Cell Cycle*, vol. 11, no. 7, pp. 1296–1300, 2012.

[56] N. Brockdorff, "Chromosome silencing mechanisms in X-chromosome inactivation: unknown unknowns," *Development*, vol. 138, pp. 5057–5065, 2011.

[57] C. Prezioso and V. Orlando, "Polycomb proteins in mammalian cell differentiation and plasticity," *FEBS Letters*, vol. 585, no. 13, pp. 2067–2077, 2011.

[58] C. L. Fisher and A. G. Fisher, "Chromatin states in pluripotent, differentiated, and reprogrammed cells," *Current Opinion in Genetics and Development*, vol. 21, no. 2, pp. 140–146, 2011.

[59] R. Margueron and D. Reinberg, "The Polycomb complex PRC2 and its mark in life," *Nature*, vol. 469, no. 7330, pp. 343–349, 2011.

[60] M. C. Trask and J. Mager, "Complexity of polycomb group function: diverse mechanisms of target specificity," *Journal of Cellular Physiology*, vol. 226, no. 7, pp. 1719–1721, 2011.

[61] K. Williams, J. Christensen, M. T. Pedersen et al., "TET1 and hydroxymethylcytosine in transcription and DNA methylation fidelity," *Nature*, vol. 473, no. 7347, pp. 343–349, 2011.

[62] H. Wu, A. C. D'Alessio, S. Ito et al., "Dual functions of Tet1 in transcriptional regulation in mouse embryonic stem cells," *Nature*, vol. 473, no. 7347, pp. 389–394, 2011.

[63] H. Wu, A. C. D'Alessio, S. Ito et al., "Genome-wide analysis of 5-hydroxymethylcytosine distribution reveals its dual function in transcriptional regulation in mouse embryonic stem cells," *Genes and Development*, vol. 25, no. 7, pp. 679–684, 2011.

[64] K. E. Davies and K. J. Nowak, "Molecular mechanisms of muscular dystrophies: old and new players," *Nature Reviews Molecular Cell Biology*, vol. 7, no. 10, pp. 762–773, 2006.

[65] K. Matsumura, F. M. S. Tome, H. Collin et al., "Expression of dystrophin-associated proteins in dystrophin-positive muscle fibers (revertants) in Duchenne muscular dystrophy," *Neuromuscular Disorders*, vol. 4, no. 2, pp. 115–120, 1994.

[66] J. M. Ervasti and K. J. Sonnemann, "Biology of the Striated Muscle Dystrophin-Glycoprotein Complex," *International Review of Cytology*, vol. 265, pp. 191–225, 2008.

[67] C. Colussi, C. Mozzetta, A. Gurtner et al., "HDAC2 blockade by nitric oxide and histone deacetylase inhibitors reveals a common target in Duchenne muscular dystrophy treatment," *Proceedings of the National Academy of Sciences of the United States of America*, vol. 105, no. 49, pp. 19183–19187, 2008.

[68] D. Cacchiarelli, J. Martone, E. Girardi et al., "MicroRNAs involved in molecular circuitries relevant for the duchenne muscular dystrophy pathogenesis are controlled by the dystrophin/nNOS pathway," *Cell Metabolism*, vol. 12, no. 4, pp. 341–351, 2010.

[69] D. Cacchiarelli, T. Incitti, J. Martone et al., "MiR-31 modulates dystrophin expression: new implications for Duchenne muscular dystrophy therapy," *EMBO Reports*, vol. 12, no. 2, pp. 136–141, 2011.

[70] G. C. Minetti, C. Colussi, R. Adami et al., "Functional and morphological recovery of dystrophic muscles in mice treated with deacetylase inhibitors," *Nature Medicine*, vol. 12, no. 10, pp. 1147–1150, 2006.

[71] S. Consalvi, V. Saccone, L. Giordani, G. Minetti, C. Mozzetta, and P. L. Puri, "Histone deacetylase inhibitors in the treatment of muscular dystrophies: epigenetic drugs for genetic diseases," *Molecular Medicine*, vol. 17, no. 5-6, pp. 457–465, 2011.

[72] M. Gomes-Pereira, T. A. Cooper, and G. Gourdon, "Myotonic dystrophy mouse models: towards rational therapy development," *Trends in Molecular Medicine*, vol. 17, no. 9, pp. 506–517, 2011.

[73] J. W. Day and L. P. W. Ranum, "RNA pathogenesis of the myotonic dystrophies," *Neuromuscular Disorders*, vol. 15, no. 1, pp. 5–16, 2005.

[74] L. A. Boyer, K. Plath, J. Zeitlinger et al., "Polycomb complexes repress developmental regulators in murine embryonic stem cells," *Nature*, vol. 441, no. 7091, pp. 349–353, 2006.

[75] A. P. Bracken, N. Dietrich, D. Pasini, K. H. Hansen, and K. Helin, "Genome-wide mapping of polycomb target genes unravels their roles in cell fate transitions," *Genes and Development*, vol. 20, no. 9, pp. 1123–1136, 2006.

[76] T. I. Lee, R. G. Jenner, L. A. Boyer et al., "Control of developmental regulators by polycomb in human embryonic stem cells," *Cell*, vol. 125, no. 2, pp. 301–313, 2006.

[77] N. Nègre, J. Hennetin, L. V. Sun et al., "Chromosomal distribution of PcG proteins during Drosophila development," *PLoS Biology*, vol. 4, no. 6, article e170, 2006.

[78] Y. B. Schwartz, T. G. Kahn, D. A. Nix et al., "Genome-wide analysis of Polycomb targets in Drosophila melanogaster," *Nature Genetics*, vol. 38, no. 6, pp. 700–705, 2006.

[79] S. L. Squazzo, H. O'Geen, V. M. Komashko et al., "Suz12 binds to silenced regions of the genome in a cell-type-specific manner," *Genome Research*, vol. 16, no. 7, pp. 890–900, 2006.

[80] B. Tolhuis, I. Muijrers, E. De Wit et al., "Genome-wide profiling of PRC1 and PRC2 Polycomb chromatin binding in Drosophila melanogaster," *Nature Genetics*, vol. 38, no. 6, pp. 694–699, 2006.

[81] P. Asp, R. Blum, V. Vethantham et al., "Genome-wide remodeling of the epigenetic landscape during myogenic differentiation," *Proceedings of the National Academy of Sciences of the United States of America*, vol. 108, no. 22, pp. E149–E158, 2011.

[82] G. Caretti, M. Di Padova, B. Micales, G. E. Lyons, and V. Sartorelli, "The Polycomb Ezh2 methyltransferase regulates muscle gene expression and skeletal muscle differentiation," *Genes and Development*, vol. 18, pp. 2627–2638, 2004.

[83] A. H. Juan, R. M. Kumar, J. G. Marx, R. A. Young, and V. Sartorelli, "Mir-214-dependent regulation of the polycomb protein Ezh2 in skeletal muscle and embryonic stem cells," *Molecular Cell*, vol. 36, no. 1, pp. 61–74, 2009.

[84] L. Stojic, Z. Jasencakova, C. Prezioso, A. Stutzer, and B. Bodega, "Chromatin regulated interchange between polycomb repressive complex 2 (PRC2)-Ezh2 and PRC2-Ezh1 complexes controls myogenin activation in skeletal muscle cells," *Epigenetics Chromatin*, vol. 4, article 16, 2011.

[85] A. H. Juan, A. Derfoul, X. Feng et al., "Polycomb EZH2 controls self-renewal and safeguards the transcriptional identity of skeletal muscle stem cells," *Genes and Development*, vol. 25, no. 8, pp. 789–794, 2011.

[86] M. Cesana, D. Cacchiarelli, I. Legnini, T. Santini, and O. Sthandier, "A long noncoding RNA controls muscle differentiation by functioning as a competing endogenous RNA," *Cell*, vol. 147, pp. 358–369, 2011.

[87] M. V. Neguembor and D. Gabellini, "In junk we trust: repetitive DNA, epigenetics and facioscapulohumeral muscular dystrophy," *Epigenomics*, vol. 2, no. 2, pp. 271–287, 2010.

[88] D. Palacios, C. Mozzetta, S. Consalvi et al., "TNF/p38α/polycomb signaling to Pax7 locus in satellite cells links inflammation to the epigenetic control of muscle regeneration," *Cell Stem Cell*, vol. 7, no. 4, pp. 455–469, 2010.

[89] A. Mattout, B. L. Pike, B. D. Towbin, E. M. Bank, A. Gonzalez-Sandoval, and et al., "An EDMD mutation in C. elegans lamin blocks muscle-specific gene relocation and compromises muscle integrity," *Current Biology*, vol. 21, pp. 1603–1614, 2011.

[90] R. Tawil and S. M. Van Der Maarel, "Facioscapulohumeral muscular dystrophy," *Muscle and Nerve*, vol. 34, no. 1, pp. 1–15, 2006.

[91] R. Tawil, D. Storvick, T. E. Feasby, B. Weiffenbach, and R. C. Griggs, "Extreme variability of expression in monozygotic twins with FSH muscular dystrophy," *Neurology*, vol. 43, no. 2, pp. 345–348, 1993.

[92] R. C. Griggs, R. Tawil, M. McDermott, J. Forrester, D. Figlewicz, and B. Weiffenbach, "Monozygotic twins with facioscapulohumeral dystrophy (FSHD): implications for genotype/phenotype correlation," *Muscle and Nerve*, vol. 18, no. 2, pp. S50–S55, 1995.

[93] J. C. T. Van Deutekom, C. Wijmenga, E. A. E. Van Tienhoven et al., "FSHD associated DNA rearrangements are due to deletions of integral copies of a 3.2 kb tandemly repeated unit," *Human Molecular Genetics*, vol. 2, no. 12, pp. 2037–2042, 1993.

[94] J. C. T. van Deutekom, E. Bakker, R. J. L. F. Lemmers et al., "Evidence for subtelomeric exchange of 3.3 kb tandemly repeated units between chromosomes 4q35 and 10q26: implications for genetic counselling and etiology of FSHD1," *Human Molecular Genetics*, vol. 5, no. 12, pp. 1997–2003, 1996.

[95] P. W. Lunt, P. E. Jardine, M. C. Koch et al., "Correlation between fragment size at D4F104S1 and age at onset or at wheelchair use, with a possible generational effect, accounts for much phenotypic variation in 4q35-facioscapulohumeral muscular dystrophy (FSHD)," *Human Molecular Genetics*, vol. 4, no. 5, pp. 951–958, 1995.

[96] E. Ricci, G. Galluzzi, G. Deidda, S. Cacurri, and L. Colantoni, "Progress in the molecular diagnosis of facioscapulohumeral muscular dystrophy and correlation between the number of KpnI repeats at the 4q35 locus and clinical phenotype," *Annals of Neurology*, vol. 45, pp. 751–757, 1999.

[97] R. Tawil, J. Forrester, R. C. Griggs et al., "Evidence for anticipation and association of deletion size with severity in facioscapulohumeral muscular systrophy," *Annals of Neurology*, vol. 39, no. 6, pp. 744–748, 1996.

[98] R. Tupler, A. Berardinelli, L. Barbierato et al., "Monosomy of distal 4q does not cause facioscapulohumeral muscular dystrophy," *Journal of Medical Genetics*, vol. 33, no. 5, pp. 366–370, 1996.

[99] R. J. L. F. Lemmers, P. De Kievit, L. Sandkuijl et al., "Facioscapulohumeral muscular dystrophy is uniquely associated with one of the two variants of the 4q subtelomere," *Nature Genetics*, vol. 32, no. 2, pp. 235–236, 2002.

[100] R. J. L. F. Lemmers, P. J. Van Der Vliet, R. Klooster et al., "A unifying genetic model for facioscapulohumeral muscular dystrophy," *Science*, vol. 329, no. 5999, pp. 1650–1653, 2010.

[101] R. J. F. L. Lemmers, M. Wohlgemuth, R. R. Frants, G. W. Padberg, E. Morava, and S. M. Van Der Maarel, "Contractions of D4Z4 on 4qB subtelomeres do not cause facioscapulohumeral muscular dystrophy," *American Journal of Human Genetics*, vol. 75, no. 6, pp. 1124–1130, 2004.

[102] N. S. T. Thomas, K. Wiseman, G. Spurlock, M. MacDonald, D. Üstek, and M. Upadhyaya, "A large patient study confirming that facioscapulohumeral muscular dystrophy (FSHD) disease expression is almost exclusively associated with an FSHD locus located on a 4qA-defined 4qter subtelomere," *Journal of Medical Genetics*, vol. 44, no. 3, pp. 215–218, 2007.

[103] R. Lyle, T. J. Wright, L. N. Clark, and J. E. Hewitt, "The FSHD-associated repeat, D4Z4, is a member of a dispersed family of homeobox-containing repeats, subsets of which are clustered on the short arms of the acrocentric chromosomes," *Genomics*, vol. 28, no. 3, pp. 389–397, 1995.

[104] E. Bakker, C. Wijmenga, R. H. A. M. Vossen et al., "The FSHD-linked locus D4F104S1 (p13E-11) on 4q35 has a homologue on 10qter," *Muscle and Nerve*, vol. 18, no. 2, pp. S39–S44, 1995.

[105] G. Deidda, S. Cacurri, P. Grisanti, E. Vigneti, N. Piazzo, and L. Felicetti, "Physical mapping evidence for a duplicated region on chromosome 10qter showing high homology with the facioscapulohumeral muscular dystrophy locus on chromosome 4qter," *European Journal of Human Genetics*, vol. 3, no. 3, pp. 155–167, 1995.

[106] R. J. L. F. Lemmers, P. de Kievit, M. van Geel et al., "Complete allele information in the diagnosis of facioscapulohumeral muscular dystrophy by triple DNA analysis," *Annals of Neurology*, vol. 50, no. 6, pp. 816–819, 2001.

[107] D. Gabellini, M. R. Green, and R. Tupler, "Inappropriate gene activation in FSHD: a repressor complex binds a chromosomal repeat deleted in dystrophic muscle," *Cell*, vol. 110, no. 3, pp. 339–348, 2002.

[108] B. Bodega, G. D. Ramirez, F. Grasser et al., "Remodeling of the chromatin structure of the facioscapulohumeral muscular dystrophy (FSHD) locus and upregulation of FSHD-related gene 1 (FRG1) expression during human myogenic differentiation," *BMC Biology*, vol. 7, article 41, 2009.

[109] D. Gabellini, G. D'Antona, M. Moggio et al., "Facioscapulohumeral muscular dystrophy in mice overexpressing FRG1," *Nature*, vol. 439, no. 7079, pp. 973–977, 2006.

[110] G. Jiang, F. Yang, P. G. M. van Overveld, V. Vedanarayanan, S. van der Maarel, and M. Ehrlich, "Testing the position-effect variegation hypothesis for facioscapulohumeral muscular dystrophy by analysis of histone modification and gene expression in subtelomeric 4q," *Human Molecular Genetics*, vol. 12, no. 22, pp. 2909–2921, 2003.

[111] R. J. Osborne, S. Welle, S. L. Venance, C. A. Thornton, and R. Tawil, "Expression profile of FSHD supports a link between retinal vasculopathy and muscular dystrophy," *Neurology*, vol. 68, no. 8, pp. 569–577, 2007.

[112] S. van Koningsbruggen, R. W. Dirks, A. M. Mommaas et al., "FRG1P is localised in the nucleolus, Cajal bodies, and speckles," *Journal of Medical Genetics*, vol. 41, no. 4, article e46, 2004.

[113] S. van Koningsbruggen, K. R. Straasheijm, E. Sterrenburg et al., "FRG1P-mediated aggregation of proteins involved in pre-mRNA processing," *Chromosoma*, vol. 116, no. 1, pp. 53–64, 2007.

[114] L. Snider, L. N. Geng, R. J. Lemmers et al., "Facioscapulohumeral dystrophy: incomplete suppression of a retrotransposed gene," *PLoS Genetics*, vol. 6, no. 10, article e1001181, 2010.

[115] M. Dixit, E. Ansseau, A. Tassin et al., "DUX4, a candidate gene of facioscapulohumeral muscular dystrophy, encodes a transcriptional activator of PITX1," *Proceedings of the National Academy of Sciences of the United States of America*, vol. 104, no. 46, pp. 18157–18162, 2007.

[116] L. Snider, A. Asawachaicharn, A. E. Tyler et al., "RNA transcripts, miRNA-sized fragments and proteins produced from D4Z4 units: new candidates for the pathophysiology of facioscapulohumeral dystrophy," *Human Molecular Genetics*, vol. 18, no. 13, pp. 2414–2430, 2009.

[117] V. Kowaljow, A. Marcowycz, E. Ansseau et al., "The DUX4 gene at the FSHD1A locus encodes a pro-apoptotic protein," *Neuromuscular Disorders*, vol. 17, no. 8, pp. 611–623, 2007.

[118] D. Bosnakovski, S. Lamb, T. Simsek et al., "DUX4c, an FSHD candidate gene, interferes with myogenic regulators and abolishes myoblast differentiation," *Experimental Neurology*, vol. 214, no. 1, pp. 87–96, 2008.

[119] L. N. Geng, Z. Yao, L. Snider, A. P. Fong, and J. N. Cech, "DUX4 activates germline genes, retroelements, and immune mediators: implications for facioscapulohumeral dystrophy," *Developmental Cell*, vol. 22, pp. 38–51, 2012.

[120] D. S. Cabianca and D. Gabellini, "FSHD: copy number variations on the theme of muscular dystrophy," *Journal of Cell Biology*, vol. 191, no. 6, pp. 1049–1060, 2010.

[121] S. M. Van der Maarel, R. Tawil, and S. J. Tapscott, "Facioscapulohumeral muscular dystrophy and DUX4: breaking the silence," *Trends in Molecular Medicine*, vol. 17, no. 5, pp. 252–258, 2011.

[122] J. E. Hewitt, R. Lyle, L. N. Clark et al., "Analysis of the tandem repeat locus D4Z4 associated with facioscapulohumeral muscular dystrophy," *Human Molecular Genetics*, vol. 3, no. 8, pp. 1287–1295, 1994.

[123] J. C. de Greef, R. J. Lemmers, B. G. van Engelen et al., "Common epigenetic changes of D4Z4 in contraction-dependent and contraction-independent FSHD," *Human Mutation*, vol. 30, no. 10, pp. 1449–1459, 2009.

[124] P. G. M. Van Overveld, R. J. F. L. Lemmers, G. Deidda et al., "Interchromosomal repeat array interactions between chromosomes 4 and 10: a model for subtelomeric plasticity," *Human Molecular Genetics*, vol. 9, no. 19, pp. 2879–2884, 2000.

[125] W. Zeng, J. C. De Greef, Y. Y. Chen et al., "Specific loss of histone H3 lysine 9 trimethylation and HP1*y*/cohesin binding at D4Z4 repeats is associated with facioscapulohumeral dystrophy (FSHD)," *PLoS Genetics*, vol. 5, no. 7, article e1000559, 2009.

[126] J. Dekker, K. Rippe, M. Dekker, and N. Kleckner, "Capturing chromosome conformation," *Science*, vol. 295, no. 5558, pp. 1306–1311, 2002.

[127] P. S. Masny, U. Bengtsson, S. A. Chung et al., "Localization of 4q35.2 to the nuclear periphery: is FSHD a nuclear envelope disease?" *Human Molecular Genetics*, vol. 13, no. 17, pp. 1857–1871, 2004.

[128] R. Tam, K. P. Smith, and J. B. Lawrence, "The 4q subtelomere harboring the FSHD locus is specifically anchored with peripheral heterochromatin unlike most human telomeres," *Journal of Cell Biology*, vol. 167, no. 2, pp. 269–279, 2004.

[129] B. Bodega, M. F. Cardone, S. Müller et al., "Evolutionary genomic remodelling of the human 4q subtelomere (4q35.2)," *BMC Evolutionary Biology*, vol. 7, article 39, 2007.

[130] A. Ottaviani, C. Schluth-Bolard, S. Rival-Gervier et al., "Identification of a perinuclear positioning element in human subtelomeres that requires A-type lamins and CTCF," *EMBO Journal*, vol. 28, no. 16, pp. 2428–2436, 2009.

[131] I. Scionti, G. Fabbri, C. Fiorillo, G. Ricci, and F. Greco, "Facioscapulohumeral muscular dystrophy: new insights from compound heterozygotes and implication for prenatal genetic counselling," *Journal of Medical Genetics*, vol. 49, pp. 171–178, 2012.

[132] G. Patrizi and M. Poger, "The ultrastructure of the nuclear periphery. The Zonula Nucleum Limitans," *Journal of Ultrasructure Research*, vol. 17, no. 1-2, pp. 127–136, 1967.

[133] F. Lin and H. J. Worman, "Structural organization of the human gene encoding nuclear lamin A and nuclear lamin C," *The Journal of Biological Chemistry*, vol. 268, no. 22, pp. 16321–16326, 1993.

[134] A. E. Rusiñol and M. S. Sinensky, "Farnesylated lamins, progeroid syndromes and farnesyl transferase inhibitors," *Journal of Cell Science*, vol. 119, no. 16, pp. 3265–3272, 2006.

[135] G. Krohne, I. Waizenegger, and T. H. Hoger, "The conserved carboxy-terminal cysteine of nuclear lamins is essential for lamin association with the nuclear envelope," *Journal of Cell Biology*, vol. 109, no. 5, pp. 2003–2011, 1989.

[136] S. G. Young, L. G. Fong, and S. Michaelis, "Prelamin A, Zmpste24, misshapen cell nuclei, and progeria—new evidence suggesting that protein farnesylation could be important for disease pathogenesis," *Journal of Lipid Research*, vol. 46, no. 12, pp. 2531–2558, 2005.

[137] J. L. V. Broers, B. M. Machiels, H. J. H. Kuijpers et al., "A- and B-type lamins are differentially expressed in normal human tissues," *Histochemistry and Cell Biology*, vol. 107, no. 6, pp. 505–517, 1997.

[138] L. Guelen, L. Pagie, E. Brasset et al., "Domain organization of human chromosomes revealed by mapping of nuclear lamina interactions," *Nature*, vol. 453, no. 7197, pp. 948–951, 2008.

[139] H. Pickersgill, B. Kalverda, E. De Wit, W. Talhout, M. Fornerod, and B. Van Steensel, "Characterization of the Drosophila melanogaster genome at the nuclear lamina," *Nature Genetics*, vol. 38, no. 9, pp. 1005–1014, 2006.

[140] J. L. V. Broers, F. C. S. Ramaekers, G. Bonne, R. Ben Yaou, and C. J. Hutchison, "Nuclear lamins: laminopathies and their role in premature ageing," *Physiological Reviews*, vol. 86, no. 3, pp. 967–1008, 2006.

[141] B. R. Johnson, R. T. Nitta, R. L. Frock et al., "A-type lamins regulate retinoblastoma protein function by promoting subnuclear localization and preventing proteasomal degradation," *Proceedings of the National Academy of Sciences of the United States of America*, vol. 101, no. 26, pp. 9677–9682, 2004.

[142] V. L. R. M. Verstraeten, J. L. V. Broers, F. C. S. Ramaekers, and M. A. M. van Steensel, "The nuclear envelope, a key structure in cellular integrity and gene expression," *Current Medicinal Chemistry*, vol. 14, no. 11, pp. 1231–1248, 2007.

[143] N. D. Willis, T. R. Cox, S. F. Rahman-Casañs et al., "Lamin A/C is a risk biomarker in colorectal cancer," *PLoS ONE*, vol. 3, no. 8, article e2988, 2008.

[144] P. Hozak, A. M. J. Sasseville, Y. Raymond, and P. R. Cook, "Lamin proteins form an internal nucleoskeleton as well as a peripheral lamina in human cells," *Journal of Cell Science*, vol. 108, no. 2, pp. 635–644, 1995.

[145] L. Vergnes, M. Péterfy, M. O. Bergo, S. G. Young, and K. Reue, "Lamin B1 is required for mouse development and nuclear integrity," *Proceedings of the National Academy of Sciences of the United States of America*, vol. 101, no. 28, pp. 10428–10433, 2004.

[146] T. Sullivan, D. Escalante-Alcalde, H. Bhatt et al., "Loss of A-type lamin expression compromises nuclear envelope integrity leading to muscular dystrophy," *Journal of Cell Biology*, vol. 147, no. 5, pp. 913–919, 1999.

[147] B. C. Capell and F. S. Collins, "Human laminopathies: nuclei gone genetically awry," *Nature Reviews Genetics*, vol. 7, no. 12, pp. 940–952, 2006.

[148] I. Landires, J. M. Pascale, and J. Motta, "The position of the mutation within the LMNA gene determines the type and extent of tissue involvement in laminopathies [1]," *Clinical Genetics*, vol. 71, no. 6, pp. 592–596, 2007.

[149] J. Scharner, V. F. Gnocchi, J. A. Ellis, and P. S. Zammit, "Genotype-phenotype correlations in laminopathies: how does fate translate?" *Biochemical Society Transactions*, vol. 38, no. 1, pp. 257–262, 2010.

[150] D. K. Shumaker, T. Dechat, A. Kohlmaier et al., "Mutant nuclear lamin A leads to progressive alterations of epigenetic control in premature aging," *Proceedings of the National Academy of Sciences of the United States of America*, vol. 103, no. 23, pp. 8703–8708, 2006.

[151] Y. Wang, A. J. Herron, and H. J. Worman, "Pathology and nuclear abnormalities in hearts of transgenic mice expressing M371K lamin A encoded by an LMNA mutation causing Emery-Dreifuss muscular dystrophy," *Human Molecular Genetics*, vol. 15, no. 16, pp. 2479–2489, 2006.

[152] V. Andrés and J. M. González, "Role of A-type lamins in signaling, transcription, and chromatin organization," *Journal of Cell Biology*, vol. 187, no. 7, pp. 945–957, 2009.

[153] S. Marmiroli, J. Bertacchini, F. Beretti et al., "A-type lamins and signaling: the PI 3-kinase/Akt pathway moves forward," *Journal of Cellular Physiology*, vol. 220, no. 3, pp. 553–564, 2009.

[154] P. L. Puri, S. Iezzi, P. Stiegler et al., "Class I histone deacetylases sequentially interact with MyoD and pRb during skeletal myogenesis," *Molecular Cell*, vol. 8, no. 4, pp. 885–897, 2001.

[155] J. H. Van Berlo, J. W. Voncken, N. Kubben et al., "A-type lamins are essential for TGF-β1 induced PP2A to dephosphorylate transcription factors," *Human Molecular Genetics*, vol. 14, no. 19, pp. 2839–2849, 2005.

[156] R. L. Frock, B. A. Kudlow, A. M. Evans, S. A. Jameson, S. D. Hauschka, and B. K. Kennedy, "Lamin A/C and emerin are critical for skeletal muscle satellite cell differentiation," *Genes and Development*, vol. 20, no. 4, pp. 486–500, 2006.

[157] Y. E. Park, Y. K. Hayashi, K. Goto et al., "Nuclear changes in skeletal muscle extend to satellite cells in autosomal dominant Emery-Dreifuss muscular dystrophy/limb-girdle muscular dystrophy 1B," *Neuromuscular Disorders*, vol. 19, no. 1, pp. 29–36, 2009.

[158] J. Wang, R. M. Kumar, V. J. Biggs, H. Lee, Y. Chen et al., "The Msx1 homeoprotein recruits polycomb to the nuclear periphery during development," *Developmental Cell*, vol. 21, pp. 575–588, 2011.

[159] A. Mattout, B. L. Pike, B. D. Towbin, E. M. Bank, and A. Gonzalez-Sandoval, "An EDMD mutation in C. elegans lamin blocks muscle-specific gene relocation and compromises muscle integrity," *Current Biology*, vol. 21, no. 19, pp. 1603–1614, 2011.

[160] L. Giacinti, P. Vici, and M. Lopez, "Epigenome: a new target in cancer therapy," *Clinica Terapeutica*, vol. 159, no. 5, pp. 347–360, 2008.

[161] C. Ptak and A. Petronis, "Epigenetics and complex disease: from etiology to new therapeutics," *Annual Review of Pharmacology and Toxicology*, vol. 48, pp. 257–276, 2008.

[162] S. Pepke, B. Wold, and A. Mortazavi, "Computation for ChIP-seq and RNA-seq studies," *Nature methods*, vol. 6, no. 11, pp. S22–S32, 2009.

[163] J. Peedicayil, "Pharmacoepigenetics and pharmacoepigenomics," *Pharmacogenomics*, vol. 9, no. 12, pp. 1785–1786, 2008.

[164] B. Claes, I. Buysschaert, and D. Lambrechts, "Pharmaco-epigenomics: discovering therapeutic approaches and biomarkers for cancer therapy," *Heredity*, vol. 105, no. 1, pp. 152–160, 2010.

14

Genome Microscale Heterogeneity among Wild Potatoes Revealed by Diversity Arrays Technology Marker Sequences

Alessandra Traini,[1] Massimo Iorizzo,[2] Harpartap Mann,[3] James M. Bradeen,[3] Domenico Carputo,[1] Luigi Frusciante,[1] and Maria Luisa Chiusano[1]

[1] *Department of Agricultural Sciences, University of Naples Federico II, Via Università 100, 80055 Portici, Naples, Italy*
[2] *Department of Horticulture, University of Wisconsin-Madison, 1575 Linden Drive, Madison, WI 53706, USA*
[3] *Department of Plant Pathology, University of Minnesota, 495 Borlaug Hall/1991 Upper Buford Circle, St. Paul, MN 55108, USA*

Correspondence should be addressed to Maria Luisa Chiusano; chiusano@unina.it

Academic Editor: Ancha Baranova

Tuber-bearing potato species possess several genes that can be exploited to improve the genetic background of the cultivated potato *Solanum tuberosum*. Among them, *S. bulbocastanum* and *S. commersonii* are well known for their strong resistance to environmental stresses. However, scant information is available for these species in terms of genome organization, gene function, and regulatory networks. Consequently, genomic tools to assist breeding are meager, and efficient exploitation of these species has been limited so far. In this paper, we employed the reference genome sequences from cultivated potato and tomato and a collection of sequences of 1,423 potato Diversity Arrays Technology (DArT) markers that show polymorphic representation across the genomes of *S. bulbocastanum* and/or *S. commersonii* genotypes. Our results highlighted microscale genome sequence heterogeneity that may play a significant role in functional and structural divergence between related species. Our analytical approach provides knowledge of genome structural and sequence variability that could not be detected by transcriptome and proteome approaches.

1. Background

The subgenus *Potatoe* of the Solanaceae family includes approximately 188 tuber-bearing species [1]. They display large ecological adaptation encompassing several traits that are lacking in the commercial potato and useful for breeding [2]. Among wild potato species, *Solanum bulbocastanum* Dun. and *S. commersonii* Dun. ex Poir. have attracted the attention of researchers and breeders. *S. bulbocastanum* is a known source of resistance to late blight disease of potato, and four late blight resistance genes have been cloned from this species to date [3–7]. *S. commersonii* ranks first among *Solanums* in terms of cold tolerance and capacity to cold acclimate, and it is also a source of resistance to pathogens such as *Ralstonia solanacearum* and *Pectobacterium carotovorum* [8, 9]. *S. bulbocastanum* and *S. commersonii* are among approximately 20 diploid potato species classified as superseries *Stellata* by Hawkes [10]. Despite their importance as sources of genes for crop improvement, relatively few genetic and genomic resources are available for these species, and little is known on their genome organization, gene function, and regulatory networks. Recently, a Diversity Arrays Technology (DArT) array was constructed for potato [11]. The array contains markers derived from various *Solanum* species, including *S. bulbocastanum* and *S. commersonii*. DArT arrays offer the potential to simultaneously survey large numbers of anonymous loci distributed throughout the genome. DArT markers are highly transferrable across populations or even across species, since the DArT array comprises a structured marker set that is surveyed in each experiment. Importantly, polymorphic DArT markers correspond to a set of DNA clones that can be sequenced for downstream applications.

The availability of the potato DArT array together with the recent release of the complete genome sequences of cultivated potato [12] and tomato [13] provide an attractive opportunity for comparative genomic studies aimed at understanding genome evolution at the species level. The genomes of potato and tomato are largely syntenic, and molecular markers and gene content are predominantly conserved [13–16]. This degree of similarity has already enabled cross species comparative genomics approaches for gene mapping and cloning, reviewed by Bradeen [17]. Bioinformatics platforms improve community access to these resources and related *omics* collections, playing an important role for data mining and genome integration [18, 19]. In contrast to this wealth of knowledge and resources for cultivated potato and tomato, very little is known about genome structure and gene content in the wild relatives of potato.

In this paper, we exploited the reference genome sequences of potato and tomato and a collection of sequences of potato DArT array markers that show polymorphic representation across the genomes of *S. bulbocastanum* and/or *S. commersonii* genotypes. Our aim was to define a preliminary collection of marker sequences informative for the two species as a starting point for investigation of genome structure. This collection was also useful to highlight microscale genome sequence heterogeneity that possibly plays a meaningful role in functional and structural divergence between related species.

2. Materials and Methods

2.1. Plant Materials and DArT Marker Analyses. Two genotypes of *Solanum bulbocastanum* and two genotypes of *Solanum commersonii* were analyzed in this study. *S. bulbocastanum* genotypes include PT29 (PI243510), a source of the late blight resistance gene *RB* [3], and G15 (PI255516), a source of the *RB* locus allele *RB-rc* [20]. The *S. commersonii* genotypes include the frost tolerant cmm1T (PI243503) [8] and cmm6-3 (PI590886), a seedling genotype selected based on its crossability with cmm1T [21]. Total genomic DNA of individual plants for molecular marker analysis was isolated from fully expanded leaves from greenhouse-grown plants, following the protocol of Doyle and Doyle [22], with minor modifications. Two grams of leaf tissue were frozen in liquid nitrogen and ground in a mortar and pestle. Ground tissue was suspended in 6 mL lysis buffer (100 mM Tris-HCl pH 8.0, 20 mM EDTA, 2% CTAB, and 1.4 M NaCl) and incubated for 20 min at 65°C with occasional mixing by inversion. One volume of chloroform was added, and the tubes were mixed well and incubated at room temperature for 20 min with occasional inversion. Tubes were then centrifuged for 15 min at 1000 g, and the supernatant was transferred to a separate tube containing 2 volumes of 100% ethanol. Contents were gently mixed by inversion. Precipitated DNA was hooked out using sterile micropipette tips and transferred to 1.5 mL microfuge tubes. The DNA was washed twice with 75% ethanol and resuspended in TE (Tris pH 8.0 + 1 mM EDTA) buffer. DNA was shipped to Diversity Arrays Technology Pty Ltd. (Canberra, Australia) for DArT marker analysis.

Construction of the potato DArT array has been previously described [11]. The potato DArT array contains markers derived from *Solanum* species representative of the secondary and tertiary genepools of potato. Hybridization of genome representations from *S. bulbocastanum* and *S. commersonii* genotypes to the potato array and automatic calling of marker states were performed by Diversity Arrays Technology Pty Ltd. using established protocols [23]. Data that passed quality standards were analyzed for polymorphisms between genotypes within each species, and polymorphic markers were selected for downstream analyses. Clone cultures corresponding to each of these markers were robotically arrayed into a Whatman EasyClone 384 well plate (Whatman plc, Kent, UK) by Diversity Arrays Technology Pty Ltd. following manufacturer's instructions. Briefly, 10 μL of each clone culture was applied to a well followed by air-drying of the plate. The FTA plates were then shipped to the University of Minnesota for PCR amplification and sequencing of clone inserts.

For clone insert PCR, 45 μL of 10 mM Tris pH 8.0 + 0.1 mM EDTA was applied to each FTA plate well for 10 min at room temperature. PCRs were conducted in a 50 μL volume that consisted of 1x PCR buffer (Applied Biosystems, Foster City, CA), 2.5 U of Amplitaq (Applied Biosystems), 200 μM of each dNTP, 1 μL of eluate from the FTA plates (as template), and 50 pmol of each primer (DArT-M13f: GTTTTCCCAGTCACGACGTTG and DArT-M13r: TGAGCGGATAACAATTTCACACAG; Integrated DNA Technologies (Coralville, IA)). Thermocycler (GeneAmp PCR System 2700 (Applied Biosystems)) conditions were 35 cycles of 94°C for 30 sec, 55°C for 30 sec, and 72°C for 30 sec followed by a single cycle of 75°C for 5 min. To each PCR, 5 μL of 3 M NaOAC and 125 μL of ice-cold ethanol were added. The PCR plates were stored at −20°C for at least one hour and then centrifuged at 2,500 g at 4°C for 30 min. The supernatant was gently poured off, and the open plates were centrifuged upside down at 800 g for 30 sec. To each tube, 175 μL of room temperature 70% ethanol was added. The plates were again stored at −20°C and centrifuged as described above. Plates were dried completely at 37°C before adding 20 μL of TE. Amplification was confirmed by agarose gel electrophoresis of 2 μL of each purified PCR, staining with ethidium bromide, and visualization under UV light.

DNA sequencing of inserts was completed at the University of Minnesota BioMedical Genomics Center using BigDye Terminator (Applied Biosystems) cycle sequencing on an Applied Biosystems 3100 or 3700 automatic sequencer. Each sequencing reaction contained 1 μL of purified PCR product and 3.2 pmol of DArT-M13f or DArT-M13r. Each insert was sequenced in both directions in separate reactions. Resulting sequences were trimmed of vector and assembled into consensus sequences using SeqMan, part of the DNASTAR (Madison, WI) Lasergene software package.

Out of 1,423 DArT marker clones sequenced, 756 hybridized in a polymorphic fashion with *S. bulbocastanum* genotypes and 550 hybridized in a polymorphic fashion with *S. commersonii* genotypes. Hereafter, these markers will be referred to as BLB- and CMM-specific markers, respectively. The remaining 117 DArT markers hybridized and were polymorphic in both species (indicated as BLB/CMM).

2.2. Sequence Analysis and Data Interpretation. The genome sequence of *Solanum phureja* [12] served as the reference genome for our analyses. The genome sequence of *Solanum lycopersicum* [13], another reference species among Solanaceae, was also employed. For both genomes, our analyses included 12 pseudomolecule sequences as well as unanchored scaffolds. We adopted gene annotations reported by the iTAG group (international Tomato Annotation Group) [24], assuring uniform annotation criteria and bioinformatics strategies and allowing coherent comparisons of the two reference genomes herein considered [13].

DArT marker sequences were aligned to the genome sequences using the splicing alignment software GenomeThreader [25] with 70% minimal nucleotide coverage and sequence identity. DArT alignments to genome sequences were grouped into six different categories (Figure 1(a)). A DArT marker sequence that aligned to a genome region independent of other DArT markers (i.e., one that does not overlap with any other marker sequences in the same genomic region) was classified as *solitary*. Each *solitary* marker was further subclassified as (1) *solitary one match*, if it aligned only once to the genome, or (2) *solitary multiple matches*, if it aligned more than once. A DArT marker whose alignment to the genome overlapped that of other DArT marker sequences was classified as an *overlapping* DArT. A DArT marker sequence having multiple matches to the genome, some of which are *solitary* and some of which are *overlapping*, was classified as subcategory (3) *mixed*. Other *overlapping* markers were further classified as *overlapping in uniform groups* when the group was composed of the same set of overlapping DArT marker sequences. This category comprised two subcategories: (4) *overlapping in uniform groups—one match* occurring only once in the genome and (5) *overlapping in uniform groups—multiple matches* appearing in two or more genome locations. DArT marker sequences which show multiple matches to the genome sequence and overlap sets of different DArT markers are defined as (6) *overlapping in heterogeneous groups*.

Fifty-three DArT marker sequences that did not align to either the potato or tomato genome sequences based on the GenomeThreader approach were assembled using CAP3 [26] (parameters: -p 40 –o 80) before a second alignment attempt based on BLASTn [27] (parameters: -e 0.003). These same DArT sequences were also aligned to the GenBank nucleotide collection (nr/nt) using BLASTn and to the nonredundant protein sequences dataset using BLASTp [28]. A BLAST2GO analysis [29, 30] was performed to classify genes associated to DArT marker sequences to show the cellular, biological, and molecular functional information of the subset annotation.

3. Results and Discussion

3.1. Dataset Description. The majority of the 1,423 DArT sequences analyzed have a length ranging between 350 and 850 nucleotides, providing a consistent dataset for subsequent bioinformatics analyses. In particular, 68% of BLB markers and 73% of CMM markers are 450 to 700 nucleotides in length (data not shown).

About 92% and 79% of all DArT sequences could be aligned the potato and tomato genomes, respectively (Table 1). These comprise 93% of BLB, 91% of CMM, and 90% of BLB/CMM DArT markers relative to the potato genome and 78% of BLB, 81% of CMM, and 76% of BLB/CMM DArT markers relative to the tomato genome. The discrepancy between the percentage of alignments to each genome is consistent with the composition of the reference potato DArT array that emphasizes markers from *Solanum* species more closely related to potato [11].

Table 1: Results of DArT alignments to potato and tomato reference genomes. For each collection, the total number of DArT markers and the number (%) of aligned DArT markers are reported.

Collection	Total no. of DArT	No. aligned (%) to	
		Potato	Tomato
BLB	756	703 (92.9)	586 (77.5)
CMM	550	499 (90.7)	446 (81.1)
BLB/CMM	117	105 (89.7)	89 (76.1)
All	1423	1307 (91.8)	1121 (79.0)

Sequence alignments were grouped into six categories, as described in Section 2. In the alignments to both potato and tomato genomes, DArT markers most frequently occurred as group (1) *solitary one match*, with 344 and 321 matches for potato and tomato, respectively, and as group (4) *overlapping in uniform groups-one match*, with 755 matches for potato and 663 matches for tomato (Figure 1(a)). For alignments to the potato genome, these two categories encompass 84% of all sequenced DArT markers: 82% for BLB, 87% for CMM, and 88% for BLB/CMM (Figure 1(b)). For alignments to the tomato genome, these same categories comprise 88% of all DArT marker sequences: 84% for BLB, 91% for CMM, and 91% for BLB/CMM. The remaining four marker alignment categories each represent less than 10% of the total number of aligned DArT marker sequences (Figure 1(b)). Briefly, groups (2) *solitary multiple-matches* and (5) *overlapping in uniform groups-multiple matches* show alignment to more than one genome region; this is probably due to repeated regions in the genome sequence; therefore, we considered these markers to be redundant. Groups (3) *mixed* and (6) *overlapping in heterogeneous groups* comprise DArT sequences with different alignment configurations probably due to intrinsic sequence properties. DArT marker sequences assigned to categories (1) and (4) localize in unique regions in both the potato and tomato genomes. Since these markers are associated unambiguously to specific genome locations, they were considered as nonredundant markers and were subjected to further analyses; DArT markers not assigned to alignment categories (1) and (4) were not considered further.

3.2. Analysis of Nonredundant DArT Markers. In total 1,099 and 984 nonredundant (i.e., group (1) and group (4)) DArT marker sequences align to the potato and tomato genome sequences, respectively. The majority of the marker sequences aligns with a sequence identity exceeding 80% and a coverage greater than 90% (Figure 2). The percentage of alignments in the highest coverage category (between 90 and 100%) is 92% for potato and 75% for tomato. Many of the alignments overlap gene regions in both genomes (Figure 2). This is not unexpected since DArT markers are obtained through digestion by *Pst*I. *Pst*I is a methylation-sensitive enzyme; therefore, it is possible that it acts mainly on hypomethylated

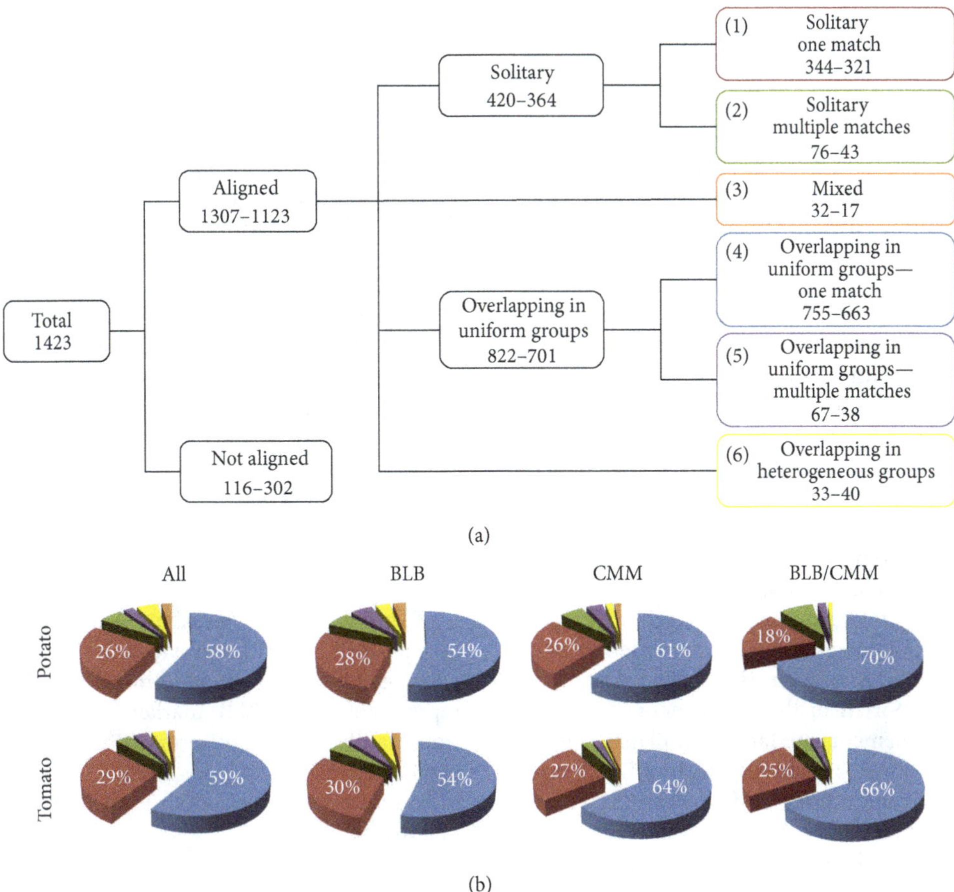

FIGURE 1: Categories of DArT markers alignments. (a) Values represent the number of alignments along the potato and tomato genome, respectively. (b) Pie charts of the percentage of aligned DArT markers, for each collection. The colour code is associated to the coloured rectangles of (a) and percentages are reported only when greater than 10%.

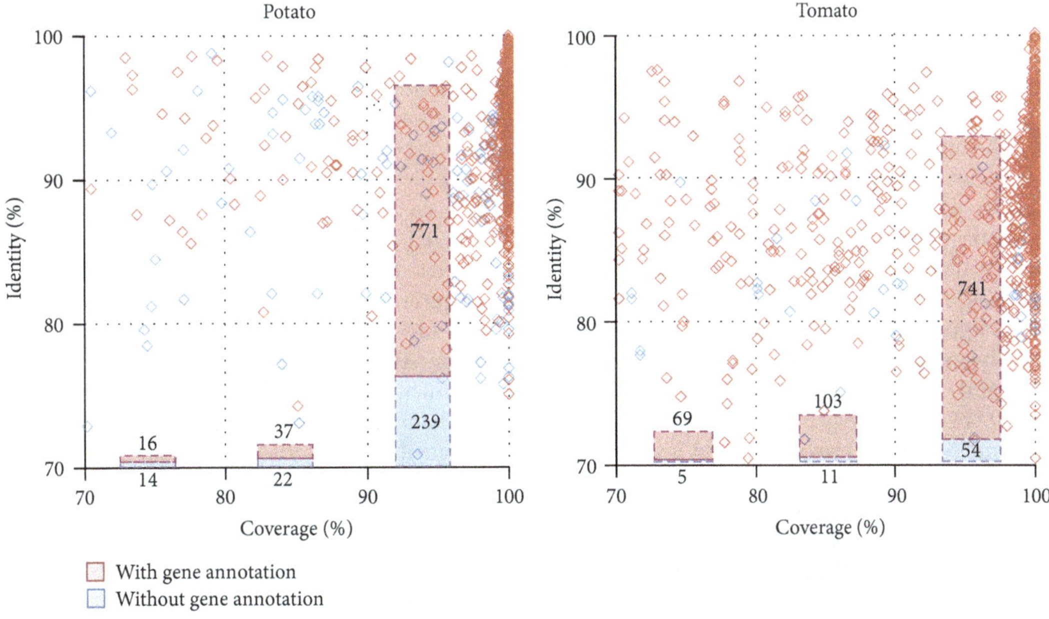

FIGURE 2: DArT marker sequences align predominantly with gene coding regions of the potato and tomato genome. The alignments associated (or not) to a gene locus along the potato and tomato genomes are highlighted in red (or blue). For each group, the number of alignments is also given.

DNA which, in turn, may correspond to gene regions, which are typically hypomethylated [31]. In Figure 3, the BLAST2GO analyses of the genes overlapping DArT marker regions are shown for both potato and tomato annotations. In particular, the figure shows the overrepresentation of genes associated with catalytic and binding activities.

In percentage, the two marker groups (1 and 4) represent 84% and 88% of all markers sequences aligned to the potato and the tomato genomes, respectively. Interestingly, in contrast with average results across all DArT sequences (Table 1) showing more matching DArT sequences to potato than to tomato, a higher proportion of the nonredundant groups align to the tomato genome than to the potato genome. This may be due to the higher contribution of ambiguous alignments (group (2) and (5)) in potato. This in turn suggests a higher sequence repetitiveness in the potato genome or better sequence quality for the tomato genome [12, 13]. Overall, nonredundant DArT marker sequences show very high coverage in potato compared to tomato (Figure 2), confirming higher phylogenetic similarity amongst potato species.

We next examined total coverage of the genome sequences from cultivated potato and tomato represented by alignments with DArT marker sequences (Table 2). Details per chromosomes are reported in the supplementary Table S1 (see Table S1 in Supplementary Material available at http://dx.doi.org/10.1155/2013/257218). In general, BLB DArT markers encompass a greater number of nucleotides in each genome than CMM or BLB/CMM markers. This is not surprising since BLB markers are the largest subset of DArT markers examined in this study. BLB DArT markers represent 208.8 Kbp of the potato genome but only 175.8 Kbp of the tomato genome. In contrast, CMM and BLB/CMM markers represent approximately equivalent regions of the potato and tomato genomes (CMM: 137.9 Kbp for potato versus 139.6 Kbp for tomato; BLB/CMM: 29.4 Kbp for potato versus 24.7 Kbp for tomato). We further divided the nonredundant DArT markers into two subclasses. *Common* markers align with the genome sequences of both potato and tomato; *specific* markers align to only one of the two genomes (Table 2). Within each sub-class, alignments were either *ungapped* (i.e., marker sequences aligned to genome sequences without disruption) or *gapped* (i.e., marker sequences aligned to genome sequences but alignments were interrupted by genome sequence not found in marker sequences). It is noteworthy that the same DArT marker sequence could be *ungapped* when aligned to the potato genome and *gapped* when aligned to the tomato genome or *vice versa*. The relative ratio of *gapped* versus *ungapped* regions of all BLB, CMM, and CMM-BLB DArT marker sequences relative to the potato and tomato genome sequences provides insight into patterns of genome evolution and species relationships. Distinction between *gapped* and *ungapped* alignments is necessary since variability in the length of *gapped* markers can complicate interpretation of the degree of genome coverage by the marker sequences. In potato, for example, the size of most of the gaps (89%) ranges from 20 to ~1000 bps. The remaining ones reach a maximum at ~5000 bps (not shown). For *common* DArT markers, the contribution of *ungapped* regions to total genome representation is higher in potato than in tomato for each marker collection. In contrast, for *common* markers, the contribution of *gapped* regions is generally lower in potato than in tomato. This again reflects higher phylogenetic similarity of the wild species to the cultivated potato. However, it is interesting to note that the relative frequency of *common gapped* regions compared to *common ungapped* ones in potato versus tomato is comparable for both BLB (14.21% in potato and 16.68% in tomato) and BLB/CMM (5.14% potato and 3.16% in tomato) DArT markers. The frequency of CMM *common gapped* and *ungapped* regions differs in potato (7.73%) with respect to tomato (15.64%). This indicates that, in contrast to BLB markers, CMM markers align with fewer gaps to the potato genome sequence than to the tomato genome sequence. This implies that the genomes of *S. commersonii* and potato are more similar at a DNA sequence level than are the genomes of *S. bulbocastanum* and potato, consistent with *S. commersonii* being phylogenetically more closely related to potato than is *S. bulbocastanum*, as the analyses based on plastid genomes previously suggested [32–34].

Considering the contribution of *specific* DArT markers, *ungapped* BLB markers provided the greatest overall genome coverage for both potato and tomato, consistent with higher representation of BLB markers in our dataset (Table 2). Importantly, the relative proportion of *gapped* regions compared to *ungapped* regions for the *specific* alignments indicates a comparable behaviour in the three marker collections in both species.

3.3. Genome Sequence Heterogeneity. We compared marker origins and alignment classifications across the potato and tomato genomes (Table 3). In general, the majority of aligned DArT markers are *ungapped* in both potato and tomato: 328 (77%) for BLB, 297 (83%) for CMM, and 65 (91%) for BLB/CMM. Eight BLB and 16 CMM markers align to both genomes in a *gapped* configuration (Table 3). Interestingly, a high percentage of aligned markers exhibit heterogeneous behaviours across the potato and tomato genomes (i.e., *gapped* versus *ungapped* in potato versus tomato and vice versa). These sequences are a source of marker variability between wild and cultivated species that can be exploited in future studies.

Seven BLB, 16 CMM, and one BLB/CMM markers aligned to the genomes of both potato and tomato in a *gapped* configuration (Table 3). As shown in Table S2, each of the seven BLB DArT markers aligned to gene regions in both species. Among these, five regions corresponded to genes with identical annotations in potato and tomato. On the other hand, among the 16 CMM DArT markers, only 10 and 14 aligned to gene coding regions in potato and tomato, respectively. Of the 10 CMM markers aligning to both potato and tomato gene coding regions, all of the 10 aligned to regions with identical gene annotations in both species (Table S2).

Nine DArT markers, three from BLB and six from CMM, aligned with the same alignment structure (i.e., number and length of *gapped* and *ungapped* regions) to homologous chromosomes in both potato and tomato and to gene loci with the same annotation (Table S2). The remaining two BLB, four

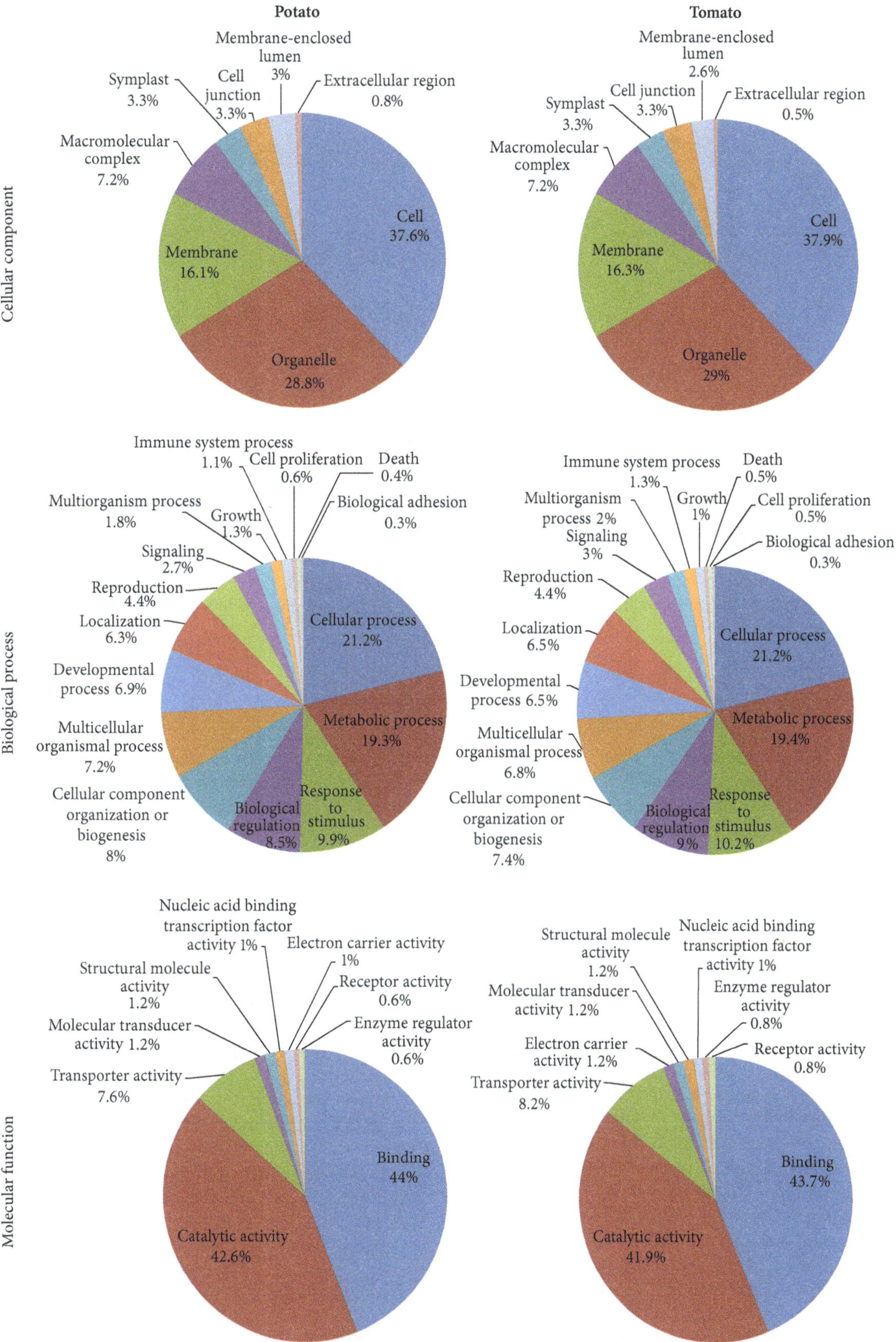

FIGURE 3: BLAST2GO analyses of the genes overlapping DArT marker regions.

Table 2: Number of nucleotides (in Kbp units) covered by DArT alignments. For details on coverage categories, see Section 2.

Coverage category	Potato			Tomato		
	BLB	CMM	BLB/CMM	BLB	CMM	BLB/CMM
Common						
Ungapped	132.3	97.7	20.5	128.2	95.3	19.1
Gapped	18.8	7.6	1.1	21.4	14.9	0.6
Specific						
Ungapped	46.1	22.3	7.7	22.5	17.3	4.4
Gapped	10.2	10.4	0.2	3.7	12.1	0.7
Total	208.8	137.9	29.4	175.8	139.6	24.7

Table 3: Comparison between DArT alignments to potato and tomato genomes. Number of DArT markers aligned along the potato (horizontal) and tomato (vertical) genomes for each collection, given in parenthesis. Each cell, within each matrix, shows the number of DArT markers per alignment type: ungapped, gapped, or not aligned.

Tomato		Potato: BLB (573)					CMM (432)					BLB/CMM (94)		
		Ungapped	Gapped	Not aligned			Ungapped	Gapped	Not aligned			Ungapped	Gapped	Not aligned
BLB (494)	Ungapped	328	43	61	CMM (407)	Ungapped	296	14	47	BLB/CMM (82)	Ungapped	65	3	9
	Gapped	50	7	5		Gapped	29	16	5		Gapped	3	1	1
	Not aligned	115	30			Not aligned	63	14			Not aligned	21	1	

CMM, and one BLB/CMM markers, although aligning to homologous chromosomes in genes of the same annotation, showed heterogeneous (i.e., number and length of *gapped* and *ungapped* regions) alignment structure (Table S2). These observations of microscale genome heterogeneity may be relevant to investigation of genome structures, functionalities, and properties of the represented *Solanum* species.

3.4. DArT Marker Sequences Not Aligned to the Reference Genomes. Some DArT markers could not be aligned to one or both genome sequences (Table 1). In particular, 116 marker sequences could not be aligned to the potato genome, 302 marker sequences could not be aligned to the tomato genome, and 51 marker sequences could be aligned to neither to potato nor to tomato. These were selected as putative wild species-specific markers and were assembled using the CAP3 software, yielding seven assembled consensus sequences comprising 20 sequences in total. The remaining 31 DArT marker sequences could not be assembled. Next, we attempted a less stringent alignment of the resulting 38 sequences (31 unassembled sequences plus seven consensus sequences) to the potato and the tomato genome sequences using the BLASTn algorithm (Table S3). Using this approach, 18 DArT marker sequences could be assigned to single locations in both the potato and tomato genomes, and only nine markers aligned to multiple genome locations in one or both species. In these cases, the less stringent alignment search performed by the BLAST software helped to confirm the presence in the potato and tomato genomes of 27 DArT marker sequences, previously unidentified in the more stringent GenomeThreader analysis. Moreover, in some cases, the BLASTn analysis confirmed matches to the same chromosome for both potato and tomato (e.g., DArT markers 472847 (chromosome 1), 537586 (chromosome 8), 473780 (chromosome 2), and 534573 (chromosome 11)). The presence of low level sequence similarity between these markers and the potato or tomato genome sequences revealed distant relationships between the wild and cultivated species and may be exploited in the study of cross-species genome heterogeneity. Twenty-two DArT markers (Table S3) showed extreme repetitive distribution along the potato and tomato chromosomes and were described by ambiguous annotations. Nevertheless, protein-based annotations (BLASTp), when present, generally confirmed homology with *Solanum* proteins or with those from more distantly related plant species. Two DArT marker sequences failed to align to the genomes of either potato or tomato even under more permissive analytical criteria.

4. Conclusions

Potato (*S. tuberosum*) and tomato (*S. lycopersicum*) belong to the subgenus *Potatoe* of the large and diverse genus *Solanum*. Although horticulturally distinct, potato and tomato share a clear evolutionary history that is well supported by molecular data [35, 36]. The species are thought to have diverged from a common ancestor approximately 6.2 to 7.3 million years ago [37, 38]. Sexual isolation and subsequent divergence of the two species were accompanied by a series of structural genomic changes including chromosome arm inversions and large-scale translocations [14, 15]. Nevertheless, the genomes of potato and tomato are largely syntenic and molecular marker and gene content are predominantly conserved [14–16]. This degree of similarity has enabled cross species comparative genomics approaches for gene mapping and cloning, reviewed by Bradeen [17], efforts that will likely be furthered by the recent release of the complete genome sequences of potato [12] and tomato [13].

In this study, we proposed a suitable methodology to exploit partial genome information from wild species in the presence of reference genomes from related species. This approach, here exploited with DArT marker sequences, can also be employed in partial genome resequencing or similar efforts. Our results also highlighted the presence of divergent sequence relationships and heterogeneous alignment structures, including the presence/absence of gaps, which are detectable thanks to appropriate, less stringent comparative methods. This divergence commonly occurred even in gene pairs with apparent orthologous relationships and presumed functional conservation, and it could often be confirmed both in potato and tomato genomes. Evidence from results supported by two reference-related species partially overcomes possible limits that may be due to the quality of first released genomes and suggests a fine microscale genome structural divergence between wild and cultivated species in the Solanaceae. Our results confirm the utility of suitable analytical approaches that could be applied when partial genome information is available, capable of highlighting genome microscale variability that, although often occurring at the gene level, is not detectable when investigating genome functionality at transcriptome and proteomic levels.

Conflict of Interests

The authors declare no conflict of interests.

Acknowledgments

This research was carried out within the project "Approcci "-omici" integrati per lo studio e l'utilizzazione della biodiversità di patata" funded by the Italian Ministry of Agriculture. It was supported in part by USDA/NIFA through an Agriculture and Food Research Initiative (AFRI) grant to JMB and by the GENHORT project funded by MIUR. The support of the Minnesota Supercomputing Institute at the University of Minnesota is gratefully acknowledged.

References

[1] D. M. Spooner and A. Salas, "Structure, biosystematics, and genetic resources," in *Handbook of Potato Production, Improvement, and Post-Harvest Management*, J. Gopal and S. M. Paul Khurana, Eds., pp. 1–39, Haworth's Press, Binghampton, NY, USA, 2006.

[2] J. E. Bradshaw, "Potato-breeding strategy," in *Potato Biology and Biotechnology: Advances and Perspectives*, D. Vreugdenhil, Ed., pp. 157–177, Elsevier, Oxford, UK, 2007.

[3] J. Song, J. M. Bradeen, S. K. Naess et al., "Gene RB cloned from *Solanum bulbocastanum* confers broad spectrum resistance to potato late blight," *Proceedings of the National Academy of Sciences of the United States of America*, vol. 100, no. 16, pp. 9128–9133, 2003.

[4] E. Van Der Vossen, A. Sikkema, B. T. L. Hekkert et al., "An ancient R gene from the wild potato species *Solanum bulbocastanum* confers broad-spectrum resistance to Phytophthora infestans in cultivated potato and tomato," *Plant Journal*, vol. 36, no. 6, pp. 867–882, 2003.

[5] E. A. G. Van Der Vossen, J. Gros, A. Sikkema et al., "The Rpi-blb2 gene from *Solanum bulbocastanum* is an Mi-1 gene homolog conferring broad-spectrum late blight resistance in potato," *Plant Journal*, vol. 44, no. 2, pp. 208–222, 2005.

[6] T. H. Park, J. Gros, A. Sikkema et al., "The late blight resistance locus Rpi-blb3 from *Solanum bulbocastanum* belongs to a major late blight R gene cluster on chromosome 4 of potato," *Molecular Plant-Microbe Interactions*, vol. 18, no. 7, pp. 722–729, 2005.

[7] T. Oosumi, D. R. Rockhold, M. M. Maccree, K. L. Deahl, K. F. McCue, and W. R. Belknap, "Gene *Rpi-bt1* from *Solanum bulbocastanum* confers resistance to late blight in transgenic potatoes," *American Journal of Potato Research*, vol. 86, no. 6, pp. 456–465, 2009.

[8] D. Carputo, T. Cardi, J. P. Palta, P. Sirianni, S. Vega, and L. Frusciante, "Tolerance to low temperatures and tuber soft rot in hybrids between *Solanum commersonii* and *Solanum tuberosum* obtained through manipulation of ploidy and endosperm balance number (EBN)," *Plant Breeding*, vol. 119, no. 2, pp. 127–130, 2000.

[9] D. Carputo, R. Aversano, A. Barone et al., "Resistance to ralstonia solanacearum of sexual hybrids between *Solanum commersonii* and *S. tuberosum*," *American Journal of Potato Research*, vol. 86, no. 3, pp. 196–202, 2009.

[10] J. G. Hawkes, *The Potato: Evolution, Biodiversity and Genetic Resources*, Smithsonian Institution Press, Washington, DC, USA, 1990.

[11] J. Sliwka, H. Jakuczun, M. Chmielarz et al., "A resistance gene against potato late blight originating from *Solanum* x *michoacanum* maps to potato chromosome VII," *Theoretical and Applied Genetics*, vol. 124, no. 2, pp. 397–406, 2012.

[12] The Tomato Genome Consortium, "Genome sequence and analysis of the tuber crop potato," *Nature*, vol. 475, no. 7355, pp. 189–195, 2011.

[13] The Tomato Genome Consortium, "The tomato genome sequence provides insights into fleshy fruit evolution," *Nature*, vol. 485, pp. 635–641, 2012.

[14] M. W. Bonierbale, R. L. Plaisted, and S. D. Tanksley, "RFLP maps based on a common set of clones reveal modes of chromosomal evolution in potato and tomato," *Genetics*, vol. 120, no. 4, pp. 1095–1103, 1988.

[15] S. D. Tanksley, M. W. Ganal, J. P. Prince et al., "High density molecular linkage maps of the tomato and potato genomes," *Genetics*, vol. 132, no. 4, pp. 1141–1160, 1992.

[16] R. C. Grube, E. R. Radwanski, and M. Jahn, "Comparative genetics of disease resistance within the solanaceae," *Genetics*, vol. 155, no. 2, pp. 873–887, 2000.

[17] J. M. Bradeen, "Cloning of late blight resistance genes: strategies and progress," in *Genetics, Genomics and Breeding of Potato*, J. M. Bradeen and C. Kole, Eds., pp. 153–183, CRC Press/Science Publishers, Enfield, NH, 2011.

[18] K. Mochida and K. Shinozaki, "Advances in omics and bioinformatics tools for systems analyses of plant functions," *Plant Cell Physiology*, vol. 52, no. 12, pp. 2017–2038, 2011.

[19] M. L. Chiusano, N. D'Agostino, A. Traini et al., "ISOL" An Italian SOLAnaceae genomics resource," *BMC Bioinformatics*, vol. 9, no. 2, article S7, 2008.

[20] M. J. Sanchez, *Allelic mining for late blight resistance in wild* Solanum *species belonging to series Bulbocastana [M.S. thesis]*, Department of Plant Pathology, University of Minnesota, St. Paul, Minn, USA, 2005.

[21] M. Iorizzo, *Uso di strumenti genomici per lo studio di linee di patata ottenute tramite ingegneria genetica e genomica [Ph.D. thesis]*, Department of Soil, Plant and Environmental and Animal Production Science, University of Naples Federico II, 2008.

[22] J. J. Doyle and J. L. Doyle, "Isolation of plant DNA from fresh tissue," *Focus*, vol. 12, pp. 13–15, 1990.

[23] D. Jaccoud, K. Peng, D. Feinstein, and A. Kilian, "Diversity arrays: a solid state technology for sequence information independent genotyping," *Nucleic Acids Research*, vol. 29, no. 4, article E25, 2001.

[24] L. Mueller, S. Tanksley, J. J. Giovannoni et al., "A snapshot of the emerging tomato genome sequence," *The Plant Genome*, vol. 2, no. 1, pp. 78–92, 2009.

[25] G. Gremme, V. Brendel, M. E. Sparks, and S. Kurtz, "Engineering a software tool for gene structure prediction in higher organisms," *Information and Software Technology*, vol. 47, no. 15, pp. 965–978, 2005.

[26] X. Huang and A. Madan, "CAP3: a DNA sequence assembly program," *Genome Research*, vol. 9, no. 9, pp. 868–877, 1999.

[27] S. F. Altschul, W. Gish, W. Miller, E. W. Myers, and D. J. Lipman, "Basic local alignment search tool," *Journal of Molecular Biology*, vol. 215, no. 3, pp. 403–410, 1990.

[28] M. Johnson, I. Zaretskaya, Y. Raytselis, Y. Merezhuk, S. McGinnis, and T. L. Madden, "NCBI BLAST: a better web interface," *Nucleic Acids Research*, vol. 36, pp. W5–W9, 2008.

[29] N. Blüthgen, K. Brand, B. Cajavec, M. Swat, H. Herzel, and D. Beule, "Biological profiling of gene groups utilizing gene ontology," *Genome Informatics*, vol. 16, pp. 106–115, 2005.

[30] A. Conesa, S. Götz, J. M. García-Gómez, J. Terol, M. Talón, and M. Robles, "Blast2GO: a universal tool for annotation, visualization and analysis in functional genomics research," *Bioinformatics*, vol. 21, no. 18, pp. 3674–3676, 2005.

[31] P. Wenzl, H. Li, J. Carling et al., "A high-density consensus map of barley linking DArT markers to SSR, RFLP and STS loci and agricultural traits," *BMC Genomics*, vol. 7, article 206, 2006.

[32] D. M. Spooner and T. Raul Castillo, "Reexamination of series relationships of South American wild potatoes (Solanaceae: *Solanum* sect. *Petota*): evidence from chloroplast DNA restriction site variation," *American Journal of Botany*, vol. 84, no. 5, pp. 671–685, 1997.

[33] D. M. Spooner, K. McLean, G. Ramsay, R. Waugh, and G. J. Bryan, "A single domestication for potato based on multilocus amplified fragment length polymorphism genotyping," *Proceedings of the National Academy of Sciences of the United States of America*, vol. 102, no. 41, pp. 14694–14699, 2005.

[34] D. Gargano, N. Scotti, A. Vezzi et al., "Genome-wide analysis of plastome sequence variation and development of plastidial CAPS markers in common potato and related *Solanum* species," *Genetic Resources and Crop Evolution*, pp. 1–12, 2011.

[35] D. M. Spooner, G. J. Anderson, and R. K. Jansen, "Chloroplast DNA evidence for the interrelationships of tomatoes, potatoes, and pepinos (Solanaceae)," *American Journal of Botany*, vol. 80, no. 6, pp. 676–688, 1993.

[36] T. L. Weese and L. Bohs, "A three-gene phylogeny of the genus *Solanum* (Solanaceae)," *Systematic Botany*, vol. 32, no. 2, pp. 445–463, 2007.

[37] Y. Wang, A. Diehl, F. Wu et al., "Sequencing and comparative analysis of a conserved syntenic segment in the solanaceae," *Genetics*, vol. 180, no. 1, pp. 391–408, 2008.

[38] F. Wu and S. D. Tanksley, "Chromosomal evolution in the plant family Solanaceae," *BMC Genomics*, vol. 11, no. 1, article 182, 2010.

15

Repertoire of Protein Kinases Encoded in the Genome of *Takifugu rubripes*

R. Rakshambikai,[1] S. Yamunadevi,[1,2,3] K. Anamika,[1,4] N. Tyagi,[1,5] and N. Srinivasan[1]

[1] *Molecular Biophysics Unit, Indian Institute of Science, Bangalore 560012, India*
[2] *Max Planck Institute for Intelligent Systems (Formerly Max Planck Institute for Metals Research), BioQuant BQ0038, Im Neuenheimer Feld 267, 69120 Heidelberg, Germany*
[3] *German Cancer Research Center, BioQuant BQ0038, Im Neuenheimer Feld 267, 69120 Heidelberg, Germany*
[4] *Department of Functional Genomics and Cancer and Department of Structural Biology and Genomics, Institut de Génétique et de Biologie Moléculaire et Cellulaire (IGBMC), CNRS UMR 7104, INSERM U 964, Université de Strasbourg, 1 rue Laurent Fries, 67404 Illkirch Cedex, France*
[5] *European Bioinformatics Institute, Wellcome Trust Genome Campus, Hinxton, CB10 1SD Cambridge, UK*

Correspondence should be addressed to N. Srinivasan, ns@mbu.iisc.ernet.in

Academic Editor: G. Pesole

Takifugu rubripes is teleost fish widely used in comparative genomics to understand the human system better due to its similarities both in number of genes and structure of genes. In this work we survey the fugu genome, and, using sensitive computational approaches, we identify the repertoire of putative protein kinases and classify them into groups and subfamilies. The fugu genome encodes 519 protein kinase-like sequences and this number of putative protein kinases is comparable closely to that of human. However, in spite of its similarities to human kinases at the group level, there are differences at the subfamily level as noted in the case of KIS and DYRK subfamilies which contribute to differences which are specific to the adaptation of the organism. Also, certain unique domain combination of galectin domain and YkA domain suggests alternate mechanisms for immune response and binding to lipoproteins. Lastly, an overall similarity with the MAPK pathway of humans suggests its importance to understand signaling mechanisms in humans. Overall the fugu serves as a good model organism to understand roles of human kinases as far as kinases such as LRRK and IRAK and their associated pathways are concerned.

1. Introduction

Takifugu rubripes is a teleost fish native to northwest pacific seas. It belongs to the family Tetraodontidae and order Tetraodontiformes. The fugu genome is rather compact with a size of ~400 Mb although the number of genes is comparable to that of higher eukaryotes indicating a considerable reduction in the intergenic regions [1]. Detailed analysis reveals that the intron-exon boundaries [2] and in certain cases alternative splicing [3], synteny [4, 5] have been conserved with respect to that of humans suggesting the possibility of conserved elements from a common ancestor. Thus, due to such features indicating close relationship, the fugu is suggested to be a good model organism and an effective way to study evolution of structure of complex vertebrate genomes [4, 6].

In 2002, the first draft sequence of the fugu genome was reported by the International fugu genome Consortium using the "whole-genome shotgun" strategy. Subsequently many versions of the genomic data have been made available at http://www.fugu-sg.org/. The latest version (v-5) was released in 2010 which covers about 392 Mb and 72% of genome being organized into chromosomes.

Response to environmental stimulus via complex signaling systems is a central feature of all living cells. Phosphorylation is one such posttranslational modification employed in signaling circuits which usually results in a functional

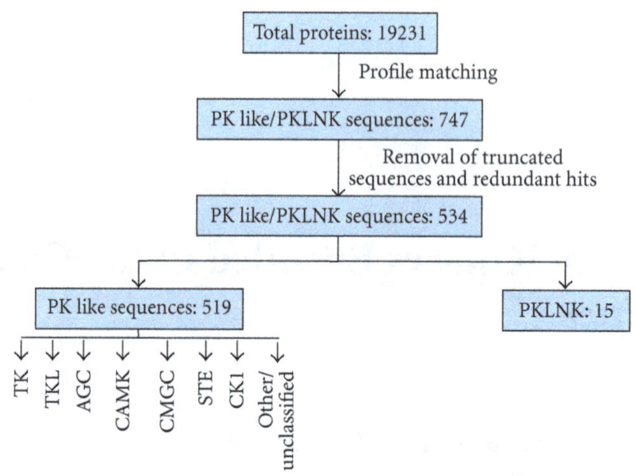

FIGURE 1: Distribution of fugu kinases into various Hanks and Hunter groups. PKLNK: protein-kinase-like nonkinases; CAMK: calcium/calmodulin-dependent protein kinase; CMGC: the group of cyclin-dependent protein kinase, mitogen-activated protein kinase, glycogen synthase kinase, casein kinase-2; AGC: the group of protein kinase A, protein kinase G, protein kinase C; STE: sterile (homologs of yeast STE); TK: tyrosine kinase; TKL: tyrosine kinase like; CK1: casein kinase 1.

change in the substrate by changing enzyme activity, cellular location, or association with other proteins. Thus, protein kinases have implication in regulation of various cellular processes encompassing metabolism, stress responses, cell cycle control, organ development, and intercellular communication [28, 29]. Abnormalities in the functioning of these kinases usually have implications in developmental disorders and malignancies [30, 31]. The eukaryotic kinases mainly constitute the Ser/Thr and Tyr kinases which share a common three-dimensional fold and the catalytic core spanning to about 300 residues [32].

The fugu genome has been used in a myriad of comparative genomic studies to elucidate the function of proteins involved in neurodegenerative diseases [33, 34], signaling systems [35, 36], and so forth and has been suggested as a method to elucidate cognate pathways in humans. In this paper, using sensitive sequence analysis [37–54] we recognize the repertoire of Ser/Thr and Tyr kinases encoded in the fugu genome. This is not trivial as homologous kinases are known to be characterized by weak sequence similarity. In addition, we classify these kinases on the basis of their catalytic domain sequence and the domains covalently tethered to the catalytic kinase domain [38]. Finally, we provide comparative analysis with distribution of the kinases in other model organisms and other proteins of the MAPK pathway especially in relation to the higher eukaryotes like human.

2. Materials and Methods

2.1. Identification of Protein Kinases. The complete set of predicted proteins from the ORF's of the *Takifugu rubripes* fifth assembly genome has been obtained from http://www.fugu-sg.org/. We have adopted sensitive sequence profile matching algorithms to identify and examine Ser/Thr and Tyr kinases encoded in the genome. The protocol used is identical to that adopted for analysis of kinases of other organisms earlier in this laboratory [43–49]. Briefly, we have employed multiple sensitive sequence search and analysis methods PSI-BLAST [37], MulPSSM [39–41] involving extensive use of RPS-BLAST [41] and HMMer [42] which match Hidden Markov Models (HMMs) to identify protein kinase catalytic domain and their co-occurring domains. The criteria used to associate a given protein kinase to a given subfamily on the basis of its primary structure include the degree of sequence identity greater than 30% with members of known subfamily of kinases and the presence of signature amino acids that are characteristics of protein kinase subfamilies [43] which include the glycine rich loop and catalytic aspartate of consensus sequence HRDLKXXN. In addition, search procedures such as PSI-BLAST have been used to detect sequences homologous to the kinase catalytic domain using an E-value cutoff of 0.0001 which is decided on the basis of previous prototypic study [50]. Truncated sequences which are less than 200 amino acids long were eliminated to arrive at a set of 534 PK- (protein-kinase-) like sequences. The data set of putative PK-like sequences has been obtained from the compilation of hits obtained during various search procedures. Out of these, 15 sequences lack aspartate in the catalytic loop and, therefore, are unlikely to function as kinases. These are referred as protein-kinase-like non-kinases (PKLNKs) [55]. These sequences were subjected to fold recognition approach PHYRE (http://www.sbg.bio.ic.ac.uk/phyre/) [51, 52] to ensure that they fold like kinases. The final number of 519 sequences is likely to function as protein kinases. The entire operation works stepwise with filtering of sequences at every stage in order to recognize kinases. The number of sequences involved at various stages is depicted in Figure 1.

2.2. Classification of Kinases into Hanks and Hunter Groups. Hanks and Hunter have proposed classification of kinases based on sequence analysis [38]. In order to classify the fugu kinases into these groups and subfamilies reverse PSI-BLAST (RPS-BLAST) was used to search each of the 519 PK-like sequences as a query against the database containing 2810 position-specific scoring matrices (PSSMs) created for the various subgroups of protein kinases corresponding to subfamilies of kinases. A query kinase sequence was associated to its subfamily based on the extent of sequence similarity. Sequences with greater than 30% identity and 70% profile coverage with at least one of the members of a kinase groups have been considered as members of the group or subfamily concerned. CLUSTALW [53] was used to generate multiple sequence alignment for the 519 kinases that were associated to specific groups. MEGA version 4 [54] was used to generate the dendrogram showing various groups of protein kinases. The sequences belonging to the group "Others" have been clustered using another dendrogram. Also MEGA 4 [54] has been used to cluster sequences of the kinase domain regions of certain families from other organisms including human, *Drosophila melanogaster, Caenorhabditis elegans*, and *Saccharomyces cerevisiae.*

2.3. Assignment of Domains to Multidomain Kinases. Domain assignments have been made for protein kinase catalytic domain containing gene products using the HMMer method by querying each of the kinase domain containing proteins against the protein family HMMs available in the Pfam database [56] and MULPSSM profiles [40, 41] of families in Pfam database. Transmembrane segments were detected using TMHMM [57].

2.4. Identification of MAPK Pathway Proteins. Sequences of MAPK proteins involved in the MAPK pathways in human were obtained from KEGG pathway database http://www.genome.jp/kegg/pathway.html [58]. The sequences were then used as a query to search using BLAST against the predicted protein sequences of fugu genome with an E-value cut off of 0.0001. The hits were obtained after pruning on the basis of coverage and percentage identity. The proteins which did not identify any homolog were then queried with an integrated dataset of fugu genome and SWISSPROT using PSI-BLAST [37] with an *E*-value cutoff of 0.001.The pathway was then generated as a network using CYTOSCAPE2.6.3 [59].

3. Results

The genome of fugu encodes 534 PK-like sequences. Of the 534 PK-like sequences 519 them possess the critical aspartate residue at the location characteristic of catalytic base and have at least one glycine conserved in the "glycine rich" loop GXGXXG present in the subdomain I of the kinase catalytic domain. A list of all these 519 sequences identified in this work along with classification and domain combinations are deposited in the KinG database [43] which was developed in this laboratory and the information is publicly available at http://king.mbu.iisc.ernet.in/. In addition, these 519 kinases are listed in the supplementary data file 1 in Supplementary Material available online at doi. 10.1155/2012/258284. The 15 sequences lacking the critical aspartate are unlikely to be functional as kinases though they are likely to adopt the kinase fold. Though the exact roles of these PKLNK's (protein kinase like non kinases) is not entirely clear, such sequences have been reported to have implications in various signaling pathways [60]. Out of the 519 putative functional kinases, 407 of them have an arginine residue preceding the critical aspartate and these are called as the "RD" kinases [48, 61]. This arginine has been indicated to be to be involved in an interaction with phosphate group of a phosphorylated sidechain in the activation loop of kinases. This interaction is known to be a critical step in switching a kinase from inactive to active state through conformational changes. Hence the switching mechanism of these RD kinases is likely to be mediated by phosphorylation at the activation segment in the kinase catalytic domain.

Among the putative PKs, 135 are likely to be tyrosine kinases and 296 are likely to be Ser/Thr kinases. 17 of the 296 Ser/Thr kinases are predicted to have membrane spanning regions. Similarly, 50 of the tyrosine kinases are predicted to be receptor tyrosine kinases. List of these kinases including the information on predicted transmembrane region is included in the supplementary data file 1.

3.1. Groupwise Distribution of Protein Kinases. The 519 kinases have been classified on the basis of the Hanks and Hunter [38] scheme of classifying protein kinases into 7 groups with clearly defined functional roles, namely, AGC (regulated by binding of second messengers), STE (proteins featuring in the MAPK signaling cascades in yeast), CMGC (include MAPK, CDK proteins), CAMK (Calcium/Calmodulin regulated kinases), CK1 (Casein kinase 1), TK (tyrosine kinase), and TKL (tyrosine kinase like). Apart from the 7 groups enlisted, there is yet another group which comprises of sequences that cannot be classified into any of the standard groups and are termed as "Other/Unclassified." The group-wise distribution of kinases in fugu has been shown in Figure 2 with the TK group being most prevalent and CK1 the least prevalent. Clustering of the sequences of the catalytic kinase domain using BLAST-CLUST was performed and 27 sequences are identified as outliers to the group/subfamily concerned. A dendrogram was constructed without these 27 sequences or the TK group and it resulted in clear distinct grouping of kinases (Figure 3). Though the members belonging to the "Other/Unclassified" group are significantly different from the classical groups, clusters are observed indicating high similarity among the sequences within a node (Figure 4). This points to the possible emergence of newer subfamilies of protein kinases that do not conform to the known groups [38] in the classification of kinases. However it is also possible that some of these "new subfamilies" represent outliers of currently known subfamilies. The group-wise distribution (percentage number of sequences in each group) has been compared with that of other model eukaryotic organisms such as human, *Drosophila melanogaster, Caenorhabditis elegans,* and *Saccharomyces cerevisiae* by performing similar analysis (Figure 5 and Table 1). The distribution of kinases in fugu is very similar to that of other eukaryotes considered and in most cases very similar to that of humans at the group level.

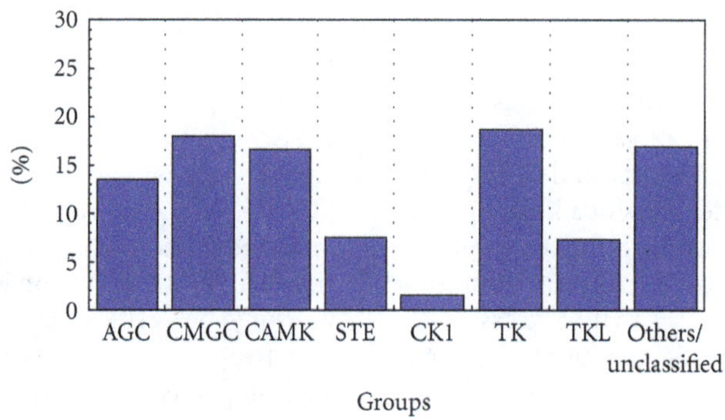

Figure 2: Percentage-wise distribution of fugu kinase groups. CAMK: calcium/calmodulin dependent protein kinase; CMGC: a group of cyclin dependent protein kinase, mitogen activated protein kinase, glycogen synthase kinase and casein kinase-2; AGC: a group of protein kinase A, protein kinase G, and protein kinase C; STE: sterile (homologs of yeast STE); TK: tyrosine kinase; TKL: tyrosine kinase like; CK1: casein kinase 1.

Figure 3: Dendrogram depicting clustering of fugu kinases. Abbreviations followed in the diagram are same as listed in the legend of Figure 2.

Table 1: Distribution of kinases of fugu, yeast, human, Drosophila, C. elegans into various Hanks and Hunter groups. Numbers indicated as percentage.

Groups	Organism				
	Fugu	Yeast	Human	*D. melanogaster*	*C. elegans*
AGC	13.48	15.23	11.72	17.35	11.04
CMGC	17.91	20.95	19.82	16.43	15.82
CAMK	16.57	22.85	16.72	15.51	12.53
STE	7.51	13.33	8.9	8.21	6.56
CK1	1.54	3.8	1.8	5.45	19.4
TK	18.68	0	20.34	15.52	20.59
TKL	7.32	0	7.2	6.39	3.28
Others/Unclassified	16.95	26.66	13.27	15.06	10.74

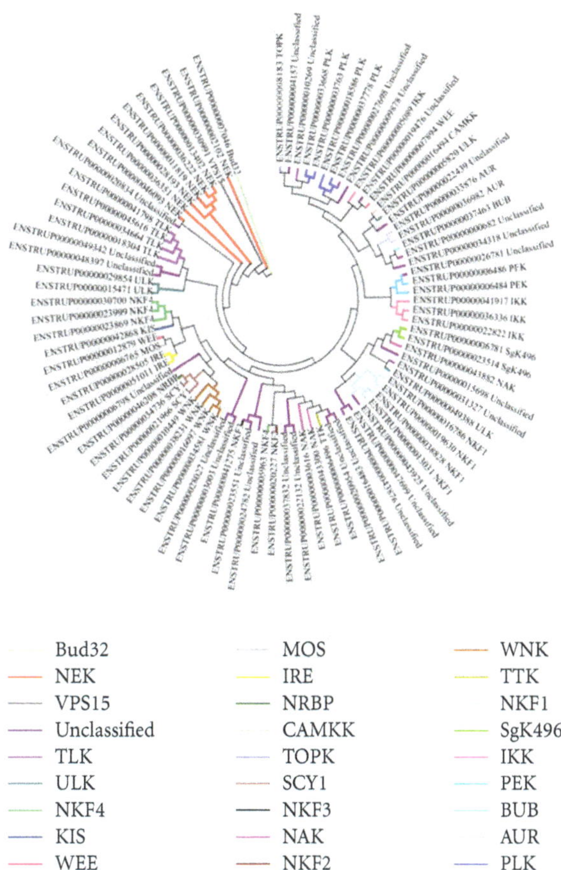

FIGURE 4: Dendrogram depicting clustering of kinases belonging to the "Other/Unclassified" group.

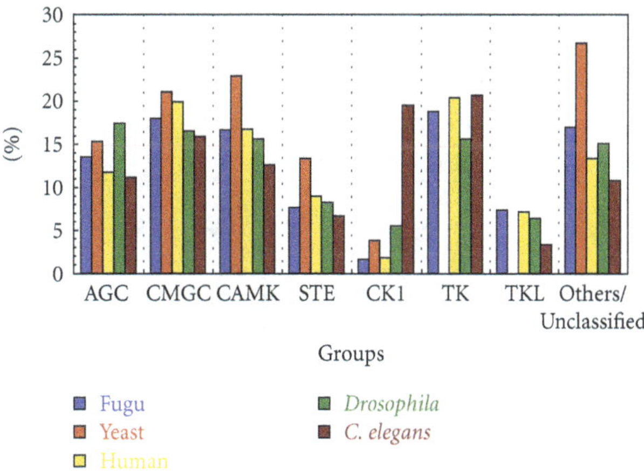

FIGURE 5: Comparison of group wise distribution of kinases from fugu, human, yeast, *Drosophila* and *C. elegans*. Abbreviations followed in the figure are same as given in the legend to Figure 2.

Though the group-wise distribution is quite comparable we wanted to explore at the level of sub-families to see if the same trend is observed. Interestingly, there are certain subfamilies in which the representation by fugu kinases is noticeably higher or lower than those of other model organisms considered. The percentage-wise distribution of these kinase subfamilies has been indicated in Figures 6 and 7. Figure 6 indicates distribution of kinases for 15 subfamilies in which fugu kinases occur in higher frequency with DYRK being highest and those of tyrosine kinase group generally being highly represented. Likewise Figure 7 shows distribution of kinases for 7 subfamilies in which fugu kinases occur at a lower frequency with CAMKK and KIS subfamilies being least. The functions of each of the

Table 2: Kinase sub-families with abnormal distribution in fugu. Overall function of each of the family is also indicated.

Family	Function
Highly represented families	
JakA	Receptor tyrosine kinase. Activates STAT involved in interferon signaling [7]
Lmr	Receptor tyrosine kinase involved in apoptosis [8]
PDGFR	Receptor tyrosine kinase activating factors for growth, differentiation, development [7]
Eph	Receptor tyrosine kinase component of developmental pathways [9]
InsR	Receptor that binds insulin and has a tyrosine-protein kinase activity [10]
PIM	Phosphorylating chromatin proteins and controlling transcription [11]
CDKL	Mediates phosphorylation of MECP2 [12]
Trio	Guanine nucleotide exchange factors that mediate cell invasiveness [13]
DYRK	Directs cellular response to stress conditions and also implicated in neuropathological characteristics of Down's syndrome [14, 15]
Trk	Receptor for neurotrophin-3 (NT-3) [16]
Src	Nonreceptor protein tyrosine kinase that plays pivotal roles in numerous cellular processes such as proliferation, migration, and transformation [17]
VEGFR	The VEGF-kinase ligand/receptor signaling system plays a key role in vascular development and regulation of vascular permeability [18]
Sgk496	Induces both caspase-dependent apoptosis and caspase-independent cell death [19]
Trbl	Interacts with MAPK kinases and regulates activation of MAP kinases [20]
Wnk	Controls sodium and chloride ion transport [21]
Underrepresented families	
KIS	Function unknown
MLCK	Calcium/calmodulin-dependent enzyme implicated in smooth muscle contraction via phosphorylation of myosin light chains [22]
MLK	Involved in the JNK pathway [23]
MAPKAPK	Integrative element of signaling in both mitogen and stress responses [24]
CDK	Involved in regulation of cell cycle by binding to cyclins [25]
CAMKK	Calcium/calmodulin-dependent protein kinase that belongs to a proposed calcium-triggered signaling cascade involved in a number of cellular processes [26]
Ste7	MAP2K homologous to yeast Ste7 [27]

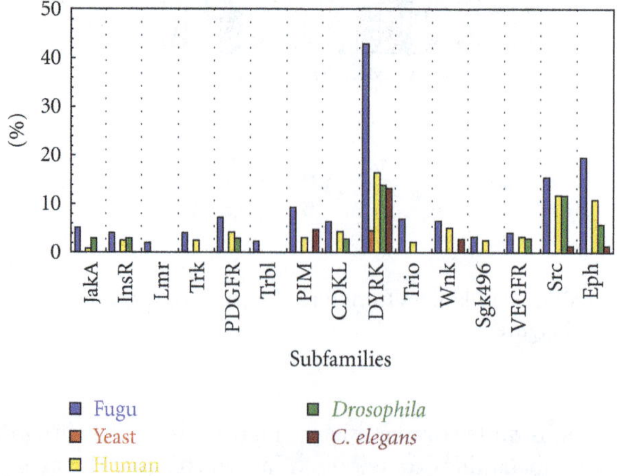

Figure 6: Percentage distribution of subfamilies that are overrepresented in fugu in comparison to various model organisms. JakA: janus kinase A; InsR: insulin receptor, Lmr: lemur kinase; Trk: neurotrophic tyrosine kinase receptor type 1; PDGFR: platelet derived growth factor receptor; Eph: ephrin receptor; VEGFR: vascular endothelial growth factor receptor; Trbl: tribbles; CDKL: cyclin-dependent kinase like; DYRK: dual specificity tyrosine regulated kinase, Wnk: with no lysine (K)' kinases.

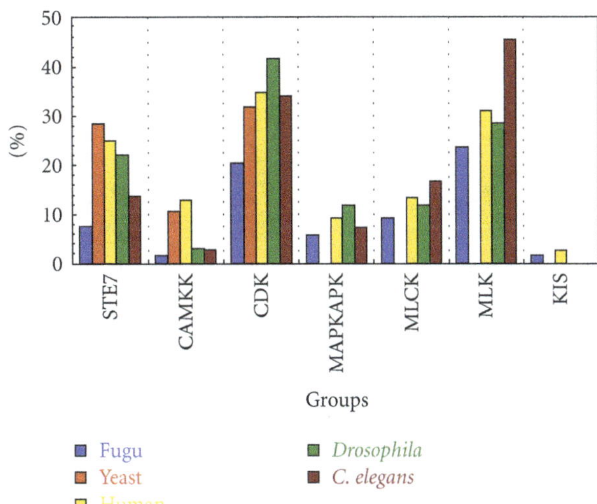

FIGURE 7: Percentage distribution of subfamilies that are under-represented in various model organisms. MLK: mixed lineage kinase; MAPKAPK: mitogen- activated protein kinase- activated protein kinase; MLCK: myosin light chain kinase; CDK: cyclin dependent kinase; CAMKK: calmodulin dependent protein kinase kinase; KIS: kinase interacting with stathmin; Ste7: sterile7.

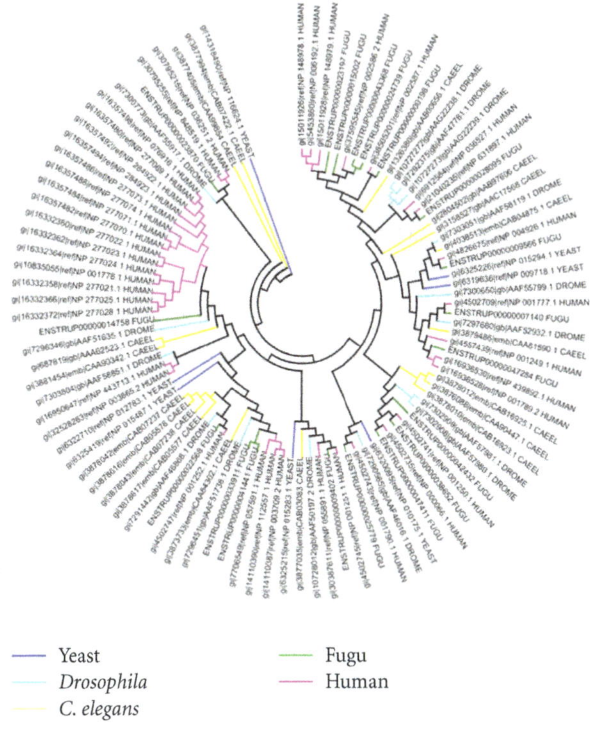

FIGURE 8: Dendrogram depicting clustering of kinases of CDK sub-family from fugu, human, *Drosophila*, and *C. elegans*.

proteins have been indicated in Table 2. Mutations or change in expression levels of these proteins has been implicated in various cancers (see the publications cited in Table 2). Also, profound differences in the number of paralogous proteins in two organisms could result in diverse outputs resulting in distinct features in the molecular processes in the two organisms [62]. The percentage distribution of each of the subfamilies for all the organisms considered can be obtained from Supplementary data file 2.

We have analyzed the nature of clustering of these fugu kinases with respect to those in other organisms. For most sub-families, fugu kinases group quite closely to those of human and in a few cases with other organisms. A similar trend is observed in the underrepresented sub-families with different clusters being observed that are not organism specific. The example of CDK subfamily is depicted in Figure 8. In each of the clusters human kinases dominate, with few representatives from fugu. Since paralogous proteins

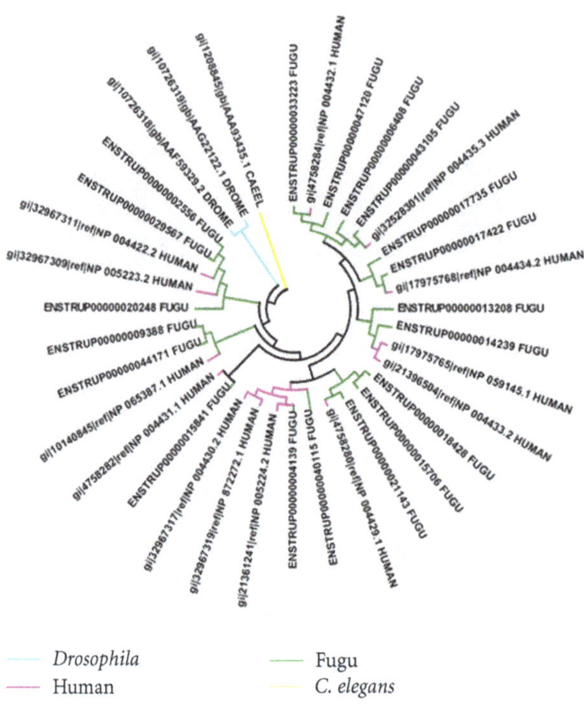

Figure 9: Dendrogram depicting clustering of kinases of ephrin receptor subfamily from fugu, human, *Drosophila*, and *C. elegans*.

perform diverse functions, their presence in large numbers is indicative of high functional diversity [62]. Therefore, it appears that paralogous CDKs in humans show larger functional diversity than fugu CDKs which are fewer in number. However, in the highly represented subfamilies of fugu kinases, apart from the above trend few kinases cluster separately suggesting higher divergence in a subset of DYRK kinases. Figure 9 shows the example of ephrin receptor family. In case of the DYRK family, 30 out of 40 kinases cluster separately (Figure 10). We then explored if this is a fish-specific trait. Fortunately another fish (zebrafish) genome has been sequenced and the list of zebrafish kinases are available in the KinG database [43]. However, the zebrafish has a large kinome (~900 kinases). Analysis on these shows that in most cases the fugu kinases do not cluster closely with those of zebrafish (supplementary data file 3). We performed a BLAST [37] search against the SWISSPROT database to identify the nearest homologs for these kinases in other organisms irrespective of the genome being sequenced fully or not. It was observed that, in the sequences which cluster close to human kinases, the percentage identity of those sequences with that of human counterpart is greater than 60%. Nevertheless, the kinases belonging to the fugu-specific cluster indicate low sequence identity with that of human and relatively higher identity of about 40% with that of certain lower eukaryotes like *Xenopus tropicalis, Dictyostelium discoideum,* and fungal species (supplementary data file 4). These may be fugu-specific sequences which are functionally divergent compared to those paralogs which show significant similarity to the human sequences. Similar expansions have been seen in other organisms as well [63] that help the organism to adapt to its environment.

3.2. Domain Combinations. The domain combinations for all the predicted kinases of fugu are provided in supplementary data file 1. Most of the domain combinations observed have been observed in various higher eukaryotes according to the Pfam database [56]. However there are 2 cases with an unusual domain combination as depicted in Figure 11.

In the first case, a galectin binding domain has been associated with protein kinase domain. The galectin domain is a carbohydrate binding domain and its function has been attributed to regulation of immunity and inflammatory responses, progression of cancer, and in specific developmental cascades [64]. These domains may function within or outside the cell. In this particular context, since there is no transmembrane component or domains which localize it to the membrane, the protein is likely to be cytosolic. This is further corroborated by the lack of any signal peptide motifs in this sequence which was analyzed using SignalP server [65]. The protein kinase domain may play a regulatory role wherein the activation state of the kinase might dictate the binding abilities of the galectin binding domain.

The second case involves a YkyA domain tethered to the protein kinase, CNH, PBD, PH, DMPK_Coil, C1_1, and the M protein repeats. Such a combination has been predicted also in another closely related fresh water species of Puffer fish, *Tetraodon nigroviridis,* however, without the YkyA domain. The YkyA domain has been reported only in bacterial species and is a putative lipoprotein binding domain occurring as a single-domain protein which aids in virulence [66]. It is likely that this protein is localized to the membrane due to the presence of the PH domain which is a reasonable indicator for membrane localization [67]. However, the role of YkyA in fugu is unclear.

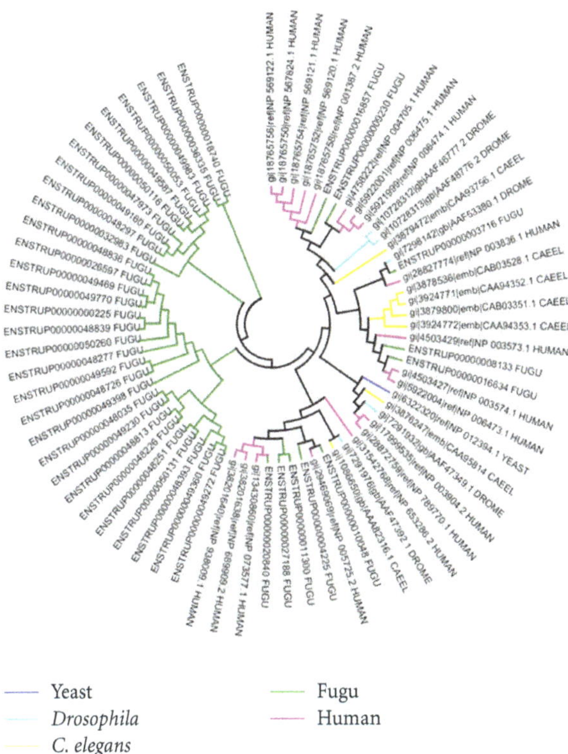

FIGURE 10: Dendrogram depicting clustering of kinases of dual specificity tyrosine regulated kinases (DYRKs) subfamily from fugu, human, yeast, *Drosophila*, and *C. elegans*.

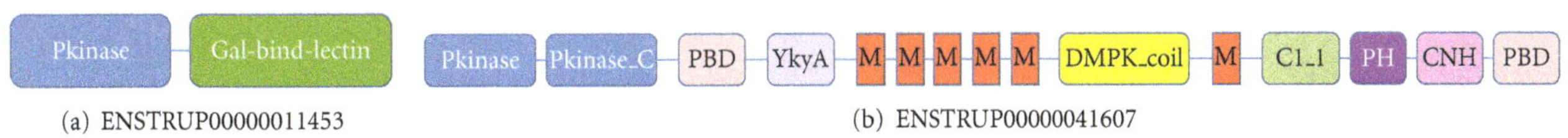

FIGURE 11: Unusual domain combinations. (a) Protein kinase tethered to a galectin binding domain. (b) Yka_A domain associated with protein kinase domain long with PBD, YkyA, M repeats, DMPK_coil, C1_1, PH, CNH. PBD: P21-Rho-binding domain; C1_1: phorbol esters/diacylglycerol binding domain; PH: pleckstrin homology.

Eight fugu kinase sequences are predicted to have two protein kinase domains each containing all the prerequisites for a functional kinase. This feature is distinct from that of Janus kinase which comprises a nonfunctional kinase domain (lacking critical aspartate) apart from a functional kinase domain. Five of these 8 cases correspond to the AGC group while the rest belong to the CAMK group. Such twin kinases have been reported in other higher eukaryotes in proteins like MAPKAP-K1 [68] wherein the C-terminal kinase domain is regulatory in nature and is involved in the activation of the N-terminal kinase domain. In all these 8 cases, both the kinase domains within a protein belong to the same group and sub-family which may be indicative of a duplication event and it may act by fine-tuning the activity levels of the protein. Interestingly, in most cases the C-terminal protein kinase domain has lower sequence identity to the respective groups than the N-terminal domain.

3.3. MAPK Signaling Pathway in Fugu. Given the slight unusual distribution of fugu kinases in terms of subfamilies we wanted to investigate the overall effect on a signaling pathway by considering distribution of all the proteins involved (including nonkinases) in the pathway. Although we did not observe any significant skewing in the clustering of the MAPK subfamily we chose to work on the MAPK pathway as the proteins involved in this pathway are extremely well characterized in other eukaryotes, especially for humans. Both kinases and non-kinases (including upstream factors and downstream effectors) were considered for the analyses. The extent of sequence similarity is an approximate indicator of the similarities of the functions of human and fugu proteins involved in MAPK pathway. Results are depicted pictorially in Figure 12. The cases with high sequence identity (>30%) and coverage (>70%) (compared to human proteins) have been represented as green boxes. These indicate the presence of functional counterparts, which are present in approximately 90% of the cases. Interestingly in a few cases (represented as yellow boxes) the sequence similarity levels reveal distant homologs suggesting differences in molecular events. These include proteins which act as ligands (FAS, TNF) and receptors (those of IL1, TNF, and certain lipopolysaccharides) which are mainly

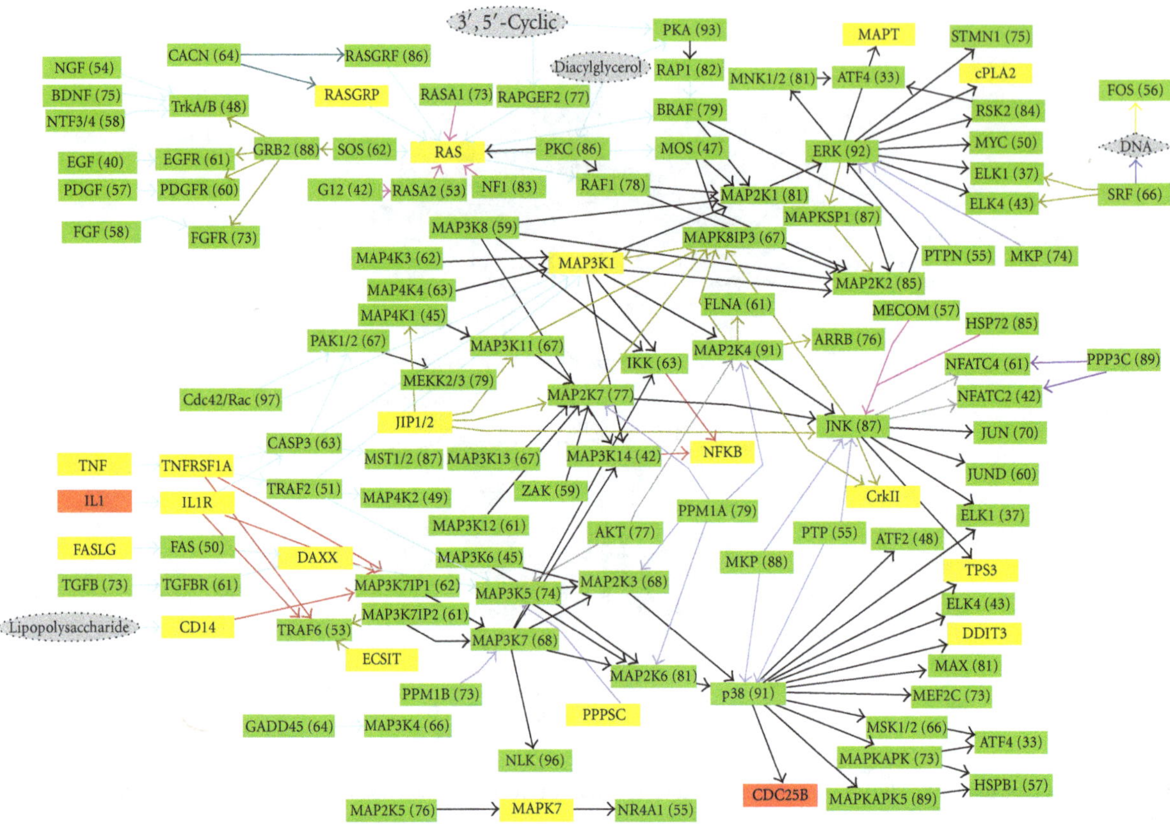

FIGURE 12: Human-fugu homologs MAPK pathway. Boxes in green represent closely related homologs of human MAPK pathway proteins present in fugu; yellow boxes indicate remote homlogs and red boxes indicate potential absence of homologs in fugu.

implicated in signaling pathways of immune system along with few proteins of the three tier MAPK (MAPKKK, MAPKK, MAPK) cascade. It also includes Ras and its activating protein RasGrp. This is especially interesting because Ras is a key component involved in GTP exchange which is a crucial step in activating the MAPK cascades. Even more glaringly, two proteins (IL1, CDC25b) do not have any identified homologs in fugu. Potential absence of IL1 is especially interesting because its absence indicates alternate ligands which are able to bind to its receptor. The other protein without any detectable homolog is CDC25b, which is a phosphatase involved in activating CDK. Absence of CDC25b may be viewed in the light of the fact that CDK subfamily is underrepresented in the fugu genome.

4. Conclusions

The current analysis on the fugu genome indicates a kinase repertoire of ~3% of the total genome which is slightly higher than an average of around 2% in the other model organisms. All the groups of eukaryotic protein kinases are found to be present in this genome with comparable numbers to that of humans. The observed distribution of few kinase subfamilies is fugu specific. The presence of unique domain combinations gives an insight into possibility of new regulatory functions of kinases. Finally, the similarity to signaling pathways in human may not only provide a platform for studying signaling systems but can also be used in kinase drug screening.

Acknowledgments

The authors are grateful to the anonymous referee for excellent suggestions in improving the presentation of this paper. This research is supported by Department of Biotechnology, Government of India.

References

[1] S. Aparicio, J. Chapman, E. Stupka et al., "Whole-genome shotgun assembly and analysis of the genome of fugu rubripes," *Science*, vol. 297, no. 5585, pp. 1301–1310, 2002.

[2] P. Gilligan, S. Brenner, and B. Venkatesh, "Fugu and human sequence comparison identifies novel human genes and conserved non-coding sequences," *Gene*, vol. 294, no. 1-2, pp. 35–44, 2002.

[3] O. Coutelle, G. Nyakatura, S. Taudien et al., "The neural cell adhesion molecule l1: genomic organisation and differential splicing is conserved between man and the pufferfish fugu," *Gene*, vol. 208, no. 1, pp. 7–15, 1998.

[4] G. Elgar, R. Sandford, S. Aparicio, A. Macrae, B. Venkatesh, and S. Brenner, "Small is beautiful: comparative genomics with the pufferfish (fugu rubripes)," *Trends in Genetics*, vol. 12, no. 4, pp. 145–150, 1996.

[5] B. Brunner, T. Todt, S. Lenzner et al., "Genomic structure and comparative analysis of nine fugu genes: conservation of synteny with human chromosome xp22.2-p22.1," *Genome Research*, vol. 9, no. 5, pp. 437–448, 1999.

[6] S. Brenner, G. Elgar, R. Sandford, A. Macrae, B. Venkatesh, and S. Aparicio, "Characterization of the pufferfish (fugu) genome as a compact model vertebrate genome," *Nature*, vol. 366, no. 6452, pp. 265–268, 1993.

[7] G. Manning, "Genomic overview of protein kinases," *Wormbook : the Online Review of C. Elegans Biology*, pp. 1–19, 2005.

[8] S. Kawa, J. Fujimoto, T. Tezuka, T. Nakazawa, and T. Yamamoto, "Involvement of brek, a serine/threonine kinase enriched in brain, in ngf signalling," *Genes to Cells*, vol. 9, no. 3, pp. 219–232, 2004.

[9] R. A. Lindberg and T. Hunter, "Cdna cloning and characterization of eck, an epithelial cell receptor protein-tyrosine kinase in the eph/elk family of protein kinases," *Molecular and Cellular Biology*, vol. 10, no. 12, pp. 6316–6324, 1990.

[10] A. Ullrich, J. R. Bell, and E. Y. Chen, "Human insulin receptor and its relationship to the tyrosine kinase family of oncogenes," *Nature*, vol. 313, no. 6005, pp. 756–761, 1985.

[11] N. Koike, H. Maita, T. Taira, H. Ariga, and S. M. M. Iguchi-Ariga, "Identification of heterochromatin protein 1 (hp1) as a phosphorylation target by pim-1 kinase and the effect of phosphorylation on the transcriptional repression function of hp1," *FEBS Letters*, vol. 467, no. 1, pp. 17–21, 2000.

[12] F. Mari, S. Azimonti, I. Bertani et al., "Cdkl5 belongs to the same molecular pathway of mecp2 and it is responsible for the early-onset seizure variant of rett syndrome," *Human Molecular Genetics*, vol. 14, no. 14, pp. 1935–1946, 2005.

[13] B. Salhia, N. L. Tran, A. Chan et al., "The guanine nucleotide exchange factors trio, ect2, and vav3 mediate the invasive behavior of glioblastoma," *American Journal of Pathology*, vol. 173, no. 6, pp. 1828–1838, 2008.

[14] K. Yoshida, "Nuclear trafficking of pro-apoptotic kinases in response to dna damage," *Trends in Molecular Medicine*, vol. 14, no. 7, pp. 305–313, 2008.

[15] J. Park, Y. Oh, and K. C. Chung, "Two key genes closely implicated with the neuropathological characteristics in down syndrome: dyrk1a and rcan1," *Bmb Reports*, vol. 42, no. 1, pp. 6–15, 2009.

[16] F. Lamballe, R. Klein, and M. Barbacid, "Trkc, a new member of the trk family of tyrosine protein kinases, is a receptor for neurotrophin-3," *Cell*, vol. 66, no. 5, pp. 967–979, 1991.

[17] J. T. Parsons, V. Wilkerson, and S. J. Parsons, "Structural and functional motifs of the rous sarcoma virus src protein," *Gene Amplification and Analysis*, vol. 4, pp. 1–19, 1986.

[18] M. A. McTigue, J. A. Wickersham, C. Pinko et al., "Crystal structure of the kinase domain of human vascular endothelial growth factor receptor 2: a key enzyme in angiogenesis," *Structure*, vol. 7, no. 3, pp. 319–330, 1999.

[19] J. Zha, Q. Zhou, L. G. Xu et al., "Rip5 is a rip-homologous inducer of cell death," *Biochemical and Biophysical Research Communications*, vol. 319, no. 2, pp. 298–303, 2004.

[20] E. Kiss-Toth, S. M. Bagstaff, H. Y. Sung et al., "Human tribbles, a protein family controlling mitogen-activated protein kinase cascades," *Journal of Biological Chemistry*, vol. 279, no. 41, pp. 42703–42708, 2004.

[21] F. Veríssimo and P. Jordan, "Wnk kinases, a novel protein kinase subfamily in multi-cellular organisms," *Oncogene*, vol. 20, no. 39, pp. 5562–5569, 2001.

[22] V. Lazar and J. G. N. Garcia, "A single human myosin light chain kinase gene (MLCK; MYLK) transcribes multiple nonmuscle isoforms," *Genomics*, vol. 57, no. 2, pp. 256–267, 1999.

[23] M. Masaki, A. Ikeda, E. Shiraki, S. Oka, and T. Kawasaki, "Mixed lineage kinase lzk and antioxidant protein-1 activate nf-κb synergistically," *European Journal of Biochemistry*, vol. 270, no. 1, pp. 76–83, 2003.

[24] M. M. McLaughlin, S. Kumar, P. C. McDonnell et al., "Identification of mitogen-activated protein (MAP) kinase-activated protein kinase-3, a novel substrate of csbp p38 map kinase," *Journal of Biological Chemistry*, vol. 271, no. 14, pp. 8488–8492, 1996.

[25] J. Pines, "The cell cycle kinases," *Seminars in Cancer Biology*, vol. 5, no. 4, pp. 305–313, 1994.

[26] S. Yano, H. Tokumitsu, and T. R. Soderling, "Calcium promotes cell survival through cam-k kinase activation of the protein-kinase-b pathway," *Nature*, vol. 396, no. 6711, pp. 584–587, 1998.

[27] B. R. Cairns, S. W. Ramer, and R. D. Kornberg, "Order of action of components in the yeast pheromone response pathway revealed with a dominant allele of the ste11 kinase and the multiple phosphorylation of the ste7 kinase," *Genes and Development*, vol. 6, no. 7, pp. 1305–1318, 1992.

[28] T. Pawson, "Introduction: protein kinases," *Faseb Journal*, vol. 8, no. 14, pp. 1112–1113, 1994.

[29] D. O. Frost, "Bdnf/trkb signaling in the developmental sculpting of visual connections," *Progress in Brain Research*, vol. 134, pp. 35–49, 2001.

[30] M. H. Lee and H. Y. Yang, "Negative regulators of cyclin-dependent kinases and their roles in cancers," *Cellular and Molecular Life Sciences*, vol. 58, no. 12-13, pp. 1907–1922, 2001.

[31] R. B. Irby, W. Mao, D. Coppola et al., "Activating src mutation in a subset of advanced human colon cancers," *Nature Genetics*, vol. 21, no. 2, pp. 187–190, 1999.

[32] J. Zheng, D. R. Knighton, L. F. ten Eyck et al., "Crystal structure of the catalytic subunit of cAMP-dependent protein kinase complexed with MgATP and peptide inhibitor," *Biochemistry*, vol. 32, no. 9, pp. 2154–2161, 1993.

[33] W. P. Yu, J. M. M. Tan, K. C. M. Chew et al., "The 350-fold compacted fugu parkin gene is structurally and functionally similar to human parkin," *Gene*, vol. 346, pp. 97–104, 2005.

[34] S. Baxendale, S. Abdulla, G. Elgar et al., "Comparative sequence analysis of the human and pufferfish huntington's disease genes," *Nature Genetics*, vol. 10, no. 1, pp. 67–76, 1995.

[35] J. N. Volff and M. Schartl, "Evolution of signal transduction by gene and genome duplication in fish," *Journal of Structural and Functional Genomics*, vol. 3, no. 1-4, pp. 139–150, 2003.

[36] Y. Niimura, "On the origin and evolution of vertebrate olfactory receptor genes: comparative genome analysis among 23 chordate species," *Genome Biology and Evolution*, vol. 1, pp. 34–44, 2009.

[37] S. F. Altschul, T. L. Madden, A. A. Schäffer et al., "Gapped blast and psi-blast: a new generation of protein database search programs," *Nucleic Acids Research*, vol. 25, no. 17, pp. 3389–3402, 1997.

[38] S. K. Hanks and T. Hunter, "The eukaryotic protein kinase superfamily: kinase (catalytic) domain structure and classification," *Faseb Journal*, vol. 9, no. 8, pp. 576–596, 1995.

[39] B. Anand, V. S. Gowri, and N. Srinivasan, "Use of multiple profiles corresponding to a sequence alignment enables effective detection of remote homologues," *Bioinformatics*, vol. 21, no. 12, pp. 2821–2826, 2005.

[40] V. S. Gowri, O. Krishnadev, C. S. Swamy, and N. Srinivasan, "Mulpssm: a database of multiple position-specific scoring

matrices of protein domain families," *Nucleic Acids Research.*, vol. 34, pp. D243–246, 2006.

[41] V. S. Gowri, K. G. Tina, O. Krishnadev, and N. Srinivasan, "Strategies for the effective identification of remotely related sequences in multiple pssm search approach," *Proteins: Structure, Function and Genetics*, vol. 67, no. 4, pp. 789–794, 2007.

[42] S. R. Eddy, "Profile hidden markov models," *Bioinformatics*, vol. 14, no. 9, pp. 755–763, 1998.

[43] A. Krupa, K. R. Abhinandan, and N. Srinivasan, "King: a database of protein kinases in genomes," *Nucleic Acids Research*, vol. 32, pp. D153–D155, 2004.

[44] A. Krupa and N. Srinivasan, "The repertoire of protein kinases encoded in the draft version of the human genome: atypical variations and uncommon domain combinations," *Genome Biology*, vol. 3, no. 12, p. RESEARCH0066, 2002.

[45] A. Krupa and N. Srinivasan, "Diversity in domain architectures of ser/thr kinases and their homologues in prokaryotes," *BMC Genomics*, vol. 6, 2005.

[46] Anamika, N. Srinivasan, and A. Krupa, "A genomic perspective of protein kinases in plasmadium falciparum," *Proteins: Structure, Function and Genetics*, vol. 58, no. 1, pp. 180–189, 2005.

[47] A. Krupa, G. Preethi, and N. Srinivasan, "Structural modes of stabilization of permissive phosphorylation sites in protein kinases: distinct strategies in ser/thr and tyr kinases," *Journal of Molecular Biology*, vol. 339, no. 5, pp. 1025–1039, 2004.

[48] A. Krupa, Anamika, and N. Srinivasan, "Genome-wide comparative analyses of domain organisation of repertoires of protein kinases of arabidopsis thaliana and oryza sativa," *Gene*, vol. 380, no. 1, pp. 1–13, 2006.

[49] K. Anamika and N. Srinivasan, "Comparative kinomics of plasmodium organisms: unity in diversity," *Protein and Peptide Letters*, vol. 14, no. 6, pp. 509–517, 2007.

[50] A. Müller, R. M. MacCallum, and M. J. E. Sternberg, "Benchmarking psi-blast in genome annotation," *Journal of Molecular Biology*, vol. 293, no. 5, pp. 1257–1271, 1999.

[51] L. A. Kelley, R. M. MacCallum, and M. J. E. Sternberg, "Enhanced genome annotation using structural profiles in the program 3d-pssm," *Journal of Molecular Biology*, vol. 299, no. 2, pp. 499–520, 2000.

[52] K. Bryson, L. J. McGuffin, R. L. Marsden, J. J. Ward, J. S. Sodhi, and D. T. Jones, "Protein structure prediction servers at university college london," *Nucleic Acids Research*, vol. 33, no. 2, pp. W36–W38, 2005.

[53] R. Chenna, H. Sugawara, T. Koike et al., "Multiple sequence alignment with the clustal series of programs," *Nucleic Acids Research*, vol. 31, no. 13, pp. 3497–3500, 2003.

[54] S. Kumar, K. Tamura, and M. Nei, "Mega: molecular evolutionary genetics analysis software for microcomputers," *Computer Applications in the Biosciences*, vol. 10, no. 2, pp. 189–191, 1994.

[55] K. Anamika, K.R. Abhinandan, K. Deshmukh et al., "Classification of nonenzymatic homologues of protein kinases," *Comparative and Functional Genomics*, vol. 2009, Article ID 365637, 17 pages, 2009.

[56] A. Bateman, E. Birney, L. Cerruti et al., "The pfam protein families database," *Nucleic Acids Research*, vol. 30, no. 1, pp. 276–280, 2002.

[57] A. Krogh, B. Larsson, G. Von Heijne, and E. L. L. Sonnhammer, "Predicting transmembrane protein topology with a hidden markov model: application to complete genomes," *Journal of Molecular Biology*, vol. 305, no. 3, pp. 567–580, 2001.

[58] M. Kanehisa and S. Goto, "Kegg: kyoto encyclopedia of genes and genomes," *Nucleic Acids Research*, vol. 28, no. 1, pp. 27–30, 2000.

[59] P. Shannon, A. Markiel, O. Ozier et al., "Cytoscape: a software environment for integrated models of biomolecular interaction networks," *Genome Research*, vol. 13, no. 11, pp. 2498–2504, 2003.

[60] L. R. Potter and T. Hunter, "Phosphorylation of the kinase homology domain is essential for activation of the a-type natriuretic peptide receptor," *Molecular and Cellular Biology*, vol. 18, no. 4, pp. 2164–2172, 1998.

[61] L. N. Johnson, M. E. M. Noble, and D. J. Owen, "Active and inactive protein kinases: structural basis for regulation," *Cell*, vol. 85, no. 2, pp. 149–158, 1996.

[62] E. V. Koonin, "Orthologs, paralogs, and evolutionary genomics," *Annual Review of Genetics*, vol. 39, pp. 309–338, 2005.

[63] G. Manning, D. S. Reiner, T. Lauwaet et al., "The minimal kinome of giardia lamblia illuminates early kinase evolution and unique parasite biology," *Genome Biology*, p. R66, 2011.

[64] H. Leffler, S. Carlsson, M. Hedlund, Y. Qian, and F. Poirier, "Introduction to galectins," *Glycoconjugate Journal*, vol. 19, no. 7-9, pp. 433–440, 2002.

[65] T. N. Petersen, S. Brunak, G. vonHeijne et al., "SignalP 4.0: discriminating signal peptides from transmembrane regions," *Nature Methods*, vol. 8, no. 10, pp. 785–786, 2011.

[66] W. Wei, Z. W. Cao, Y. L. Zhu et al., "Conserved genes in a path from commensalism to pathogenicity: comparative phylogenetic profiles of staphylococcus epidermidis rp62a and atcc12228," *Bmc Genomics*, vol. 7, article no. 112, 2006.

[67] M. Saraste and M. Hyvonen, "Pleckstrin homology domains: a fact file," *Current Opinion in Structural Biology*, vol. 5, no. 3, pp. 403–408, 1995.

[68] M. Deak, A. D. Clifton, J. M. Lucocq, and D. R. Alessi, "Mitogen- and stress-activated protein kinase-1 (mskl) is directly activated by mapk and sapk2/p38, and may mediate activation of creb," *Embo Journal*, vol. 17, no. 15, pp. 4426–4441, 1998.

16

Characterization of the OmyY1 Region on the Rainbow Trout Y Chromosome

Ruth B. Phillips,[1,2] Jenefer J. DeKoning,[1] Joseph P. Brunelli,[2,3] Joshua J. Faber-Hammond,[1] John D. Hansen,[4] Kris A. Christensen,[2,3] Suzy C. P. Renn,[5] and Gary H. Thorgaard[2,3]

[1] *School of Biological Sciences, Washington State University Vancouver, 14204 NE, Salmon Creek Avenue, Vancouver, WA 98686-9600, USA*
[2] *Center for Reproductive Biology, Washington State University, Pullman, WA 99164-7520, USA*
[3] *School of Biological Sciences, Washington State University, Pullman, WA 99164-4236, USA*
[4] *US Geological Survey, Western Fisheries Research Center, 6505 NE 65th Street, Seattle, WA 98115, USA*
[5] *Biology Department, Reed College, 3203 SE Woodstock Boulevard, Portland, OR 97202-8199, USA*

Correspondence should be addressed to Ruth B. Phillips; phillipsr@vancouver.wsu.edu

Academic Editor: G. Pesole

We characterized the male-specific region on the Y chromosome of rainbow trout, which contains both sdY (the sex-determining gene) and the male-specific genetic marker, OmyY1. Several clones containing the OmyY1 marker were screened from a BAC library from a YY clonal line and found to be part of an 800 kb BAC contig. Using fluorescence *in situ* hybridization (FISH), these clones were localized to the end of the short arm of the Y chromosome in rainbow trout, with an additional signal on the end of the X chromosome in many cells. We sequenced a minimum tiling path of these clones using Illumina and 454 pyrosequencing. The region is rich in transposons and rDNA, but also appears to contain several single-copy protein-coding genes. Most of these genes are also found on the X chromosome; and in several cases sex-specific SNPs in these genes were identified between the male (YY) and female (XX) homozygous clonal lines. Additional genes were identified by hybridization of the BACs to the cGRASP salmonid 4x44K oligo microarray. By BLASTn evaluations using hypothetical transcripts of OmyY1-linked candidate genes as query against several EST databases, we conclude at least 12 of these candidate genes are likely functional, and expressed.

1. Introduction

Salmonid fishes have the XX/XY system of sex determination [1]. Most rainbow trout have morphologically distinguishable sex chromosomes [2], with the short arm being longer in the X chromosome than the Y, primarily due to presence of 5S rDNA sequences on the X chromosome (reviewed in [3]). Some wild fish and several male clonal lines, including the Swanson YY and Arlee YY clones have a Y chromosome that resembles the X chromosome in having a longer short arm including the 5S rDNA sequences [4]. This fact and the viability seen in rainbow trout and Chinook salmon YY individuals [5, 6] suggest that the X and Y chromosomes share considerable genetic content.

The SEX locus is not found on a common linkage group in Pacific salmon and trout, but rather each species has the SEX locus on a different linkage group as shown by genetic mapping and localization of male-specific markers using in situ hybridization with clones to GH-Y [7–9]. (Rainbow and cutthroat trout are an exception to this [10], and hybrids between these two species are interfertile.) Although it is possible that each species has a different master sex determining gene, this is unlikely because there are several male-specific markers shared by the *Oncorhynchus* species. These include OmyY1, found in all Pacific trout and salmon [11], OtY2 [12] which is related to OmyY1, and GH-Y [13], which are found in most species of Pacific salmon, and OtY1, found only in Chinook salmon [14]. These results support the hypothesis that a small chromosomal segment containing the SEX locus, or the small short arm containing the SEX locus, is transposing to a new chromosome in each of these species (reviewed in [9, 15]).

The OmyY1 genetic marker, which has been used to sex fish from all of the *Oncorhynchus* species, was isolated from lambda libraries prepared from male fish of both rainbow trout (OSU × Hotcreek) and Chinook salmon (where the region is referred to as OtY3) [11]. Recently, evidence has been presented supporting the hypothesis that an immune function related-gene, sdY, which is found within the rainbow trout OmyY1 lambda clone, may be the master sex determining gene for many species in the Salmoninae [16]. We present evidence based on sequence information of BAC clones isolated from the OmyY1 region of rainbow trout that there are a number of sex-linked genes shared between the X and Y chromosomes, implying the sex chromosomes are in the early stages of differentiation. Some of the genes in the region adjacent to sdY show male-specific expression.

2. Materials and Methods

2.1. Doubled Haploid Rainbow Trout Clonal Lines and Crosses. Type I loci and other markers were sequenced in four doubled haploid parental lines produced by androgenesis (female Oregon State University (OSU), female Whale Rock (WR), male Swanson (SW), and male Arlee (AR)). Markers were mapped in a cross of (OSU × AR) to determine concordance with phenotypic sex. These fish were previously used to generate a dense genome-wide linkage map [17–19] to analyze quantitative trait loci for a variety of traits [20, 21] and to characterize linkage of markers on the Y chromosome [4, 22]. In all crosses, sex phenotype was determined by internal examination of gonads.

2.2. BAC Library Screening for OmyY1 Marker Associated BACs. High density filters corresponding to the 5X BAC DH YY male Swanson (SW) library (EcoR1 set) for rainbow trout were screened using [^{32}P] dCTP-labeled amplicons from the OmyY1 region. For screening the filters, two probes were generated by PCR amplification of SW genomic DNA using previously described primers to the OmyY1 region [11]. These amplicons were TOPO TA cloned and transformed in JM107 *E. coli*. After identification and isolation of OmyY1+clones, plasmid DNA was extracted and used as a probe to screen the filters. The product was purified (Qiagen) and then labeled with ^{32}P-dCTP (Amersham) using Ready-to-Go beads (GE Healthcare) according to the manufacturer's suggestions. The labeled probe was purified using G-50 columns (ProbeQuant, GE Healthcare) and then used for hybridization. Filters were hybridized (5X SSC, 1% SDS, 0.5% sodium pyrophosphate, 0.5% nonfat milk and 10% dextran sulfate at 65°C) and then washed (0.5% SSC/0.5% SDS, 65°C) under stringent conditions. Positive clones were confirmed by PCR using primer sets corresponding to OmyY1. This resulted in many hits, but only the strongest signals were further evaluated by PCR and FISH. BACs were named according to the coordinates on the original plate. Clones were obtained from the National Center for Cool and Cold Water Aquaculture, ARS-USDA as stab cultures. BAC DNA for PCR confirmation was prepared from selected clones grown in 300 mls LB broth + chloramphenicol (12.5 μg/mL) overnight at 37C with shaking at 250 rpm, and DNA was isolated using a Qiagen Plasmid Maxi Kit.

2.3. PCR Screening of OmyY1 Marker Associated BACs. Further PCR screening of the selected clones was performed in a PTC100 or 200 Thermalcycler (MJ Research) using primers noted above. Reactions were carried out in 10 μL volumes containing 0.5 U Taq DNA Polymerase (GenScript), .4 uM primers, 1 μL 10x PCR buffer containing 1.5 mM $MgCl_2$, 200 μM dNTPs, and 1 μL of 40 ng BAC DNA template. PCR conditions comprised of an initial denaturation step of 95°C for 5 min; 35 cycles with a denaturation step of 95°C for 30 sec, 45 sec at specific annealing temperature, and extension at 72°C for 45 sec followed by a final extension at 72°C for 5 minutes. PCR products were separated on a 1.2% agarose gel containing .5x TBE and GelRed Nucleic Acid stain (Phenix). The DNA fragments were visualized using a Gel Logic 100 system and UV trans-illuminator (Fisher Biotech).

2.4. Assembly of the OmyY1 BAC Contig. BAC clones RE223F6 and RE143K8 were confirmed using OmyY1-specific PCR primers and sent to Ming Chen Luo at University of California, Davis for assembly of a contig with the Swanson (SW) physical map [23]. Contig #6256 was identified as encompassing the OmyY1 BACs and included 60 additional BACs. Eleven BACs spanning the contig were ordered to use for FISH and end-sequencing and included RT399M07, RT278E18, RT423A24, RT399C14, RT439A05, RT578J04, RT450D22, RT015B10, RT492A21, RT004L12, and RT225N07. After BAC DNA isolation as described above, and determination of DNA concentration, aliquots were sent to the WSU Pullman Molecular Biology Core for end-sequencing using either T7, SP6, or M13 primers.

2.5. Localization of OmyY1 BAC Clones on Rainbow Trout Chromosomes. Blood was cultured from rainbow trout using standard methods [4]. Rainbow trout BAC DNA was labeled with Spectrum Orange using a nick translation kit (Abbott Molecular). Human placental DNA (2 μgs) and Cot-1 DNA (1 μg, prepared from rainbow trout) were added to the probe mixture for blocking. Hybridizations were carried out at 37°C overnight and posthybridization washes were as recommended by the manufacturer (Abbott Molecular) with minor modifications [24]. Antibodies to Spectrum Orange (Molecular Probes) were used to amplify the signal. Slides were counterstained with 4,6-diamidino-2-phenylindole (DAPI) at a concentration of 125 ng DAPI in 1 mL antifade solution (Abbott Molecular). Images were captured with a Jai camera and analyzed with Cytovision Genus (Applied Imaging, Inc.) software. Chromosomes were arranged according to size within the metacentric/submetacentric and acrocentric groups.

2.6. Characterization of the OmyY1 BAC Contig: Illumina Sequencing. A minimum tiling path of BAC clones from the OmyY1 region was sequenced by Amplicon Express, Inc. Pullman, WA by Focused Genome Sequencing (FGS). FGS is a next-generation sequencing (NGS) method developed at Amplicon Express that allows very high quality assembly of BAC clone sequence data using the Illumina HiSeq (San Diego, CA). The proprietary FGS process makes NGS tagged libraries of BAC clones and generates a consensus sequence of the BAC clones.

Five OmyY1 BACs spanning the OmyY1 physical map contig (RT578J04, RT450D22, RT015B10, including BAC clones RE223F06, and RE143K08 containing the OmyY1 locus) were sequenced using Roche 454 GS FLX next-generation sequencing technologies, by the WSU Pullman bioinformatics genomics core lab. This sequence evaluation yielded 94325 total reads, assembling 188 contigs ranging in size from 50.4 kb to <100 bp.

2.7. Characterization of the OmyY1 BAC Contig: Comparative Genomic Hybridization. We used a 4 x 44 k Agilent salmonid microarray [25] to determine gene homologies within the OmyY1 contig through Comparative Genomic Hybridization (CGH). This microarray was designed using 60 bp oligos from the 3′ end of most known cDNAs from Rainbow Trout or Atlantic salmon. We pooled six clones from the contig (RT492A21, RT450D22, RT015B10, RT423A24, RT004L12, and RE143K08) and labeled with both Alexa Fluor 5 and and Alexa Fluor 3 using the BioPrime Total Genomic Labeling System (Invitrogen). Samples were hybridized to the salmonid microarray in duplicate versus unrelated BAC clones using the manufacturer's protocol for CGH (Agilent). Using the R statistical software package, microarray data were floored at 2 standard deviations above the local background then normalized with a LOESS normalization method. Color ratios were fit to a Linear Model and then analyzed using empirical Bayes statistics [26]. Positive hits on the array features with significant P values and higher fluorescence intensity for OmyY1 contig samples than the competing samples from unrelated BAC clones.

False positives were removed from the initial list of positive hits by conducting a stand-alone BLAST of the 60 bp oligo sequences from the array against a first draft rainbow trout genome (M. R. Miller, unpublished) and removing any repetitive elements. A second BLAST inquiry was conducted in GenBank, and additional repetitive elements were removed from the list. Additionally, vector (*E. coli*) genomic DNA was hybridized to the microarray, and overlapping hits with those from the OmyY1 contig positive hit list were removed. The final positive hit list represented possible candidates for functional genes closely linked to OmyY1. Genes of interest were confirmed by searching the Illumina assembly and/or by PCR using primers designed to the EST target in the OmyY1 BAC clones and rainbow trout genomic DNA, followed by Sanger sequencing on an ABI 3130xl.

Once possible candidates for functional genes were confirmed within the OmyY1 BAC contig, the corresponding BAC sequences were compared to EST sequences from the 4 x 44 k salmonid microarray annotation file, the cGRASP rainbow trout EST database (http://web.uvic.ca/grasp/), or cDNA sequences from GenBank. In addition, the list of possible functional genes was BLAST-analyzed against both the rainbow trout and Atlantic salmon [27] draft genome assemblies to identify potential paralogs and eliminate additional repetitive sequences. Candidate genes were proposed to be functional within the OmyY1 contig if a full-length hypothetical mRNA could be identified within the BAC contig sequence. Criteria for functionality included presence of start and stop codons, with no in-frame stop codons. Gene identity required corresponding BAC sequence to be 100% identical to candidate genes, excluding compressions found in 454 and Illumina sequencing of homopolymers. For microarray hits not found in the BAC contig assembly, corresponding rainbow trout-specific expressed sequences were compared to Sanger sequenced PCR products from OmyY1 contig BAC clones or to OmyY1-linked scaffolds from the draft rainbow trout genome assembly.

2.8. PCR Confirmation and Sanger Sequencing of BAC Clones Containing Loci Identified in BLAST Analysis of Illumina Sequencing Results and SNP Identification. GenBank BLASTn and BLASTx analyses were performed on contig sequences obtained from the Illumina/454 OMY BACs assembly of the 6 OmyY1 BAC clones spanning the minimum tiling path of the physical map contig #6256. Following identification of regions of homology within contigs to various teleost annotated gene loci, the resulting sequences were subsequently screened for suspected and known repetitive elements in salmonids using a salmonid-specific repeat masker (http://lucy.ceh.uvic.ca/repeatmasker/cbr_repeatmasker.py).

PCR primers were designed either to the OMY BAC contig sequence itself, or to the GB cDNA or EST sequence available (see Table 3). Additionally, subsequent to the Comparative Genomic Hybridization (CGH) results, ESTs identified as positive hits were also used for template in target design. ESTs identified by CGH were aligned with the rainbow trout draft genome sequence, and scaffolds with significant homology provided another source for primer design.

PCR amplified products were treated with 2 U of FastAP (Fermentas) and 1 U Exonuclease I (USB) and sequenced directly using 2.3 uls of the sequencing RR mix from the Big Dye Terminator v3.1 cycle sequencing kit (ABI), 2 μls of the provided 5x buffer, .4 μm primer, and .5–2 μL of the PCR reaction. Reactions were run for 29 cycles with 4 minute extension at 72C , then cleaned by CleanSEQ magnetic bead separation (Agencourt) and run on an ABI 3130xl sequencer. Sequence reads were aligned using DNASTAR Seqman II v5 sequence analysis software.

3. Results

3.1. The OmyY1 BAC Contig. Three BAC clones containing the OmyY1 sequence were confirmed following screening of the Swanson rainbow trout YY RE library. These clones overlapped with a contig from the Swanson RT library which originally contained 30 clones. Most recently a third generation BAC physical map has been prepared with BACs from the three Swanson BAC libraries: RE (EcoRI), RB (BamHI), and RT (HindIII) (Y. Palti, pers. com.). This updated map expands the previous OmyY1 BAC-containing contig, now

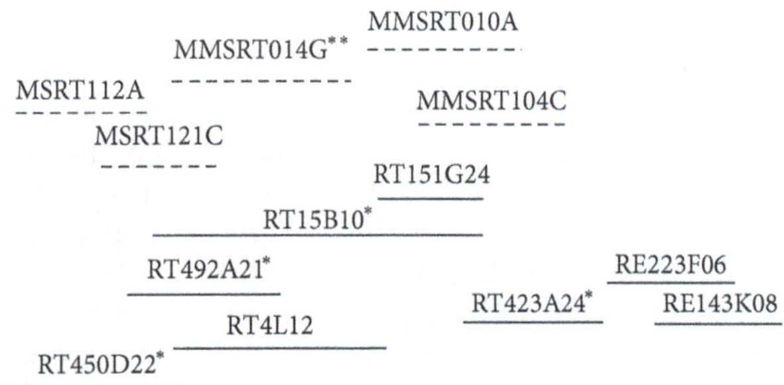

FIGURE 1: Diagram of the OmyY1 BAC contig showing clones that we examined in this project. The dashed lines above the BAC clones indicate Scaffolds from Mike Miller's rainbow trout genome assembly. *Indicates that the clone was sequenced by Illumina.**Indicates scaffolds that did not include BACs known to be in the OmyY1 contig (6256), but which apparently had BACs that are in this region because sequences of genes identified in the OmyY1 BACs matched the sequence of genes in these scaffolds exactly.

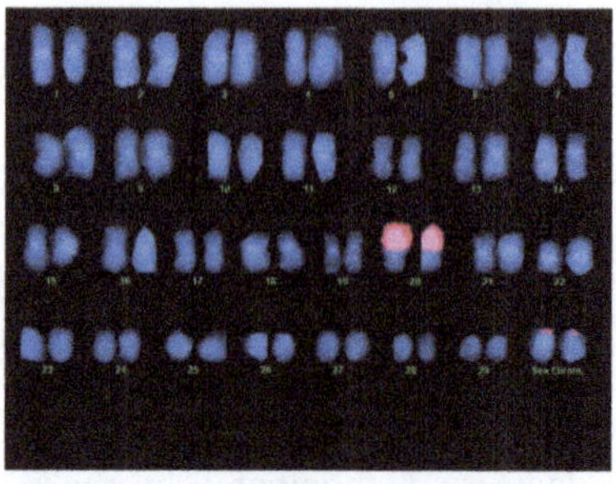

(a)

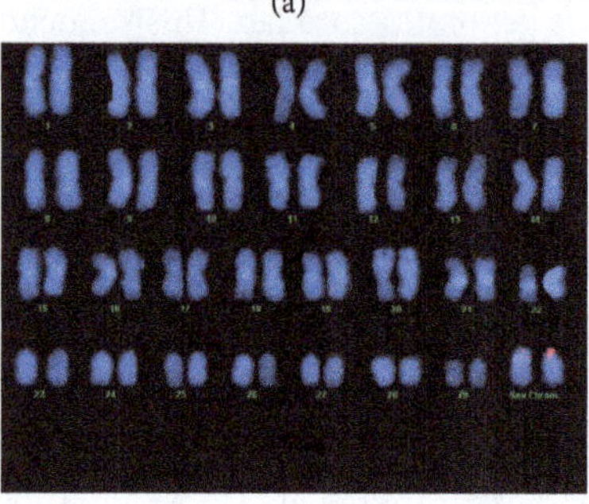

(b)

FIGURE 2: (a) Hybridization of the BAC clone RE143K08 (which is found near the 3′ end of the OmyY1 contig and contains the OmyY1 marker, the sdY gene and ribosomal DNA) to male rainbow trout chromosomes. Note the signals on the sex chromosome pair and chromosome pair 20, the site of the 18S rDNA locus in rainbow trout. (b) Hybridization of the BAC clone RT429A21 to male rainbow trout chromosomes. This BAC is close to the 5′ end of the OmyY1 contig (6256) (see Figure 1). Note the larger signal on the Y chromosome compared to the X.

contains 60 clones, and is designated as #6256. Although a PCR product was obtained following amplification with OmyY1 primers in most of the clones, we were able to sequence the OmyY1 product from only two of the RE BAC clones. The proposed sex determining gene, sdY [16], is found in the same two RE BAC clones and is contained within the 22 kb OmyY1 GenBank deposition. A diagram showing the clones from physical map contig #6256 analyzed in this project is shown in Figure 1.

3.2. Localization of OmyY1 BAC Clones on Rainbow Trout Chromosomes. A dozen BAC clones from physical map contig #6256 were used as probes in FISH (fluorescence *in situ*) experiments for hybridization to rainbow trout chromosomes. All of these clones hybridized to the end of the short arm of the Y chromosome of rainbow trout with some signal observed on the end of the short arm of the X chromosome in many cells (Figures 2(a) and 2(b)). Hybridization was done on chromosomes prepared from OSU × Hot Creek hybrids. The Hot Creek strain has a Y chromosome which is morphologically distinct from the X, so the X and Y chromosomes can be easily distinguished, with the short arm of the Y being smaller than the short arm of the X chromosome [2]. Some of the BAC clones contained ribosomal DNA and those clones also hybridized to the NORs on chromosome 20 in rainbow trout (Figure 2(a)). BAC clones in the middle of the contig hybridized only to the Y chromosome in many cells (Figure 2(b)). These clones did not contain ribosomal DNA. The BAC clones containing OmyY1 and sdY did not hybridize to the X or Y chromosomes in Chinook Salmon (data not shown). As mentioned above, these clones contained 18S ribosomal DNA and hybridized only to known sites of rDNA on Chinook chromosomes.

3.3. Characterization of the OmyY1 BAC Contig: Illumina and 454 Sequencing Results. Five BACs spanning the OmyY1 physical map contig #6256 (RT450D22, RT015B10, RT492A21, RT004L12, and RT423A24) were Illumina sequenced and assembled by Amplicon Express (http://www.ampliconexpress.com/). The BAC clones were sequenced by Focused Genome Sequencing (FGS). FGS is a next-generation sequencing (NGS) method developed at Amplicon Express that allows very high quality assembly of BAC clone sequence data using the Illumina HiSeq (San Diego, CA). The proprietary FGS process makes NGS tagged libraries of BAC clones and generates a consensus sequence of the BAC clones. Illumina sequencing yielded an average of 2,412,973 reads per BAC clone assembling into 56 contigs ranging in size from approximately 245 kb to 1 kb. Additionally, five OmyY1 BACs spanning the OmyY1 physical map contig (RT578J04, RT450D22, RT015B10, including BAC clones RE223F06, and RE143K08 containing the OmyY1 locus) were sequenced using Roche 454 GS FLX next-generation sequencing technologies, by the WSU Pullman bioinformatics genomics core lab.

Assembled DNA sequence contigs were evaluated directly for gene coding content using NCBI-BLASTx [28] and also redundantly masked for repetitive elements using the salmonid-specific repeat masker prior to NCBI-BLASTx evaluation, to identify candidate gene sequences. BLASTx coding content homologies provided access to GenBank gene sequence data depositions, which were then used in NCBI distributed stand-alone BLAST-2.2.25 application

Table 1: List of predicted protein coding genes, pseudogenes, and unknown transcripts found in rainbow trout physical map contig #6256 sequence assemblies. Pseudogenes were differentiated from protein coding genes based on the presence of premature stop codons in open reading frames. The BAC clone column denotes the clone in the contig that gave a PCR product for the gene in question.

Gene name	Predicted gene function	PCR amplified in BAC clone	Annotation
Oncorhynchus mykiss sdY	Protein coding	143K08/223F06	Genbank AB626896.1
DENN domain-containing protein 4B-like (dennd4b-like)	Protein coding	RT015B10	Genbank KC686345
Immunoglobulin superfamily DCC subclass member 3-like (Igdcc3-like) [partial]	Protein coding	RT015B11	Genbank KC686344
cAMP-responsive element-binding protein 3-like protein 4 (cr3l4)	Protein coding	RT015B12	Genbank KC686342
F-box/WD repeat-containing protein 2 (FBXW2)	Protein coding	RT423A24	Genbank KC686346
Na/K/2Cl co-transporter (nkcc1a)	Protein coding	RT450D22	Genbank KC686348
Oocyte protease inhibitor-1 (opi-1)	Protein coding	RT004L12	Genbank KC686349
Zinc/iron-regulated protein (zip1)	Protein coding	RT004L12	Genbank KC686350
CREB-regulated transcription coactivator 2-like (TORC2-like)	Protein coding		Genbank KC686350
Zinc finger CCHC domain-containing protein 3-like	Protein coding		Genbank KC686343
General transcription factor II-I repeat domain-containing protein 2-like	Protein coding		Genbank KC686351
rRNA promoter binding protein	Protein coding		Genbank KC686343
Interleukin enhancer-binding factor 2 (ILF-2)	Pseudogene	RT015B10	Genbank KC686347
Unknown (Some homology to laminin subunit beta-1-like)	Unknown		cGRASP EST omyk-BX860091
Unknown	Unknown		cGRASP EST omyk-CB488087
Unknown (Some homology to Sec16A)	Unknown		cGRASP EST Contig22207
Unknown	Unknown		cGRASP EST omyk-DV199273
Unknown	Unknown		cGRASP EST Contig19333

evaluations. The contigs containing the OmyY1 BAC clone illumine sequence data were put into the Standalone BLAST databases format and then BLASTn was conducted on this OmyY1 database with downloaded candidate gene sequences for comprehensive gene sequence content homology ([16, 29] and http://web.uvic.ca/grasp/). When assembled, sequence contigs were found to sequentially contain gene coding-region content, and these adjacent assemblies were evaluated for flanking DNA sequence homology to allow assembly of larger continuous sequence reads spanning the candidate gene sequence. Candidate gene sequences were characterized for intron-exon boundaries by sequence alignment with GenBank mRNA or predicted gene sequence depositions.

By this method the following 12 predicted protein coding gene sequences have been identified (including sdY), partially annotated and deposited to GenBank (Table 1). Five of these genes have confirmed transcripts in at least one of five different databases that were searched and these are reported in Table 2. Additionally, one predicted pseudogene was identified in the contig #6256 sequence assemblies (Table 1), determined by presence of premature stop codons relative to annotated predicted proteins from other species (data not shown). Further BLAST analyses within the cGRASP EST cluster rainbow trout databases yielded an additional 5 OmyY1 linked transcripts for unknown genes: omyk-BX860091, omyk-CB488087, Contig22207, omyk-DV199273, and Contig19333 (Tables 1 and 2). BLASTx in GenBank revealed that omyk-BX860091 shows poor homology to laminin subunit beta-1-like (best homology is 63% identity to *Oreochromis niloticus*) and Contig22207 has partial homology to Sec16A (all homology <55% identity for <25% of the transcript). No predicted protein sequences for these unknown genes could be analyzed.

3.4. Characterization of the OmyY1 BAC Contig: Comparative Genomic Hybridization Results. A 4 x 44 k Agilent salmonid microarray [25] was used to identify gene homologies within the OmyY1 contig. The preliminary list of hits from the OmyY1 contig contained 124 expressed sequences. Following the removal of duplicates within the array and false positive hits by removing overlapping hits with unrelated BAC clones, we were left with 48 ESTs (see Supplemental File 1 of the Supplementary Material available online at http://dx.doi.org/10.1155/2013/261730). This list was compared to genes found in the Illumina and 454 assemblies to confirm that the CGH yielded true positives, with six genes found on both lists: cAMP-responsive element-binding protein 3-like protein 4, F-Box/WD repeat containing protein 2, Na/K/2Cl co-transporter, CREB-regulated transcription coactivator 2, DENN domain-containing protein 4B, and Zinc transporter ZIP1. No other genes confirmed in the

Table 2: List of predicted nonrepetitive transcribed genes in rainbow trout physical map contig #6256 with corresponding EST's in at least one database. EST designations in bold had the highest homology to OmyY1 linked sequence, and those sequences were used for homology and coverage analyses.

Gene name	4 × 44 K salmonid oligo array designation	cGRASP EST cluster contig annotation	GenBank Annotation	Miller draft rainbow trout genome scaffold	Miller scaffold homology (base pair identities)	Coverage in Miller scaffold	OmyY1 contig (Illumina unless noted)	OmyY1 contig homology (base pair identities)	Coverage within OmyY1 contig	PCR amplification in BAC clone	Transcribed gene in OmyY1 contig?
Oncorhynchus mykiss sdY			**AB626896.1**	n/a	n/a	n/a	454 contig_3	3270/3273	100%	143K08/223F06	Yes
F-Box/WD repeat containing protein 2 (FBXW2)	C262R066/ C179R024	**Contig53026**		MMSRT112A/ 121C (overlap)	2319/2326	100%	Contig_2	2319/2326	100%	RT450D22/ RT492A21	Yes
Unknown (Some homology to laminin subunit beta-1-like)		**omyk-BX860091**		MMSRT112A/ 121C (overlap)	747/778	100%	Contig_2	747/778	100%		Likely
Na/K/2Cl co-transporter (nkcc1a)	C029R137	**omyk-CR375193**		MMSRT112A	832/843	100%	Contig_2	832/843	100%	RT450D22	Yes
Oocyte protease inhibitor-1 (opi-1)		**Contig59765**		MMSRT010A/ 121C	919/940, 883/906	97%, 94%	Contig_2	883/906	94%		Likely in MMSRT010A scaffold
Unknown		**omyk-CB488087**		MMSRT010A/ 121C	315/315, 311/315	100%, 100%	Contig_2	311/315	100%		Likely in MMSRT010A scaffold
Unknown (Some homology to Sec16A)		**Contig22207**		MMSRT104C	1193/1197	100%	Contig_28	1193/1197	100%		Yes
cAMP-responsive element-binding protein 3-like protein 4 (cr3l4)	C264R086	**Contig51133**		MMSRT014G	1173/1181	99%	Contig_121	1173/1181	99%	RT15B10	Yes
CREB-regulated transcription coactivator 2 (TORC2)	**C117R024**			MMSRT069D*	657/726	97%	Contig_107	698/748	100%	RT15B10	Possibly
Zinc/iron-regulated protein (zip1) (splice A)	**C053R110**			MMSRT069D*	391/415	39%	Contig_108, 110	1058/1065	100%	RT15B10	Yes
Zinc/iron-regulated protein (zip1) (splice B)	C014R029	**CLUSTER_ID#5374769**		MMSRT069D*	916/1137	99%	Contig_108	1125/1148	100%	RT15B10	Likely
Unknown		**omyk-DV199273**		MMSRT069D*	451/518	74%	Contig_135, 153	696/700	100%		Yes
Unknown		**Contig19333**		MMSRT069D*	637/755	87%	Contig_119	815/816	94%		Yes

*Denotes scaffolds from Michael Miller's draft genome that are paralogous to OmyY1 linked sequence.

TABLE 3: Primer information for markers in the OmyY1 BAC contig (#6256).The parental lines are the doubled haploid male (AR/SW) and female (OSU/WR) rainbow trout clonal lines in which the products were amplified. Clonal lines listed on either side of "/" denote which clonal lines have the SNP or indel. The last column shows the base pair location of the SNP(s) or indel in the amplified product.

Mapping primers	Gene	Primer name	Tm	BAC clone	Product size	Parental line	SNP or indel
F1:GCCTATCAACCTCCTGTACTT R1:CCTGCGAAATAACCAATG	Fbxw2	Fbxw2-F/R	58–60	450D22/492A1	703 bp	AR, SW/OSU, WR	261bp
F2:CTACAGGAAGTCCAAACGAGG R2:TGGGGATGGTAGTTGACAGTC	Fbxw2	Fbxw2-F2/R2	58–64	450D22/492A21	740 bp	AR, SW/OSU, WR	162(A/G) 525(T/A)
F3:TAGACATTTCCCCTGTATTACC R3:AAAAGGGTCTGCTACTCATT	Fbxw2	Fbxw2-F3/R3	58	450D22/492A21	818 bp	AR, SW/OSU, WR	378(G/T), 530(G/C) 756(G/A), 642–46 bp
F4:CATGTCTTCTTTCCATTGGCC R4:TTGTGACCTCCATCTTGTAGG	Intergenic	c451-grp2-F/R	62	450D22	392 bp	AR, SW/OSU, WR	147(G/T), 166(G/T)
F5:CGACAGCACCAAACTAATCTTT R5:GATCATCAGTTGCACGGAGG	Intergenic	c35-450D22-F2-R2	62	450D22	978 bp	AR, SW/OSU, WR	male-specific amplification
F6:TTACAGTGACATTCGGTTCAA R6:TTTGGGTGTGTGATGCTATTT	Intergenic	MM121C-5′14kb	62	450D22/492A21	687 bp	AR/SW/OSU/WR	168(G/A)
F7:CGGAATGCACCAAACCCTAA R7:GTATGTTGTCCTTGGCTCCGAA	Intergenic	1M42E-14201-F5/14944-R	65	423A24	743 bp	AR/OSU	AFLP in AR

OmyY1 contig assemblies were represented on the 4 x 44 k oligo array by exact name or alias.

Stand-alone BLAST analysis was conducted with the 48 array hits using a draft *Oncorhynchus mykiss* genome assembly as the database. This draft genome assembly included sequence from 7 BAC clones within the rainbow trout physical map contig #6256; 5 of which were chosen for Illumina and 454 sequencing in our study (Figure 1). Microarray sequences matching any corresponding assembled genome scaffolds were confirmed as OmyY1-linked. Of these 7, scaffolds MMSRT112A (containing sequence from RT450D22) and MMSRT121C (containing sequence from RT492A21) showed significant sequence homology and coverage to 5 microarray hits each, 3 of which fall in an overlapping region between both BAC RT450D22 and BAC RT492A21clones (supplemental file 1).

These microarray results identified additional sex-linked or paralogous scaffolds that contained a significant number of microarray hits. Stand-alone BLAST results showed MMSRT069D had significant homology to 4 hits (Supplemental File 1). To check whether this scaffold was sex-linked or paralogous, primers were designed to the microarray EST sequences then partial gene sequences were amplified from the OmyY1 BAC clones and sequenced. Sequences from MMSRT069D showed 90% homology to OmyY1 BAC sequence indicating the scaffold was most likely a paralogous scaffold, amplifying in BAC clone RT15B10. There were also five rDNA and/or repetitive ESTs that contained homology to similar sets of draft genome scaffolds, including MMSRT092E, 100C, 059D, 036A, 103G, 136F, and 084G (Supplemental File 1). The 5 EST sequences had similar rates of homology with both the draft genome and OmyY1 contig assemblies, but none show with 100% homology with any assembly. Due to the difficulty in amplifying clean rDNA and repetitive sequences in the BACs, and because a great number of scaffolds showed identical homology to the ESTs, we could not determine whether these scaffolds were paralogous or identical to the OmyY1 region. It is worth noting, however, that MMSRT100C also contains IRF9 (interferon regulatory factor 9) which is the gene from which *sdY* was originally derived [16]. This suggests that even though these rDNA and repetitive sequences are found in many locations throughout the genome, there are specific sex-linked versions surrounding OmyY1 and sdY.

The 48 CGH hits queried in stand-alone BLAST evaluations of the OmyY1 contig sequence and draft genome assemblies were checked against GenBank (using BLASTn and BLASTx) for homology with known salmon and trout repetitive elements and checked against the rainbow trout draft genome. This search showed 17 hits represented transposable elements, 5 contained ribosomal DNA, and 12 more were repetitive, leaving 14 ESTs showing significant nonrepetitive sequence homology to the OmyY1 region (Supplemental file 1). Although repetitive hits were not included in Tables 1 and 2, many of these repetitive ESTs from the microarray seem to be expressed within the OmyY1 contig and may still be functionally and evolutionarily important (details of both BLAST searches are included in Supplemental File 1). It is also important to note that for some ESTs, repetitive elements only made up a small portion of the entire transcript.

Some ESTs within the 4 x 44 k annotation file were derived from Atlantic Salmon (*Salmo salar*), and for those ESTs with less than 99-100% homology to either the *O. mykiss* draft genome assembly or the OmyY1 contig Illumina assembly, rainbow trout-specific EST databases were searched to find expressed orthologs or paralogs. These databases include the cGRASP *O. mykiss* EST cluster databases, GenBank, Salem's 454 database [29], and the *O. mykiss* sex-specific gonad EST databases at 35dpf from Yano et al. [16]. Found orthologs and paralogs were evaluated for strong homology and comprehensive gene coverage within OmyY1 contig sequence and for the lack of premature stop codons when predicted protein sequences were available to determine functionality of genes.

The final list of predicted functional, nonrepetitive genes within the OmyY1 contig contained 11 hits from the 4 x 44 k microarray (Table 2). Six of these ESTs (cr3l4, FBXW2, nkcc1a, TORC2-like, and ZIP1 (with 2 predicted splice variants)) were found in the Illumina assembly for the OmyY1 contig.

3.5. Analysis of SNPs from the OmyY1 BACs. For further confirmation of these loci within the BAC clones, Sanger sequencing was performed. PCR primers were designed either to the OMY BAC Illumina/454 contig sequence, GB cDNA, or EST sequence available, or to the *O. mykiss* draft genome homologous scaffolds (see Figure 1). The F-box and WD repeat domain containing 2 (Fbxw2) gene comprises a 7.6 kb region within BAC RT450D22. It was evaluated for sequence polymorphisms in the DH RT clonal lines AR, OSU, SW, and WR, [30, 31] yielding 10 SNPs; 4 of which were population specific and 6 of which appeared sex specific. All SNPs are referenced against the Fbx numbering system of the GenBank deposition and locations are tabulated (see Table 3).

Gene-specific amplification attempts of the *Oncorhynchus mykiss* oocyte protease inhibitor-1 gene, BAC RT492A1 (GenBank KC686349), failed to amplify a single locus, except within the BAC itself. Sequencing of PCR products of genomic DNA from the OSU female DH line yielded the cleanest reads, with primers F1/R1 showing discernible indels and SNPs and an overall homology of 94% as compared to the Y sequence, but it could not be determined whether this was a related sequence on the X chromosome or an autosomal homolog.

Sex-linkage and SNP evaluations of the cAMP-responsive element-binding protein 3-like protein 4 (cr3l4) gene, BAC RT15B10 (GenBank KC686342) of DH RT genomic DNAs preferentially amplified paralogous loci; no primer combinations yielded sex-linked polymorphisms for this gene.

Also within BAC RT15B10, the Zinc transporter ZIP1 (zip1) gene was evaluated by PCR and amplified product sequencing of RT clonal lines for sex-linked polymorphisms, which are referenced against the ZIP1 GenBank KC686350 deposition. The CREB-regulated transcription coactivator 2 (TORC2-like) gene was PCR evaluated for sex-linked polymorphisms in the RT clonal line panel, identifying 4 SNPs between the clonal lines, 1 sex-linked SNP, and a 17 bp deletion in SW only.

Population and RT clone-specific polymorphisms were found in the DENN domain-containing protein 4B-like gene amplified in BAC RT15B10. All polymorphisms are referenced against the GenBank KC686345 sequence. RT15B10 also contains an interleukin enhancer-binding factor 2 pseudogene (GenBank KC686347); however the 2 primer sets used to confirm sequence within RT15B10 amplified more than one locus in all the DH genomic DNA analyzed.

In addition to these EST templates and annotated loci, a number of primers were designed to other unannotated regions within the Illumina/454 assembly contigs as well as to the homologous MMSRT draft genome scaffolds. Of these, there were 2 primer sets that yielded sex-specific SNPs and these are included in Table 3.

4. Discussion

In this study we describe sequencing and CGH results from BAC clones forming a contig with the OmyY1 marker isolated by Brunelli and Wertzler [11] and the sdY sex determining gene isolated by Yano et al. [16]. Primers for the OmyY1 marker and the sdY gene amplified products from two of the BACs on the 3′ end of the contig. These BACs hybridized to the end of the short arm of the Y chromosome with some signal also on the end of the short arm of the X chromosome. Illumina sequencing of the BACs, because of the large number of repetitive sequences present, produced many small contigs. A number of putative genes were identified, and all genes PCR amplified appear to be present in both males and females, with only a small portion showing sex-specific sequence differences. Large contiguous sequences with over 95% homology to each other were found in nonoverlapping BACs suggesting that duplications of a number of regions, some of which containing putative genes, are present in contig #6256. Taken together these results suggest that the X and Y are in an early stage of differentiation, perhaps because infrequent episodes of crossing over have partially homogenized the content of the X and Y.

Because the rainbow trout X and Y chromosomes are usually morphologically distinct, it has been assumed that they are in a later stage of differentiation compared to species such as medaka where the X and Y are not distinguishable. In the case of medaka, the only difference between the X and Y is the insertion of the sex-determining gene DMY into the Y chromosome by transposition [32]. DMY was found in only one BAC clone. We analyzed a contig composed of a minimum tiling path of six BACs from the Swanson YY library. Sex-specific SNPs were identified in genes from BACs in the 5′ and middle of the contig, but few genes were found on the sdY on the 3′ end other than rDNA and transposable elements. Because the ends of contig #6256 contain very repetitive sequences, we were not able to extend the contig in either direction. It is possible that Y-specific genes are found in regions outside of the region that was sequenced. Future work could involve isolation of BACs and sequencing of these adjacent regions in both directions.

The expression of all of the genes in the OmyY1 region could be examined to determine if any are expressed preferentially in males during sex determination or sexual differentiation. Isolation of the corresponding region from the OSU BAC library prepared from an XX female clonal line and comparison of the content and expression of genes on the X and the Y in the region adjacent to the proposed sex determination gene, sdY would provide information about the degree of divergence between the X and Y chromosomes.

In the Swanson YY clonal line, the Y chromosome is not morphologically distinct from the X, with rDNA on the short arm adjacent to the centromere. This undifferentiated Y chromosome is possibly the result of crossing over between the X and Y in the population that gave rise to this clonal line. Future work on the male-specific region of the Y chromosome in YY clonal lines with the morphologically differentiated Y chromosome should reveal whether greater variation is present in the coding regions of the Y chromosome in such rainbow trout strains.

Finally, we would like to compare the male-specific content of the Y chromosome in the other *Oncorhynchus* species and determine the size of the region that is transposing to form new Y chromosomes in this genus. A BAC clone has been isolated that contains the sdY gene from a male Chinook salmon (RH Devlin, Fisheries and Oceans, pers. com.) and it will be interesting to compare the sequences of the adjacent regions in the two species. We are currently performing CGH experiments to compare the genes present in YY and XX genomic DNA from rainbow trout, coho salmon, and sockeye salmon which should also provide information on differences in gene content between the X and Y chromosomes of these species.

5. Conclusions

We analyzed sequence data from a contig of rainbow trout BAC clones from the Swanson YY clonal line that includes the male-specific OmyY1 genetic marker and the sex-determining gene, sdY. Our data shows that this region is rich with rDNA and repetitive elements, but also has many predicted protein coding genes. Most genes in the region seem to be present on both the X and Y chromosomes, although sex-specific SNPs are found in a small portion of them, suggesting that the X and Y are at an early stage of differentiation. None of these genes were found in the cosmid or BAC clones from the Y chromosome of Chinook salmon (Phillips and Devlin, unpublished) and the two BAC clones containing OmyY1 and sdY did not hybridize to the Y chromosome of Chinook salmon, suggesting that the shared region containing the sex-determining gene between these two species is very small.

Acknowledgments

This project was supported by Agriculture and Food Research Initiative Competitive Grant no. 2009-35205-05067 from the USDA National Institute of Food and Agriculture. The following individuals from Amplicon express worked on the sequencing: Evan Hart, Amy Mraz, Keith Stormo, Jason Dobry, and Travis Ruff. The authors would also like to thank Michael R. Miller for providing the rainbow trout draft genome assembly which greatly improved our analysis. The use of trade, firm, or corporation names in this publication is for the information and convenience of the reader. Such use does not constitute an official endorsement or approval by the U.S. Department of Interior or the U.S. Geological Survey of any product or service to the exclusion of others that may be suitable.

References

[1] R. H. Devlin and Y. Nagahama, "Sex determination and sex differentiation in fish: an overview of genetic, physiological, and environmental influences," *Aquaculture*, vol. 208, no. 3-4, pp. 191–364, 2002.

[2] G. H. Thorgaard, "Heteromorphic sex chromosomes in male rainbow trout," *Science*, vol. 196, no. 4292, pp. 900–902, 1977.

[3] R. B. Phillips, M. A. Noakes, M. Morasch, A. Felip, and G. H. Thorgaard, "Does differential selection on the 5S rDNA explain why the rainbow trout sex chromosome heteromorphism is NOT linked to the SEX locus?" *Cytogenetics and Genome Research*, vol. 105, pp. 122–125, 2004.

[4] A. Felip, A. Fujiwara, W. P. Young et al., "Polymorphism and differentiation of rainbow trout Y chromosomes," *Genome*, vol. 47, no. 6, pp. 1105–1113, 2004.

[5] J. E. Parsons and G. H. Thorgaard, "Production of androgenetic diploid rainbow trout," *Journal of Heredity*, vol. 76, no. 3, pp. 177–181, 1985.

[6] R. H. Devlin, C. A. Biagi, and D. E. Smailus, "Genetic mapping of Y-chromosomal DNA markers in Pacific salmon," *Genetica*, vol. 111, no. 1–3, pp. 43–58, 2001.

[7] J. Stein, R. B. Phillips, and R. H. Devlin, "Identification of the Y chromosome in chinook salmon (*Oncorhynchus tshawytscha*)," *Cytogenetics and Cell Genetics*, vol. 92, no. 1-2, pp. 108–110, 2001.

[8] R. B. Phillips, M. R. Morasch, L. K. Park, K. A. Naish, and R. H. Devlin, "Identification of the sex chromosome pair in coho salmon (*Oncorhynchus kisutch*): lack of conservation of the sex linkage group with chinook salmon (*Oncorhynchus tshawytscha*)," *Cytogenetic and Genome Research*, vol. 111, no. 2, pp. 166–170, 2005.

[9] R. B. Phillips, J. DeKoning, M. R. Morasch, L. K. Park, and R. H. Devlin, "Identification of the sex chromosome pair in chum salmon (*Oncorhynchus keta*) and pink salmon (*Oncorhynchus gorbuscha*)," *Cytogenetic and Genome Research*, vol. 116, no. 4, pp. 298–304, 2007.

[10] M. A. Alfaqih, R. B. Phillips, P. A. Wheeler, and G. H. Thorgaard, "The cutthroat trout Y chromosome is conserved with that of rainbow trout," *Cytogenetic and Genome Research*, vol. 121, no. 3-4, pp. 255–259, 2008.

[11] J. P. Brunelli, K. J. Wertzler, K. Sundin, and G. H. Thorgaard, "Y-specific sequences and polymorphisms in rainbow trout and Chinook salmon," *Genome*, vol. 51, no. 9, pp. 739–748, 2008.

[12] J. P. Brunelli and G. H. Thorgaard, "A new Y-chromosome-specific marker for Pacific salmon," *Transactions of the American Fisheries Society*, vol. 133, no. 5, pp. 1247–1253, 2004.

[13] Shao Jun Du, R. H. Devlin, and C. L. Hew, "Genomic structure of growth hormone genes in chinook salmon (*Oncorhynchus tshawytscha*): presence of two functional genes, GH-I and GH-II, and a male- specific pseudogene, GH-Y," *DNA and Cell Biology*, vol. 12, no. 8, pp. 739–751, 1993.

[14] R. H. Devlin, B. K. McNeil, T. D. D. Groves, and E. M. Donaldson, "Isolation of a Y-chromosomal DNA probe capable of determining genetic sex in Chinook salmon (*Oncorhynchus tshawytscha*)," *Canadian Journal of Fisheries and Aquatic Sciences*, vol. 48, pp. 1606–1612, 1991.

[15] R. A. Woram, K. Gharbi, T. Sakamoto et al., "Comparative genome analysis of the primary sex-determining locus in Salmonid fishes," *Genome Research*, vol. 13, no. 2, pp. 272–280, 2003.

[16] A. Yano, R. Guyomard, B. Nichol et al., "An immune-related gene evolved into the master sex-determining gene in rainbow

trout, *Oncorhynchus mykiss*," *Current Biology*, vol. 22, pp. 1–6, 2012.

[17] W. P. Young, P. A. Wheeler, V. H. Coryell, P. Keim, and G. H. Thorgaard, "A detailed linkage map of rainbow trout produced using doubled haploids," *Genetics*, vol. 148, no. 2, pp. 839–850, 1998.

[18] K. M. Nichols, W. P. Young, R. G. Danzmann et al., "A consolidated linkage map for rainbow trout (*Oncorhynchus mykiss*)," *Animal Genetics*, vol. 34, no. 2, pp. 102–115, 2003.

[19] Y. Palti, S. A. Gahr, J. D. Hansen, and C. E. Rexroad, "Characterization of a new BAC library for rainbow trout: evidence for multi-locus duplication," *Animal Genetics*, vol. 35, no. 2, pp. 130–133, 2004.

[20] K. M. Nichols, J. Bartholomew, and G. H. Thorgaard, "Mapping multiple genetic loci associated with *Ceratomyxa shasta* resistance in *Oncorhynchus mykiss*," *Diseases of Aquatic Organisms*, vol. 56, no. 2, pp. 145–154, 2003.

[21] A. M. Zimmerman, J. P. Evenhuis, G. H. Thorgaard, and S. S. Ristow, "A single major chromosomal region controls natural killer cell-like activity in rainbow trout," *Immunogenetics*, vol. 55, no. 12, pp. 825–835, 2004.

[22] R. B. Phillips, J. J. Dekoning, A. B. Ventura et al., "Recombination is suppressed over a large region of the rainbow trout y chromosome," *Animal Genetics*, vol. 40, no. 6, pp. 925–932, 2009.

[23] Y. Palti, M. C. Luo, Y. Hu et al., "A first generation BAC-based physical map of the rainbow trout genome," *BMC Genomics*, vol. 10, article 462, 2009.

[24] R. B. Phillips, K. M. Nichols, J. J. DeKoning et al., "Assignment of rainbow trout linkage groups to specific chromosomes," *Genetics*, vol. 174, no. 3, pp. 1661–1670, 2006.

[25] B. F. Koop, K. R. Von Schalburg, J. Leong et al., "A salmonid EST genomic study: genes, duplications, phylogeny and microarrays," *BMC Genomics*, vol. 9, article 545, 2008.

[26] G. K. Smyth, "Linear models and empirical Bayes methods for assessing differential expression in microarray experiments," *Statistical Applications in Genetics and Molecular Biology*, vol. 3, pp. 1–26, 2004.

[27] W. S. Davidson, B. F. Koop, S. J. Jones et al., "Sequencing the genome of Atlantic salmon (*Salmo salar*)," *Genome Biology*, vol. 11, no. 9, article 403, 2010.

[28] S. F. Altschul, T. L. Madden, A. A. Schäffer et al., "Gapped BLAST and PSI-BLAST: a new generation of protein database search programs," *Nucleic Acids Research*, vol. 25, no. 17, pp. 3389–3402, 1997.

[29] M. Salem, C. E. Rexroad III, J. Wang, G. H. Thorgaard, and J. Yao, "Characterization of the rainbow trout transcriptome using Sanger and 454-pyrosequencing approaches," *BMC Genomics*, vol. 11, no. 1, article 564, 2010.

[30] W. P. Young, P. A. Wheeler, V. H. Coryell, P. Keim, and G. H. Thorgaard, "A detailed linkage map of rainbow trout produced using doubled haploids," *Genetics*, vol. 148, no. 2, pp. 839–850, 1998.

[31] M. R. Miller, J. P. Brunelli, P. A. Wheeler et al., "A conserved haplotype controls parallel adaptation in geographically distant salmonid populations," *Molecular Ecology*, vol. 21, pp. 237–249, 2012.

[32] M. Kondo, U. Hornung, I. Nanda et al., "Genomic organization of the sex-determining and adjacent regions of the sex chromosomes of medaka," *Genome Research*, vol. 16, no. 7, pp. 815–826, 2006.

17

De Novo Transcriptome Assembly and Differential Gene Expression Profiling of Three *Capra hircus* Skin Types during Anagen of the Hair Growth Cycle

Teng Xu, Xudong Guo, Hui Wang, Xiaoyuan Du, Xiaoyu Gao, and Dongjun Liu

The Key Laboratory of Mammalian Reproductive Biology and Biotechnology of the Ministry of Education, Inner Mongolia University, Hohhot, Inner Mongolia Autonomous Region 010021, China

Correspondence should be addressed to Dongjun Liu; liudjimu@yahoo.cn

Academic Editor: Soraya E. Gutierrez

Despite that goat is one of the best nonmodel systems for villus growth studies and hair biology, limited gene resources associated with skin or hair follicles are available. In the present study, using Illumina/Solexa sequencing technology, we *de novo* assembled 130 million mRNA-Seq reads into a total of 49,115 contigs. Searching public databases revealed that about 45% of the total contigs can be annotated as known proteins, indicating that some of the assembled contigs may have previously uncharacterized functions. Functional classification by KOG and GO showed that activities associated with metabolism are predominant in goat skin during anagen phase. Many signaling pathways was also created based on the mapping of assembled contigs to the KEGG pathway database, some of which have been previously demonstrated to have diverse roles in hair follicle and hair shaft formation. Furthermore, gene expression profiling of three skin types identified ~6,300 transcript-derived contigs that are differentially expressed. These genes mainly enriched in the functional cluster associated with cell cycle and cell division. The large contig catalogue as well as the genes which were differentially expressed in different skin types provide valuable candidates for further characterization of gene functions.

1. Introduction

Inner Mongolia Cashmere Goat (*Capra hircus*, IMCG) is a diploid ($2n = 60$) mammal that belongs to the family of *Bovidae*. It plays an important role in the world animal fiber industry because it can produce high quality underhair (cashmere is the commercial name) and is one of the world's largest breeding groups. Cashmere produced by IMCG, which is of a small diameter (14–18 μm) and is soft to touch, is grown from the secondary hair follicle (HF) of the body skin [1, 2]. Fiber diameter and length determine both the quality and the amount of cashmere produced by an animal. The longer the length and the smaller the diameter of the cashmere fibers, the higher the price becomes. IMCGs exhibit seasonal rhythm and annual cycle of cashmere growth that are controlled by daylength. During the period from the summer solstice to midwinter, when the length of day is reduced, cashmere fiber has a high growth speed; in contrast, it becomes low during the period from midwinter to the next summer solstice [3, 4]. This photoperiodic characteristic of cashmere fiber growth is convenient for cashmere harvest and formulating a management strategy of cashmere production. During the past decades, many mammalian genomic and transcript sequences have become available, including *Homo sapiens*, *Mus musculus*, and *Bos taurus* which play important roles in understanding HF formation and hair growth. However, only a total of 561 ESTs are pertinent to goat skin or hair follicle indicating that few studies focused on understanding the gene expression pattern of goat skin or hair follicle. On the other hand, almost all the deposited goat skin associated ESTs were sequenced by the traditional approach from randomly picked cDNA clones which did not guarantee that the less-abundant transcripts could be efficiently detected. In addition, we also observed that cashmere fibers of IMCG are mainly produced by the back (BK) and side of the body (BS) skin of the trunk coat, but few are produced by the belly (BL) skin. This indicated that the gene expression patterns of BK and BS skin are different from BL. Therefore, the gene expression patterns and differential gene expression profiling of three skin sites during active hair growth phase (anagen) are important to understanding the underlying molecular mechanism associated with cashmere fiber growth.

The advents of the ultra-high-throughput, cost-effective next-generation sequencing (NGS) technologies make the whole transcriptome sequencing and analysis feasible, even in the absence of genomic data [5]. In the past few years, NGS has been widely used in RNA sequencing (RNA-Seq) which provided researchers with more information about gene expression, regulation, and networks under specific physiological conditions or developmental stages in both model and nonmodel organisms [6–8]. It offers accurate quantitative and digital gene expression profiles of sequenced transcripts [5]; moreover, it has very low background noise and a large dynamic range of gene expression levels compared with DNA microarray [5]. In the present study, we utilized Illumina/Solexa paired-end mRNA-Seq approach to sequence and *de novo* assemble the goat skin transcriptome during anagen phase, and further investigated the gene expression profiles of three different skin types based on the assembled contigs, skin from the BL, BK, and BS of the trunk coat, which have a large discrepancy in cashmere production.

2. Materials and Methods

2.1. Goat Skin Tissue Collection and RNA Extraction. Generally, according to the different skin types of trunk coat, skin from the back and body side of IMCG produce either wool or cashmere, whereas belly skin produces mainly wool fiber and few cashmere [9]. A breeding, two-year-old female goat was sampled from Yi Wei White Cashmere Goat Breeding Farm at Ulan Town of Erdos in Inner Mongolia Autonomous Region, China. During anagen phase (Nov, 2010), the hair (wool and cashmere) of belly, back, and body sides of the goat were sheared and further shaved. After sterilization with 70% alcohol, full-thickness skin sections of each body part were excised and immediately frozen in liquid nitrogen for storage and transport until RNA isolation. Goat skin tissue collections were carried out in accordance with the guidelines of Inner Mongolian Animal Society Ethics Committee. This study has been checked and approved by Inner Mongolian Animal Society which is responsible for Animal Care and Use in Inner Mongolia Autonomous Region, China. Skin excision was performed after Xylazine hydrochloride anesthesia, and all efforts were made to minimize suffering. Before RNA extraction, each sample was washed with 10 mL PBS (pH 7.2) and 0.5 mM EDTA. Total RNA of each sample was isolated by using a TRIzol plus RNA purification kit according to the manufacturer's protocol (Invitrogen). Total RNA quality and concentration were determined by a 2100 bioanalyzer nanochip (Agilent). Enrichment of mRNA from total RNA was performed with a RiboMinus RNA-Seq Kit according to the manufacturer's instructions (Invitrogen).

2.2. Paired-End Library Construction and Illumina Sequencing. It has been reported that mRNA fragmentation can result in a more even coverage along the entire gene body, whereas cDNA fragmentation is more biased towards the 3′ end of the transcript [8]. Therefore, each ploy(A)-enriched RNA sample was chemically fragmented into small pieces by using divalent cations at 94°C for 5 min. The fragmented RNA was reverse-transcribed into cDNA by using random hexamer primers containing a tagging sequence at the 3′ end with the use of a superscript III double-stranded cDNA synthesis kit according to the manufacturer's protocol (Invitrogen). The double-stranded cDNA was subjected to end-repair and further 3′ terminal tagging by the addition of 5′ DNA adaptors and T4 DNA ligase with overnight incubation at 16°C for 16 hours. The targeted di-tagged cDNA was purified by polyacrylamide gel electrophoresis (PAGE) and gel excision (200 ± 25 bp). The clean di-tagged cDNA was enriched by limited-cycle PCR amplification (18 cycles) with primer pairs that annealed to the tagging sequences of the di-tagged cDNA. Library purification by PAGE removed any residual nucleotides, PCR primers, and small amplicons. Three independent paired-end libraries were sequenced on a HiSeq2000 system. The initial Illumina short reads from this study have been submitted to the NCBI sequence read archive (SRA, http://www.ncbi.nlm.nih.gov/Traces/sra/sra.cgi/) with the accession number SRA055764.

2.3. De Novo Assembly of mRNA-Seq Reads, CDS Prediction, and Validation. Due to the fact that the base quality requirement for *de novo* assembly is more strict than that of the resequencing project, customized Perl scripts were used to remove reads which contained adaptor contamination, low quality bases (>2% base smaller than Q20 per read), and undetermined bases (>2% "N"s per read) from each dataset generated from different skin types. Three datasets were concatenated in a left-to-left and right-to-right manner. Next, the clean high-quality reads dataset was *de novo* assembled with default parameters by using Inchworm assembler which is a component of Trinity software [10]. All of the sequence reads were initially trimmed to 50 bp (nucleotides 21 to 70 of each read) in length and then used in mapping experiments and statistical analysis. Mapping short reads uniquely back to the contigs was performed by SOAPaligner 2.20 with two mismatched bases per read permitted [11]. Coding sequence prediction was performed by GENSCAN [12]. The contigs which contain two or three predicted CDSs may attribute to a small proportion of false positive discovery of predicting coding sequences from mature transcript sequences. Ten randomly selected putative full length transcripts which did not assign with known protein functions and another ten transcripts that are associated with hair cycling were subjected to perform reverse transcription polymerase chain reaction (RT-PCR) and Sanger sequencing. Primer sequences used for RT-PCR are available upon request. The assembled transcriptome sequences (49,115 contigs greater than 300 bp in length) in this study have been deposited to the NCBI's transcriptome shotgun assembly (TSA) database under the consecutive accession numbers from KA304519 to KA353633.

2.4. Functional Annotation and Classification of Assembled Contigs. Mouse and Cow RefSeq protein sequences

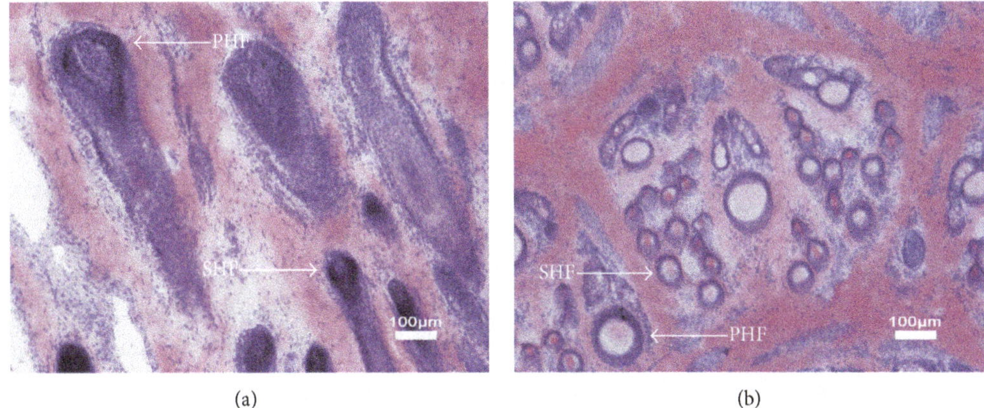

(a) (b)

FIGURE 1: Frozen sections of cashmere goat skin stained by hematoxylin. White arrows indicate the primary hair follicles (PHFs) and secondary hair follicles (SHFs) in the sample.

were downloaded from the NCBI RefSeq database (ftp://ftp.ncbi.nih.gov/refseq/). Homology searches against the Swiss-prot and NR databases were performed by using BLASTX algorithms with an E-value cutoff of 10^{-10}. Mouse and Cow RefSeq RNA data were also downloaded as reference sequences for short reads mapping to calculate hit numbers compared with *de novo* assembled contigs. BLASTX was used to perform KOG and KEGG annotation [13, 14], followed by retrieval of the functional proteins and assignment to each of the classification entries (E-value 10^{-10}). Gene ontology (GO) against the NR database was conducted by Blast2GO (E-value 10^{-10}) [15]. WEGO [16] and GO terms classifications counter (http://www.animalgenome.org/tools/catego/) were used for assignment of each GO ID to the related ontology terms.

2.5. Digital Profiling of Differentially Expressed Transcripts and qRT-PCR Validation. According to the AC statistical framework [17], the p value of differential expression significance of each transcript-derived contigs between two samples was calculated by using the following equation:

$$p(y \mid x) = \left(\frac{N_2}{N_1}\right)^y \frac{(x+y)!}{x!y!(1+N_2/N_1)^{(x+y+1)}}. \quad (1)$$

N_1 and N_2 represent the total uniquely mapped reads in sample one and sample two, respectively. x is the number of reads mapped to a certain gene in sample one, and y represents the number of reads mapped to the same gene in sample two. After the calculation of p value, multiple hypothesis testing was performed to correct p value by using Phyper function of the R tool (http://www.r-project.org/). RPKM values of each contig were estimated by aligning trimmed reads back to the contigs [18]. Both a q-value of less than $10^{-3}$and a RPKM value with at least 2-fold difference between the two samples were used as criteria to determine significant DEGs. KOG enrichment analysis was conducted by hypergeometric distribution test by using the Phyper function in the R software package. Bonferroni correction was further used to adjust the p values, respectively. The significantly enriched functional clusters were selected when the corrected p value (q-value) was less than 10^{-3}. Quantitative real-time PCR (qRT-PCR) was performed on the same individual of the corresponding body part skin. Total RNA was firstly treated with DNase I before reverse transcription by superscript III double-stranded cDNA synthesis kit (Invitrogen). Each cDNA sample was used as template for qRT-PCR by using the SYBR Premix Ex Taq II kit (TaKaRa) on a 7300 real-time PCR system (Applied Biosystems), and at least three technical repeats were performed for all the genes within each template. Acetyl-CoA carboxylase 1 (a7431; 102) was used to normalize gene expression quantities between samples. Gene expression fold difference between two samples was calculated by the $2^{-\Delta\Delta Ct}$ method [19]. PCR primer sequences are available upon request.

3. Results

3.1. Illumina Paired-End Sequencing, De Novo Transcriptome Assembly, and Ab Initio CDS Prediction. To obtain comprehensive transcripts of skin tissue that provide an overview of gene expression profile during anagen in the cashmere goat, skin tissues from the belly (BL), back (BK), and the side of the body (BS) during anagen were sampled. The skin sections were made to show primary hair follicles and secondary hair follicles (Figure 1). Then, total RNA from each sample was isolated, respectively. Three RNA-Seq libraries were constructed and sequenced by using Illumina/Solexa technology. As a result, a total of approximately 130 million raw reads (65 million paired-end reads, 2 × 100 bp) that represented roughly 13 GB of sequence data were generated from three independent 200 bp insert libraries. The initial read quantities of three libraires were listed in Table 1. After the removal of reads with low quality and containing ambiguous bases (>2% "N" per reads), we employed Inchworm assembler to *de novo* assemble high quality reads which were generated from three different skin types. *De novo* assembly mRNA-Seq reads yielded 49,115 contigs over 300 bp comprising 45.4 MB of total sequence length, with an average length of 924 bp and N50 length of 1380 bp. Of the 49,115 contigs, there were 12,892 (26.2%) contigs greater than 1 Kb, 13,768

TABLE 1: Read number and mapping result from three independent libraries.

Sample	Total reads	Clean reads	Mapped reads	Mapped/clean
BL	41232601	40731588	35593595	87.4%
BK	39363238	38876298	34033557	87.5%
BS	49962602	49337308	42383188	85.9%

TABLE 2: Length distribution of 49,115 assembled contigs.

Contig length (bp)	Number of contigs	Cumulative size
301 ~ 500	22455	8.5 Mb
501 ~ 700	7987	4.7 Mb
701 ~ 1000	5781	4.8 Mb
1001 ~ 1500	4908	6.0 Mb
1500 ~ 2000	2914	5.1 Mb
>2000	5070	16.3 Mb
Total	49115	45.4 Mb

(28.1%) varying from 501 bp to 1 Kb, and the remaining 22,455 (45.7%) ranging from 301 bp to 500 bp in length (Table 2). To identify the protein encoding regions, we used GENSCAN to perform the *ab initio* prediction of the coding sequence (CDS) of 49,115 contigs. We found that 23,039 putative CDSs were identified from 22,734 (46.3%) assembled contigs. Of the 22,734 CDS-contained contigs, 22,440 have one putative CDS, 283 have two, and 11 contain three CDSs. Further analysis indicated 8,184 out of 23,039 contained a putative full-length CDSs (i.e., containing start and stop codons). Further, 6,889 CDSs contained a start but no stop codon, 3,171 predicted to have a stop but no start codon, and 4,795 have neither. The average length of putative 8,184 full-length CDSs reached 1,326 bp, while the partial CDSs have an average length of 605 bp. Among 8,184 predicted full-length CDSs, 127 of them cannot be annotated by known proteins. To validate sequence assemblies, we randomly selected ten predicted full-length CDSs that did not possess BLASTX hits in the NR database and ten genes which have been demonstrated that are specific to hair cycling and hair growth to perform RT-PCR and Sanger sequencing. These genes include the signaling molecules such as Wnt, insulin-like growth factor 1 (IGF-1), members of fibroblast growth factor (FGF), and their receptors encoding genes such as Frizzled and IGF-1R. The result showed that 19 PCRs are positive and the Sanger sequencing results all showed higher than 97% identities with *de novo* assembled transcripts indicating the relatively high credibility of sequences assemblies (Table S1 and Table S2 in Supplementary Material available online at http://dx.doi.org/10.1155/2013/269191). To further validate the quality and sequencing depth of the assembled contigs, we mapped the total short reads back to the assembled contigs. The sequencing depth ranged from 2- to 164,789-fold, with an average sequencing depth of 249-fold. Specifically, 87.4%, 87.5%, and 85.9% corresponding to the BL, BK, and BS skin of high-quality reads, respectively, can uniquely be realigned to the 49,115 contigs (Table 1). The relatively small proportion of unmapped reads may be involved in comprising the contigs which are shorter than 300 bp. This suggests that the majority of short reads in our RNA-Seq data were efficiently assembled into relatively larger contigs.

3.2. Codon Usage and SSR Marker Identification. Examining the codon usage of 23,039 predicted CDSs showed that the most abundant amino acids encoded by triplet codons are nonpolar (hydrophobic) amino acids (44.1%), and then the uncharged polar amino acids (29.5%), while the acidic and basic amino acids accounted for 12.3% and 14.1%, respectively (Table S3). Among the 23,039 predicted CDSs, the average GC content reaches 54.9%, while the maximal GC content is 86.7% and minimum is 31.3%. Scanning the stop codon of 11,355 CDSs (8,184 predicted full lengths CDSs plus 3,171 stop codon containing CDSs) indicated that the stop codon most frequently used in goat is TGA which account for 53.8%, whereas TAG (23.2%) and TAA (23.0%) have the approximately equal utility frequency.

We used MISA (http://pgrc.ipk-gatersleben.de/misa/misa.html) to search the potential simple sequence repeats (SSRs) existed in our assembled transcripts. In this study, the repeat sequence which consists of dinucleotide, trinucleotide, tetranucleotide, pentanucleotide, and hexanucleotide tandem repeats with at least 18 bp in size was considered as an SSR. We found a total of 2,011 transcripts-derived SSRs that represent 158 unique repeat motifs scattered in 1,850 contigs of which 141 contigs contain at least 2 SSRs. The frequency of SSR occurrence is 4.09% and the average distance is 22.6 kb in our assembled 49,115 large contigs. 43.6% of the total SSRs are trinucleotide repeats, followed by dinucleotide (26.3%), hexanucleotide repeats (20.4%), and only a small proportion of them are pentanucleotide and tetranucleotide repeats (5.5% and 4.2%), respectively. The AC/GT (24.1%) motif comprises the highest frequency among all the identified SSRs, followed by CCG/CGG (18.4%), AGC/CTG (11.5%), and AGG/CCT (7.9%). The types of SSRs and their occurrence frequencies that we found in the goat are similar to the findings in other mammals but different from the results in plants. For example, the AC/GT repeat type, most abundant in the goat transcriptome sequence, is also very abundant in the human genome and in other vertebrates, whereas this repeat type is rarely observed in rice and sweet potato [20, 21]. The microsatellites identified in this work will become a valuable resource for goat genetic mapping.

3.3. Functional Annotation of Transcript-Derived Contigs. To functionally annotate the assembled contigs, a sequence similarity search was performed against *Bos taurus* RefSeq protein sequences (32,242 sequences), the Swiss-Prot protein database, and the nonredundant (NR) protein database by

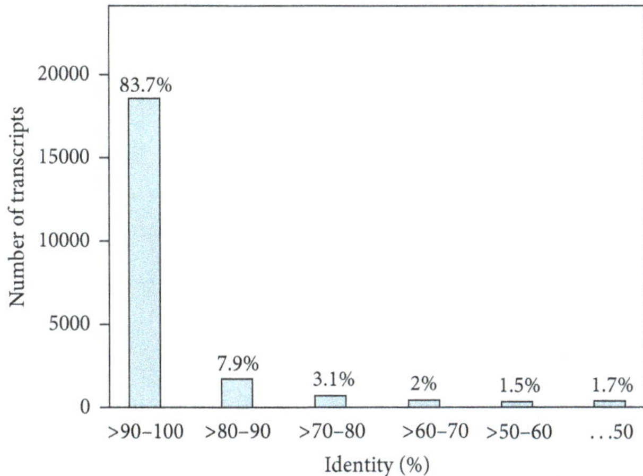

FIGURE 2: Sequence identity distribution. All BLASTX-hit transcripts were calculated. Vertical light blue bar shows the number of transcripts-derived contigs with which the range of percentage hit by BLASTX.

using BLASTX algorithm with a constant E-value (10^{-10}). With this approach, 21,104, 21,193, and 22,146 contigs were annotated by *Bos taurus* RefSeq protein sequences, Swiss-Prot protein database, and the NR database, respectively. 20,984 out of 22,146 contigs showed ≥70% sequence identity with NR top hit at matched region (Figure 2). Among the 22,146 BLASTX-hit possessing contigs, it is worth to note that only 103 (0.47%) contigs corresponding to the NR database top hits match goat itself, which could explain the limited number of goat gene and protein sequences currently available in the public database. Examining the 22,146 (45.1%) contigs with a high similarity to NR, we found that 19,040 contigs also harbored a predicted CDS, demonstrating that many putative CDSs cannot be annotated by known functions and thus may indicate some new genes existed in our assembled contigs. The remaining 26,969 contigs which had no significant hits in the NR database are more likely 5′ and 3′-untranslated regions (UTRs) or previously uncharacterized ESTs (or genes specifically expressed in *Capra hircus*). These transcripts are shorter in average length and relatively less abundant in their sequencing depth compared with those contigs which significantly hit to the NR database (data not shown). However, we also noted that many contigs with high sequencing depth showed no hits with the NR database. In the top 1,000 most reads-abundant contigs, 69 contigs with an average length of 893 bp showed no hits with known proteins. In fact, when we aligned sequenced reads to the 36,442 *Mus musculus* and 34,573 *Bos taurus* RefSeq mRNA, only 8,717 (25.2%) and 14,836 (43.0%) sequences were mapped to each, respectively. This suggested that *Capra hircus* is not phylogenetically close to the other two mammals, even though *Capra hircus* belongs to the same family as *Bos taurus*.

Since hair (wool and cashmere) mainly consists of highly compressed dead keratinocytes, we elucidated the relative abundance by calculating the values of reads per kilobase per million reads (RPKM) of 49,115 contigs which enabled us examine the expression level of the keratin-encoding genes relative to other genes. Through BLAST searching, we found a total of 126 keratin or keratin-associated protein (KAP) encoding sequences presented in our contig database. As expected, of the total 49,115 contigs, the top ten most abundant exclusively encode keratin or KAP, including K5, K14, KAP3.1, K33B, and KAP1.1. The remaining 116 keratins or KAP associated contigs also showed a greater average abundance than other genes. We also noted that some of the keratin-related contigs exhibited relatively higher amino acid diversity (6 out of 126 with <80% identities and 11 with <90% identities at BLAST matched region) when compared with the NR database top hits, suggesting a series of novel keratin variants may be synthesized in skin tissue undergoing rapid hair (anagen phase) growth. The expression level of keratin or KAP may also give an insight into the promoter efficiency and selection while performing exogenous gene expression in the skin tissue.

3.4. Functional Classification of Assembled Contigs by KOG, KEGG, and GO. We used BLASTX to search against functional proteins from the KOG (euKaryotic Orthologous Groups) database which is a component of the clusters of orthologous groups (COG) database [13]. 16,036 contigs had significant hits (E-value 10^{-10}), and these were classified into 4 groups and 25 functional clusters. Apart from 4,750 poorly characterized genes, cellular process, and signaling appeared as the largest group, which consisted of three highly abundant clusters including signal transduction mechanisms (3,106 genes), intracellular trafficking, secretion, vesicular transport (906 genes), and cytoskeleton (864 genes). Furthermore, to analyze pathway-based biological activities of genes which were expressed in goat skin, we annotated the 49,115 contigs against the KEGG (http://www.genome.jp/kegg/) protein database. From our contig database, 15,020 (30.6%) contigs were assigned to the 291 KEGG pathways. Among them, 3,948 genes can be further assigned to 23 signaling pathways, including the MAPK, Wnt, Insulin, Hedgehog, TGF-beta, VEGF, and Notch pathways which have been previously demonstrated to play various important roles in HF development and hair shaft differentiation.

In addition, gene ontology (GO) (http://www.geneontology.org/) analysis was also performed by using the Blast2GO program to further classify the transcript-derived contigs [15]. 18,069 contigs were cataloged into three main GO domains with a total of 129,669 GO IDs, and further subdivided into 47 subcategories (Figure 3). Of these assigned GO terms, biological process was the predominant domain followed by molecular function and cellular component. Under the biological process category, we found that cellular process and metabolic process are prominently represented, as they were in the KOG and KEGG classification, suggesting complicated metabolic activities occurred in anagen phase goat skin. A high correlation among KOG, KEGG, and GO classifications may reflect that goat hair growth is mediated by the complicated metabolic processes.

3.5. Differential Gene Expression Profiling, Functional Enrichment Analysis, and qRT-PCR Validation. A popular method

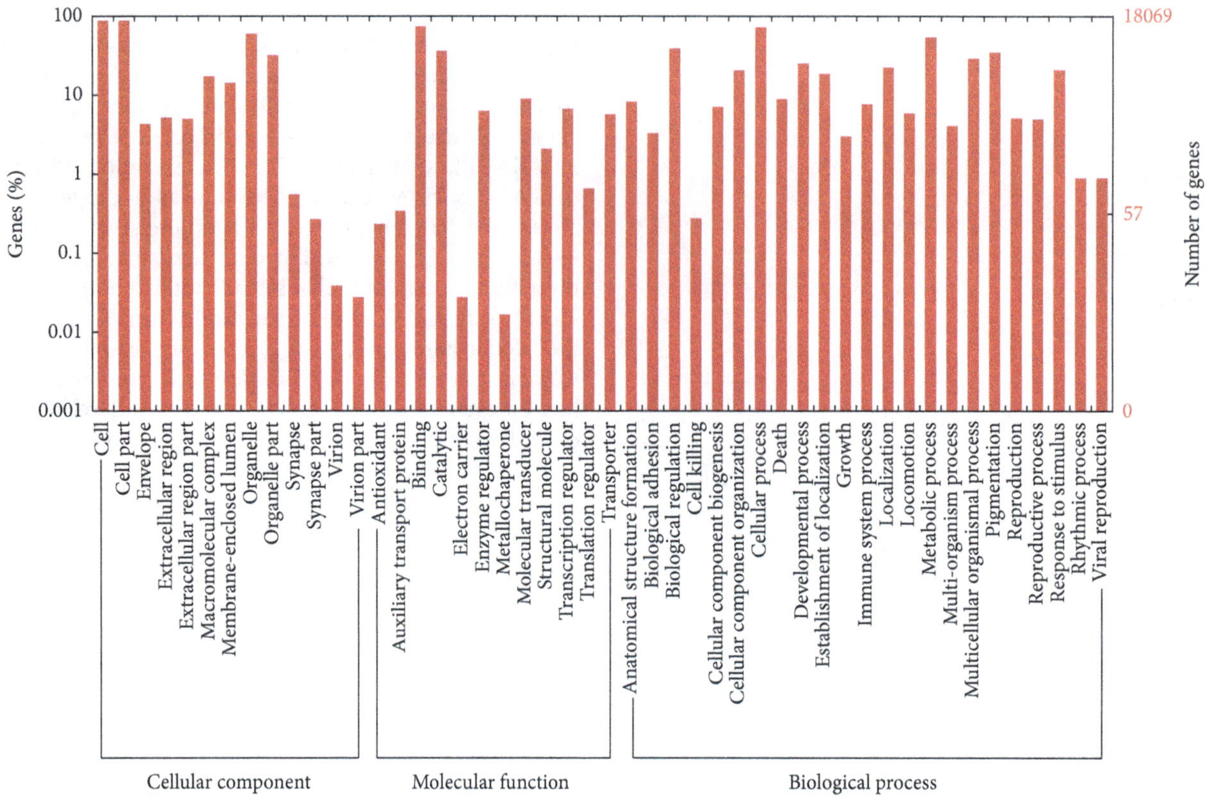

FIGURE 3: Gene Ontology classification of the transcriptome sequences. 18,069 transcriptome sequences can be annotated by the GO database. The classification results are displayed in the three main ontologies: cellular component, molecular function, and biological process.

of global measurement of differentially expressed genes is to take the quantity of NGS reads as an indicator for calculating transcript abundance [5–7]. Quantitative measurement of gene expression by NGS technologies has been suggested to be accurate and highly correlated with other methods of detecting gene expression levels, such as qRT-PCR and DNA microarray [5]. To identify differentially expressed genes (DEGs) among the three different skin types reflected in our short reads dataset, we mapped the short reads datasets from three libraries to the 49,115 contigs by SOAPaligner with the seed length of 50 bp. After mapping skin type-specific short reads to the reference, we calculated RPKM values for all contigs which can be used to quantify the expression of contigs both within and between samples [18]. An AC statistical framework was applied to calculate the p value of each transcript-derived contig expression difference by two-sample comparison [17]. Then, we performed multiple hypothesis testing by controlling the false discovery rate (FDR, q-value) to correct the p-value. In this study, both the FDR was less than 10^{-3} and the RPKM values were greater than 2-fold (or less than 1/2-fold) different between two samples, then they were considered as a statistically significant DEGs (Table S4). From BL-BK comparison, we observed that 3,532 transcript-derived contigs were upregulated expression from BK when compared to BL skin, and 9,927 were downregulated (Figure S1). Similarly, 3,128 upregulated and 6,811 downregulated contigs were detected in the comparison of BS skin with BL skin. Further analysis revealed that 1,360 transcript-derived contigs were consistently upregulated and 4,973 were downregulated in BK and BS compared to BL skin (Figure 4). However, the number of DEGs was sharply reduced to 5,367, of which 3,338 were upregulated and 2,029 were downregulated from BS to BK skin, indicating that the genes expression patterns between BK and BS are more similar than those two comparisons.

The BK and BS skin of the Cashmere goat mainly produce cashmere fiber, whereas BL skin mostly grows wool fiber but few cashmere fibers. Therefore, to investigate differences between these two kinds of skin types, we annotated the 6,333 consistent DEGs and further performed KOG enrichment analysis compared with transcriptome background. We found that the clusters involved in the cell cycle control, cell division, and chromosome partitioning in KOG classification were overrepresented by these DEGs (Table S5). The evidence is that 71 out of 1442 (4.92%) annotated DEGs (KOG database annotation) which derived from 538 counterparts of total 17,594 (2.97%) annotated transcriptome sequences ($p < 8.47e{-6}$, q-value $< 2e{-4}$). Furthermore, ten significant DEGs that enriched in this cluster were selected to perform qRT-PCR and to investigate gene expression difference among three skin types. Although the exact fold difference for each DEG by qRT-PCR were different from the RNA-Seq method

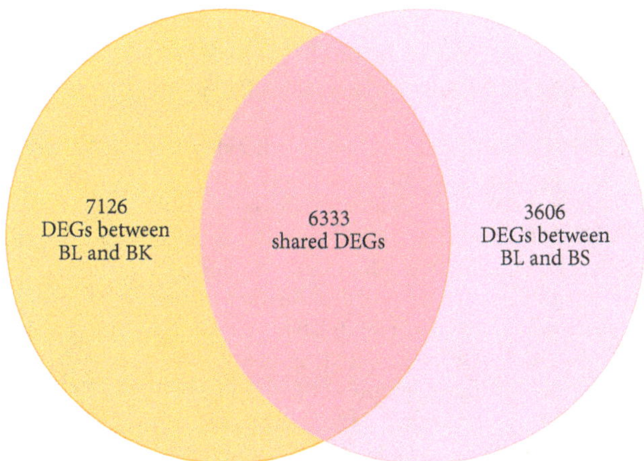

FIGURE 4: Venn diagram of shared DEGs between BL versus BK and BL versus BS comparisons.

but all the comparison pairs had the similar trends with RNA-Seq approach suggesting the relative high consistency between RNA-Seq and qRT-PCR (Table S6).

4. Discussion

To obtain the comprehensive transcripts of goat and its gene expression profiles reflected in different skin types during anagen phase, we sequenced and assembled mRNA from BL, BK, and BS skin of body coat. The assembler used in this work is Inchworm which is a major component of Trinity software [10]. Initially, the assembler generated 265,169 contigs over 100 bp (average contig length 299 bp and N50 length 417 bp) corresponding to ~79.3 MB sequence length, 87,962 contigs of which are longer than 200 bp (average contig length 622 bp and N50 length 1001 bp) representing ~54.8 MB nonredundant sequence length. When we use these smaller contigs for functional annotation, more redundant hit accessions were obtained, and the functional classification of transcripts-derived contigs would be more redundantly represented in each functional cluster, which gave us biased interpretation and overview of the transcripts functions. In this study, we used the contigs over 300 bp for annotation, which yielded 22,146 hits and represented 17,472 nonredundant accessions, indicating that the large proportion of the contigs belonged to UniGene clusters. On the other hand, approximately 87% of the total short reads can uniquely be mapped to the 49,115 contigs also suggesting that the majority of reads contributed to comprising those larger contigs.

As the different skin-regions of the Cashmere goat body coat, cashmere fibers are mainly produced from skin on the BK and BS, with few growing from the BL part. The molecules that are differentially expressed among these skin types may have underlying or potential roles associated with cashmere growth. Through calculating the short reads mapped on each contig from different libraries, we identified 6,333 consistently differentially expressed transcripts from BK and BS responded to BL skin (Figure 4). These DEGs were mainly enriched in the cell cycle control, cell division, and chromosome partitioning in the KOG functional clusters, indicating that the gene expression pattern associated with cell cycle and the cell division between two types of skin are significantly different. For instance, a kinetochore-bound protein kinase, named budding uninhibited by benzimidazoles 1 (Bub1), was identified 4- and 3-fold downregulation from BK and BS compared with BL skin (contig ID a91887;15). This kinase functions in part by phosphorylating a member of the miotic checkpoint complex and activating the spindle checkpoint [22, 23]. Similarly, we found another important kinetochore protein known as NDC80 (a132791;11) which is responsible for chromosomes segregation during M phase of the cell division that was also identified as a significant downregulated gene from BK and BS compared with BL skin [24]. In addition to the genes involved in the spindle checkpoint during cell division, we also noted that some important regulators which directly participate in the regulation of cell cycle progress are differentially expressed. For example, we found that cyclin-dependent kinase 7 (CDK7, a53471;29) as well as its partner, Cyclin-H (a65131;18), which form a complex to directly regulate cycle division were both identified as downregulated DEGs in BK and BS libraries compared within BL part [25]. Furthermore, profiling of cell cycle associated DEGs by using qRT-PCR method also showed relatively high consistent result with RNA-Seq. This may suggest that the hair synthesis rates between cashmere- and wool-producing skins are significantly different.

A large body of literature has focused on revealing the molecular mechanisms of HF initiation, patterning, and hair cycling in mammalian model organisms such as *Mus musculus*. Many significant studies mainly focused on the function of upstream molecules of the signal transduction pathway such as Wnt/β-catenin, TGF-β, Eda, Hedgehog, IGF, and their receptors [26–42]. Generally, conditional knockout or skin tissue-specific overexpression of ligand, receptor, or adaptor molecules from these pathways during the embryogenesis or postnatal stages usually has diverse effects on HF and hair shaft formation. One example is β-catenin which serves as an important adaptor molecule in embryogenesis. Sustained epithelial β-catenin activation by a transgenic approach caused excessive induction and fusion of HF with severely impaired fiber shaft formation [43]. However, the factors which directly promote hair growth are rarely characterized. The most compelling molecule discovered to promote hair growth is FGF-5, a secreted protein that when ablated from mice leads to abnormally long hair (~1.5-fold longer than wild type) by either the elongation of the anagen phase or retardation of catagen initiation [44]. IGF-1, another important mitogen associated with HF development, has been reported as an elongation factor of mouse whiskers [31] and was recently demonstrated to promote body hair growth by overexpression of IGF-1 in mouse skin through transgenic approach [30]. But this growth was accompanied by an absence of two types of body coat hair and disorientation of a small proportion of HF [30]. Nevertheless, we did not identify that these molecules encoding genes were significantly differentially expressed. This may ascribe to the three samples that were all derived from anagen phase of the hair cycle, because many previous studies have demonstrated that these signaling molecules and their receptors function as important regulators during the transition from telogen (resting phase) to anagen, such as Wnt, Sonic hedgehog, and TGF-β family members [45–47], suggesting that the expression levels of these signals are different between two periods. In our DEG catalogue, only the genes associated with cell division and cell cycle are significantly enriched, which might indicate that the efficiency of hair synthesis is different between two skin types. The function of these enriched DEGs involved in the hair cycling should be further characterized.

5. Conclusions

Taking into consideration the read abundance, average contig length, N50 length, and total contig size, our assembled contig catalogue provided a relatively complete and comprehensive dataset which could reflect the goat skin transcriptome during the anagen phase. The identification of numerous genes, including those showing differential expression in three skin types, especially those DEGs which are enriched in the functional cluster, will provide us good launching points and resources for further characterizing gene functions associated with hair growth. Our dataset was generated solely with the use of an Illumina HiSeq2000 platform, demonstrating that this ultra-high-throughput sequencing technology is a suitable tool for investigation of the large eukaryotic organism transcriptome and global measurement of gene expression profile. Finally, the extremely abundant paired-end reads generated from anagen phase will be very useful for subsequent studies, such as comparisons with gene expression patterns from catagen or telogen phase, which will be very helpful for further identifying genes associated with hair follicle development and fiber growth.

Authors' Contribution

Teng Xu and Xudong Guo contributed equally to this work.

Acknowledgments

This work was supported by the Genetically Modified Organisms Breeding Major Projects of China (Grant no. 2011ZX08008-002), the National Natural Science Foundation of China (Grant no. 31160228), and the National High Technology Research and Development Program of China (Grant no. 2013AA102506). The authors thank Yongbin Zhang at Yi Wei White Cashmere Goat Breeding Farm for kindly helping with samples preparation and collection. They also thank Guiming Liu at Beijing Institute of Genomics for helpful discussions in data analysis.

References

[1] M. Ibraheem, H. Galbraith, J. Scaife, and S. Ewen, "Growth of secondary hair follicles of the Cashmere goat in vitro and their response to prolactin and melatonin," *Journal of Anatomy*, vol. 185, part 1, pp. 135–142, 1994.

[2] D. Allain and C. Renieri, "Genetics of fibre production and fleece characteristics in small ruminants, Angora rabbit and South American camelids," *Animal*, vol. 4, no. 9, pp. 1472–1481, 2010.

[3] A. J. Nixon, M. P. Gurns, and K. Betterid, "Seasonal hair follicle activity and fibre growth in some New Zealand Cashmere-bearing goats (Caprus hircus)," *Journal of Zoology*, vol. 224, pp. 589–598, 1991.

[4] B. J. McDonald, W. A. Hoey, and P. S. Hopkins, "Cyclical fleece growth in cashmere goats," *Australian Journal of Agricultural Research*, vol. 38, pp. 597–609, 1987.

[5] Z. Wang, M. Gerstein, and M. Snyder, "RNA-Seq: a revolutionary tool for transcriptomics," *Nature Reviews Genetics*, vol. 10, no. 1, pp. 57–63, 2009.

[6] S. Guo, Y. Zheng, J. G. Joung et al., "Transcriptome sequencing and comparative analysis of cucumber flowers with different sex types," *BMC Genomics*, vol. 11, no. 1, article 384, 2010.

[7] S. Wickramasinghe, G. Rincon, A. Islas-Trejo, and J. F. Medrano, "Transcriptional profiling of bovine milk using RNA sequencing," *BMC Genomics*, vol. 13, article 45, 2012.

[8] U. Nagalakshmi, Z. Wang, K. Waern et al., "The transcriptional landscape of the yeast genome defined by RNA sequencing," *Science*, vol. 320, no. 5881, pp. 1344–1349, 2008.

[9] Z. Wenguang, W. Jianghong, L. Jinquan, and M. Yashizawa, "A subset of skin-expressed microRNAs with possible roles in goat and sheep hair growth based on expression profiling of mammalian microRNAs," *OMICS A Journal of Integrative Biology*, vol. 11, no. 4, pp. 385–396, 2007.

[10] M. G. Grabherr, B. J. Haas, M. Yassour et al., "Full-length transcriptome assembly from RNA-Seq data without a reference genome," *Nature Biotechnology*, vol. 29, no. 7, pp. 644–652, 2011.

[11] R. Li, C. Yu, Y. Li et al., "SOAP2: an improved ultrafast tool for short read alignment," *Bioinformatics*, vol. 25, no. 15, pp. 1966–1967, 2009.

[12] C. Burge and S. Karlin, "Prediction of complete gene structures in human genomic DNA," *Journal of Molecular Biology*, vol. 268, no. 1, pp. 78–94, 1997.

[13] R. L. Tatusov, N. D. Fedorova, J. D. Jackson et al., "The COG database: an updated vesion includes eukaryotes," *BMC Bioinformatics*, vol. 4, article 41, 2003.

[14] M. Kanehisa, M. Araki, S. Goto et al., "KEGG for linking genomes to life and the environment," *Nucleic Acids Research*, vol. 36, no. 1, pp. D480–D484, 2008.

[15] A. Conesa, S. Götz, J. M. García-Gómez, J. Terol, M. Talón, and M. Robles, "Blast2GO: a universal tool for annotation, visualization and analysis in functional genomics research," *Bioinformatics*, vol. 21, no. 18, pp. 3674–3676, 2005.

[16] J. Ye, L. Fang, H. Zheng et al., "WEGO: a web tool for plotting GO annotations," *Nucleic Acids Research*, vol. 34, pp. W293–W297, 2006.

[17] S. Audic and J. M. Claverie, "The significance of digital gene expression profiles," *Genome Research*, vol. 7, no. 10, pp. 986–995, 1997.

[18] A. Mortazavi, B. A. Williams, K. McCue, L. Schaeffer, and B. Wold, "Mapping and quantifying mammalian transcriptomes by RNA-Seq," *Nature Methods*, vol. 5, no. 7, pp. 621–628, 2008.

[19] K. J. Livak and T. D. Schmittgen, "Analysis of relative gene expression data using real-time quantitative PCR and the 2-$\Delta\Delta$CT method," *Methods*, vol. 25, no. 4, pp. 402–408, 2001.

[20] X. Tao, Y. H. Gu, H. Y. Wang, W. Zheng, X. Li et al., "Digital gene expression analysis based on integrated de novo transcriptome

assembly of sweet potato [Ipomoea batatas (L.) Lam]," *PLoS One*, vol. 7, Article ID e36234, 2012.

[21] U. Lagercrantz, H. Ellegren, and L. Andersson, "The abundance of various polymorphic microsatellite motifs differs between plants and vertebrates," *Nucleic Acids Research*, vol. 21, no. 5, pp. 1111–1115, 1993.

[22] F. Marchetti and S. Venkatachalam, "The multiple roles of Bub1 in chromosome segregation during mitosis and meiosis," *Cell Cycle*, vol. 9, no. 1, pp. 58–63, 2010.

[23] C. Klebig, D. Korinth, and P. Meraldi, "Bub1 regulates chromosome segregation in a kinetochore-independent manner," *Journal of Cell Biology*, vol. 185, no. 5, pp. 841–858, 2009.

[24] Y. Chen, D. J. Riley, P. L. Chen, and W. H. Lee, "HEC, a novel nuclear protein rich in leucine heptad repeats specifically involved in mitosis," *Molecular and Cellular Biology*, vol. 17, no. 10, pp. 6049–6056, 1997.

[25] R. P. Fisher and D. O. Morgan, "A novel cyclin associates with MO15/CDK7 to form the CDK-activating kinase," *Cell*, vol. 78, no. 4, pp. 713–724, 1994.

[26] J. Huelsken, R. Vogel, B. Erdmann, G. Cotsarelis, and W. Birchmeier, "β-Catenin controls hair follicle morphogenesis and stem cell differentiation in the skin," *Cell*, vol. 105, no. 4, pp. 533–545, 2001.

[27] M. Yuhki, M. Yamada, M. Kawano et al., "BMPR1A signaling is necessary for hair follicle cycling and hair shaft differentiation in mice," *Development*, vol. 131, no. 8, pp. 1825–1833, 2004.

[28] V. A. Botchkarev and M. Y. Fessing, "Edar signaling in the control of hair follicle development," *The Journal of Investigative Dermatology Symposium Proceedings*, vol. 10, no. 3, pp. 247–251, 2005.

[29] C. Chiang, R. Z. Swan, M. Grachtchouk et al., "Essential role for Sonic hedgehog during hair follicle morphogenesis," *Developmental Biology*, vol. 205, no. 1, pp. 1–9, 1999.

[30] N. Weger and T. Schlake, "IGF-I signalling controls the hair growth cycle and the differentiation of hair shafts," *Journal of Investigative Dermatology*, vol. 125, no. 5, pp. 873–882, 2005.

[31] H. Y. Su, J. G. H. Hickford, P. H. B. The, A. M. Hill, C. M. Frampton, and R. Bickerstaffe, "Increased vibrissa growth in transgenic mice expressing insulin-like growth factor 1," *Journal of Investigative Dermatology*, vol. 112, no. 2, pp. 245–248, 1999.

[32] H. Kulessa, G. Turk, and B. L. M. Hogan, "Inhibition of Bmp signaling affects growth and differentiation in the anagen hair follicle," *EMBO Journal*, vol. 19, no. 24, pp. 6664–6674, 2000.

[33] D. J. Headon and P. A. Overbeek, "Involvement of a novel Tnf receptor homologue in hair follicle induction," *Nature Genetics*, vol. 22, no. 4, pp. 370–374, 1999.

[34] L. Guo, L. Degenstein, and E. Fuchs, "Keratinocyte growth factor is required for hair development but not for wound healing," *Genes and Development*, vol. 10, no. 2, pp. 165–175, 1996.

[35] V. A. Botchkarev, N. V. Botchkareva, A. A. Sharov, K. Funa, O. Huber, and B. A. Gilchrest, "Modulation of BMP signaling by noggin is required for induction of the secondary (nontylotrich) hair follicles," *Journal of Investigative Dermatology*, vol. 118, no. 1, pp. 3–10, 2002.

[36] V. A. Botchkarev, N. V. Botchkareva, W. Roth et al., "Noggin is a mesenchymally derived stimulator of hair-follicle induction," *Nature Cell Biology*, vol. 1, no. 3, pp. 158–164, 1999.

[37] S. Vauclair, M. Nicolas, Y. Barrandon, and F. Radtke, "Notch1 is essential for postnatal hair follicle development and homeostasis," *Developmental Biology*, vol. 284, no. 1, pp. 184–193, 2005.

[38] A. Rangarajan, C. Talora, R. Okuyama et al., "Notch signaling is a direct determinant of keratinocyte growth arrest and entry into differentiation," *EMBO Journal*, vol. 20, no. 13, pp. 3427–3436, 2001.

[39] A. G. Li, M. I. Koster, and X. J. Wang, "Roles of TGFβ signaling in epidermal/appendage development," *Cytokine and Growth Factor Reviews*, vol. 14, no. 2, pp. 99–111, 2003.

[40] L. P. Sanford, I. Ormsby, A. C. Gittenberger-de Groot et al., "TGFβ2 knockout mice have multiple developmental defects that are non-overlapping with other TGFβ knockout phenotypes," *Development*, vol. 124, no. 13, pp. 2659–2670, 1997.

[41] T. Andl, S. T. Reddy, T. Gaddapara, and S. E. Millar, "WNT signals are required for the initiation of hair follicle development," *Developmental Cell*, vol. 2, no. 5, pp. 643–653, 2002.

[42] M. Ito, Z. Yang, T. Andl et al., "Wnt-dependent de novo hair follicle regeneration in adult mouse skin after wounding," *Nature*, vol. 447, no. 7142, pp. 316–320, 2007.

[43] K. Närhi, E. Järvinen, W. Birchmeier, M. M. Taketo, M. L. Mikkola, and I. Thesleff, "Sustained epithelial β-catenin activity induces precocious hair development but disrupts hair follicle down-growth and hair shaft formation," *Development*, vol. 135, no. 6, pp. 1019–1028, 2008.

[44] J. M. Hebert, T. Rosenquist, J. Gotz, and G. R. Martin, "FGF5 as a regulator of the hair growth cycle: evidence from targeted and spontaneous mutations," *Cell*, vol. 78, no. 6, pp. 1017–1025, 1994.

[45] K. S. Stenn and R. Paus, "Controls of hair follicle cycling," *Physiological Reviews*, vol. 81, no. 1, pp. 449–494, 2001.

[46] S. E. Millar, "Molecular mechanisms regulating hair follicle development," *Journal of Investigative Dermatology*, vol. 118, no. 2, pp. 216–225, 2002.

[47] L. Alonso and E. Fuchs, "Stem cells in the skin: waste not, Wnt not," *Genes and Development*, vol. 17, no. 10, pp. 1189–1200, 2003.

18

The Expression, Purification, and Characterization of a Ras Oncogene (Bras2) in Silkworm (*Bombyx mori*)

Zhengbing Lv,[1,2] Tao Wang,[1,2] Wenhua Zhuang,[1,2] Dan Wang,[1,2] Jian Chen,[1,2] Zuoming Nie,[1,2] Lili Liu,[1,2] Wenping Zhang,[1,2] Lisha Wang,[3] Deming Wang,[3] Xiangfu Wu,[1,2] Jun Li,[4] Lian Qian,[5] and Yaozhou Zhang[1,2]

[1] *College of Life Sciences, Zhejiang Sci-Tech University, Hangzhou 310018, China*
[2] *Zhejiang Provincial Key Laboratory of Silkworm Bioreactor and Biomedicine, Hangzhou 310018, China*
[3] *School of Pharmacy, Xuzhou Medical College, Xuzhou 221004, China*
[4] *Department of Obstetrics and Gynecology, Yong Loo Lin School of Medicine, National University Health System, Singapore 119228*
[5] *Agilent Technologies Singapore, Singapore 117681*

Correspondence should be addressed to Yaozhou Zhang; yaozhou@chinagene.com

Academic Editor: Graziano Pesole

The Ras oncogene of silkworm pupae (Bras2) may belong to the Ras superfamily. It shares 77% of its amino acid identity with teratocarcinoma oncogene 21 (TC21) related ras viral oncogene homolog-2 (R-Ras2) and possesses an identical core effector region. The mRNA of *Bombyx mori* Bras2 has 1412 bp. The open reading frame contains 603 bp, which encodes 200 amino acid residues. This recombinant BmBras2 protein was subsequently used as an antigen to raise a rabbit polyclonal antibody. Western blotting and real-time PCR analyses showed that BmBras2 was expressed during four developmental stages. The BmBras2 expression level was the highest in the pupae and was low in other life cycle stages. BmBras2 was expressed in all eight tested tissues, and it was highly expressed in the head, intestine, and epidermis. Subcellular localization studies indicated that BmBras2 was predominantly localized in the nuclei of Bm5 cells, although cytoplasmic staining was also observed to a lesser extent. A cell proliferation assay showed that rBmBras2 could stimulate the proliferation of hepatoma cells. The higher BmBras2 expression levels in the pupal stage, tissue expression patterns, and a cell proliferation assay indicated that BmBras2 promotes cell division and proliferation, most likely by influencing cell signal transduction.

1. Introduction

Ras family small GTPases play essential roles in a variety of cellular responses including cell proliferation, differentiation, survival, transformation, and tumor development [1–4]. The family has approximately 20 members in mammals [5, 6], and as many as 36 Ras family genes have been identified in humans [7] with evolutionarily conserved orthologs in *Drosophila*, *C. elegans*, *S. cerevisiae*, *S. pombe*, *Dictyostelium*, and plants [8]. This family includes the classical Ras proteins (H-Ras, N-Ras, K4A-Ras, and K4B-Ras), the R-Ras proteins (R-Ras, TC21/R-Ras2, and M-Ras/R-Ras3), the Rap proteins (Rap1A, Rap1B, Rap2A, and Rap2B), and the Ral proteins (RalA and RalB) [9]. The first identified classical Ras (hereafter simply referred to as Ras) proteins have been studied most intensively. It was shown that Ras transduces signals from receptor-type tyrosine kinases to downstream effectors and thereby controls the proliferation and differentiation of various cell types [10]. Ras proteins function as molecular switches and are controlled by a regulated GDP/GTP cycle. Guanine nucleotide exchange factors (GEFs, e.g., SOS and mCDC25/GRF) promote the formation of active, GTP-bound Ras, whereas GTPase-activating proteins (GAPs, p120 and NF1 GAP) promote the formation of inactive, GDP-bound Ras [11].

Like Ras, other members of the Ras superfamily are also believed to function as molecular regulatory switches that control a spectrum of diverse cellular processes [12]. Ras-related proteins share significant similarities in molecular weight (20 to 25 kDa) and sequence identity (30 to 55%)

with Ras proteins [13]. Despite possessing strong structural and biochemical similarities with Ras proteins, only a limited number of Ras-related proteins have been shown to exhibit transforming potential [14]. Apart from the classical Ras proteins, the only other member of the Ras subfamily of GTPases found to be mutated in human cancers was TC21 (also called R-Ras2) [15].

The aim of this study is to elucidate ras oncogene (*Bras2*) properties and to characterize the *Bras2* present in the silkworm *Bombyx mori*. The cDNA of *BmBras2* from *Bombyx mori* consists of 1,412 bp. The open reading frame (ORF) contains 603 bp, encoding 200 amino acid residues with a predicted molecular weight of 22.9 kDa and theoretical isoelectric point (pI) of 6.62. Its accession number in GenBank is AB206960. In this paper, we report the cDNA cloning, expression, purification, and characterization of silkworm Bras2 for the first time. We found that silkworm Bras2 shares 77% of its amino acid identity with TC21 and may be involved in the regulation of normal cell growth. This study lays a good foundation for further research on the function of this protein.

Bombyx mori has a well-studied genetic background and high developmental synchronization. Because it is susceptible to nuclear polyhedrosis virus and easy to breed at a large scale, *B. mori* has been used as a bioreactor to produce recombinant proteins with the *B. mori* nucleopolyhedrovirus (BmNPV) expression system [16]. One of the major advantages of the BmNPV expression system is that it can be used to produce relatively large quantities of posttranslationally modified heterologous proteins. This expression system is inexpensive, convenient and has a high production level, so it has been widely used to express recombinant proteins.

2. Materials and Methods

2.1. Animals, Tissues, and Bm5 Cells. The *B. mori* strain in this study was the progeny of Qiufeng × Baiyu. Silkworms were reared on mulberry leaves under standard conditions. Heads, intestines, epidermises, silk glands, fat bodies, malpighian tubules, ovaries, and testes were dissected from the fifth instar larvae, frozen immediately in liquid nitrogen, and stored at −80°C. The fifth instar larvae, pupae, moths, and nascent eggs were also frozen in liquid nitrogen and stored at −80°C. Bm5 cells were seeded at 1×10^5 cells per flask in culture flask and cultured for three days at 37°C in a 5% CO_2 incubator. The medium was removed and the Bm5 cells were collected for western bolt with anti-BmBras2.

2.2. Construction of a Recombinant Plasmid. The cDNA, which was previously constructed from metaphase pupae by our laboratory [17], was used as a template to amplify the coding region of BmBras2 by polymerase chain reaction (PCR). We designed gene-specific primers on the basis of the *BmBras2* cDNA sequence, which included restriction enzyme sites for *Eco*RI and *Hind*III. The sequences of the gene-specific primer (GSP) were 5′-GGGAATTCATGT-CTCGAGCAGGCGACAGAC-3′ and 5′-GGGAAGCTT-TTACAGGATGGTGCACTTC-3′. The PCR cycle conditions were one cycle at 94°C for 5 min, then 30 cycles of denaturing at 94°C for 1 min, followed by an annealing step at 53°C for 50 s, then an extension step at 72°C for 1 min and one additional extension cycle at 72°C for 10 min using a *Taq* DNA polymerase Kit (Promega, USA). The PCR products were purified using a PCR Rapid Purification Kit (BioDev-Tech, China). After digestion with *Eco*RI and *Hind*III, the amplicons were subcloned into expression vector pET-28a using T4 DNA ligase (Promega, USA) and then transformed into *E. coli* TG1 cells (which are maintained in our laboratory) for screening purposes. Positive colonies in which the *BmBras2* gene was successfully integrated into the plasmid were identified by double plasmid digestion and subsequently sequenced by an ABI PRISM 3130-XL/A automated sequencer (applied biosystems).

2.3. Sequence Analysis. A nucleotide and protein sequence similarity analysis was carried out at GenBank using BLASTN (in the EST, other database) and BLASTP (in all nonredundant databases) algorithms. The deduced amino acid sequence was analyzed with the Expert Protein Analysis System (http://www.expasy.org/). Multiple sequence alignments of the Bras2 and R-Ras family were conducted using the Clustal W program in Bioedit software. The protein conformation was modeled by SWISS-MODEL (http://swissmodel.expasy.org/) and viewed in the Swiss PDB Viewer [18].

2.4. Expression and Purification of BmBras2 . The recombinant expression plasmid pET-28a (+)-*BmBras2* was transformed into *E. coli* Rosetta (DE3) (which is maintained in our laboratory). Bacterial expression cultures were incubated at 37°C in LB medium containing kanamycin (50 μg/mL) and chloramphenicol (50 μg/mL) until an A_{600} of 0.5 was reached. Recombinant protein expression was induced by the addition of IPTG (Sanland-chem, USA) at a final concentration of 0.1 mM. Following 4 h of incubation at 37°C, bacterial culture was harvested by centrifugation and frozen at −20°C. Frozen bacterial pellets were thawed and resuspended in lysis buffer and then lysed by pulsed sonication with a cell disruptor (SCIENTZ-IID) on ice. The lysates were centrifuged at 14,000 × g for 20 min at 4°C. The supernatant was collected and filtered through a 0.45 μm filter (Millipore, USA). Nickel metal affinity resin columns were used for rBmBras2 single-step purification [19], and the protein purity was examined by SDS-PAGE as described by Laemmli [20]. The purity was also examined by HPLC, which showed a fusion protein purity of over 98%.

2.5. Molecular Weight Determination. The molecular weight of rBmBras2 was assessed by SDS-PAGE using 12% SDS gels. The molecular weight was also determined by 4700 MALDI-TOF/TOF mass spectrometry [21] (ABI, USA) using default settings for laser energy and TOF parameters.

2.6. Protein Extraction. The fifth instar larvae, pupae, moths, eggs, and tissues isolated from fifth instar larvae were ground to a powder in liquid nitrogen. The powders were suspended in protein dissolution buffer and then incubated for 30 min on ice. Homogenates were centrifuged at 12,000 × g for 15 min at 4°C. The protein concentrations of all samples were quantified before SDS-PAGE analysis using a Bradford assay, and total protein loading was normalized to an equal amount with alpha tubulin.

2.7. Western Blot Analyses. Whole protein extracts from each tissue and Bm5 cells were separated by 12% SDS-PAGE and then electrotransferred to PVDF membranes (Millipore). Membranes were incubated in blocking buffer containing 5% skim milk and 0.05% Tween 20 in TBS (TBST; pH 7.5) at room temperature for 1 h or 4°C overnight and were then incubated with polyclonal antibody (diluted to 1/1000) in blocking buffer at room temperature for 1 h or 4°C overnight, then washed with 0.05% TBST three times for 5 min per wash and detected with HRP-labeled anti-rabbit IgG (Dingguo Biotechnology) and DAB. The preparation of polyclonal antibody for anti-rBmBras2 was described by Sheng et al. [22].

2.8. RNA Extraction. Total RNA was extracted from the various silkworm developmental stages and the tissues of fifth instar larvae using Trizol reagent (Invitrogen) according to the manufacturer's instructions. Contaminating genomic DNA (gDNA) was removed by adding DNase I (Invitrogen). RNA purity was determined with a UV spectrophotometer. The UV_{260}/UV_{280} ratios were between 1.8 and 2.1 for all analyzed RNA samples. The concentration of total RNA was determined by measuring the absorbance at 260 nm with an ND-1000 spectrophotometer (Bio-Rad, USA).

2.9. Primer Design and Real-Time PCR. Real-time PCR primers were designed using Primer Premier 5.0 software. The primer pairs for BmBras2 were as follows: upstream primer, 5′-TCCTGCTGGTCTTCTCCGTG-3′ downstream primer, 5′-GACACCACTCGCTGCGTTTC-3′; 18S rRNA, upstream primer, 5′-CGATCCGCCGACGTTACTACA-3′; and downstream primer, 5′-GTCCGGGCCTGGTGAGA-TTT-3′. Real-time PCR was performed with an ABI Prism 7300 Sequence Detection System (Applied Biosystems) under the following PCR conditions: one cycle at 95°C for 30 s, followed by 40 cycles at 95°C for 5 s and 60°C for 31 s. A SYBR PrimeScript RT-PCR Kit II was used for the real-time PCR. Each reaction was performed in triplicate in a 96-well plate along with the endogenous 18S rRNA control gene. At the end of the real-time PCR cycles, a dissociation curve was performed to check for the presence of nonspecific dsDNA SYBR Green hybrids, such as primer dimers. Data analysis was performed with ABI Prism 7300 SDS Software V1.3.1 (Applied Biosystems, USA). The target gene expression levels were normalized against the 18S rRNA gene expression level. The relative expression level was calculated using $2^{-\Delta\Delta CT}$. (Where $\Delta CT = CT$ (*BmBras2*) − CT (18S rRNA) for different stages or tissues; $\Delta\Delta CT = \Delta CT$ for different stages or tissues − ΔCT maximum).

2.10. Subcellular Localization. Bm5 cells were seeded in a special confocal microscope dish (Bio-line Instruments). After 12 h, the culture medium was removed. Cells were rinsed twice with 1 mL phosphate-buffered saline (PBS), and then they were fixed in 3.7% formaldehyde at 25°C for 10 min. The cells were blocked with 3% BSA at 37°C for 2 h. After that, the cells were incubated with anti-BmBras2 IgG antibody (dilution 1/1000) at 4°C for 12 h (cells incubated with negative serum without the antibody served as the control). After three washes in PBST (PBS+0.05% Tween-20, 10 min each), cells were incubated with goat anti-rabbit antibody (dilution, 1/2000, Cy3 labeled, Promega) and DAPI (4′-6-diamidino-2-phenylindole, at a dilution of 1/2,000) at 37°C for 2 h. Following three washes with PBST (10 min each), the cells were analyzed with a Nikon ECLIPSE TE2000-E confocal microscope with EZ-C1 image analysis software.

2.11. Cell Proliferation Assay. One hundred microliters of cell suspension (approximately 2×10^3 hepatoma cells) was added to each well of a 96-well plate and cultured for one day at 37°C in a 5% CO_2 incubator. Ten microliters of rBmBras2 protein was pipetted into each well, and then the hepatoma cells were divided into a control group (0 mg/mL) and test groups of rBmBras2 at different concentrations (1.5, 5, 20, 40 μg/mL) with 6 replicates per group. The hepatoma cells were then cultured for 3 more days in the cell incubator. Ten microliters of Cell Counting Kit-8 (CCK-8) solution was transferred to each well in the 96-well plate and incubated for approximately 1 hour. All samples were measured with a Universal Microplate Reader ELX-800 (Bio-Tek, USA) at a wavelength of 450 nm for measurements and 630 nm for references. All results were presented as $\overline{x} \pm s$, and a Student's t-test was used for statistical analysis, with statistical significance defined as $P < 0.05$ or $P < 0.01$.

3. Results

3.1. Biological Information about BmBras2 . The complete mRNA of the *BmBras2* gene was 1412 bp in length. This length includes a 5′-terminal untranslated region (UTR) of 122 bp, a 3′-terminal UTR of 687 bp with a canonical polyadenylation signal sequence of AATAAA and a poly(A) tail, and an open reading frame (ORF) of 603 bp encoding a polypeptide of 200 amino acids. The BmBras2 protein had a predicted molecular weight of 22.9 kDa and a theoretical isoelectric point of 6.62. The conserved domain of BmBras2 shared specific similarities with the M-Ras/R-Ras-like subfamily, and this subfamily contains R-Ras2/TC21, M-Ras/R-Ras3, and related members of the Ras family. The R-Ras family of Ras-related proteins contains R-Ras, TC21 (R-Ras2), and M-Ras (R-Ras3). We employed a Clustal W program in Bioedit software to generate multiple alignments of the R-Ras family members and BmBras2, and significant similarities were detected between members of the R-Ras family and BmBras2. BmBras2 shares 77% of its amino acid identity with TC21 (R-Ras2) and also includes the identical core effector regions (Figure 1(a)). BmBras2's homology with ras-related proteins from other species is very high, with 76% shared amino acid identity with a Ras-related protein R-Ras2 precursor of

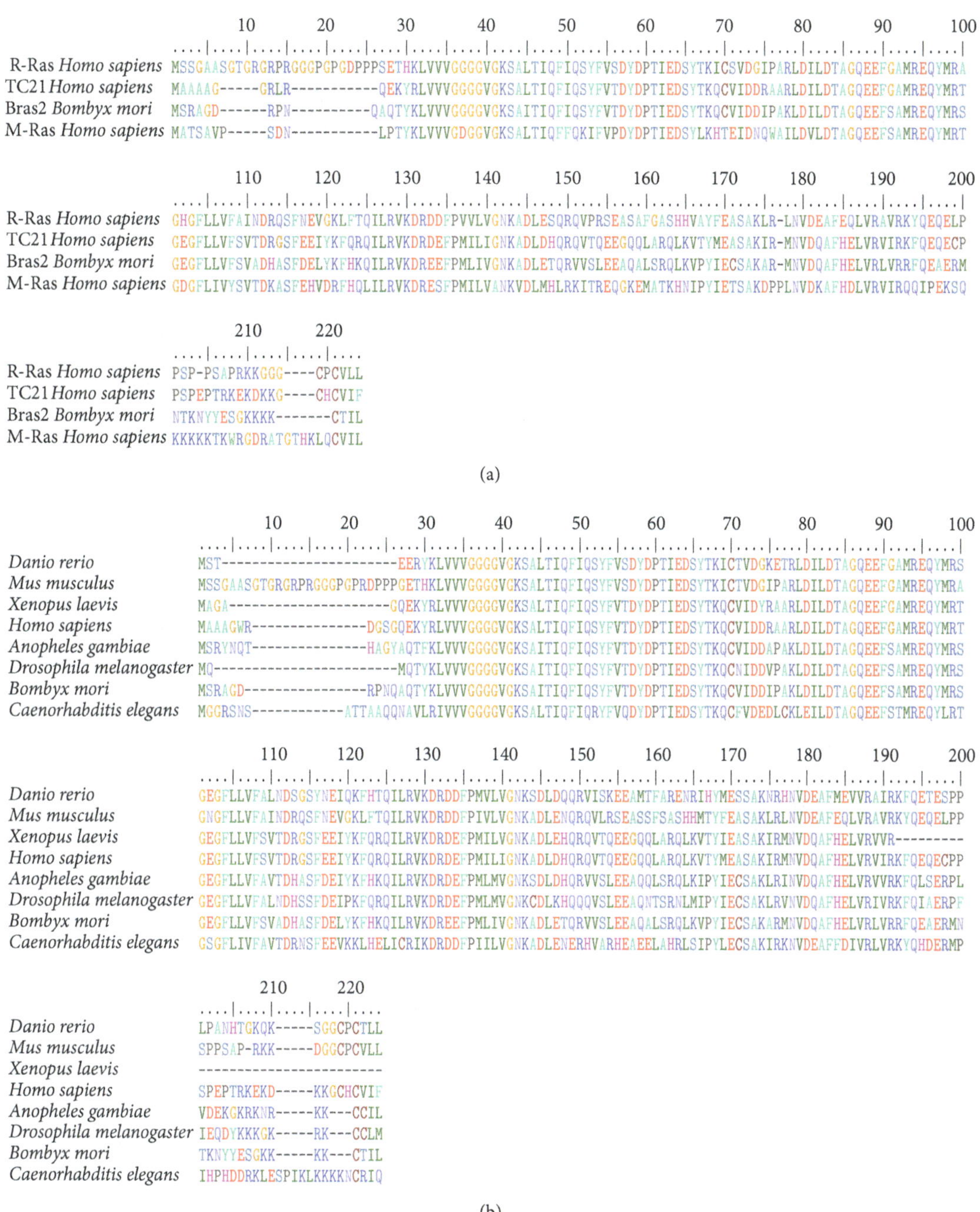

FIGURE 1: (a) Alignment of amino acid sequence of BmBras2 with members of the R-Ras superfamily from *Homo sapiens*. BmBras2 shares 69% of its amino acid identity with R-Ras, 77% with TC21 (R-Ras2) and 59% with M-Ras (R-Ras3). (b) Alignment of BmBras2 amino acid sequence with homologous proteins from different species.

Mus musculus and 78% amino acid identity with the Ras oncogene at 64B of *Drosophila melanogaster* (Figure 1(b)). These analyses indicated that the BmBras2 of silkworm pupae may belong to the Ras superfamily.

3.2. rBmBras2 Expression and Purification in E. coli. A PCR fragment containing the complete open reading frame of mature BmBras2 was inserted into pET-28a by ligation at the *Eco*RI and *Hind*III restriction enzyme sites. The expressed product was a recombinant protein with a 6 × His tag. The fusion protein His-BmBras2 was successfully expressed in *E. coli* and purified using nickel metal affinity resin columns. A molecular weight of 26,690.0 D was determined by MALDI-TOF/TOF analysis (data not shown) and is in agreement with the calculation using amino acid composition (22879.1 D + 3825.2 D = 26,704.3 D). The polyclonal antibody was prepared by subcutaneously immunizing male New Zealand white rabbits. The titer of the polyclonal antibody was more than 1:12800 when measured by indirect ELISA (data not shown). In Western blot analyses (Figure 2), the polyclonal antibody recognized recombinant His-BmBras2 protein.

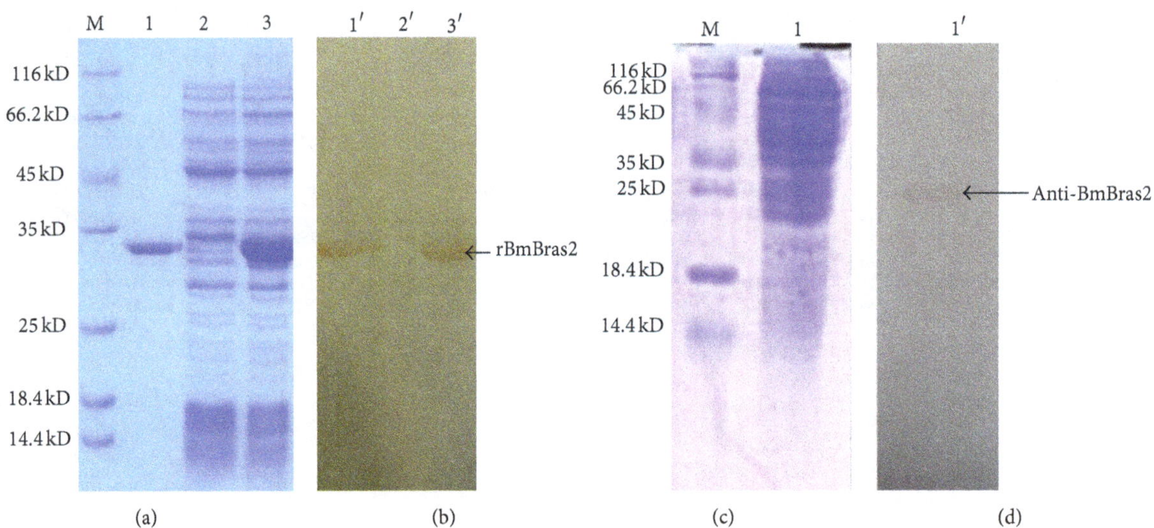

Figure 2: (a) The expression and purification of recombinant BmBras2 were analyzed by SDS-PAGE. (b) The expression and purification of recombinant BmBras2 were analyzed by Western blotting. M, protein mass marker; 1, purified recombinant protein by Ni-NTA superflow cartridges; 2, the lysate of *E. coli* Rosetta with pET-28a (+)-*BmBras2* without induction; 3, the lysate of *E. coli* Rosetta with pET-28a (+)-*BmBras2* after IPTG induction. (c) The lysate of Bm5 was analyzed by SDS-PAGE. (d) The Bm5 cells were collected for western bolt with anti-BmBras2. (Bm5 cells were seeded at 1×10^5 cells per flask in culture flask and cultured for three days at 37°C in a 5% CO_2 incubator.) M, protein mass marker; 1, the lysate of Bm5 cells.

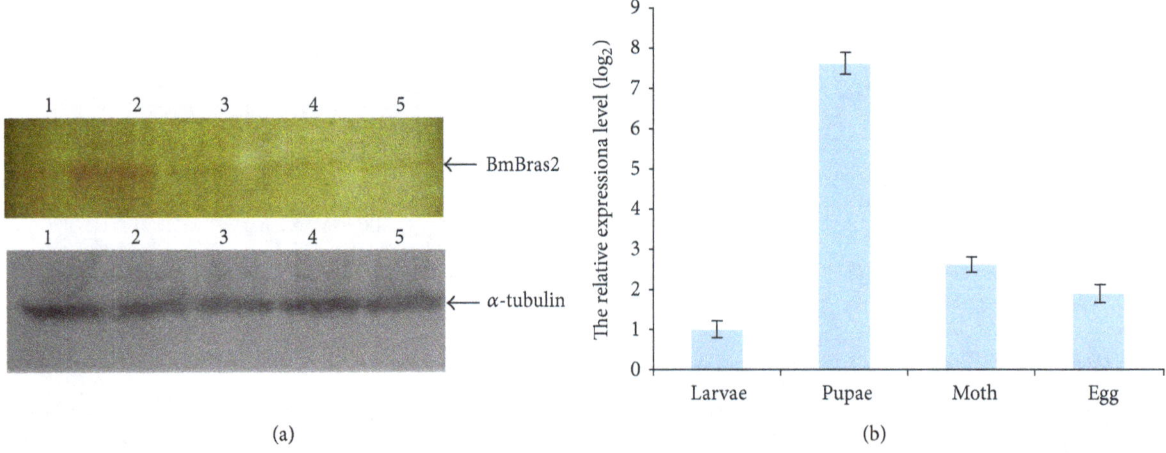

Figure 3: The expression analysis of BmBras2 during different silkworm developmental stages. (a) The expression levels of BmBras2 by western blotting; 1, fifth instar larvae; 2, pupae; 3, moth; 4, egg; 5, purified rBmBras2. (b) The relative expression levels of BmBras2 analyzed by real-time PCR. The relative expression level was calculated by using $2^{-\Delta\Delta CT}$, where $\Delta\Delta CT$ = (CT, *BmBras2*-CT, 18S rRNA) for different stages and (CT, *BmBras2*-CT, 18S rRNA) pupae.

3.3. Expression Analysis and Tissue Distribution of BmBras2. There is no information regarding the expression patterns of BmBras2 from EST sources. We therefore employed SYBR Green real-time PCR and western blot analyses to quantify *BmBras2* gene expression levels during different silkworm developmental stages and tissue distribution in fifth instar larvae [23]. For Western blot analyses, we used the polyclonal rabbit antibody (Anti-rBmBras2) to analyze the protein expression levels throughout four silkworm developmental stages and from eight tissues of the fifth instar larvae. For the real-time PCR analyses, we used a constitutively expressed gene, namely, 18S rRNA, as an internal control. For the *BmBras2* gene and 18S rRNA of silkworm, dissociation curves indicated proper amplification of the intended targets at the corresponding melting temperatures. *BmBras2* expression levels during different silkworm developmental stages are shown in Figure 3. BmBras2 was expressed throughout four developmental stages, and its expression was higher in the pupal stage and lower in others such as the fifth instar larvae, moth, and nascent egg. The real-time PCR results were consistent with the western blot analyses. Tissue distributions of BmBras2 in the fifth instar larvae are shown in Figure 4(a), with alpha-tubulin as an internal control. According to real-time PCR analysis, BmBras2 was expressed in all eight tissues (Figure 4(b)), and it was highly expressed in the head, intestine, and epidermis. Real-time PCR results were basically consistent with the western blot analyses (Figure 4(a)). Western blot analysis of Bm5 lysate illustrated that there was a

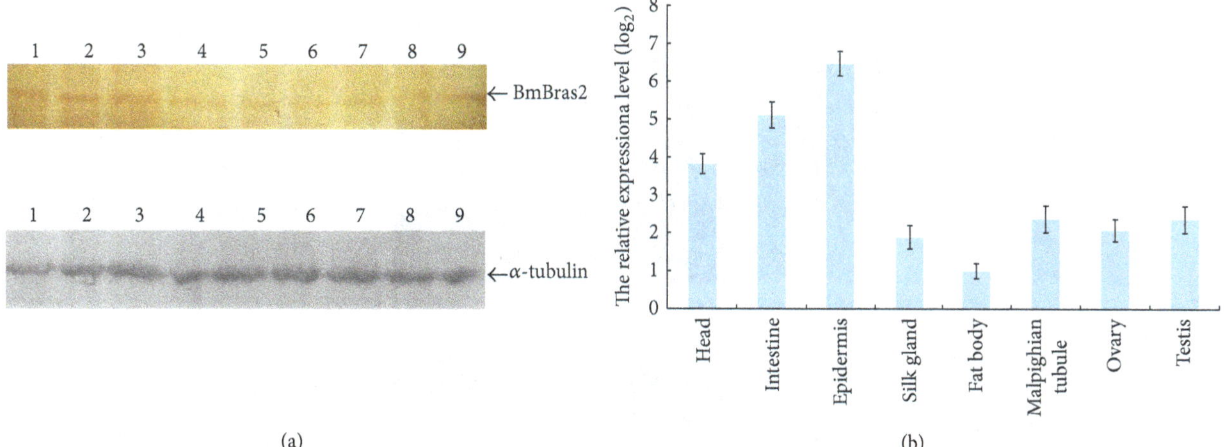

FIGURE 4: Distribution of BmBras2 in different fifth instar larvae tissues. (a) tissue distributions of BmBras2 by western blotting; 1, head; 2, intestine; 3, epidermis; 4, silk gland; 5, fat body; 6, malpighian tubule; 7, ovaries; 8, testis; 9, purified rBmBras2. (b) tissue distributions of BmBras2 analyzed by real-time PCR. Sg, silk gland; Fb, fat body; Mt, malpighian tubule. The relative expression level was calculated by using $2^{-\Delta\Delta CT}$ here $\Delta\Delta CT$ = (CT, *BmBras2*-CT, 18S rRNA) for different tissues and (CT, *BmBras2*-CT, 18S rRNA) fat body.

clear target band near 35 kDa band of the maker, while others are not obvious, which show good specificity of the prepared antibodies.

3.4. Subcellular Localization of BmBras2 . Bm5 cells were used to investigate the subcellular localization of endogenous BmBras2. Immunostaining with polyclonal rabbit anti-rBmBras2 antibody indicated that BmBras2 localized in both the cytoplasm and nucleus. The fluorescence intensity was predominantly stronger in the nucleus than in the cytoplasm, which indicated that BmBras2 is mainly localized in the nuclei of Bm5 cells (Figure 5).

3.5. rBmBras2 Effects on Cell Proliferation. PLC cells were treated with rBmBras2 for approximately 3 days and then measured at a wavelength of 450 nm with CCK-8. According to Figure 6, rBmBras2 proliferated on PLC in a dose-dependent manner. In other words, the effect of cell proliferation increased with an increasing concentration of rBmBras2 ($P < 0.05$). Meanwhile, the same assay was performed using HepG2 cells. The result showed that rBmBras2 can also promote the proliferation of HepG2 cells with the same effect as the PLC cells (data not shown).

4. Discussion

Ras superfamily GTPases function as GDP/GTP-regulated molecular switches [24]. They share a set of conserved G box GDP/GTP-binding motif elements beginning at the N-terminus as follows: G1, GXXXXGKS/T; G2, T; G3, DXXGQ/H/T; G4, T/NKXD; and G5, C/SAK/L/T [25]. We aligned the amino acid sequence of BmBras2 in the NCBI database and found that its conserved domain also contained G box GDP/GTP-binding motif elements. This protein showed specific similarities with an M-Ras/R-Ras-like subfamily, and it belongs to the Ras-like GTPase superfamily. Many members of the Ras superfamily of GTPases have been implicated in hematopoietic cell regulation, with roles in growth, survival, differentiation, cytokine production, chemotaxis, vesicle trafficking, and phagocytosis [6]. However, it is becoming increasingly evident that different members of the Ras subfamily may have different biological functions that depend not only on differences in their affinities to regulators or effectors but also on their precise subcellular localization [6].

The silkworm undergoes a complete metamorphosis, and its life cycle has four developmental stages, that is, egg, larvae, pupae, and moth. During these four stages, there are substantial changes in external morphology, physiological function, and biological characteristics. Although the insects seemed to be superficially very quiet during periods such as the egg stage, pupal stage, silking stage, and the molting stage, there are drastic internal changes going on; at the surface in the pupal stage, however, there is intensive internal organizational dissociation and histogenesis. These changes prepare the silkworm for the mating and oviposition of the adult stage. We analyzed the expression of BmBras2 during different developmental stages, and we found that the expression level of our target protein was the highest during the pupal stage and lower in the fifth instar larvae, moths, and nascent eggs. Because intense tissue differentiation and dissociation happen during the pupal stage, we predicted that BmBras2 might play an important role in this process.

Silkworm growth is a comprehensive embodiment of system and organ growth, and the growth of systems and organs is the embodiment of cell growth. There are three ways for cells to grow, as follows: (1) cell division, in which the growth pattern mainly relies on increasing the number of similar cells, but the basic cell size remains unchanged, as in sperm cells; (2) cell volume increase, with cell division mainly appearing during the embryonic development stage, but in the larval stage, the growth of entire organs and tissues is implemented only by increasing the cell size, as in silk glands; (3) cell division and cell volume increase, which include

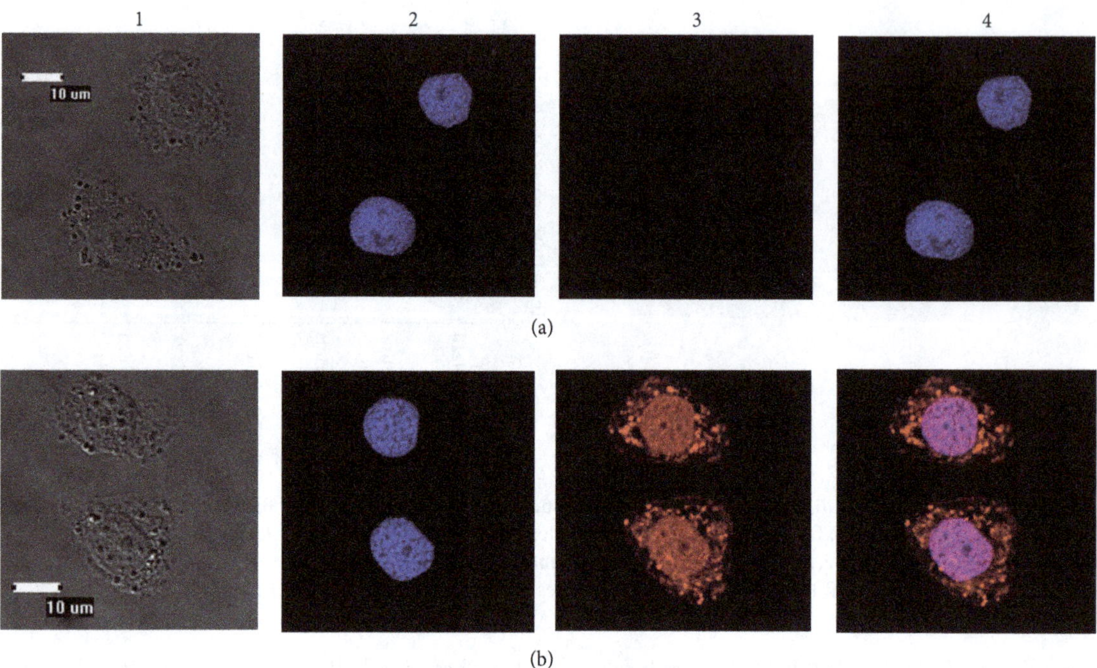

FIGURE 5: Subcellular localization of BmBras2. (a) Negative result amplified by 40 × 2.5; (b) positive result amplified by 40 × 2.5; 1, visible light images; 2, DAPI fluorescence images; 3, Cy3 fluorescence images; 4, mixed images.

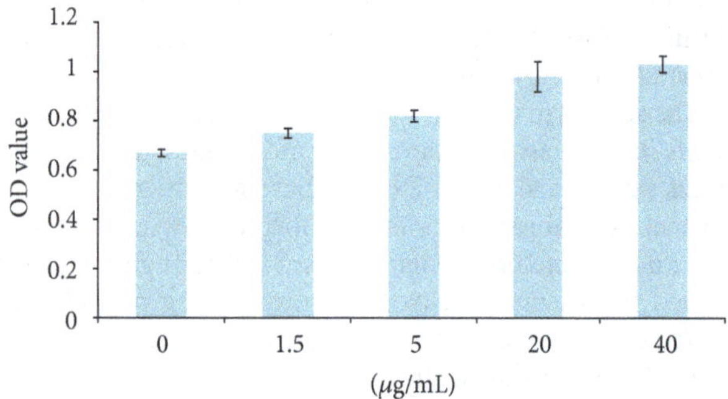

FIGURE 6: rBmBras2 effects on PLC proliferation promotion.

most of tissues and organs, such as epithelial cells. To learn more about the distribution and related functions of BmBras2 protein in various tissues, we measured mRNA and protein expression levels. The results of both experiments are largely consistent with one another, and BmBras2 is highly expressed in the head, intestine, and epidermis. High expression levels of tissue BmBras2 are essential for silkworms to complete their growth and development (that is, for cell division and proliferation). We have therefore proposed that BmBras2 protein is involved in cell cycle regulation, and it can promote cell division and proliferation through cell signal transduction, which is needed for tissue differentiation and dissociation.

Subcellular localization is closely related to protein function. Immunofluorescence analysis indicated that BmBras2 is mainly localized in the nuclei of Bm5 cells and partly in the cytoplasm. It was thus hypothesized that BmBras2 may be involved in cell cycle regulation, and this result will lay a foundation for further study of BmBras2 protein function.

TC21 is known to be a powerful oncogene [7], and constitutive TC21 activity induces cell proliferation and transformation [9]. TC21 overexpression in hepatocellular carcinoma (HCC) has been positively correlated to tumor size. Sequence analysis results suggest that Bmbras2 shares a high homology with TC21. To investigate whether BmBras2 can stimulate hepatoma cell proliferation, we conducted a cell proliferation assay. The results showed that rBmBras2 protein can promote PLC and HepG2 cell proliferation, but the cell proliferation mechanism caused by rBmBras2 protein is unclear and requires further study.

The silkworm is a widely used model organism. Basic research on the silkworm Bras2 protein location and biological functions will provide an important basis for further study of this protein's physiological role as well as that of other silkworm protein families.

Conflict of Interests

The authors have declared that no conflict of interests exists.

Acknowledgments

This work was supported by financial grants from the National High Technology Research and Development Program (no. 2012ZX09102301-009, no. 2011AA100603), the National Basic Research Program of China (no. 2012CB114600), and the Singapore National Medical Research Council (R-174-000-137-275).

References

[1] P. Sun, H. Watanabe, K. Takano, T. Yokoyama, J. I. Fujisawa, and T. Endo, "Sustained activation of M-Ras induced by nerve growth factor is essential for neuronal differentiation of PC12 cells," *Genes to Cells*, vol. 11, no. 9, pp. 1097–1113, 2006.

[2] R. B. Blasco, S. Francoz, D. Santamaría et al., "c-Raf, but not B-Raf, is essential for development of K-Ras oncogene-driven non-small cell lung carcinoma," *Cancer Cell*, vol. 19, no. 5, pp. 652–663, 2011.

[3] S. García-Silva, O. Martínez-Iglesias, L. Ruiz-Llorente, and A. Aranda, "Thyroid hormone receptor $\beta 1$ domains responsible for the antagonism with the ras oncogene: role of corepressors," *Oncogene*, vol. 30, no. 7, pp. 854–864, 2011.

[4] A. L. Kennedy, J. P. Morton, I. Manoharan et al., "Activation of the PIK3CA/AKT pathway suppresses senescence induced by an activated RAS oncogene to promote tumorigenesis," *Molecular Cell*, vol. 42, no. 1, pp. 36–49, 2011.

[5] G. W. Reuther and C. J. Der, "The Ras branch of small GTPases: Ras family members don't fall far from the tree," *Current Opinion in Cell Biology*, vol. 12, no. 2, pp. 157–165, 2000.

[6] A. Ehrhardt, G. R. A. Ehrhardt, X. Guo, and J. W. Schrader, "Ras and relatives—job sharing and networking keep an old family together," *Experimental Hematology*, vol. 30, no. 10, pp. 1089–1106, 2002.

[7] J. Colicelli, "Human RAS superfamily proteins and related GTPases," *Science's STKE*, vol. 2004, no. 250, p. RE13, 2004.

[8] K. Wennerberg, K. L. Rossman, and C. J. Der, "The Ras superfamily at a glance," *Journal of Cell Science*, vol. 118, no. 5, pp. 843–846, 2005.

[9] M. Rosário, H. F. Paterson, and C. J. Marshall, "Activation of the Ral and phosphatidylinositol 3′ kinase signaling pathways by the Ras-related protein TC21," *Molecular and Cellular Biology*, vol. 21, no. 11, pp. 3750–3762, 2001.

[10] T. Gotoh, Y. Niino, M. Tokuda et al., "Activation of R-Ras by Ras-guanine nucleotide-releasing factor," *Journal of Biological Chemistry*, vol. 272, no. 30, pp. 18602–18607, 1997.

[11] S. M. Graham, A. D. Cox, G. Drivas, M. G. Rush, P. D'Eustachio, and C. J. Der, "Aberrant function of the Ras-related protein TC21/R-Ras2 triggers malignant transformation," *Molecular and Cellular Biology*, vol. 14, no. 6, pp. 4108–4115, 1994.

[12] A. Valencia, P. Chardin, A. Wittinghofer, and C. Sander, "The ras protein family: evolutionary tree and role of conserved amino acids," *Biochemistry*, vol. 30, no. 19, pp. 4637–4648, 1991.

[13] G. M. Bokoch and C. J. Der, "Emerging concepts in the Ras superfamily of GTP-binding proteins," *The FASEB Journal*, vol. 7, no. 9, pp. 750–759, 1993.

[14] S. M. Graham, S. M. Oldham, C. B. Martin et al., "TC21 and Ras share indistinguishable transforming and differentiating activities," *Oncogene*, vol. 18, no. 12, pp. 2107–2116, 1999.

[15] K. T. Barker and M. R. Crompton, "Ras-related TC21 is activated by mutation in a breast cancer cell line, but infrequently in breast carcinomas in vivo," *British Journal of Cancer*, vol. 78, no. 3, pp. 296–300, 1998.

[16] Y. Zhang, J. Chen, Z. Lv, Z. Nie, X. Zhang, and X. Wu, "Can 29 kDa rhGM-CSF expressed by Silkworm pupae bioreactor bring into effect as active cytokine through orally administration?" *European Journal of Pharmaceutical Sciences*, vol. 28, no. 3, pp. 212–223, 2006.

[17] Y. Z. Zhang, J. Chen, Z. M. Nie et al., "Expression of open reading frames in silkworm pupal cDNA library," *Applied Biochemistry and Biotechnology*, vol. 136, no. 3, pp. 327–343, 2007.

[18] T. Schwede, J. Kopp, N. Guex, and M. C. Peitsch, "SWISS-MODEL: an automated protein homology-modeling server," *Nucleic Acids Research*, vol. 31, no. 13, pp. 3381–3385, 2003.

[19] T. Maniatis, E. F. Fritsch, and J. Sambrook, *Molecular Cloning: A Laboratory Manual*, Cold Spring Harbor Laboratory, Cold Spring Harbor, NY, USA, 1982.

[20] U. K. Laemmli, "Cleavage of structural proteins during the assembly of the head of bacteriophage T4," *Nature*, vol. 227, no. 5259, pp. 680–685, 1970.

[21] J. A. Falkner, M. Kachman, D. M. Veine, A. Walker, J. R. Strahler, and P. C. Andrews, "Validated MALDI-TOF/TOF mass spectra for protein standards," *Journal of the American Society for Mass Spectrometry*, vol. 18, no. 5, pp. 850–855, 2007.

[22] Q. Sheng, J. Xia, Z. Nie, and Y. Zhang, "Cloning, expression, and cell localization of a novel small heat shock protein gene: BmHSP25.4," *Applied Biochemistry and Biotechnology*, vol. 162, no. 5, pp. 1297–1305, 2010.

[23] L. Kong, Z. Lv, J. Chen et al., "Expression analysis and tissue distribution of two 14-3-3 proteins in silkworm (*Bombyx mori*)," *Biochimica et Biophysica Acta*, vol. 1770, no. 12, pp. 1598–1604, 2007.

[24] I. R. Vetter and A. Wittinghofer, "The guanine nucleotide-binding switch in three dimensions," *Science*, vol. 294, no. 5545, pp. 1299–1304, 2001.

[25] H. R. Bourne, D. A. Sanders, and F. McCormick, "The GTPase superfamily: conserved structure and molecular mechanism," *Nature*, vol. 349, no. 6305, pp. 117–127, 1991.

Permissions

The contributors of this book come from diverse backgrounds, making this book a truly international effort. This book will bring forth new frontiers with its revolutionizing research information and detailed analysis of the nascent developments around the world.

We would like to thank all the contributing authors for lending their expertise to make the book truly unique. They have played a crucial role in the development of this book. Without their invaluable contributions this book wouldn't have been possible. They have made vital efforts to compile up to date information on the varied aspects of this subject to make this book a valuable addition to the collection of many professionals and students.

This book was conceptualized with the vision of imparting up-to-date information and advanced data in this field. To ensure the same, a matchless editorial board was set up. Every individual on the board went through rigorous rounds of assessment to prove their worth. After which they invested a large part of their time researching and compiling the most relevant data for our readers. Conferences and sessions were held from time to time between the editorial board and the contributing authors to present the data in the most comprehensible form. The editorial team has worked tirelessly to provide valuable and valid information to help people across the globe.

Every chapter published in this book has been scrutinized by our experts. Their significance has been extensively debated. The topics covered herein carry significant findings which will fuel the growth of the discipline. They may even be implemented as practical applications or may be referred to as a beginning point for another development. Chapters in this book were first published by Hindawi Publishing Corporation; hereby published with permission under the Creative Commons Attribution License or equivalent.

The editorial board has been involved in producing this book since its inception. They have spent rigorous hours researching and exploring the diverse topics which have resulted in the successful publishing of this book. They have passed on their knowledge of decades through this book. To expedite this challenging task, the publisher supported the team at every step. A small team of assistant editors was also appointed to further simplify the editing procedure and attain best results for the readers.

Our editorial team has been hand-picked from every corner of the world. Their multi-ethnicity adds dynamic inputs to the discussions which result in innovative outcomes. These outcomes are then further discussed with the researchers and contributors who give their valuable feedback and opinion regarding the same. The feedback is then collaborated with the researches and they are edited in a comprehensive manner to aid the understanding of the subject.

Apart from the editorial board, the designing team has also invested a significant amount of their time in understanding the subject and creating the most relevant covers. They scrutinized every image to scout for the most suitable representation of the subject and create an appropriate cover for the book.

The publishing team has been involved in this book since its early stages. They were actively engaged in every process, be it collecting the data, connecting with the contributors or procuring relevant information. The team has been an ardent support to the editorial, designing and production team. Their endless efforts to recruit the best for this project, has resulted in the accomplishment of this book. They are a veteran in the field of academics and their pool of knowledge is as vast as their experience in printing. Their expertise and guidance has proved useful at every step. Their uncompromising quality standards have made this book an exceptional effort. Their encouragement from time to time has been an inspiration for everyone.

The publisher and the editorial board hope that this book will prove to be a valuable piece of knowledge for researchers, students, practitioners and scholars across the globe.

List of Contributors

Pavel Dobrynin and A. P. Kozlov
The Biomedical Center, Saint Petersburg 194044, Russia
Dobzhansky Center for Genome Bioinformatics, Saint Petersburg State University, Saint Petersburg 190004, Russia

Ekaterina Matyunina
The Biomedical Center, Saint Petersburg 194044, Russia

S. V. Malov
Dobzhansky Center for Genome Bioinformatics, Saint Petersburg State University, Saint Petersburg 190004, Russia

Jungnam Lee and Kyudong Han
Department of Nano biomedical Science & WCU Research Center, Dankook University, Cheonan 330-714, Republic of Korea

Seyoung Mun
Department of Microbiology, College of Advance Science, Dankook University, Cheonan 330-714, Republic of Korea

Thomas J. Meyer
Department of Biological Sciences, Louisiana State University, Baton Rouge, LA 70803, USA

Shunzhao Sui, Jianghui Luo, Jing Ma and Mingyang Li
College of Horticulture and Landscape, Chongqing Engineering Research Center for Floriculture, Key Laboratory of Horticulture Science for Southern Mountainous Regions of Ministry of Education, Southwest University, Chongqing 400715, China

Qinlong Zhu
College of Life Science, South China Agricultural University, Guangzhou 510642, China

Xinghua Lei
Department of Botony, Chongqing Agricultural School, Chongqing 401329, China

Rosemary Jagus and Allen R. Place
Institute of Marine and Environmental Technology, University of Maryland Center for Environmental Science, 701 E. Pratt Street, Baltimore, MD 21202, USA

Tsvetan R. Bachvaroff
Smithsonian Environmental Research Center, 647 Contees Wharf Road, Edgewater, MD 21037, USA

Bhavesh Joshi
Bridge Path Scientific, 4841 International Boulevard, Suite 105, Frederick, MD 21703, USA

Chiara Gamberi and Paul Lasko
Department of Biology, McGill University, 3649 Promenade Sir William Osler, Montreal, QC, Canada

Wei Yu, Meihui Wang, Hanming Zhang, Yanping Quan and Yaozhou Zhang
Institute of Biochemistry, College of Life Sciences, Zhejiang Sci-Tech University, Hangzhou, Zhejiang 310018, China
Zhejiang Provincial Key Laboratory of Silkworm Bioreactor and Biomedicine, Hangzhou, Zhejiang 310018, China

Hongli Cui and Yinchu Wang
The Coastal Zone Bio-Resource Laboratory, Yantai Institute of Coastal Zone Research, Chinese Academy of Sciences, Yantai 264003, China
Yantai Institute of Coastal Zone Research, Graduate University of the Chinese Academy of Sciences, Beijing 100049, China

Song Qin
The Coastal Zone Bio-Resource Laboratory, Yantai Institute of Coastal Zone Research, Chinese Academy of Sciences, Yantai 264003, China

Sang-Je Park
National Primate Research Center, Korea Research Institute of Bioscience and Biotechnology, Ochang, Chungbuk 363-883, Republic of Korea
Department of Biological Sciences, College of Natural Sciences, Pusan National University, Busan 609-735, Republic of Korea

Young-Hyun Kim and Kyu-Tae Chang
National Primate Research Center, Korea Research Institute of Bioscience and Biotechnology, Ochang, Chungbuk 363-883, Republic of Korea
National Primate Research Center, Korea Research Institute of Bioscience and Biotechnology, University of Science & Technology, Ochang, Chungbuk 363-883, Republic of Korea

Jae-Won Huh
National Primate Research Center, Korea Research Institute of Bioscience and Biotechnology, Ochang, Chungbuk 363-883, Republic of Korea

Heui-Soo Kim
Department of Biological Sciences, College of Natural Sciences, Pusan National University, Busan 609-735, Republic of Korea

Laurence Wurth
Gene Regulation Programme, Center for Genomic Regulation (CRG) and UPF, 08003 Barcelona, Spain

Y. Q. He
The Laboratory Animal Research Center, Jiangsu University, Zhenjiang 212013, China
College of Food Science and Biological Engineering, Jiangsu University, Zhenjiang 212013, China

R. Zhao
School of Clinical Medicine, Jiangsu University, Zhenjiang 212013, China

Y. Pan and L. J. Ying
The Laboratory Animal Research Center, Jiangsu University, Zhenjiang 212013, China

Debbie Laudencia-Chingcuanco
Genomics and Gene Discovery Unit, USDA-ARS WRRC, 800 Buchanan Street, Albany, CA 94710, USA

D. Brian Fowler
Department of Plant Sciences, University of Saskatchewan, 51 Campus Drive, Saskatoon, SK, Canada S7N 5A8

Greco Hernandez
Division of Basic Research, National Institute for Cancer (INCan), Avenida San Fernando No. 22, Col. Seccion XVI, Tlalpan, 14080 Mexico City, Mexico

Christopher G. Proud
Centre for Biological Sciences, University of Southampton, Life Sciences Building (B85), Southampton SO17 1BJ, UK

Thomas Preiss
Genome Biology Department, The John Curtin School of Medical Research, The Australian National University, Building 131, Garran Road, Acton, Canberra, ACT 0200, Australia

Armen Parsyan
Goodman Cancer Centre and Department of Biochemistry, Faculty of Medicine, McGill University, 1160 Pine Avenue West, Montreal, QC, Canada
Division of General Surgery, Department of Surgery, Faculty of Medicine, McGill University Health Centre, Royal Victoria Hospital, 687 Pine Avenue West, Montreal, QC, Canada

Chiara Lanzuolo
CNR Institute of Cellular Biology and Neurobiology, IRCCS Santa Lucia Foundation, Via Del Fosso di Fiorano 64, 00143 Rome, Italy

Alessandra Traini, Domenico Carputo, Luigi Frusciante and Maria Luisa Chiusano
Department of Agricultural Sciences, University of Naples Federico II, Via Universita 100, 80055 Portici, Naples, Italy

Massimo Iorizzo
Department of Horticulture, University of Wisconsin-Madison, 1575 Linden Drive, Madison, WI 53706, USA

Harpartap Mann and James M. Bradeen
Department of Plant Pathology, University of Minnesota, 495 Borlaug Hall/1991 Upper Buford Circle, St. Paul, MN 55108, USA

R. Rakshambikai and N. Srinivasan
Molecular Biophysics Unit, Indian Institute of Science, Bangalore 560012, India

S. Yamunadevi
Molecular Biophysics Unit, Indian Institute of Science, Bangalore 560012, India
Max Planck Institute for Intelligent Systems (Formerly Max Planck Institute for Metals Research), BioQuant BQ0038, Im Neuenheimer Feld 267, 69120 Heidelberg, Germany
German Cancer Research Center, BioQuant BQ0038, Im Neuenheimer Feld 267, 69120 Heidelberg, Germany

K. Anamika
Molecular Biophysics Unit, Indian Institute of Science, Bangalore 560012, India
Department of Functional Genomics and Cancer and Department of Structural Biology and Genomics, Institut de Genetique et de Biologie Moleculaire et Cellulaire (IGBMC), CNRS UMR 7104, INSERM U 964, Universite de Strasbourg, 1 rue Laurent Fries, 67404 Illkirch Cedex, France

N. Tyagi
Molecular Biophysics Unit, Indian Institute of Science, Bangalore 560012, India
European Bioinformatics Institute, Welcome Trust Genome Campus, Hinxton, CB10 1SD Cambridge, UK

Ruth B. Phillips
School of Biological Sciences, Washington State University Vancouver, 14204 NE, Salmon Creek Avenue, Vancouver, WA 98686-9600, USA
Center for Reproductive Biology, Washington State University, Pullman, WA 99164-7520, USA

Jenefer J. DeKoning and Joshua J. Faber-Hammond
School of Biological Sciences, Washington State University Vancouver, 14204 NE, Salmon Creek Avenue, Vancouver, WA 98686-9600, USA

Joseph P. Brunelli, Kris A. Christensen and Gary H. Thorgaard
Center for Reproductive Biology, Washington State University, Pullman, WA 99164-7520, USA
School of Biological Sciences, Washington State University, Pullman, WA 99164-4236, USA

John D. Hansen
US Geological Survey, Western Fisheries Research Center, 6505 NE 65th Street, Seattle, WA 98115, USA

Suzy C. P. Renn
Biology Department, Reed College, 3203 SE Woodstock Boulevard, Portland, OR 97202-8199, USA

Teng Xu, Xudong Guo, Hui Wang, Xiaoyuan Du, Xiaoyu Gao and Dongjun Liu
The Key Laboratory of Mammalian Reproductive Biology and Biotechnology of the Ministry of Education, Inner Mongolia University, Hohhot, Inner Mongolia Autonomous Region 010021, China

Zhengbing Lv, Tao Wang, Wenhua Zhuang, Dan Wang, Jian Chen, Zuoming Nie, Lili Liu, Wenping Zhang, Xiangfu Wu and Yaozhou Zhang
College of Life Sciences, Zhejiang Sci-Tech University, Hangzhou 310018, China
Zhejiang Provincial Key Laboratory of Silkworm Bioreactor and Biomedicine, Hangzhou 310018, China

Lisha Wang and Deming Wang
School of Pharmacy, Xuzhou Medical College, Xuzhou 221004, China

Jun Li
Department of Obstetrics and Gynecology, Yong Loo Lin School of Medicine, National University Health System, Singapore 119228

Lian Qian
Agilent Technologies Singapore, Singapore 117681

www.ingramcontent.com/pod-product-compliance
Lightning Source LLC
Chambersburg PA
CBHW080700200326
41458CB00013B/4926

* 9 7 8 1 6 3 2 3 9 3 5 4 8 *